P9-ELV-260

Remediation and Beneficial Reuse of Contaminated Sediments

Editors

Robert E. Hinchee, Augusto Porta, and Marco Pellei

The First International Conference on Remediation of Contaminated Sediments

Venice, 10–12 October 2001

Library of Congress Cataloging-in-Publication Data

International Conference on Remediation of Contaminated Sediments (1st : 2002: Venice, Italy.)

Proceedings of the First International Conference on Remediation of Contaminated Sediments, Venice Italy, October 10-12, 2001 / edited by Marco Pellei, Augusto Porta, Robert E. Hinchee.

p. cm.

Includes bibliographical references and index.

Contents: v 1. Characterization of contaminated sediments (S1-1) -- v. 2. Management of contaminated sediments (S1-2) -- v. 3. Remediation and beneficial reuse of contaminated sediments (S1-3).

ISBN 1-57477-125-6 (3-volume set casebound : alk. paper) -- ISBN 1-57477-127-2 (v. 1 : alk. paper) -- ISBN 1-57477-128-0 (v. 2 : alk. paper) -- ISBN 1-57477-129-9 (v. 3 : alk. paper)

1. Contaminated sediments--Management--Congresses. 2. Soil remediation--Congresses. I. Hinchee, Robert E. II. Porta, A. (Augusto) III. Pellei, Marco, 1972- . IV. Title.

TD878.I563 2002

628.5'5--dc21

2002019143

Printed in the United States of America

Battelle Press
505 King Avenue
Columbus, Ohio 43201, USA
614-424-6393 or 1-800-451-3543
Fax: 1-614-424-3819
Internet: press@battelle.org
Website: www.battelle.org/bookstore

For information on future environmental conferences, write to:

Battelle
Environmental Restoration Department
505 King Avenue
Columbus, Ohio 43201-2693
Phone: 614-424-7604
Fax: 614-424-3667
Website: www.battelle.org/conferences

CONTENTS

Phytoremediation

Electrokinetic Remediation

Physical-Chemical Remediation

Monitored Natural Recovery

Environmental Dredging

FOREWORD

Contaminated sediments, in both freshwater and marine systems, are a significant issue worldwide. Many steps have been taken to restrict or ban from use some contaminants (e.g., DDT, PCBs, antifouling paints) and reduce discharge of others. However, contaminants can persist for many years in sediments, where they have the potential to adversely affect human health and the environment. Some chemicals continue to be released to surface waters from industrial and municipal sources, and polluted runoff streams from urban and agricultural areas continue to build up harmful levels of contamination in sediments. When contaminated sediments are dredged to maintain and improve navigation, major financial costs can be incurred. Estimates are that as much as 10% of dredged material is contaminated, and the costs of proper disposal of contaminated dredge material are substantially higher than for open-ocean dumping. Costs for confined disposal facilities and in situ confinement of contaminated sediments are 3 to 6 times higher than for ocean dumping. Alternative treatment technologies are emerging, but to date the costs have been reported to be 10 to 100 times those associated with dumping.

The need is clear for a better understanding of the nature of contaminated sediments and the risks they pose, better technology for dealing with sediments, and reasonable approaches to management of the problem. It is apparent also that international cooperation among site managers, decision makers, and scientists and engineers is needed to share information and develop workable, cost-effective solutions. Therefore, Battelle Geneva Research Center, in collaboration with Azienda Multiservizi Ambientali Veneziana (AMAV), organized the First International Conference on Remediation of Contaminated Sediments (Venice, 10-12 October 2001). Despite the adverse conditions of international travel during that period, nearly 500 sediments and remediation experts came to Venice from approximately 40 countries around the world to attend the Conference.

Each person who made a presentation at the Conference was invited to submit a paper for the proceedings. Following editorial review, 129 papers were accepted for publication and organized into three volumes:

Characterization of Contaminated Sediments (S1-1). Editors: Marco Pellei, Augusto Porta, and Robert E. Hinchee.
Management of Contaminated Sediments (S1-2). Editors: Augusto Porta, Robert E. Hinchee, and Marco Pellei.
Remediation and Beneficial Reuse of Contaminated Sediments (S1-2). Editors: Robert E. Hinchee, Augusto Porta, and Marco Pellei.

The papers in this volume, *Remediation and Beneficial Reuse of Contaminated Sediments*, are concerned with approaches for treating and detoxifying sediments and for isolating and stabilizing contaminated sediments by capping or by placement in confined disposal facilities. Authors describe their experiences with monitored natural attenuation of sediments and with a wide variety of treatment technologies (e.g., physical, chemical, thermal, electrokinetic, enhanced biodegradation, phytoremediation) to remediate sediments and wetlands. Several papers focus on options for beneficial reuse of sediments and on containment/immobilization approaches.

We appreciate the assistance of the Conference co-sponsors—Agenzia Nazionale per la Protezione dell'Ambiente (ANPA, the Italian environmental protection agency) and the U.S. Army Corps of Engineers Center for Contaminated Sediments (CCS). In addition, several international organizations and periodicals concerned with harbor and waterways maintenance and environmental quality participated by helping to publicize the Conference among their members and subscribers:

Autorità Portuale di Venezia
Central Dredging Association (CEDA)
Comune di Venezia
DHI Water and Environment (Formerly Danish Hydraulic Institute and VKI Institute for Water Environment)
Environment Canada
International Association of Ports and Harbors (IAPH)
Journal of Soils and Sediments
Magistrato alle Acque di Venezia
Ministero dell'Ambiente
PIANC—International Navigation Association
Provincia di Venezia
Regione Veneto
U.S. Environmental Protection Agency

In addition, we would like to thank the Battelle staff in Geneva and Columbus who worked with us on the details of organizing and conducting the Conference and preparing the proceedings. Carol Young provided support and insight into the organization of the Conference and coordinated publicity, correspondence with speakers, and production of printed materials. Klaus Müller helped review abstracts and papers and provided suggestions on design of the technical program. Gina Melaragno, Noemi Pinna, Brenda George, and Veronica Smith assisted with the many tasks required before and during the Conference to ensure operational efficiency. Sam Yoon, Davide Tamburini, and David Pala provided computer support for the presentation files used by nearly 100 platform speakers, ensuring that the inevitable technical difficulties presented by variations in software and graphics design did not cause the tightly paced program to deviate from the schedule. Lori Helsel devoted many hours to standardizing the format of proceedings papers and assembling the three volumes of the proceedings. Joseph Sheldrick, manager of Battelle Press, provided production-planning advice and coordination with the printer; Richard Reed designed the covers.

Although the technical review provided guidance to the authors as needed to help clarify their presentations, the materials in these volumes ultimately represent the authors' results and interpretations. The support provided to the Conference by Battelle and other sponsoring organizations should not be construed as endorsement by these organizations of the content of these volumes.

Editors
Robert E. Hinchee (Farmington, Utah, USA)
Augusto Porta (Geneva, Switzerland)
Marco Pellei (Geneva, Switzerland)

A STATE-OF-THE-ART OVERVIEW OF CONTAMINATED SEDIMENT REMEDIATION IN THE UNITED STATES

Michael R. Palermo (U.S. Army Corps of Engineers R&D Center, USA)

ABSTRACT: Remediation of contaminated sediment has received growing attention in the United States in recent years. Contaminated sediments may be viewed as a "fourth environmental medium", with concerns over sediment impacts equal to those for water, air, and land-disposed waste. Options for managing contaminated sediments include monitoring natural processes which may gradually improve conditions, restricted use of a contaminated area, treatment or isolation of the contaminated sediments in-place, and dredging or excavation followed by treatment or disposal of the sediments at another location. Technical guidance for evaluating each of these options and criteria for selecting among the options is available, but the selection of a final remedy for many sites may be complex, expensive and contentious. This paper provides a state-of-the-art overview of the options for sediment remediation; the mechanisms used in the U.S. for assessment, evaluation and implementation; and a status summary of major U.S. projects.

EXTENT OF SEDIMENT CONTAMINATION IN THE US

Contaminated sediments, defined here as sediments containing chemical concentrations that pose a known or suspected threat to the environment or human health, is "a pervasive national problem" in the U.S. (NRC 1997, 2001).

Improved control and treatment of discharges of wastes into U.S. waters beginning in the 1970s has resulted in remarkable improvements in water quality nationwide. However, sediments are sinks for many contaminants, and historical releases of contaminants in rivers, lakes, and coastal waters remains as a significant problem. In 1997, the U.S. Environmental Protection Agency (USEPA) completed the first comprehensive inventory of sediment contamination nationwide. The results of the inventory indicated sediment contamination present in over 70% of U.S. watersheds, with the total estimated volume of surficial contaminated sediment over 1 billion cubic meters (USEPA 1997). Although the percentage of U.S. waters and associated volumes of sediment with contamination at levels posing a significant environmental risk is much lower, the extent of the problem in the U.S. is massive in both its logistical scope and the overall projected costs of remediation. The map in Figure 1 shows areas of potential concern as defined by the inventory, and it is evident that the widespread areas of contamination correspond to areas with concentrated industrial activity across the nation, primarily in the East and Great Lakes areas of the U.S. But contaminated sediment sites exist in all regions of the U.S., including freshwater lakes and rivers, coastal and estuarine waters, and ocean waters. Further, it is apparent that only a fraction of the specific contaminated sediment sites in the

U.S. have been remediated or are presently being addressed through remedial investigations or actions.

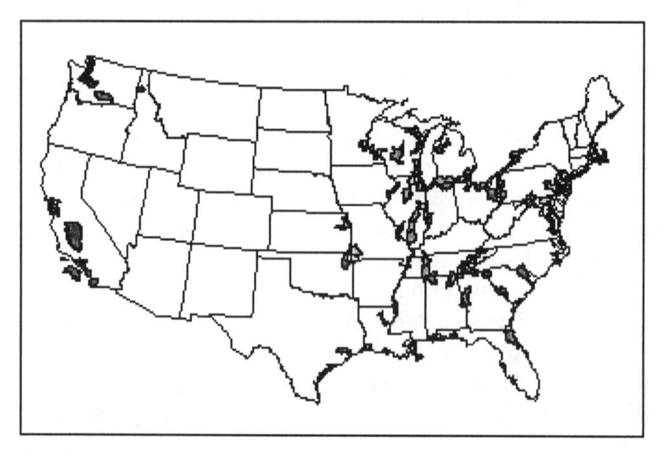

Figure 1. Probable Areas of Concern for Contaminated Sediments in the U.S. (from USEPA 1997).

ROLES AND RESPONSIBLITIES

Superfund and Other Cleanup Authorities. The USEPA is the Federal agency with the overall responsibility for environmental quality in the U.S., and so plays a major role in sediment cleanup activities nationwide. Authorities related to contaminated sediments stem from several environmental laws for which USEPA has the primary responsibility, and the agency has developed a strategy (USEPA 1998) to develop and apply consistent technical approaches and closely coordinate activities related to the problem.

Under the U.S. Federal Water Pollution Control Act Amendments of 1972, also called the Clean Water Act, several provisions relate to contaminated sediment remediation, such as those for water quality standards, effluent limits, management of dredged and fill material, etc. Specific authority was also given to identify and remove contaminated sediments from critical port and harbor areas. However, actual cleanup dredging under this authority was limited to one project, a sediment contaminated by a PCB spill in 1976 in the Duwamish Waterway in Puget Sound. The Assessment and Remediation of Contaminated Sediments (ARCS) program was also authorized to identify and assess the extent and magnitude of sediment contamination and to demonstrate bench scale treatment technologies. ARCS resulted in perhaps the most comprehensive technical guidance in the U.S. published to date for evaluation of sediment remediation technologies (USEPA 2001).

The 1980 U.S. Comprehensive Environmental Response, Compensation and Liability Act (CERCLA), along with the 1986 U.S. Superfund Amendments and Reauthorization Act (SARA), both commonly referred to as the Superfund law, provide perhaps the most far reaching Federal regulatory authority under which sediment remediation is conducted in the U.S. Superfund gives the USEPA a range of authorities to pursue the remediation of hazardous substance releases to the environment and cleanup of inactive hazardous waste disposal sites with the objective of human health and ecological risk reduction. Sites may be placed on the Superfund National Priorities List (NPL) based on a hazard ranking system, which is the start of a formal process of remedial investigations, feasibility studies, etc., prior to decisions on the remedial actions to be taken. This process is essentially a linear one, and some complex projects have taken well over a decade to reach completion. A more streamlined approach is also available under Superfund, using USEPA's "removal" authority (removal here refers to removal of the environmental risk as opposed to actual sediment removal), but even this process can be lengthy. The funding of projects under Superfund is based on the "polluter pays" principle. USEPA can fund a cleanup under the law, then seek to recover the costs from the responsible parties. However, in most cases, the parties seek an active role and may play the lead role in conducting the required studies and actual remedial actions. Under the Superfund program, the vast majority of work is paid for by the parties responsible for the contamination, not the U.S. government.

The Superfund Act also provides authority for Federal agencies such as the U.S. National Oceanic and Atmospheric Administration or the U.S. Fish and Wildlife Service, State agencies, and tribal governments to act as "Natural Resource Trustees" to access and claim damages for injuries to natural resources due to release of hazardous substances. Many Superfund sediment cleanup projects have integrated remedies with restoration and natural resource damages actions by trustee agencies, for example, the Palos Verdes Shelf project near Los Angeles and the Lavaca Bay project near Galveston, Texas. Under the natural resource provisions of CERCLA, the trustees must first prove damages, and seek compensation from the responsible parties through settlement or litigation, before restoration can be implemented. In some cases, the natural resource damages related to contaminated sediments are addressed through activities such as habitat restoration which are not directly related to cleanup of sediments.

The 1976 U.S. Resource Conservation and Recovery Act (RCRA) regulates the disposal of solid and hazardous wastes. Since the intent of the RCRA law is to regulate releases from active disposal facilities, a cleanup of releases under this law is described as a "RCRA corrective action." The number of such cleanups involving sediments is small compared to the Superfund program. RCRA may also come into play regarding the disposal of materials from sediment cleanup projects in licensed landfills.

Several States have initiated their own sediment cleanup programs or are addressing sediment sites under State-led cleanup programs. The States may also take a lead role in evaluations of cleanup under Superfund and RCRA.

Dredged Material Management. In the U.S., dredging activities are regulated by the U.S. Army Corps of Engineers (USACE) under the Rivers and Harbors Act, the Clean Water Act, and/or the Marine Protection Research and Sanctuaries Act, also known as the Ocean Dumping Act. The USACE is the primary water resources development agency in the U.S., and is considered the nation's "dredging agency." About 250 million cubic meters of sediments are dredged annually for maintenance of the U.S. navigation system, and although most of this material is clean, the management of contaminated dredged material has a major impact on the navigation dredging program. Consequently the USACE has devoted a considerable effort to increase the knowledge base, investing over $200 million in research and development on dredged material management, with much of this effort focused on issues related to contaminated sediments (USACE 2001).

Because of its navigation dredging mission and technical capabilities, the USACE plays a major support role for USEPA, other Federal agencies, and the States, regarding contaminated sediment management (Palermo and Wilson 2000). For example, the USACE acts as a principal design and contract administration agency for the USEPA Superfund program. Various provisions of the U.S. Water Resource Development Acts, also give the USACE specific authorities related to contaminated sediments, including authority to dredge for cleanup or environmental purposes outside the boundaries of and adjacent to a Federal navigation project as part of the operation and maintenance of the project.

OVERVIEW OF REMEDIAL PROJECTS IN THE US

The basic options for sediment remediation are well established. These include: non removal options such as monitored natural recovery and in-situ capping; and removal options, such as dredging with containment, and dredging with sediment treatment. Many remedies involve combinations of these options. Each of these options has been applied to sediment remediation projects in the U.S., and considerable field experience has been gained in execution of these projects. However, the effectiveness of the options and principles for selecting a given option for large and complex sites is still the subject of much debate.

The actual numbers of projects in the U.S. involving remediation of contaminated sediments and the options used are not easy to determine, because there is no common database for all the Federal and state regulatory programs which may result in remedial actions. Historically, sediment sites were not given a high priority in the Superfund program and were not identified as a specific project category. Recently, increased emphasis is being placed on sediment sites or sediment components of sites. USEPA now estimates that the Superfund program has taken an action or made a decision for an action at about 140 to 150 contaminated sediment sites nationwide since 1980 (Evison 2001). Many of these are marsh or stream sites involving small volumes of sediment.

Three published project surveys and data base reviews indicate that approximately 90 major contaminated sediment sites (those with sediment volumes exceeding about 2000 cubic meters) have been remediated or are underway in the U.S. (GE/AEM/BBL 2001, Cleland 2000, ThermoRetec 2001). Figure 2 illustrates the locations, geographic distributions, and types of remedial actions implemented at these sites. Similar to the USEPA sediment inventory, the concentration of remediation projects is in the Great Lakes, the East, and the Pacific Northwest regions of the country. Clearly, dredging and excavation, used as the sole option or in combination with other options, is by far the most common approach used to date nationwide.

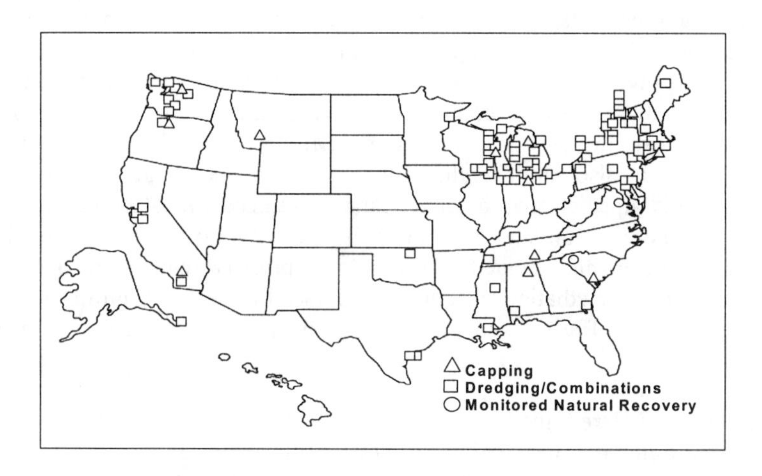

Figure 2. Distribution of completed and on-going projects in the U.S. with greater than 2000 cubic meters of contaminated sediments.

Monitored Natural Recovery. Monitored Natural Recovery (MNR) is a remedial option which relies on natural processes to contain or reduce the bioavailability or toxicity of sediments left in-place. Processes important to MNR include burial and in-place dilution following deposition of clean sediment and biodegradation or abiotic transformation processes which convert the contaminants to less toxic forms. MNR should not be confused with the "no action" alternative because by definition it must include source control and an appropriate monitoring program to insure the processes are effective. Institutional controls such as fish consumption advisories, fishing bans, or waterway or land use restrictions are also commonly a component of an MNR remedy. Only a limited number of contaminated sediment sites in the U.S. have been addressed solely be MNR. Examples include the Sangamo-Lake Hartwell site in South Carolina and the James River in Virginia. But MNR is a common component for remedies with a combination of actions, i.e., sites addressed by capping or dredging areas of higher contamination, with MNR for areas of lower contamination. Examples include the Burnt Fly Bog site in New Jersey and the Puget Sound Naval Shipyard site in Bremerton, Washington. The major advantages of MNR are its low cost, limited primarily to monitoring costs, and the avoidance of disruptions to the waterbody. The major disadvantages are the fact that contaminated sediment is left in the aquatic environment for the time period during which the natural processes act to reduce the risks, and the potential for future disruption of buried contaminants by storms, floods, or other episodic events.

In-Situ Capping. In situ capping (ISC) is an active remediation option in which a layer of clean isolating material (usually a clean sediment or soil) is placed to contain and stabilize the contaminated sediment in place. A variety of capping materials and cap placement techniques are available, and monitoring data collected for a number of projects has indicated capping can be an effective remedy. However, there is considerable debate regarding the permanence of a capping remedy for highly contaminated material and concern regarding the potential for extreme events such as storms, floods, or earthquakes to disrupt a cap. There is also an inherent disadvantage in that contaminated material remains in the aquatic environment. Capping remedies must be designed fully considering these factors and with an appropriate level of conservatism. In-situ capping has been implemented as the sole or primary remedy for a relatively small number of remediation sites in the U.S. (see Figure 2). Significant projects include the Eagle Harbor site in Washington and the Koppers site in South Carolina.

Environmental Dredging. Removal, including both dredging and dry excavation following dewatering or stream diversion, is the most common approach for sediment remediation in the U.S. Removal of the contaminated material from the aquatic environment is an obvious advantage, and this approach is usually viewed most favorably by regulatory agencies and, in some cases, by the public. The removal process for dry excavation uses conventional earth moving equipment, and the removal efficiency or effectiveness of such operations is not debated. However, the effectiveness of a dredging operation is often an important and controversial issue. The major considerations for effectiveness of dredging include resuspension of sediment during the dredging process, release of contaminants due to resuspension, and residual contaminated sediment left in place following the operation. Removal of the sediment mass is straightforward, but resuspension and related processes are complex and difficult to evaluate for all project conditions. Consequently, the definition of success for environmental dredging projects has been the subject of much debate between responsible parties and the agencies. In many cases, extensive measures are taken to isolate the dredging area from the waterbody using silt curtains, and in some cases, sheet pile enclosures.

The selection of appropriate dredging equipment and the compatibility of equipment with the selected disposal option is also an important factor, and may conflict with goals related to resuspension. Equipment normally used for navigation dredging can be used for remediation projects, but U.S. dredge designers, manufacturers, and dredging contractors are making significant contributions and have fielded a variety of innovative hydraulic and mechanical dredges especially designed for environmental work. Also, many international dredging companies have now formed partnerships with U.S. companies, allowing for use of specialty equipment from a variety of countries. Most of the remedial dredging projects in the U.S. now use some type of environmental dredging equipment.

Containment in CDFs, CADs, and Landfills. Containment refers to placement of dredged material in a disposal site with associated design features to hold the contaminants within the site. Confined disposal facilities (CDFs) and Contained Aquatic Disposal (CAD) sites are commonly used for contaminated sediments from navigation dredging and have also been used for remediation projects. But, by far, the most common containment option in the U.S. for contaminated sediments dredged for purposes of remediation has been disposal in licensed landfills.

CDFs are diked upland or nearshore areas which may be used in the remediation context both as final disposal sites and as temporary rehandling sites for storage or processing prior to sediment treatment. At Commencement Bay, Washington, one of the largest Superfund sites in the U.S., several projects have utilized CDFs created by filling areas within the harbor with contaminated materials, with the sites later used for harbor development. CAD refers to a capped underwater facility with lateral confinement, such as an excavated pit. CAD has only been used once for Superfund, at the Puget Sound Naval Shipyard site, where contaminated material dredged for both navigation and remediation was placed in the CAD facility.

Landfills are regulated under the RCRA law in the U.S., and RCRA regulations establish specific design requirements for both non-hazardous solid waste and hazardous waste landfills. One of the requirements for placement of wastes in landfills is related to "no excess water", which dictates that contaminated sediments must be essentially dewatered prior to disposal in the landfills. A common scenario for many U.S. sediment remediation projects therefore involves dredging, followed by pre-treatment to dewater the sediments, and rehandling and transport of the material for landfill disposal. This scenario has proven to be an effective option for sediments, especially for those projects involving relatively low volumes of material. A major advantage is that a dedicated disposal site need not be constructed, and the long-term liability of the disposed material is effectively shifted to the landfill operator. However, costs can be high, since the location of landfills can be many miles from the dredging sites, and the "tipping" fees charged for disposal of each ton of material can be expensive. Truck haul from pre-treatment sites to landfills is common, but material from the Marathon Battery Superfund project in New York was transported by rail directly from the site to a landfill in Ohio.

Sediment treatment. Sediment treatment refers to the use of physical, chemical, or biological processes to destroy or degrade contaminants or immobilize the contaminants within the sediment. Destruction of contaminants is an attractive option for agencies and the public, and much attention has been focused in the U.S. on evaluating innovative technologies for sediment treatment. The available processes and approaches for design and implementation of treatment have been well documented (PIANC 1996, USEPA 2001, Thompson and Francingues 2001). In addition, field pilot studies of promising treatment technologies have been conducted under USEPA programs including the ARCS program, and authorities under the Water Resources Development Act. However, no "silver bullet" solution has been found, i.e. one that is both environmentally and economically effective across a broad range of sediment chemical and physical characteristics.

Many treatment technologies have proven effective, but cost is the major constraint to wider application of treatment for sediment remediation projects. Treatment costs can range from around $50US per cubic meter for process such as stabilization to over $1000US per cubic meter for high temperature thermal processes. Nonetheless, treatment is the selected remedy for the most highly contaminated fraction of sediments at many sites in the U.S. Examples include the Marathon Battery Superfund site in New York and the Bayou Bonfouca site near New Orleans.

Future Projects. Even though many contaminated sediment sites have been addressed in the U.S., much work remains to be done. For example, around 1300 sites are presently on the Superfund National Priorities List (NPL), and about 20% percent of current NPL sites involve contaminated sediment as at least a significant component. Since many of these projects are in the remedial investigation or feasibility study stage, not all of them will necessarily need active remediation. But USEPA expects to list about 20 to 30 new NPL sites each year in the coming years and expects that a few of these will be sediment sites (Evison 2001). Presumably, state led cleanup programs and voluntary cleanups conducted by responsible parties would also add to the yearly total of new sediment cleanups nationwide.

Even as work progresses in future years, several very large and complex Superfund projects involving contaminated sediments promise to intensify the level of controversy in the U.S. over these projects. For example, there is heated debate regarding the need to dredge large volumes of PCB-contaminated sediment from the Hudson River in New York state, and this site has received much attention at the national level. The Fox River in Wisconsin, another large and complex PCB-contaminated site, promises to add to the level of scrutiny on the nationwide problem of contaminated sediments.

REMEDY EVALUATION AND SELECTION

The experiences gained in the U.S. have provided much insight on how options for sediment remediation should be evaluated and conducted. Selection of the most appropriate and potentially effective remedy for a given project remains the most vexing issue. The basic components of an evaluation should include a through characterization and assessment of the problem and/or project requirements, site and sediment characterization, implementation of source controls, screening of alternatives, detailed assessment of alternatives, selection of a preferred alternative, and provisions for monitoring and management. All these evaluations should be conducted within a risk-based framework

There is universal agreement that risk assessment is a critical component in evaluating options for sediment remediation, and the recent U.S. National Research Council report emphasized this principle (NRC 2001). In the remediation context, risk assessments for sediment projects involve complex relationships for both exposure and effects, and many of the specific data desirable for assessments are newly developed. This is especially true for ecological risk assessment as compared to human health risk assessment.

Risk assessments for the baseline condition (without a remedy in place) are routinely conducted in the U.S. for sediment remediation projects. However, comparative risk assessments, in which the risks for each of the remedial options under consideration are evaluated and compared to the baseline, are not commonly done. Such comparative assessments are potentially complex and evaluations are difficult to make on an equitable basis, because the human health and ecological receptors and exposures may differ among the options. But, it is now recognized that comparative assessments in some form will be needed in the future to ensure informed decisions are made. Fortunately, many of the specific assessment tools, models, etc. frequently used for sediment evaluations are inherently risk-based.

Remedy approaches should be evaluated fully considering: 1) site specific considerations such as hydrodynamics, adjacent resources and infrastructure, water depths, and other factors which may influence the risks and costs of a given approach; 2) project specific considerations such as contaminated volumes or areas to be addressed, the regulatory framework under which the project will be evaluated, and other factors which may dictate the feasible solutions; and 3) sediment specific considerations such as the contaminants of concern, contaminant concentrations relative to background, physical properties of the sediments, and presence of debris in the sediments that may limit the ability to dredge, cap, contain or treat the sediments.

Ultimately, the selection of a management or remedial approach must consider a balance between environmental protection and economic efficiency. Experience has shown that, for large or complex sites, combinations of options are often the most efficient remedies.

ACKNOWLEDGEMENTS

This paper summarizes results of studies conducted by the U.S. Army Engineer Research and Development Center (ERDC), Environmental Laboratory, Waterways Experiment Station. Permission to publish this material was granted by the Chief of Engineers.

REFERENCES

Evison, L. 2001. Personal communication, Leah Evison, U.S. Environmental Protection Agency, Office of Emergency and Remedial Response.

General Electric Corp., Applied Environmental Management Inc., and Blasland, Bouck, and Lee, Inc. (GE/AEM/BBL). 2001. Major Contaminated Sediment Sites Database, Release 3.0, 2001. http://www.hudsonwatch.com/mcss/index.html

National Research Council (NRC), 1997. Contaminated Sediments in Ports and Waterways, National Academy Press, Washington, D.C.
(Available from National Academy Press http://www.nap.edu/bookstore)

National Research Council (NRC) 2001. A Risk-Management Strategy for PCB Contaminated Sediments, National Academy Press, Washington, D.C. (Available from National Academy Press http://www.nap.edu/bookstore)

Palermo, M.R. and J.R. Wilson. 2000. "Corps of Engineers Role In Contaminated Sediment Management And Remediation," Proceedings, American Bar Association, Section of Environment, Energy, and Resources, Panel on Contaminated Sediments: Science, Law, and Politics, 8th Section Fall Meeting, 22 September, 2000, New Orleans, Louisiana.

Permanent International Association of Navigation Congresses. 1996. "Handling and Treatment of Contaminated Dredged Material from Ports and Inland Waterways - CDM," Report of Working Group No. 17 of the Permanent Technical Committee I, Supplement to Bulletin No. 89, 1996, Permanent International Association of Navigation Congresses, Brussels, Belgium.

Thompson, D. and N. Francingues. 2001. "A Special Report on PIANC's Innovative Dredged Sediment Decontamination and Treatment Technologies Specialty Workshop," Journal of Dredging Engineering, Vol. 3, No. 2, Official Journal of the Western Dredging Association.

Cleland, J. 2000. "Results Of Contaminated Sediment Cleanups Relevant To The Hudson River - An Update To Scenic Hudson's Report *Advances In Dredging Contaminated Sediment*," Report prepared for Scenic Hudson, October, 2000. http://www.scenichudson.org/pcb_dredge.pdf

ThermoRetec. 2001. "Sediment Technologies Memo for the Fox River," Report prepared for the Wisconsin Dept. of Natural Resources by ThermoRectec Consulting Corp., Seattle Washington.

USACE. 2001. Dredging Operations Technical Support and Center for Contaminted Sediments websites. www.wes.army.mil/el/dots/ccs

USEPA. 1997. "The Incidence and Severity of Sediment Contamination in Surface Waters of the United States," EPA-823-R-97-006, 007, and 008, U.S. Environmental Protection Agency Office of Science and Technology. http://www.epa.gov/waterscience/cs/congress.html

USEPA. 1998. "EPA's Contaminated Sediment Management Strategy," USEPA 823-R-98-001, USEPA Office of Water, Wash. D.C.

USEPA. 2001. Guide to Great Lakes Program Office Assessment and Remediation of Contaminated Sediments (ARCS) Program. http://www.epa.gov/glnpo/arcs/arcs-home.html

BENEFICIAL USE OF DREDGED MATERIAL FOR URBAN REDEVELOPMENT: OENJ CHEROKEE'S CASE STUDIES

Linda P. Morgan, Farhad Jafari, and ***Irving E. Cohen***
OENJ Cherokee Corporation, Bayonne. New Jersey, USA

ABSTRACT: The successful redevelopment of OENJ Cherokee's (OENJ) waterfront brownfields properties, using amended dredged material from the New York-New Jersey Harbor (NY-NJ Harbor), proves the beneficial use of dredged material for urban redevelopment is a practical, cost-effective, and environmentally sound means of reclaiming properties that otherwise would have lain dormant and undeveloped. OENJ's redevelopment of the Elizabeth Metro Center and the Bayonne Golf Club underscore opportunities for, and obstacles to, the continued success of beneficial use of dredged material for urban redevelopment. Technologies used for amendment of dredged material (pug mill system and in-barge mixing) are also discussed in this paper.

ELIZABETH METRO CENTER (JERSEY GARDENS MALL)

Introduction. Upland beneficial use of dredged material in the New York-New Jersey region is a recent phenomenon. In 1997, when the federal government banned ocean dumping of contaminated-dredged material (CDM) in the historic "mud dump" off the shore of Sandy Hook, regulators and policymakers faced a dilemma about how to dispose of significant volumes of maintenance-dredging materials. To maintain adequate depths for oceangoing vessels to offload cargo at the Port of New York and New Jersey, each year the Army Corps of Engineers (ACOE) contracts dredging of about 3 to 4 million cubic meters of material. Therefore, when the Sandy Hook mud dump closed in 1997, the ACOE had few choices: bury the CDM in underwater, secure cells; experiment with decontaminating it using various heat-treating, bioremediation or sediment-washing processes; treat it, transport it and dispose of it out of state; or amend the material to bind up any contamination and beneficially reuse it as a cap on brownfield sites.

Project Overview. The Elizabeth Metro Center, location of the Jersey Gardens mall, was the State's first upland project that beneficially used CDM from the NY-NJ Harbor. This 67 hectares (166-acre) former municipal landfill site was purchased in 1992 by OENJ. Conveniently located adjacent to Newark International Airport, Newark Bay, the Port of New York and New Jersey, and the New Jersey Turnpike, the site was ripe for redevelopment. However, due to a lack of adequate utilities, road connections, and perceived environmental permitting restrictions, the property had lain dormant for almost 20 years. The developers of the Elizabeth site believed the site would be successful as a retail outlet mall because the nearby IKEA store was the largest grossing IKEA location in America. After spending more than $20 million in remediation and permitting costs, the developers faced a price tag of $15 million for clean fill, an expenditure that would place a severe financial burden on the deal.

When, in 1997, the United States Environmental Protection Agency announced a ban on ocean disposal of CDM, OENJ proposed amending the material with pozzolonic reagents including cement, cement kiln dust, lime kiln dust and fly ash, to bind up any contaminants in the dredged material preventing them from being leached out. Working with its consultants, OENJ perfected a formula that met the State Department of Environmental Protection's (DEP) Non-Residential Direct Contact Soil Cleanup Criteria. This allowed the dredged material to be beneficially used for commercial redevelopment. Working closely with the DEP, OENJ also devised a protocol to allow for diverse recyclable materials such as glass, masonry, and construction and demolition debris to be deposited on the site as structural fill. Over a one-year period, the site accepted .75 million cubic meters (1 million cubic yards) of recyclable materials, including amended-dredged material. OENJ avoided spending $15 million on a clean soil cap by using the CDM. OENJ was paid by the ACOE to accept the material. This offsets the costs of remediation, processing and placement, capping and dynamic compaction of the site for the eventual mall development, and enabled the project to proceed.

After installing a leachate collection and methane gas venting system, part of the site was paved and sold to Glimcher Realty Trust, a major mall developer. In less than one year Glimcher completed construction of the 120,000 square meter (1.3 million square foot) retail outlet Jersey Gardens mall. Currently, the mall contributes over $3.5 million in taxes to the City of Elizabeth and Union County, and employs over 5,000 people. The project has become a paradigm for innovative project management, engineering, and policymaking. Other New Jersey municipalities as well as other states have emulated this model in developing comprehensive brownfields programs.

Alternate Stabilization Techniques: Pugmill vs. In-Barge Processing. The 1997 ACOE contract enabling property to receive dredged material from Port Newark was OENJ's first experience managing and reusing dredged material as structural fill at an upland site. Over a two-year period, the site received dredged material from other areas in the NY-NJ harbor, including different reaches of Port Newark, Liberty State Park, and Union Dry Dock.. By the end of 1998, approximately 440,000 cubic meters of dredged material were placed on the property.

Two different methods were utilized to amend the dredged material: a pug mill system and in-barge mixing of dredged material. In the former, dredged material and an admixture (admix) was simultaneously introduced to a pug mill and blended with mixing blades or paddles. Materials to be mixed enter the pug mill from one end and exit from the other end with a retention time of approximately 30 to 45 seconds. The amended material is transferred from the plant to the site for final placement (Figure 1). The in-barge mixing method utilizes a mixing head attached to the arm of a long-reach hydraulic excavator. Cement slurry is injected into the dredged material in the barge through hoses attached to the excavator. The mixing head is submerged into the dredged material, and its revolutions blend the cement and dredged material (Figure 2). Before the mix is

FIGURE 1. First Generation Pugmil Processing Plant at Elizabeth NJ Site.

FIGURE 2. Typical Equipment for In-Barge Mixing of Dredged Material.

unloaded and transported to the final destination, enough time is allowed for hydration of cement.

Of 440,000 cubic meters of dredged material placed at the site, approximately half was stabilized on barges in Port Newark and imported to the site. The remainder was processed through the pug mill system installed at the site.

For the portion processed on the site, initially the raw dredged material was pumped from dredge scows approximately one kilometer off the shore through a pipeline into the pug mill. Problems arose when the pumping of dredged material, often mixed with miscellaneous debris and scrap metal, clogged the pipes. As a result, this method was abandoned.

The second generation of stabilization plant operated on gravity feed using hoppers and screw feeders. The plant was erected at another location, the Sealand property, approximately 1.5 kilometers to the north of the site. The dredged-material scows, were docked at Sealand and were unloaded using a crane and clamshell bucket. The crane fed the hopper and dredged material was pushed into the pug mill by screw conveyors. As with the previous plant, admix silos were mounted on top of the pug mill with a continuous feed. Once mixed, the amended-dredged material was pushed into a surge hopper, dumped into off-road trucks and transported to the site for final placement. That plant successfully processed 2,200 to 3,000 cubic meters in eight-hour shifts. The dredged material was mixed with 8%-10% Portland cement.

There are pros and cons with each method. In-barge mixing is less vulnerable to mechanical breakdowns than the pug mill because debris is frequently mixed with dredged material. In-barge mixing can mix clay chunks more easily than a pug mill. However, the amended material is more uniform and consistent when mixed in a pug mill.

Stabilizing the Dredged Material for Structural Suitability. Portland cement was the preferred admix used for stabilization/solidification for its consistency and availability. During the first few months of operation with the first plant, the contractor experimented with several pozzolanic admixtures such as cement kiln dust, lime kiln dust and fly ash using as much as 25% on a wet-weight basis. Using these by-products creates inconsistency in their chemical constituents and pozzolanic properties. The mixed material might not achieve the strength requirements or fail the leachate test criteria. Because of the potential drawback, Portland cement only was used at ratios approximating 8% on a wet-weight basis.

Utilizing pozzolanic admixtures increases the dredged material's workability. Maintenance dredge is mostly silt (60% to 70%) with various percentages of clay and fine sand (usually equal percentages of each). Its initial moisture content (weight of water/weight of solids) generally ranges from 150 to 200%. Immediately after mixing, the moisture content decreases to approximately the 120% to 130% range. The optimum moisture content as determined by standard proctor tests varies mostly from 40% to 50%. Therefore, significant moisture reduction is required to make dredged material compactable.

To lower the moisture content of the amended material, the material placement contractor spread the material over large areas in thin layers and exposed it to air and sun. (Figure 3) A set of farming disks pulled by a bulldozer

FIGURE 3. Placement of Dredged Material at the Elizabeth, NJ Site

constantly displaced the material until it could be compacted. In cold seasons, placement of dredged material is difficult and not economically feasible. The contractor attempted unsuccessfully to compact the dredged material to the specified density (90% of the modified maximum dry density), and had to stop until the following spring. From mid-April to mid-October, placement and compaction of dredged material was successfully accomplished by the contractor.

DOT Embankment Pilot Project. During preparation of the Elizabeth Metro Center, the New Jersey Department of Transportation (NJDOT) sponsored a pilot project to assess amended-dredged material performance in roadway embankment applications. Two embankments were constructed using dredged material stabilized with 8% Portland cement. The material was placed in 30 to 45 centimeter thick layers to a maximum height of 4 meters. Each layer was compacted to 86% of the modified proctor maximum dry density. The two embankments were monitored during an eighteen-month period for dredged material's compressibility, shear strength and durability. Samples of dredged material were also tested in laboratory for shear strength, consolidation, gradation, permeability and resilient modulus. The results indicated that compressibility and strength of dredged material once compacted was satisfactory to sustain embankment loads; however, the material had durability problems. Freeze-thaw tests indicated that the material should be protected against frost. The conclusion of the study was that the dredged material could be used as roadway embankment with certain limitations. The dredged material has to be covered by at least three feet of granular soil if it is to be used in New Jersey as sub-grade fill for roadway embankments.

In addition to geotechnical monitoring and testing of the two embankments, the material was environmentally monitored. A percolated water collection system was installed at the base of the two embankments to collect/test the percolated water from the dredged material. The study also monitored airborne particulates and surface water run-off from the dredged material. The key parameters measured in the percolated water were for PP+40, PCBs, and, Dioxins. Modified Multiple Extraction Procedure (MMEP) testing indicated few exceedences of metals above New Jersey Non-Residential criteria.

A second phase of the embankment project is contemplated to better address the performance of dredged material under actual vehicular loads. A portion of a secondary road will be constructed using dredged material as sub-grade material. This material will be covered by at least three feet of granular cover material to protect it against frost action. Parameters such as soil modulus, shear strength and compressibility will be periodically monitored. Upon completion of the pilot project, recommendations will be made regarding utilization of dredged material in roadway application.

BAYONNE GOLF COURSE CASE STUDY

Project Overview. Following on the success of the Elizabeth Metro Center project, OENJ moved across the bay to redevelop a former landfill and vacant contaminated property in Bayonne New Jersey. This 51 hectares (125-acre) upland site is located on Upper New York Bay, south of the Military Ocean Terminal. It has an excellent view of the Statue of Liberty. Surrounded by oil refineries, warehousing and industrial establishments, this brownfield property seemed an unlikely choice for a golf course. However, the adjacent Military Ocean Terminal was soon to be decommissioned by the Federal Government and returned to the City, which targeted the facility for port and mixed-use development. In addition, a light rail line recently had been built through the City, with a stop a few blocks from the property. The Mayor was supportive of developing the site for recreational uses. Structural properties of dredged material indicated that the site could be filled to 15 meters (50 feet) in the air with 2.8 million cubic meters of material.

Once the dredged material is sculpted into a golf course, it will provide excellent views of Manhattan. In addition, OENJ will design a challenging course for golfers in a county that now has no golf courses. OENJ is compensated to take the contaminated dredged material, and the city realizes a host community fee for recyclable materials, including dredged material brought to the site.

The Bayonne property posed significant remedial challenges. Half the property was a former municipal landfill that had not been properly closed. The other half, previously purchased by a public utility, had been historically contaminated with oil products many years prior to its acquisition. Although the property had been created at the turn of the century by historic filling, its cleanup required excavation and offsite disposal of known areas of drums and oil basins, full perimeter containment of groundwater, installation of leachate collection and management systems, piping of two stormwater ditches and installation of a methane gas venting system. As an innocent purchaser, OENJ undertook this complex remediation as part of the voluntary cleanup program offered by the

DEP. A key to the economic success of this project's development was maximizing placement of recyclable materials, including the amended-dredged material.

Delay in Receipt of Dredged Material. In the New York/New Jersey area, the dredging process is directed by three entities. The Federal Army Corps of Engineers prepares the technical specifications, drawings, and request for proposals (RFP) for each dredging project. The Port Authority of New York and New Jersey, a bi-state agency, is the local sponsor of the dredging contracts and is responsible for 30% of the total cost of each contract, while the Federal government funds the balance of the contracts. The third entity is the State of New Jersey, through its Office of Maritime Resources in the Department of Transportation and the NJDEP's Office of Dredging and Sediment Technology, which coordinates the permitting for beneficial use of dredged material on upland sites. In the late 1990s, anticipating developers' competitive bidding for the dredged material, the Office of Maritime Resources created incentives and a lottery system to direct the ACOE-awarded dredging contracts to those upland sites that were successful in becoming permitted to accept the dredged material.

Once the Elizabeth Metro Center project was recognized as a profitable beneficial use project, other firms entered into the dredge material management and placement business. Between 3 and 5 million cubic meters per year of maintenance dredged material were expected to be available for upland sites. This was a tremendous opportunity for those developers who could successfully secure upland sites and could navigate the permitting process in preparing a site to receive material. Over a three-year period, developers rushed to prepare their sites to accept the dredged material.

No significant dredging contracts were awarded until the U.S. Congress passed the Water Resources Development Act in 2000. This authorized funding for the deepening of the New York/New Jersey Harbor channels. The first RFP was released to the dredging companies in late spring 2000. After several amendments to the RFP, bids were finally opened in September 2000. The outcome of the bid was challenged, and a Federal judge ruled in March 2001. The contract was awarded in April 2001 and dredged material was finally delivered to the Bayonne site in May 2001.

Dredged Material Processing Technique. To prepare the site for material, in 1999 OENJ dredged approximately two acres of open water and filled them with coarse stone and rock creating an off-loading platform for dredged material scows to dock. Approximately 140,000 cubic meters of bay sediments were dredged to create an access channel for dredged material scows to dock at the platform. Two hundred and six meters (700 feet) of steel sheet piles with a tie back system were installed at the bulkhead. The stabilization plant was erected over a concrete platform supported on 160 timber piles.

A third generation of stabilization plant with two separate pug mills and two silos was constructed at the platform. The previous design was modified by lowering the loading hopper to a level at which a hydraulic excavator with clamshell bucket could unload the dredged material into the hopper. A hydraulic excavator has more cycles per minute than a crane. The modification increased the production rate. A bucket elevator was designed to convey the dredged material from the bottom of the loading hopper into the pug mill. The plant is rated to process 450 cubic meters per hour (Figures 4 and 5).

Within the first month of operation, the plant experienced mechanical problems and was retrofitted while still processing material. In a three month period from mid May 2001 to early August, 2001, approximately 200,000 cubic meters of dredged material was processed with 8% Portland cement and placed at the site. The amended dredged material was placed in cells in 1 to 1.3 meter thick layers. Once solidified, one foot of recycled masonry or soil was placed on top of the dredged material to allow off-road trucks to move over the placed dredged material. Within two to three days, another layer of dredged material was placed on top of previously placed dredged material and the 30 centimeter cover. This process will continue until the site is graded according to the plan.

This placement method at the Bayonne site is different from the Elizabeth Metro Center's. During placement of the dredged material in the Metro Center project, OENJ learned that constant displacement of the dredged material breaks the bonds between soils and cement crystals and result in strength loss. The above-mentioned method of placing dredged material in cells and allowing it to cure in-situ relies on strength gain through cementation of dredged material parti cles. Since the live loads of a golf course are minimal, no additional compaction effort will be required to further solidify the dredged material.

FIGURE 4. Unloading of Dredged Material into Processing Plant at Bayonne, NJ Site.

FIGURE 5. Third Generation Stabilization Plant-Dual Pugmill at the Bayonne, NJ Site.

Port Authority Clay Pilot Project. Concurrent with using dredged contaminated silt sediments, OENJ Cherokee is in the process of conducting a pilot study sponsored by the Port Authority of NY/NJ to demonstrate the possibility of beneficial reuse of dredged clay material for landfill caps or liners. Dredged clay will be spread in 20 centimeter layers and compacted. The clay will be then be monitored for workability, shear strength, compressibility and permeability. The study's findings will determine whether beneficial reuse of dredged clay is technically and economically feasible.

FUTURE OUTLOOK FOR BENEFICIAL USE OF DREDGED MATERIAL

Many opportunities exist for redeveloping properties by beneficially using dredged material. Capping brownfield sites with amended dredged material is an excellent way to recycle and reclaim damaged sites. There are a multitude of waterfront sites in the New York-New Jersey region that can be successfully remediated using either amended contaminated-dredged material, or clays from harbor-dredging maintenance projects. The dredged material is a long-term, cost-effective source of fill material for reclaiming abandoned and underutilized sites. Given that there are currently over 100 landfills in New Jersey that have not been properly closed, this is one opportunity for using the material. In addition, dredged material may be used to fill in old quarries or mines, or for roadway embankments, as demonstrated by the NJDOT project in Elizabeth. OENJ is currently experimenting with mixing dredged material with other recyclables such as automobile tires and construction and demolition debris because those recyclables are more structurally stable or more economically disposable material for landfill.

Unfettered use of dredged material for upland placement is limited by the fairly complex permitting process necessary to obtain an approved disposal site, and the lack of certainty in the schedule for delivery of dredged material. The current process for releasing maintenance dredging contracts is problematic because of the requirements set by the dredge processing contractor for acceptance of dredged material (e.g. dewatering of dredged material, debris removal) and the need to mesh the operations of the dredging contractor (sea-based operation) with the upland facility operator (land-based operation). Also, it is complicated to plan for site remediation and site redevelopment if the source or schedule of fill (and income) is not reliable. In addition, there are competing regional interests in securing the fill for upland brownfields redevelopment sites, or for open water cells as a federal engineering project. Finally, the process of managing the material itself will continue to challenge developers and engineers wanting to work in the field of dredged material management.

PROPOSED FRAMEWORK FOR EVALUATING BENEFICIAL USES OF DREDGED MATERIAL IN NY/NJ HARBOR

Nancy Bonnevie (Battelle Memorial Institute, Duxbury, MA)
Thomas Gulbransen (Battelle Memorial Institute, Great River, NY)
Jerry Diamantides (David Miller & Associates, Providence, RI)
James Lodge (US Army Corps, Planning Division, NY)

ABSTRACT: The U.S. Army Corps of Engineers has developed and is in the process of implementing a Dredged Material Management Plan (DMMP) for the Port of New York and New Jersey. The DMMP mission is to provide economically cost-effective and environmentally sound dredged material management that satisfies the Port's need for safe navigation. The Corps has included numerous stakeholders in the planning process, such as Federal agencies, non-Federal agencies, non-Federal sponsors, private industry, and citizen groups. These stakeholders acknowledge ongoing and future demands for removal of dredged material due to naturally occurring sedimentation and changes in Port commerce. In addressing this common need, the DMMP defines and provides a preference ranking for numerous dredged material management options that may be implemented during the next 40 years. The DMMP presents a strong preference for dredged material management options that result in beneficial use of dredged material. Beneficial uses of dredged material may include habitat creation, enhancement, and restoration as well as land remediation.

Despite advances in the economists' ability to place a dollar value on changes in environmental attributes, many potential benefits and costs of beneficial use management options cannot be adequately valued in dollar terms, either because of the prohibitively high cost of applying the appropriate evaluation method, or the presence of potentially significant non-use values that cannot be appropriately included in the standard benefit/cost analysis, or because the willingness-to-pay approach may not identify the total value of project impacts. Furthermore, many of the benefits and costs identified in this expanded framework are not applicable to a benefit/cost analysis.

This paper describes development of a single, systematic framework consistent with Corps planning guidance that evaluates and compares various beneficial use options. This framework is intended to incorporate economic, environmental, and policy related information that would be supplemental to a standard benefit/cost analysis. It extends the standard federal benefit/cost analytical structure by identifying and evaluating the full range of benefits and costs that occur at the national, regional, and local level, including those benefits and costs that cannot be adequately measured in a monetary metric. It increases the number of factors to be considered and compared and creates a transparent means of evaluating the decision making process. The presentation will examine use of this framework in group consensus building sessions, as well as in executive planning discussions. A need for such a framework exists because decisions concerning dredged material management options are affected by considerations outside of the classic benefit/cost analytical structure. An interactive case study will be presented that illustrates the potential utility of this approach.

INTRODUCTION

The U.S. Army Corps of Engineers has developed and is in the process of implementing a Dredged Material Management Plan (DMMP) for the Port of New York and New Jersey. The DMMP mission is to provide economically cost-effective and environmentally sound dredged material management that satisfies the Port's need for safe navigation. The Corps has included numerous stakeholders in the planning process, such as Federal agencies, non-Federal agencies, non-Federal sponsors, private industry, and citizen groups. These stakeholders acknowledge ongoing and future demands for removal of dredged material due to naturally occurring sedimentation and changes in Port commerce. In addressing this common need, the DMMP suggests a preference ranking for numerous dredged material management options that may be implemented during the next 40 years. The DMMP presents a strong preference for dredged material management options that result in beneficial use of dredged material such as but not limited to habitat creation, enhancement, and restoration as well as land remediation.

Each of the potential beneficial uses is associated with a wide variety of 'benefits' ranging from reduced operational costs to increased habitat for key species or a reduction in human health risk. One of the difficulties faced by the Corps is how best to compare and contrast these benefits to determine the most appropriate option for a given project. Due to the broad spectrum of potential benefits, there is no common 'metric' for conducting this evaluation. For example, it is difficult to compare the benefit gained from habitat restoration with that gained from reduced operational costs. Despite advances in economists' ability to place a dollar value on changes in environmental attributes, many potential benefits and costs of beneficial use management options cannot be adequately valued in dollar terms, either because of the prohibitively high cost of applying the appropriate evaluation method, the presence of potentially significant non-use values that cannot be appropriately included in the standard benefit/cost analysis, or because the willingness-to-pay approach may not identify the total value of project impacts.

The objective of this analysis was the development of a single, systematic framework consistent with Corps planning guidance that evaluates and compares proposed beneficial use options. This framework is intended to incorporate economic, environmental, and policy related information that would be supplemental to a standard benefit/cost analysis. It extends the standard federal benefit/cost analytical structure by identifying and evaluating the full range of benefits and costs that occur at the national, regional, and local level, including those benefits and costs that cannot be adequately measured in a monetary metric.

One of the primary challenges in developing this framework was the need to 'normalize' the various parameters considered for each benefit to one common unit of measurement. Similar problems are encountered in ecological risk assessment, where data available for evaluating the potential risks at a single site may include sediment chemistry, toxicity tests, community structure analyses, and tissue chemistry evaluating bioaccumulation and the potential for trophic transfer of contaminants. To address this issue in ecological assessments, an approach referred to as 'weight of evidence' was developed to balance and integrate multiple lines of evidence to achieve a more objective, unbiased conclusion (Menzie et al., 1996). A modification of the concepts described by Menzie et al., 1996, has been applied to develop a framework for evaluating the potential beneficial uses of dredged material.

MATERIALS AND METHODS

As described by Menzie et al, 1996, the weight of evidence approach is an attempt to normalize the various types of information generated from an ecological risk assessment to a common metric so that all can be equally evaluated with respect to the individual strengths and weaknesses of the different data results. As stated by Menzie et al., 1996, one of the primary objectives of this approach was to provide an objective and transparent approach for incorporating various professional judgements into the evaluation of data that might otherwise not be comparable.

The weight of evidence approach employs a quantitative methodology based on three basic characteristics of each measurement endpoint: an assigned weight, the magnitude or response, and the concurrence of the among outcomes of multiple measurement endpoints (Menzie et al., 1996). For a specific measurement endpoint, each of these characteristics is defined according to pre-determined criteria, to ensure objectivity. For example, the assigned weight is determined based on attributes related to the strength of the association between the measurement endpoint and the assessment endpoint, the quality of the data and the study design (Menzie et al., 1996).

For the purpose of developing a framework for evaluating beneficial uses, the concepts of weight, magnitude and concurrence were applied to the evaluation of possible benefits associated with each use. Following an ecological risk assessment paradigm, assessment and measurement endpoints were identified and weighted based on attributes developed to address the various concerns of the Corps when considering possible uses of dredged material. The text below walks through the four primary steps of the weight of evidence approach, using hypothetical examples to illustrate how the method could be applied. It is important to emphasize that the examples provided are hypothetical and for demonstration purposes only. Additional research will be required to more accurately define the various components of the framework.

Step 1: Identification of Assessment and Measurement Endpoints. The first step in the evaluation is to identify the assessment and measurement endpoints (Table 1). Assessment endpoints (AE) are typically defined as the explicit expressions of the actual environmental values that are to be protected or evaluated. They reflect social and environmental priorities and should be expressed in a manner than can be evaluated through an objective scientific process. An example of an assessment endpoint for an ecological risk assessment would be the community structure and reproductive success of songbird populations within a contaminated area.

For the purpose of evaluating beneficial uses, the following assessment endpoints were identified:

- AE 1 - Consideration of all economic impacts associated with the proposed use;
- AE 2 - Evaluation of potential environmental effects resulting from proposed use; and
- AE 3 - Consideration of resource management issues associated with the proposed use.

TABLE 1. Representative endpoints to weight of evidence of dredged material beneficial uses.

	Assessment Endpoints			
		Potential for Environmental Risk		
	Economics	**Human Health**	**Ecological Health**	**Resource Management**
Measurement Endpoints	Human Health Benefits	Evidence of Possible Exposure	Evidence of Possible Exposure	Biological Resources
	Regional Economic Benefits	Evidence of Potential Risk	Evidence of Potential Risk	Habitat Creation/Restoration
	Reduced Cost of Sediment Disposal	Exceedance of Applicable Criteria	Exceedance of Applicable Criteria	Recreational Use
	Real Estate Valuations	Direct Evidence of Toxicity	Direct Evidence of Toxicity	Aesthetic Value
	Operational Costs	Reduction in physical hazard	Risk to Endangered Species	Endangered Species
	Ecosystem Productivity Benefits		Potential for Bioaccumulation	
	Enhanced Aesthetics		Evidence of physical hazard	
	Recreational Benefits			

These endpoints were intended to encompass the diverse array of potential benefit types. Ultimately each assessment endpoint will be assigned a site-specific rank or score which can be compared among proposed sites to classify the potential beneficial uses.

Measurement endpoints (ME) are the specific lines of evidence used to evaluate the identified assessment endpoints. Multiple measurement endpoints may be associated with a single assessment endpoint, and they may vary depending on the project evaluated. Measurement endpoints provide the basis for structuring the analysis and serve as the actual measurements used to evaluate the potential effects or benefits associated with each assessment endpoint. Therefore they should be explicitly related (either directly or indirectly) to the specific assessment endpoints. Further they should include metrics (e.g., degree or response, space and/or time) that can be used as a basis for estimating risks. Examples of measuring for the identified assessment endpoints include:

- AE 1 - Economic Effects
  - ME 1: Human Health Benefits
  - ME 2: Regional Economic Benefits
  - ME 3: Reduced Cost of Sediment Disposal
- AE 2 - Environmental Effects
  - ME 1: Reductions in physical hazards
  - ME 2: Reduction in possible exposures
- AE 3 - Resource Management
  - ME 1: Habitat creation
  - ME 2: Recreational Use
  - ME 3: Improved aesthetic value for human use

It is important to note that the measurement endpoints identified above are only examples; additional measurement endpoints may be identified for each proposed project based on the available site-specific information. Table 1 identifies a broader set of measurement endpoints for economic effects related to the land remediation management option currently being implemented at Bayonne, NJ.

Step 2: Determination of Measurement Endpoint Weights. Once the individual measurement endpoints are identified, it is necessary to determine the relative importance or 'weight' of each to the overall assessment. Relative weighting creates the foundation for ensuring any metrics and comparisons reflect the values of the stakeholders a priori, that is weightings are assigned before revealing case-specific scores. Weighting factors enable the group to emphasize considerations that reflect their priorities. For example, data variability or quality may be especially important to evaluate when using certain measurement endpoints. Other endpoints may warrant weighting based on how definitively the measurement links to the assessment concern. The nature and extent of these factors, referred to as "attributes" are adjustable to each case. For the purpose of this evaluation, eight preliminary attributes were identified:

- **Strength of Association**: Refers to the extent to which the measurement endpoint is representative of, correlated with, or applicable to the assessment endpoint.
- **Feasibility**: Refers to the likelihood that the proposed beneficial use will succeed. For example, while benefits associated with wetland creation/restoration may be significant, there are data to indicate that the success rate for such projects is relatively low.
- **Strength of data/uncertainty**: Refers to the quality and reliability of the data.
- **Consistency with regional planning**: Evaluates consideration of regional goals as outlined in the NY/NJ DMMP, CCMP or other planning documents.

- **Spatial application**: Relates to the ability of the measurement endpoint to evaluate potential benefits across areas larger than the specific site.
- **Temporal application**: Relates to the ability of the measurement endpoint to evaluate potential benefits over time.
- **Sensitivity**: Relates to the ability to detect a response in the measurement endpoint.
- **Availability of Objective Measure**: Relates to the ability to judge site-specific information against predetermined and well-accepted standards, criteria or objective measures, in the case of economic valuation this relates to the ability to apply a monetary metric to the impact in question.

Each of these attributes is assigned a numerical value (i.e., scaling value) that designates its relative importance to the managers (i.e., USACE) independent of any specific proposed projects. The scaling values assigned for this preliminary framework (Table 2) were assigned based on the general values assigned to similar attributes by Menzie et al, 1966, with further adjustments, such as the "applicability of a monetary metric scale" which was assigned a value of 0.5 by the authors. It is important to note that the final scaling values should be derived based on consensus among the various stakeholders.

TABLE 2. Calculating final averages for resource management. End point weight = avg.(scores * importance)

		Habitat Creation/ Restoration		Recreational Use		Aesthetic Value		Endangered Species	
Attribute	Importance	Score (1-5)	Attribute Weight	Score (1-5)	Attribute Weight	Score (1-5)	Attribute Weight	Score (1-5)	Attribute Weight
Strength of Association	1	1	1	3	3	3	3	1	1
Strength of data/Uncertainty	0.8	4	3.2	3	2.4	3	2.4	4	3.2
Consistency with Regional Planning	0.8	5	4	5	4	5	4	5	4
Spatial Application	0.2	3	0.6	5	1	5	1	3	0.6
Temporal Application	0.2	4	0.8	2	0.4	2	0.4	4	0.8
Sensitivity	0.5	3	1.5	2	1	2	1	3	1.5
Availability of Objective Measure	0.2	2	0.4	2	0.4	2	0.4	2	0.4
Measurement Endpoint Weight			**1.74**		**2.03**		**2.03**		**1.74**

As part of the assessment of a specific project, each attribute will also be assigned a score. The score is defined as a number from 1 to 5 indicating the extent to which the specific attribute is addressed by each measurement endpoint. A value of 5 indicates a high correlation, while a score of 1 would imply little relationship.

Once the individual scaling values and scores are defined, the measurement endpoint weight is defined as follows. First, the weight of each individual attribute is calculated by multiplying the scaling value and the score:

$$AW = SV \times S \quad (1)$$

Where:

AW = Attribute weight
SV = Attribute Scaling Value
S = Attribute Score

The total weight of the measurement endpoint is then defined by averaging the individual attribute weights. Table 2 presents the total weighting for a sub-set of measurement endpoints for the Resource Management measurement endpoints of a land remediation management option at an abandoned landfill/brownfield site in Bayonne, NJ. The information used to develop the total weights of the measurement endpoints comes from a case study of beneficial use of dredged material at the OENJ – Bayonne currently being performed by the New York District of the Corps.

Step 3: Determining Finding and Magnitude for Each Measurement Endpoint. For each measurement endpoint evaluated, it is necessary to determine the finding and magnitude based on the available project-specific data. For the purpose of this evaluation, the finding indicates whether the proposed project would have a positive, negative, neutral or undetermined impact on the individual measurement endpoint. For example, in evaluating the measurement endpoint "Economic Impacts", data from the Bayonne case study indicate that approximately 76 permanent jobs would be created as the result of implementing the land remediation management option and proposed redevelopment of 138 acres at the site. This increase in permanent employment yields a positive finding, while data indicating that jobs would be lost would result in a negative finding. Once the finding has been determined, the magnitude (i.e., high or low) is determined, indicating the relative significance of each effect. In this example, creation of 76 jobs is determined to be a low, positive effect because the ratio of jobs created to acres developed (76/138 = 0.55) is very low. In the NY/NJ metropolitan region, the creation of one job for every two acres developed is reasonably considered a low level impact.

This step emphasizes the need to identify consistent metrics for determining the findings and magnitudes of a response. Table 3 presents examples of 'criteria' to be used for determining the findings and magnitudes for the evaluation. Again, the criteria listed for the economic effects are based on the Bayonne case study. However, it is important to note that these criteria are intended as examples only; additional research is required to develop appropriate benchmarks for evaluating the proposed measurement endpoints. It is important that these benchmarks be developed independently of specific projects to ensure an objective, unbiased evaluation.

Step 4: Weight of Evidence Results. As the final step of the evaluation, the measurement endpoint weights, finding, and magnitude of response are integrated to determine the overall benefits of the proposed project (Table 4). Measurement endpoints are placed on a matrix and bar charts (Figures 1 and 2) which allow normalized comparison across measurement endpoints and between projects.

TABLE 3. Criteria used to define magnitude of two resource management endpoints for NJ landfill case.

Measurement Endpoint	Finding	Magnitude	Criteria Examples
Habitat Creation/ Restoration	Positive	High	Creation of more than 10 acres of habitat
	Positive	Low	Creation of up to 10 acres of habitat
	Undetermined/Neutral	--	Data are unavailable
	Negative	Low	Loss of up to 1 acres of habitat
	Negative	High	Loss of more than 1 acres of habitat
Recreational Use	Positive	High	Greater than 10% increase in recreational use
	Positive	Low	No change or increase in recreational use of 10%
	Undetermined/Neutral	--	Data are unavailable
	Negative	Low	Loss of up to 10% of recreational use
	Negative	High	Greater than 10% decrease in recreational use

TABLE 4. Portion of weight of evidence for NJ landfill redevelopment.

Measurement Endpoint	Finding (+ or - or U)	Magnitude (1=low, 2-high)	Average Weight	WOE Result
Habitat Creation/Restoration	+	2	1.74	+3.48
Recreational Use	+	1	2.03	+2.03
Aesthetic Value	+	2	2.03	+4.05
Endangered Species	+	1	1.74	+1.74

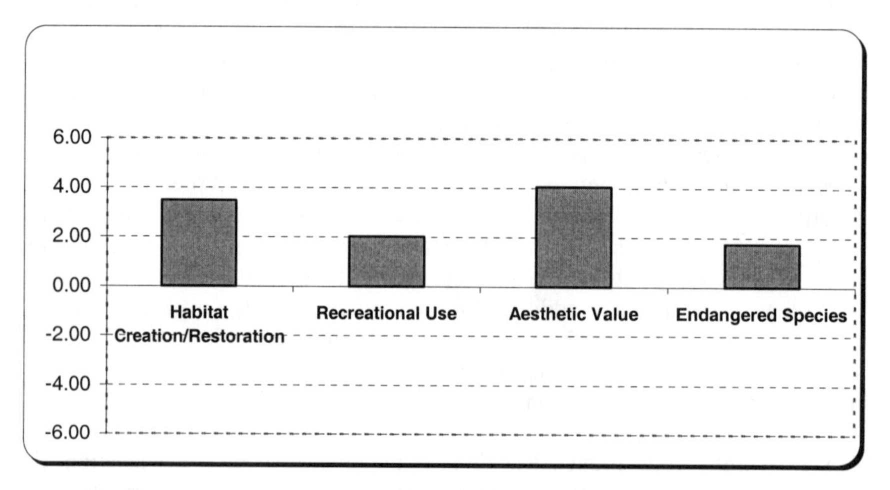

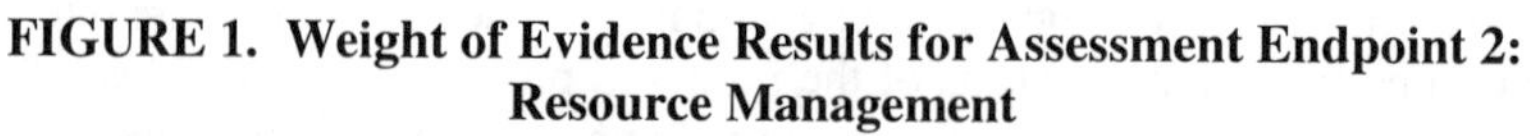
FIGURE 1. Weight of Evidence Results for Assessment Endpoint 2: Resource Management

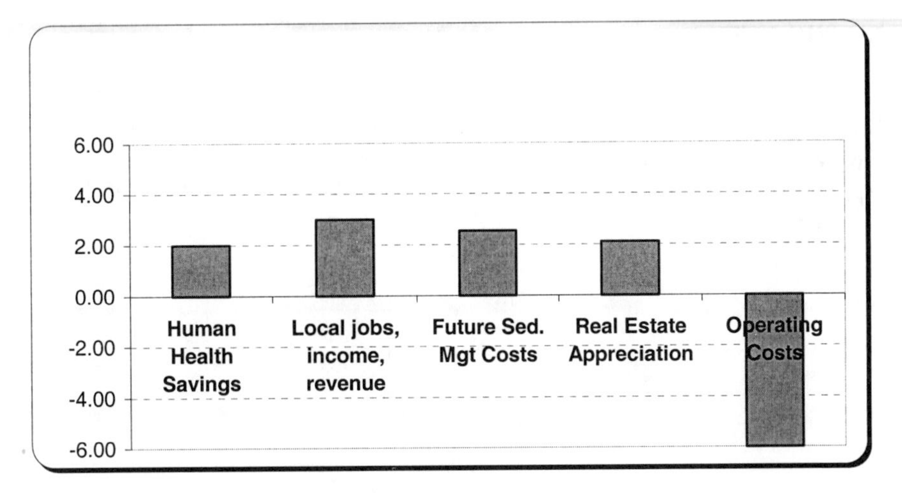

FIGURE 2. WOE Results for Assessment Endpoint 1: Economics

RESULTS AND CONCLUSIONS

The framework presented is intended to provide an unbiased, objective approach for integrating varied and conflicting pieces of information. The need for such a framework is clear, because decisions concerning dredged material management options are affected by so many considerations outside of the classic benefit/cost analytical structure. The proposed approach provides a methodology for incorporating and comparing different types of information. In addition, this approach provides a mechanism for establishing regional, consensus-based priorities for beneficial use decisions.

Building the framework upon a structured consensus development method serves two purposes. First, stakeholders are given clear opportunity to ensure that their constituent interests are evaluated. Second, the framework enables the broad spectrum of stakeholders to capture and debate the relative importance assigned to the varied factors based on systematic and consistent rationales. The landfill example illustrates the utility of applying this consensus development approach for highlighting diverse considerations and inviting decision makers to discuss the relative importance of various factors. The framework demonstrated numerous conflicts amongst stakeholder opinions and perspectives. Highlighting such conflicts is a central value of this transparent process.

It is important to note that additional work will be required to fully develop the framework to address all of the potential concerns of individual stakeholders associated with the wide array of possible beneficial uses. The example presented here was developed for presentation purposes only and many of the parameters would require additional data collection and research to adequately quantify. However, this example clearly demonstrates the utility of this approach for dredged material management decisions.

REFERENCES

Menzie, C., M.H. Henning, J. Cura, K. Finkelstein, J. Gentile, J. Maughan, D. Mitchell, S. Petron, B. Potocki, S. Svirsky, and P. Tyler. 1996. Special Report of the Massachusetts Weight-of-Evidence Workgroup: "A Weight-of-Evidence Approach for Evaluating Ecological Risks." *Human and Ecol Risk Assess* 2(2): 277-304

BENEFICIAL USE OF DREDGED MATERIALS IN PORTLAND CEMENT MANUFACTURE

Kevin H. Gardner, University of New Hampshire, Durham NH, USA; ***Bryan J. Magee***, TRL Limited, Berkshire, UK; Jennifer Dalton, Mindy Weimer, University of New Hampshire, Durham NH, USA

Abstract: Reported in this paper are findings of a research project being carried out to develop a beneficial use management strategy for contaminated dredged materials. As dredged materials inherently contain valuable resources – namely Al_2O_3, SiO_2 and Fe_2O_3 – the premise of this work is to capitalize on these properties to manufacture Portland cement, a construction material requiring considerable quantities of these three elements for its manufacture. Laboratory-scale investigations have been carried out to investigate the ability to incorporate dredged material as an alternative feedstock while maintaining product quality. Simultaneously, the fate of organic and inorganic contaminants present in the dredged material are being investigated to ensure a safe process and final product.

INTRODUCTION

Many of the ports and harbors in the United States are experiencing a sediment management crisis, and nowhere is this more apparent than in NY/NJ Harbor. Dredging required for maintenance and deepening of shipping channels creates large volumes of sediments, most of which are no longer acceptable for open ocean disposal due to the presence of contaminants and more stringent testing guidelines. New York State needs to find solutions for the management of dredged materials in order to conduct the dredging operations it needs to maintain its status as an important national and international port (Stern et al., 1998). NY/NJ harbor contains the largest port complex on the east coast of North America, handling some two million loaded containers each year. This activity creates approximately 200,000 port-related or port-dependent jobs and annually generates $30 billion in revenues and $620 million in state and local taxes.

One of the most promising management options for contaminated sediment is to find a 'beneficial use' for the material. A number of promising beneficial uses have been developed and proposed recently, including the manufacture of lightweight aggregte and the production of cement from a dedicated kiln (Rehmat et al., 1997, 1998; Fuller Company, 1998, Stern et al., 1998). The premise of the 'beneficial use' technology investigated in this work and presented here is to capitalize on the inherent oxide composition of dredged material to produce a valuable and marketable commodity – Portland cement – while taking advantage of existing infrastructure for cement production. Portland cement is an extremely fine, gray powder manufactured from some of the earth's minerals. After mixing with water, Portland cement is the glue that binds sand and gravel together into the rock-like mass we know as concrete. The principle oxides present in dredged sediments applicable to cement manufacture include Al_2O_3,

SiO_2, and Fe_2O_3. The capacity of this technology to provide a solution for the management of dredged material from the NY/NJ Harbor depends on both the capacity of accessible cement plants and the quantity of dredged material that can be incorporated, as partial feedstock, into their production processes. It also depends on the ability to incorporate this alternative material safely and without adverse impact on human health and the environment (Uchikawa et al., 1992).

In terms of accessibility, three major cement plants exist in NY State, all of which are accessible from NY/NJ harbor by barge via the Hudson River. Lafarge's plant in Ravena, NY is the largest, with a capacity of approximately 3.2 million metric tons per year. The St. Lawrence Cement plant in Catskill N.Y. and Glens Falls Cement plant in Glens Falls, NY both have a capacity of approximately 600,000 metric tons per year. This brings the current total capacity along the Hudson River to approximately 4.4 million metric tons per year. As preliminary results indicate that dredged material may be used to replace 3-12% of the raw materials by mass, this implies a sediment management capacity of 132,000 to 528,000 metric tons per year of dry material. With a typical water content of dredged material from NY Harbor of 127% (weight of water/weight of solids), this converts to 300,000 to 1,200,000 metric tons of actual dredged material. In addition, St. Lawrence Cement is currently constructing a new facility in Greendale, NY to replace the Catskill, NY facility; total production capacity will increase to over 2 million metric tons per year (significantly increasing total capacity for sediment).

The economics of 'beneficial use' management strategies for dredged material is a further potential benefit. As an example, the State of New Jersey has used significant volumes of dredged materials in a beneficial manner as fill materials at Brownfield and Superfund sites by amending them with stabilizing agents such as cement, cement kiln dust, and fly ash. The cost of sediment managed in this way has been approximately \$30-\$58/yd^3. This is in comparison to landfill tipping fees for dredged materials in the range \$60-118. In terms of cement manufacture, preliminary economic analysis has indicated that the potential for substantial cost savings exist for both port/facility operators and the cement producer. Savings would result from reduced or minimal tipping fees, whereas for the cement company, savings would be due to reduced raw material costs. With no tipping fees charged, a cement company with production capacity of 3.2 million metric tons per year may expect savings of approximately \$400,000 per year for a 1.5% by mass material replacement (this does not consider any capital costs or operating cost increases required to handle the alternative material). Using the maximum replacement that has been calculated by the UNH research team of 12%, a plant with production capacity of 3.2 million metric tons per year would realize an annual savings of \$3 million on the cost of bauxite, iron, and fly ash (again assuming no tipping fees and no capital improvement costs).

MATERIALS AND METHODS

The process of manufacturing Portland cement consists of grinding raw materials, mixing them intimately in certain proportions, and burning them in a large rotary kiln at a temperature of up to ~1450°C – at which point the material sinters and fuses into balls known as clinker. The clinker is then cooled, ground to a fine powder with some gypsum added, and the resulting product is the commercial product Portland cement (Neville, 1995).

The principle raw materials used in the manufacture of Portland cement are oxides: lime, silica, alumina and ferric oxide. While lime (i.e. calcium oxide) is not found in nature, it is produced during cement manufacture by heating calcium carbonate, which is found in raw materials such as limestone or calcite. In conventional cement manufacture, raw materials such as clay or shale are used as sources of the oxides silica, alumina and ferric oxide (Neville, 1995).

For this research project, dredged materials were sampled from the Harlem River in the New York harbor using a standard ponar sampler; three five-gallon samples were obtained from three different locations, all of which were typical depositional environments. Samples were place in plastic 5-gal. buckets and stored at 4°C for future use in cement production and analysis. Raw materials used by Blue Circle Cement (now Lafarge), Ravena, NY in their manufacturing process (i.e. limestone, shale, fly ash, bauxite, gypsum, iron) were also collected for characterization and use in laboratory research.

Analytical techniques used to characterize sediment, clinker, and final products include x-ray diffraction and x-ray fluorescence for crystal phase identification and total elemental analysis, respectively. Inorganic contaminant availability is being investigated using a number of types of leaching tests followed by analysis using inductively-coupled plasma – optical emission spectroscopy and graphite furnace atomic absorption spectroscopy.

RESULTS AND DISCUSSION

The oxide composition of the raw feed into the rotary kiln and the three dredged materials collected from the Harlem River were analyzed by X-ray fluorescence (XRF), and are shown in Table 1 below. These results demonstrate the similarity in major oxide composition between the two types of materials, the obvious difference being the very high calcium content of the slurry. However, it is important to note that the slurry is typically comprised of material from many different geologic formations, which are mixed to achieve the desired composition. Two of the most expensive components for cement manufacturers to acquire are iron and aluminum: two abundant elements in typical dredge materials.

TABLE 1 Oxide concentrations in the dredged material and slurry

Material	SiO_2	Al_2O_3	Fe_2O_3	CaO	MgO	SO_3	Na_2O	K_2O	TiO_2	P_2O_5	MnO	Cl	LOI
DM A	54.7	10.9	8.2	5.2	2.2	2.6	3.0	1.3	0.5	0.2	0.1	1.8	9.3
DM B	61.1	11.0	7.9	4.7	1.9	1.1	2.3	1.3	0.5	0.2	0.1	0.9	7.0
DM C	66.2	8.7	6.9	5.4	1.9	0.9	2.4	1.0	0.4	0.2	0.0	0.8	5.1
SLURRY	13.9	3.2	2.1	41.3	2.5	0.6	0.2	0.5	0.1	0.0	0.0	0.0	35.4

Using the dredged material compositions above, and the major elemental composition of the other raw materials at this particular cement manufacturing plant, mix proportions were established to maximize usage of dredged material while satisfying the conditions required for proper quality assurance. Results indicate that a maximum of approximately 12% of the entire raw feedstock may be dredged material, with iron being the controlling major oxide. With a plant with the capacity of the Lafarge plant in Ravena, NY, (3.2 million metric tons/year) this translates into approximately 384,000 metric tons of dry sediment per year, or 850,000 tons of wet sediment at a typical water content of 125% (weight of water / weight of solids).

A single major issue that has arisen in these calculations is the chloride content. There are two problems with chloride. First, very little available chloride is permitted to be present in the final cement product. Even if free chloride is expected to volatilize in the kiln, however, subsequent precipitation of chloride phases in the kiln is problematic. This remains an unresolved issue in this research, and is a major focus of this research effort at this time. Depending on the outcome of these efforts, the practical implications of this are that free chloride (either in the final product or in the slurry feed) may limit the extent to which traditional materials may be replaced by dredged material, at least from marine and estuarine locations.

Three options for dealing with chloride are being investigated at this time. First, sediment pre-treatment is being investigated. Table 2 shows the effect of various treatments on chloride removal. The first row shows the effect of dewatering the sediment (in this case by allowing it to drain through a filter); results show approximately 22% of total chloride is removed. The other four rows show the effect of adding 50 or 100 ml DI water directly to the filter or of suspending the dewatered sediment in 50 or 100 ml of DI water for a short time followed by filtration. All of these subsequent four treatments had very similar removal efficiency and removed approximately 53% of total chloride. These results also indicate that only approximately 55% of the total chloride is available for leaching and that the remaining chloride may not be considered to be available. The two other, and perhaps more practical, options being considered for chloride are: 1. addition of a complexing chemical that will allow chloride to volatilize at very low temperatures, and 2. limiting dredged material concentration in the feed stream.

TABLE 2. Chloride removal treatments to 50g of sediment

Sample Treatment	Cl^- removed (% of total Cl^-)
Filtered (dewatered)	22
filtered with 50mL of water	53
filtered with 100mL of water	54
mixed with 50mL of water and filtered	50
mixed with 100mL of water and filtered	55

The environmental performance of the clinker and cement produced using this alternative material source is of primary concern and is being investigated currently. Leaching data have been developed for heavy metal leaching from clinker (without dredged material added) and for dredged material, but results for clinker made with dredged material are not yet available. Tables 3 and 4 below show the leaching of various elements from dredged material under two different extraction conditions, 1M HNO_3 and pH=4, respectively. These results demonstrate a significant difference exists in soluble metal concentration between what could considered to be environmentally available (pH=4) and that resulting from a more rigorous extraction. Although data for clinker manufactured using dredged material are not yet available, a worst-case scenario can be developed from the extraction data that is available. Dredged material and clinker (results not tabulated) were leached aggressively using 1M nitric acid, and resulting lead concentrations were 140 ppm and 14 ppm, respectively. If the maximum amount of dredged material were added to produce the clinker (12%), and no changes in leaching availability were effected in the kiln, the worst-case lead concentrations leaching from the clinker would be 29 ppm. However, it is expected that lead will be much less available in the clinker (where lead will be incorporated into the amorphous and crystalline matrices) than it was in the dredged material (where it is primarily sorbed on surfaces) (Uchikawa et al., 1992).

TABLE 3. Element leaching in 1M HNO3 for 24hours (ppm)

Ca	Cd	Cu	Fe	Pb	Zn	Cl^-
3,924	0.975	51.4	12,516	140	85.7	10,260

TABLE 4. Metal leaching at pH 4 for 24 hours (ppm)

Ca	Cd	Cu	Fe	Pb	Zn
215	0.213	1.00	11.6	3.46	0.765

CONCLUSIONS

Research has been conducted investigating the major elemental composition of fine-grained sediments from NY Harbor which typify dredged material in this region. Results indicate the sediments are enriched in iron and aluminum, two essential and relatively expensive ingredients for the manufacture of Portland cement. Based on major element chemistry, up to 12% (by mass, dry weight basis) of dredged material could be used to replace traditional sources of iron, aluminum, and silicon. Early results have identified chloride as a potential problem contaminant; this is an aspect of on-going research. The fate of trace metals and organic contaminants in the manufacturing process and the leachability of these contaminants in the final product are also continuing.

ACKNOWLEDGEMENTS

The authors would like to acknowledge the valuable assistance of Lorraine Roberts and Jack Dwyer at Blue Circle Cement (now Lafarge) for x-ray fluorescence analysis, sharing their knowledge of cement manufacturing, and providing us will samples of their various raw materials. We would also like to acknowledge the support of the New York State Department of Environmental Conservation, particularly Elyse Peterson, and Frank Estabrooks, Jim Swart, and Dennis Wolterding for assistance with collecting background information and for collecting sediment samples for us. This research was supported by the Cooperative Institute for Coastal and Estuarine Environmental Technology.

REFERENCES

FULLER COMPANY [1998], "Rotary Kiln Production of Lightweight Aggregate from an Extruded Mix of Harbor Dredgings, Plant Dust, Overburden Clay, and Shale Fines", Fuller Company, Bethlehem, PA.

NEVILLE, A.M. [1995], "Properties of Concrete - Fourth Edition", Longman Group Limited, England.

REHMAT, A., LEE, A., GOYAL, A, MENSINGER, M.C., BHATTY, J.I., and BARONE, S.P. [1997], "Production of Cement from Contaminated Concrete Using Cement-Lock™ Technology", Proceedings of International Conference on Solid Waste Technology and Management.

REHMAT, A., LEE, A., GOYAL, A, and MENSINGER, M.C. [1998], "Construction-Grade Cement Production from Contaminated Sediment Using Cement-Lock™ Technology", Forwarded by authors, March 2000.

STERN, E.A., K.R. DONATO, N.L. CLESCERI, and K.W. JONES [1998], "Integrated Sediment Decontamination from the New York/New Jersey Harbor". Proceedings of the National Conference on Management and Treatment of Contaminated Sediments, Cincinnati, OH. EPA/625/R-98/001.

UCHIKAWA, H., HANEHARA, S., SHIRASAKA, T., and HASHIMOTO, S. [1992], "Influence of Burning Atmosphere on Distribution of Minor and Trace Element in Clinker and Formation of Brown Color Clinker"' JCA Proceeding of Cement and Concrete, Vol.46, pp.32-37.

THE USE OF DREDGED MATERIALS IN ABANDONED MINE RECLAMATION

Andrew S. Voros (New York / New Jersey Clean Ocean And Shore Trust)
J. Paul Linnan (Pennsylvania Department of Environmental Protection)

ABSTRACT: Three mega-volume features of American industrial output have been combined for their mutual beneficial use: Abandoned Mine Lands, coal fly ashes, and material dredged from navigation channels. Half-a-million tons of sediments dredged from New York Harbor are being amended with pozzolonic residues from coal-burning to form a soil cement for the reclamation of a strip mine feature in west-central Pennsylvania. The nature of the trace contaminants in dredged sediments is here described, along with the physical and chemical effects of amending them with pozzolonic wastes. The project operations, including materials transport, handling, and testing are described along with preliminary results from the demonstration site.

INTRODUCTION

Dredging. On average, the United States annually dredges and disposes 450 million tons of sediments from its ports, dams, reservoirs and navigational channels. However, individual projects exist with like volumes, such as the deepening of channels to accommodate new generation shipping vessels, and the restoration of silted up hydropower reservoirs. The Port of New York/New Jersey alone is expected to generate 250,000,000 tons of dredged material from the deepening and maintenance of its navigation channels over the next 40 years; 100 million tons of it over the next 9. Dredged material is significant not only for the remarkable volumes generated (for general purposes, a ton of dredged material is equal to one cubic yard of volume) but for the fact that the means of their disposal- ocean dumping- is ceasing to become an option because of concern over the potential bioaccumulation of trace agricultural and industrial contaminants in the marine food chain. As a result, the maritime community saw the cost of dredging and disposal go from $5 per cubic yard to $125 practically overnight. Annual maintenance dredging went undone, while container ships and tankers had to carry partial loads or transfer cargoes to off shore barges. The 165,000 port jobs and annual $20 billion contribution to the region's economy was threatened. Relocation of major shipping companies to deeper ports would have impacted the entire northeast region whose imports and exports are served by the Port of NY/NJ. This regulatory change has impacted the regions economy and employment, imports/exports and transportation patterns, requiring the development of an affordable, long term upland disposal option.

Abandoned Mine Lands (AML). In 1995, The Bureau of Abandoned Mine Reclamation and the Bureau of Solid Waste and Land Recycling in

Pennsylvania's Department of Environmental Protection and NY/NJ COAST began to examine the possibility of the beneficial use of dredged material for mine reclamation. Pennsylvania is noted for the magnitude of its AML problems, estimated as requiring $15 billion for the basic remediation of the 5600 sites deemed human health hazards out of their 9000 abandoned mines. Pennsylvania has over 800 incidents of mine subsidence each year; 145 active underground mine fires (including the famous Centralia Mine Fire which has burned since 1962) and most importantly, 3000 miles of Acid Mine Drainage (AMD) impacted waterways. The fill requirements of these features are immense. Several sites (like the Jeddo Mine Tunnel and the twin 32-mile-long crop falls in the anthracite region) individually require in excess of one billion cubic yards of fill. The state's fill requirement is in excess of 10 billion cubic yards. Having successfully experimented with the sealing of AMD-causing formations with fly-ash grouts PaDEP suggested the examination of the use of the dredged material as an aggregate in a manufactured fill bound with pozzolonic fly ash, another waste disposal problem.

Coal Fly Ash. The physical and chemical properties of coal fly ashes have been studied for over 20 years and their pozzolonic (cementitious) binding properties are well known, being a significant additive in Portland cement. The United States produces an estimated 115 million tons of fly ash annually in the burning of coal that generates half the nation's electricity. However, less than one third of it is used in manufacturing while the rest is disposed of in over 750 sites around the country, from lined landfills to massive open stockpiles. Coal ash disposal has been relatively free of federal regulation, but recently, however, coal ash only narrowly escaped significant regulatory oversight from the US Environmental Protection Agency, with the likelihood of increased scrutiny over the next year.[1] The amount of ash generated will only increase due to energy demand and Clean Air Act requirements for the addition of more scrubbing materials (like lime) during the combustion process. Coal ash's well known property of expansion during chemical hydration also limits the amount that can be used in cement manufacture.

The Concept. Individually taken, consideration of any of these three problems is clearly daunting. Viewed in concert, however, a pattern of complementarity emerges: massive volumes of disposal materials / massive voids requiring fill; acid generating voids / alkaline materials requiring confined disposal; a problematic wet material / a binder that initiates hydration reactions. It was decided to test the use of dredged material as an aggregate in a fly-ash bound manufactured fill, to restore the original contours of a hill whose side had been striped away into a 120 foot high-wall extending for several thousand feet. Since the issues of abandoned mine lands and coal fly ash used in their remediation are well known, we here emphasize the dredged material portion of the project and touch on public attitudes towards such projects, and the means of their financing.

METHODS AND MATERIALS

Dredged material. *Dredged material* is the generic term for the wet sediments removed, or dredged, from navigational channels, reservoirs and so on.[2] Because they are in aquatic environments, they are mostly water (usually 65%) with the remaining solid portion being the constituent sediments of natural erosion processes and storm water runoff. Depending on their location along watercourses, the solid potion of dredged materials may range from sand, to a mixture of sand, silt and clay, and up to 7% decaying organic plant matter (common mud). We are here largely concerned with the later characterization, since clean sand is generally used for beach replenishment or aggregate, and particle size is a key factor in contaminant adsorption. This mud is often described as black mayonnaise, its small particle size making it readily adhere to surfaces and its decomposing organic material giving off the smell of hydrogen sulfide. This is true of pristine wetlands sediments as well and is not an indicator of their relative environmental health as sediments. Being mostly water, the material sloshes readily and must be transported carefully, even in barges. Their physical properties are important in considering their handling and transportation, and in their perception by the public. Coming from areas adjacent to human activities, these materials contain a certain amount of natural and man-made debris, and trace contamination of agricultural and industrial runoff, including metals and organic compounds including dioxins and PCBs. The issue of sediment contamination is complex but will be presented here in brief.

Contaminated Sediments. Many of the chemical substances of concern came into question recently with the ability to detect TRACE concentrations at parts per trillion and quadrillion. The difficulty arises when societal concerns meet (or ignore) science. The toxicity of a substance involves a complex set of circumstances including species, exposure, pathways, age, sex, condition of subject and so on. Controversy rages over the health impacts of even the more notorious contaminants like dioxin. The issue is further complicated by risk assessments extrapolated from lab tests of non-human subjects exposed to enormous concentrations of subject contaminants. It is not known what trace concentrations of contaminants pose a threat, to humans or the marine environment, if any at all.

Five major types of pollutants are found in sediments[3]:

Nutrients, including phosphorous and nitrogen compounds such as ammonia. Elevated levels of phosphorous can promote the unwanted growth of algae. This can lead to the amount of oxygen in the water being lowered when the algae die and decay. High concentrations of ammonia can be toxic to benthic organisms.

Bulk Organics, a class of hydrocarbons that includes oil and grease.

Halogenated Hydrocarbons or **Persistent Organics,** a group of chemicals that are very resistant to decay. DDT and PCBs are in this category.

Polycyclic Aromatic Hydrocarbons (PAHs), a group of organic chemicals that includes several petroleum products and byproducts.

Metals, including iron, manganese, lead, cadmium, zinc, and mercury, and **metalloids** such as arsenic and selenium.

Interestingly, the banning of ocean disposal of dredged materials is at least partly the result of the continuing momentum of anti-ocean disposal actions over the last century involving far more obnoxious substances. Ocean disposal of garbage was banned in the 1930's, disposal of acid waste some time after that, treated sewage sludge in the 1980's, and the burning of wooden marine debris out at sea was stopped in the 90's. Now, ocean disposal of even mildly contaminated sediments has been virtually phased out, and it seems that even virgin clay material from the last glaciation will not be eligible for ocean disposal.

PaDEP has established a pass/fail standard for maximum levels of contaminants acceptable for this demonstration. A complete list of over 300 analytical parameters sampled for the dredge material project is available through http://nynjcoast.org . Analytical parameters are general chemistry, inorganics, organics, pesticides and PCB's.

Perhaps most easily understood, Pennsylvania will not accept material categorized as hazardous. Ceilings for dioxin levels are at 530 parts per trillion and for PCBs at 4 parts per million (mg/l).

Coal Ash. The manufacture of cementitious grouts from ashes for large applications relies upon alkali activation to initiate hydration reactions. Structures dating from ancient Roman and Greek cultures, many still standing today, were constructed from this type of cementitious material. A pozzolan can be defined as a siliceous or siliceous and alumnus material, which possesses little or no cementitious value but will, in finely divided form and in the presence of moisture, react with calcium hydroxide at ordinary temperatures to form compounds possessing cementitious properties.

From the point of view of dredged material disposal, this project amounts to contaminant sequestration. Fine particle sizes in dredged solids provide immense surface areas. A gram of mud has 760 square meters of surface area, providing myriad sites for the attachment of contaminants and the cause of concern when the material is digested by benthic organisms. When the material is hardened into a cement matrix, the same gram of material has a surface area of about 1 square inch, about one-millionth that of the unconsolidated material. The reduction in surface area available to chemical attack alone would be a significant safeguard to the leaching of contaminants. But Toxicity Characteristic Leachate Testing (TCLP) show that the material is chemically bound in the cement matrix, as described below.

PROJECT OPERATIONS

. The dredging is performed utilizing a dredge-plant mounted on a spud-barge and equipped with various types of clamshell buckets, and two hopper barges. Local tug service is used to transport the hopper barges to and from the transfer facility.

Off-loading of the dredged material at the facility is accomplished utilizing a 50-ton crane equipped with a clamshell bucket. As necessary, the loaded barges are moored at the facility to allow the sediment to settle for dewatering of the barges prior to off-loading. Water decanted from the loaded barges is pumped through a particulate filter to portable frac-tanks. After an adequate period of settling, the water in the tanks are tested for compliance with permitted discharge criteria contained. Upon confirmation, the decanted water is discharged from the tanks to the local waterway.

The raw dredged material is off-loaded into a large receiving hopper and through a series of screens to separate debris from the sediment. This debris includes tires, pilings, timbers, large metal objects, concrete, and similar unsuitable materials and is staged for transport and disposal at an alternate approved disposal facility. The dredged material is placed into a pugmill where it is mixed with coal fly ash to bulk (physically stabilize) the material for transport. The raw material is solidified to ensure that no free liquid is present that may leak out or shift the load during transport. From the pugmill, the pre-amended material is discharged via a radial-stacking conveyor to a temporary storage area. The material is loaded from the stockpile into 110 ton gondola rail cars and covered with tarps for transport to the Bark Camp Mine Reclamation site in Pennsylvania's Clearfield County. Unloading at Bark Camp is accomplished using an excavator located on an elevated structure over the rail spur. The material is placed directly into off-road trucks and transferred to the final processing area. There the material is blended with cement and lime kiln dusts and coal fly ashes in a pugmill system according to a pre-determined mix design, and discharged onto a radial stacking conveyor. The final manufactured fill is transported utilizing off-road trucks to the reclamation area of the highwall. The fill is spread in two-foot thick lifts with a low-ground-pressure bulldozer and compacted using a vibratory roller.

Sampling And Testing Protocol. This project is conducted pursuant to all local, state, and federal regulations applicable to the dredging, processing, transport, and beneficial use of the dredged sediments and additives. Certain permits were required for each location and/or operation of the project. The testing protocols include full Toxicity Characteristic Leaching Procedure (TCLP), organics, inorganics, metals, pesticides, herbicides, PCBs and dioxin/furans. Composites taking during shipping undergo TCLP and analysis for PCBs and dioxins. The final product undergoes the Synthetic Precipitation Leaching Procedure (SPLP).

Quality Control measures for this project included characterizing the chemical and/or physical properties of the raw dredged sediment and any additives utilized in the treatment process prior to commencing dredging. The physical and chemical properties of the additives were determined from testing and analytical results provided by the generators of these materials. Vendors supplying coal ash products were required to demonstrate that the ash materials meet the PADEP Module 25 chemical criteria for ash placement in mine reclamation. Additionally, Material Safety Data Sheets (MSDS) that are available for these materials are kept on file in the administrative office at the Bark Camp facility.

The chemical analysis protocol for the dredged sediment intended for Bark Camp consists of three stages. These include the core sampling and analysis of in situ dredged material (Stage I); the sampling and analysis of the dredge material at the portside offloading facility, which is intended to confirm that the material being shipped is similar to the in situ materials (Stage II); and the sampling and analysis of the treated materials at the Bark Camp facility to assure that the manufactured materials comply with the criteria established in the Beneficial use Order (BUO), (Stage III).

The sampling and analysis is conducted in accordance with the requirements of the BUO and the New Jersey Department of Environmental Protection (NJDEP) guidance manual entitled The Management and Regulation of Dredging Activities and Dredged Material in New Jersey's Tidal Waters; October 1997. The New Jersey manual (the "Guidance Document") specifies sampling and analytical requirements for upland disposal and beneficial use of dredged materials in the State of New Jersey. The manual specifies sampling procedures and frequency requirements, target analyte lists, analytical test methods to be used, and acceptable method detection limits for in-situ sediment samples.

A Sampling and Analysis Plan (SAP) for the in situ sediment is prepared for this project and submitted for the NJDEP and PaDEP's review and approval. Individual core samples of the in situ sediment is taken to the proposed project depth plus allowable over-dredge. Composite samples are prepared from the individual core samples. The individual core and composite samples are subjected to the analysis specified in the Guidance Document and the approved SAP.

Bench tests utilizing the sediment from the in-situ testing and various percentages of additives are performed to simulate the creation of the manufactured fill. The bench test product samples are analyzed in order to chemically and physically characterize the manufactured fill and to determine the ability of the fill from each mix design tested to stabilize chemical constituents found in the in-situ sediment. The analytical and test results for the Stage I in-situ sediment samples are submitted to the NJDEP and PaDEP for their respective review and approval. NJDEP issues the Waterfront Development Permit allowing dredging and PaDEP issues written approval for use of the dredged materials at Bark Camp.

Stage II testing occurs at the portside facility and is performed to confirm that the pre-amended dredged material is physically and chemically characteristic of the material sampled in Stage I. This confirmatory sampling is performed pursuant to the BUO, at a frequency of one composite per 25,000 cubic yards of dredged material. Accordingly, one (1) composite sample is chemically analyzed and geotechnically tested pursuant to the requirements specified in the BUO. The analytical and test results are reported to the PADEP for its review and information.

The final stage (Stage III) of the QA process is performed after the final amendment of the dredged sediment at the Bark Camp facility. One (1) composite sample of the manufactured fill is obtained pursuant to the BUO requirement of one composite sample per every 25,000 cubic yards of material. The composite sample is chemically analyzed and geotechnically tested pursuant to the specific requirements specified in the BUO for the manufactured fill. The analytical and test results are reported to the PADEP for its review and information.

The analytical results for a set of in situ sediment samples, the pre-amended (portside) dredge material, and the manufactured fill placed at the Bark Camp facility are presented in the cited website, as are the geotechnical test results of the manufactured fill.

Quality Assurance measures (Stages II and III) for this project are implemented to confirm that the chemical and/or physical properties of the pre-amended dredged material transported to Bark Camp and the manufactured fill are similar to that of the in situ sediment sample properties.

CONCLUSIONS

The following conclusions can be made from monitoring the material first placed at Bark Camp three years ago and monitored since, and the material placed since then.

These data demonstrate that the technology producing the manufactured fill derived from the dredged material has effectively stabilized the chemical constituents present in the in-situ dredge material, and that the manufactured fill is physically competent as an engineered fill material suitable for use in mine reclamation.

By adding and mixing the dredged material with specific quantities and sources of coal ash, the dredge material was successfully pre-amended at the portside facility, rendering a material that could be handled and transported by common earth handling techniques and equipment.

Laboratory chemical analyses and geotechnical test results indicate that the processes creating the recyclable fill manufactured at Bark Camp have successfully solidified and chemically stabilized the dredge material. Synthetic Precipitation Leaching Procedure (SPLP) testing of the final product emplaced in the high wall shows non-detects for metals and organics

Water quality testing from six deep wells below the site and surface water collection points all pass drinking water quality standards. Water quality test points, maps and results are available at: http://www.dep.state.pa.us/dep/DEPUTATE/MINRES/BAMR/bark_camp/WaterQdata/WaterQ.htm

Geophysical tests were conducted in October 1999 by Lamont Dougherty Earth Observatory, including Ground Penetrating Radar, resistivity, conductivity and seismic imaging. These tests indicate that the placed material is uniform, solid, and has no water moving below or through it.

The stream below the reclaimed area has seen the return of over-wintering trout populations, which have not been observed since mining.

Risk Communication. It is difficult to understate the complexity of communicating the benefit of shipping mud from New York Harbor into central Pennsylvania. Given that the Commonwealth already accepts much of New York and New Jersey's solid waste, and that dredged material was vilified beyond any real threat to human or environmental health in the campaign to cease its ocean disposal, this task was especially sensitive. PaDEP is to be commended for an exhaustive effort in community outreach that preceded the project. Literally dozens of visits were undertaken to gain background and to understand sensibilities. PaDEP held several open houses regarding the project and worked closely with its Citizen's Advisory Committee to gain support for an honest assessment of this concept's potential.

Future Projects. This project was largely funded by a bond act passed in the State of New Jersey specifically to find solutions for the dredging issue, and by the Port Authority of New York/New Jersey. Bark Camp was chosen for its scientific merits, and the operation was built around it, not the other way around. A future project would have to be sited in an area that balanced several considerations including: accessibility to rail, degree of impact on the local environment that is significant and remediable, availability of ancillary funding (like reclamation bonds); nearby sources of admixture and acceptance by the local community. The greatest cost-effectiveness and economies of scale are achieved by piggy-backing multiple missions onto single projects, such as AMD abatement, wetlands restoration, mine reclamation and dredged material and coal ash disposal.

A particularly beneficial application of this technology may involve the grouting of acid drainage-causing deep mine features, which we hope to demonstrate at Bark Camps existing deep mine structures.

The ability to beneficially use the hundreds of millions of tons of dredged materials and coal fly-ash produced world wide has unprecedented implications not only for the 560,000 abandoned mine sites in the US, but for the millions of abandoned mine features and the Acid Mine Drainage they generate in North America, Europe, Africa and China.

NOTES AND REFERENCES

[1]A Plan to Dump Coal Ash Adds Salt to a Wound, *New York Times*. Business Section p. 1, April 14, 2000.
[2]Dredged Materials should be distinguished from sludge (treated human waste) and other materials, the names of which are often inaccurately applied.
[3]EPA Office of Water website: http://www.epa.gov/OST/cs/

CONTAMINATED SEDIMENTS: RAW MATERIAL FOR BRICKS

Jan Peter Ulbricht, Managing Director, HANSEATEN-STEIN Ziegelei, Hamburg, Germany

ABSTRACT: Since 1996 the HANSEATEN-STEIN GmbH (HZG) operates a brick-factory in Hamburg, Germany, which utilizes contaminated sediments from the Port of Hamburg as main raw material for the production of facing bricks (HANSEATEN-STEIN®) for the North-German building industry. The factory is running at an industrial scale, producing up to 5 million bricks per year utilizing up to 35.000 t of drained port sediments. It has been proven that it is technically possible, ecologically reasonable and economically attractive to utilize contaminated port sediments as the main ingredient for a facing brick. The know-how accomplished serves as a base for a planned large scale industrial plant in Hamburg, Germany, utilizing annually 300.000 t of drained port sediments.

INTRODUCTION

More than 2 million m³ of sediments have to be dredged out of the Port of Hamburg every year in order to ensure safe and efficient ship traffic. The River Elbe on her way from the Czech Republic through former East-Germany to the North-Sea leaves those sediments within the port basins of Hamburg. Like many other port cities, Hamburg has developed an advanced system of dredging and handling the sediments. With a world-wide unique technical draining and separation plant, all sediments are separated into gravel, sand, fine-sand and sludge, the latter forming the clay-fraction and thus containing all contamination.

After draining and separation, the gravel- and sand-fractions (250.000 t, dry matter) are used within the port for construction activities or are sold to the building industry. App. 200.000 t (dry matter) of the sludge are used for sealing disposal sites and the remaining 260.000 t (dry matter) harbor sludge are disposed at special mono-disposal sites in Hamburg (Umweltbehörde Hamburg, 2001).

In order to decrease the amounts which are disposed annually at the disposal-site, HZG examined in a specially built brick-factory during a four-year trial period the possibilities to utilize harbor sludge for brick-making.

Objectives. To find out about the technical possibilities, the ecological impacts and the economical attractiveness of making bricks from contaminated harbor sludge were the three main questions, when production was started in 1996.

"Technical possibilities" means finding the right mix of raw materials and the right production methods in order to produce a facing brick that complies with the German Norm DIN 105 (Deutsches Institut für Normung e.V., 1989) and thus can be sold in the German construction market in competition with traditional bricks.

"Ecological impacts" describes the consequences during the separate steps of the production process and the produced bricks (emissions and immissions) during their life cycle.

"Economical attractiveness" describes the cost issue from the sludge owner's point of view. The underlying question is, whether brick production as an advanced technical solution may compete with its costs with alternative solutions.

Site description. The HANSEATEN-STEIN® brick factory in Hamburg goes back to a traditional brick making facility, which started production in 1888. During the years 1995 and 1996 the developers of the HANSEATEN-STEIN® technology built a completely new factory using the old foundations and installing a custom-made machinery for each production step. The total capacity installed is for 35.000 t of drained harbor sludge or 5 million facing bricks.

MATERIAL AND METHODS

Material. Approximately 70% of the raw materials used to produce the HANSEATEN-STEIN® is drained harbor sludge from the Port of Hamburg. The following table shows typical contaminants of the sludge:

TABLE 1. Typical contamination of drained harbor sludge from the Port of Hamburg (HZG, 2001)

Compound	Unit	Concentration (median)
TOC (C)	wt.- % dry matter	3
Sulfur, total	mg/kg dry matter	2980
Chlorine, total	mg/kg dry matter	592
Fluorine, total	mg/kg dry matter	453
Arsenic (As)	mg/kg dry matter	22
Lead (Pb)	mg/kg dry matter	76
Cadmium (Cd)	mg/kg dry matter	2
Chromium (Cr)	mg/kg dry matter	37
Copper (Cu)	mg/kg dry matter	74
Nickel (Ni)	mg/kg dry matter	29
Mercury (Hg)	mg/kg dry matter	1
Zinc (Zn)	mg/kg dry matter	516
TBT (tri-butyl-tin)	µg/kg dry matter	1120

The remaining 30% is natural clay from Germany's Westerwald region, where a large variety of natural clays in different qualities are mined for brick factories worldwide. It is essential to use a "matching" clay that fits well with the characteristics of the sludge. Since the sludge - for clay standards – contains a lot of sand and thus creates more pores in the brick than desired, a matching clay must burn very densely.

Drying. As the harbor sludge arrives from the draining facility it still contains more than 40% of water that need to be dried out to 20% residual water. At temperatures of not more than 50°C / 122°F a fluidized-bed-dryer brings the water-content down to 20%, which is perfect for the subsequent shaping.

Shaping. After the dried sludge is mixed thoroughly with natural clay, a vacuum-extruder shapes the mix into the desired brick format. Vacuum-extrusion is the classical and - due to optimized machines - the most effective and powerful way of shaping in modern brick making.

Drying. Before the freshly shaped, "green" brick, can be baked in the tunnel-kiln it needs to be dried completely. This is achieved in a tunnel dryer at temperatures of app. 200°C / 392°F. All flue-gases from this process are reused in the following tunnel-kiln.

Baking. Modern tunnel-kilns ensure a ceramic careful and yet cost efficient firing method. Depending on the desired face of the brick the HANSEATEN-STEIN® is baked at temperatures of 1060°C / 1940°F. All flue-gases originating from the baking process are burnt in an auto-thermal combustion chamber, making sure that all gases are finally burnt at more than 850°C / 1562°F.

Purification. All flue-gases coming out of combustion are cleansed in a flue-gas purification system using a mix of lime hydrate ($Ca(OH)_2$) and active carbon.

RESULTS AND DISCUSSION

Technical possibilities. Like many other industrial countries Germany regulates its construction materials with industrial norms. Such norm regulates basically the physical quality of the material (e.g. gross density, compression strength, water absorption). The HANSEATEN-STEIN® meets all requirements of the corresponding norm DIN 105 since 1997 and is listed (Güteschutz Nord-West e.V., 1997) with the following data (median):

compression strength: 35 N/mm^2
gross density: 1.61 – 1.80 kg/dm^3
water absorption: 12 %

These data classify the HANSEATEN-STEIN® into one category together with most of Germany's traditional facing bricks.

Ecological impacts. The smallest fraction (< 20 μm) of the sediments dredged from the Port of Hamburg is the fraction containing most contaminants. This harbor sludge contains a variety of heavy metals such as arsenic, zinc, lead and mercury as well as several organic substances and several organic-metallic combinations such as TBT. It is considered as waste under the European Waste Catalogue (European Commission, 1994).

Therefore throughout the whole factory the sludge has to be treated in a closed system. Especially when thermal treatment is involved it has to be ensured that no gaseous emissions are set free before proper purification. Full emission control has been in effect from the first day the factory started to operate.

The initial drying of the sludge takes place at no more than 50°C / 122°F, thus no emission treatment has to established. The main drying of the shaped brick is effected at 200°C / 392°F, where many PAH's and several metals (for example Hg) become gaseous. All flue-gas from this drying is therefore reused in the kiln.

During the thermal process of baking the brick at temperatures up to 1060°C / 1940°F all organic matter is burnt completely, whereas heavy metals partly become gaseous and partly are vitrified with the clay-minerals. Since the hot flue-gas is exhausted through the first quarter of the kiln and thus is pulled hot from the burning area, it mixes with chlorides, sulfides, fluorides and unburned organic elements. This gas mix is burnt again in an auto-thermal combustion chamber.

The final flue-gas purification is subject to the strictest emission control regulation in Germany, which also applies to waste incinerators and power plants.

A very important question is the emission characteristics of the brick product. Many tests have been carried out in order to simulate the "life-cycle" of a facing brick from using it in the facing wall to its destruction and maintenance as building rubble on the disposal site. Predicting the life-cycle of a facing brick is very new to the brick industry and thus no data for comparison is available. However, it has been proven in several tests that no future mobilization of contaminants may be expected from HANSEATEN-STEIN® bricks, which may collide with any corresponding regulation (Karius, et.al., Bremen, 1999; LAGA, Berlin, 1996). Table 2 below shows an example of the latest leaching test performed with HANSEATEN-STEIN® bricks, produced from harbor sludge described in Table 1:

TABLE 2. Results of leaching test (LAGA) of HANSEATEN-STEIN® bricks made from sludge from Port of Hamburg (IZF, Essen, 2001)

Compound	Unit	Concentration
Chloride (Cl^-)	mg/l	< 10
Sulfate (SO_4^{-2})	mg/l	140
Fluoride (F^-)	mg/l	0.4
Arsenic (As)	mg/l	0.01
Lead (Pb)	mg/l	< 0.005
Cadmium (Cd)	mg/l	< 0.0003
Chromium (Cr), total	mg/l	0.011
Copper (Cu)	mg/l	< 0.005
Nickel (Ni)	mg/l	< 0.005
Mercury (Hg)	mg/l	< 0.0002
Zinc (Zn)	mg/l	< 0.005

A direct comparison of the results in Table 1 with the results in Table 2 is hardly possible, since Table 2 contains results from leaching tests opposed to

Table 1's total contents. But it shows the dimension of transforming contaminants into ceramic ingredients. Of course, all values included in Table 2 in a facing brick are considered to constitute no harm for either environment, animals or humans (LAGA, Berlin, 1996).

Economical attractiveness. Making bricks from contaminated harbor sludge is especially valuable for rather highly contaminated sludges since it operates with temperatures above 1000°C / 1832°F. At the same time the finished product is a ready to sell construction material, equal to traditional bricks.

Thus there are two economical aspects to be considered: First the port city/authority makes use of a technology, which allows for very low cost in high decontamination at the same time. Second a construction material is produced, which enables the city to use a recycled product in its truest definition.

Giving precise cost figures for the technology presented is impossible, since factors like quantity of contaminated sludge, its physical contents and condition of retail market for bricks are important variables.

Examples from Hamburg and Bremen – where brick-factories with capacities of annually 100000 t and 300000 t are planned - have shown, however, that fees around € / $ 20.00 / t drained sludge are very well achievable.

REFERENCES

Deutsches Institut für Normung e.V.. 1989. *DIN 105, Teil 1 und Teil 3.* Berlin, Germany.

Hamer, K., V. Karius, J. Schröter, H.D. Schulz. 1997 - 2000. *Bremer Baggergut in der Ziegelherstellung.* Bremen, Germany

Institut für Ziegelforschung e.V.. 2001. *Prüfungsbericht Nr. CH 6326 – Eluatgehalte gemäß LAGA-Programm.* Essen, Germany.

Länderarbeitsgemeinschaft Abfall LAGA. 1996. *Anforderungen an die stoffliche Verwertung von mineralischen Abfällen. Mittlg. LAGA 20, Stand 5.9.1995.* Berlin, Germany.

Umweltbehörde Hamburg. 2001. *Abfallwirtschaftsplan Baggergut.* Hamburg, Germany.

TECHNICAL AND ECONOMIC ASPECTS OF USING CEMENT-LOCK^{SM} TECHNOLOGY FOR SEDIMENT TREATMENT

Michael Mensinger (ENDESCO Services, Inc., Des Plaines, Illinois)
Anil Goyal (ENDESCO Services, Inc., Des Plaines, Illinois)
Amir Rehmat (ENDESCO Clean Harbors, L.L.C., Des Plaines, Illinois)
Tony Lee (Cement-Lock Group, L.L.C., Des Plaines, Illinois)

ABSTRACT: The Cement-Lock^{SM} Technology thermo-chemically transforms contaminated sediment into construction-grade cement, which has properties similar to those of portland cement. During the thermo-chemical transformation, the mixture of sediment and proprietary modifiers is imparted with latent cementitious properties that allow it to be converted into construction-grade cement. Further, all organic contaminants are destroyed and converted to innocuous carbon dioxide and water. Heavy metals are locked within the molten matrix to completely immobilize them. This paper describes the Cement-Lock Technology, its targeted applications, development status, commercial availability, beneficial use applications, logistical and regulatory requirements, and estimated costs.

INTRODUCTION

The Cement-Lock Technology is a versatile, cost-effective, and environmentally friendly manufacturing technology for converting sediment into construction-grade cement, steam, and electric power (Rehmat *et al.*, 1998, 1999, 2000). The technology can be implemented as a central processing facility or it can be customized for the needs of specific localities. This paper focuses specifically on the treatment of contaminated sediment dredged from ports, harbors, and rivers.

The development of the Cement-Lock Technology has been supported through the federal Water Resources Development Act (WRDA) of 1992, 1996, and 1999. WRDA authorized funding for the purpose of demonstrating on a commercial scale (up to 500,000 yd^3 or 382,300 m^3) the capability to decontaminate contaminated dredged material from the harbors of New York and New Jersey. The WRDA Sediment Decontamination Program has resulted in the development and evaluation of both thermal and non-thermal technologies from the laboratory and bench scale to the pilot and field scale. The goal of the WRDA program is the construction and implementation of one or more sediments decontamination facilities that result in a treatment train capable of producing a viable end product for beneficial use applications such as construction or fill materials and agricultural grade topsoil (Jones *et al.*, 1999). The New Jersey Department of Transportation (NJ-DOT) Office of Maritime Resources is also sponsoring the Cement-Lock demonstration project via technical and financial support.

Technology Description. In the Cement-Lock Technology (Figure 1) the sediment is brought together in a reactive melter with suitable modifiers in proportions required for producing an environmentally friendly material called Ecomelt® with latent cementitious properties. These properties are utilized to convert the Ecomelt into construction-grade cement, which has compressive strength properties exceeding those required for ordinary portland cement. The quantity of modifiers added, which depends upon the sediment composition, is typically less than about 20 weight percent of the sediment.

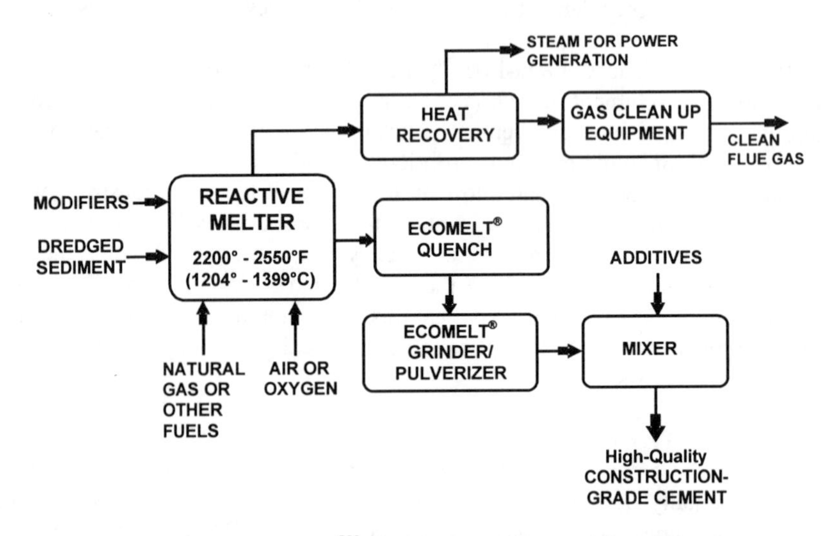

FIGURE 1. Cement-Lock℠ Technology Process Flow Diagram

The melter for carrying out this process is operated at temperatures up to about 2550°F (1399°C), sufficient to melt the sediment-modifiers mixture. In the presence of excess air/oxygen at these temperatures, organic components originally present in the sediment are completely utilized for their fuel value and converted to innocuous carbon dioxide and water. Inorganic components of the waste material are transformed into the cement matrix (predominantly calcium-alumino silicates), which are subsequently pulverized and mixed with an additive to yield construction-grade cement. The sensible heat contained in the flue gas is recovered to produce steam that can then be used for power generation.

Targeted Applications. Some of the targeted applications for the Cement-Lock Technology are listed below.

- Decontamination and conversion of dredged sediments from polluted harbors and rivers into construction-grade cement.
- Decontamination and conversion of contaminated soils and hazardous wastes from Superfund sites into construction-grade cement.
- Co-production of electricity, steam, and construction-grade cement from coal via integration of Cement-Lock Technology with coal burning power plants.

- Co-processing of wastes from a variety of industries such as automobile, food, petroleum, chemical, steel, aluminum, etc. to produce power and construction-grade cement.

Some advantages of Cement-Lock Technology deployment are:

1. Cement produced from sediment possesses strength properties similar to those of portland cement and therefore commands similar value in the market place.
2. A Cement-Lock plant emits considerably less CO_2 into the atmosphere compared to a typical cement plant. Therefore, it promotes the Kyoto Protocol for reduction of greenhouse gases from cement manufacturing.
3. Cost of cement production is drastically reduced due to receipt of tipping fees associated with the wastes.
4. Cement-Lock can operate equally effectively on sediment or a mixture of sediment and other wastes, thereby assuring a continuous supply of raw material needed for a successful business.

Development Status. Cement-Lock has been thoroughly evaluated in bench-scale studies with contaminated sediment from the Newtown Creek estuary in New York and the Trenton Channel of the Detroit River in Michigan. Pilot-scale studies (~1 ton/day scale) have also been conducted with Newtown Creek sediment. Both of these sediments are contaminated with polynuclear aromatic hydrocarbons, polychlorinated biphenyls, pesticides, insecticides, polychlorinated dioxins and furans, and heavy metals.

The results of the bench- and pilot-scale test programs for evaluating the Cement-Lock Technology have been very favorable and are summarized in Tables 1, 2, and 3 below. In summary,

- All hazardous organic contaminants, including oil and grease, PAHs, PCBs, pesticides, insecticides, chlorinated dioxins and furans, were destroyed to well below regulatory limits (Table 1).

- All heavy metals were immobilized within the cement matrix. The construction-grade cement (the end product) passed the U.S. EPA TCLP (Toxicity Characteristic Leaching Procedure) test (Table 2). The leachability of metals is several orders of magnitude below the regulatory limits.

- Samples of mortar were prepared from the cement product, sand, and water according to ASTM (American Society for Testing and Materials) C 109 procedures. The compressive strengths of these mortar samples were determined according to the ASTM procedure after 3, 7, and 28 days of curing (Table 3). The compressive strength tests were conducted by Construction Technology Laboratories, Inc. (CTL - the research arm of the Portland Cement Association, USA). The mortar produced from the cement product exceeded ASTM compressive strength requirements. Therefore, it can replace portland cement for general construction applications.

- The process did not generate any secondary hazardous waste streams.

- The pilot-plant data were consistent with the laboratory-scale data in terms of organic destruction, leachability, and the quality of the cement generated from the estuarine sediments.
- The total metal contents in the cement product from the Cement-Lock Technology are usually within the range of total metals found in ordinary portland cement.

TABLE 1. Organic Contaminant Destruction Achieved With Sediment

	----- Estuarine Sediments -----			----- River Sediments -----		
Organic	Untreated Sediment	Cement Product	DRE*	Untreated Sediment	Cement Product	DRE
	--- mg/kg (dry) ---		-- % --	--- mg/kg (dry) ---		- % -
Oil & Grease	--	--	--	18,000	< D.L.**	> 99.99
SVOCs[+]	370	0.22	99.93	51.2	< D.L.	> 99.99
	---- μg/kg (dry) ----			---- μg/kg (dry) ----		
PCBs	8,585	< D.L.	> 99.99	1,100	< D.L.	> 99.99
	---- ng/kg (dry) ----			---- ng/kg (dry) ----		
2,3,7,8-TCDD/F[+]	262	< D.L.	> 99.99	--	--	--
TCDD/Fs	2,871	< D.L.	> 99.99	--	--	--
PeCDD/Fs	4,363	< D.L.	> 99.99	--	--	--
Hx/Hp/ OCDD/Fs	34,252	< D.L.	> 99.9	--	--	--

* Destruction and removal efficiency.
** Less than detection limit of the analytical procedure used.
+ SVOCs – Semivolatile organic contaminants;
TCDD/F – Tetrachloro-dibenzo dioxins and furans, etc.

TABLE 2. Metal Immobilization in Construction-Grade Cement Produced From Sediment

			TCLP*		
	-- Untreated Material --		-- Cement Produced from --		
Metal	Estuarine Sediment	River Sediment	Estuarine Sediment	River Sediment	Regulatory Limit
	------- mg/kg (dry) -----		---------- mg/L ----------		
Arsenic	39	7.8	< 0.005**	< 0.01	5
Cadmium	27	9.5	< 0.001	< 0.002	1
Chromium	298	138	0.15	< 0.072	5
Cu	--	180	--	< 0.01	--
Lead	542	218	< 0.002	< 0.01	5
Mercury	2.9	0.55	< 0.0004	< 0.0004	0.2
Selenium	6.2	--	< 0.003	--	1
Silver	13	--	< 0.001	--	5

* Toxicity Characteristic Leaching Procedure.
** Less than the detection limit of the analytical procedure used.

TABLE 3. Compressive Strength of Construction-Grade Cement Produced from Sediment

Test	Cement-Lock Cement From			ASTM Requirements	
	Estuarine Sediment		River Sediment	Blended Cement	Portland Cement
Period	Lab-Scale	Pilot-Scale	Lab-Scale	C 595	C 150
Days	psi (MPa)				
3	1,950 (13.4)	2,230 (15.4)	2,245 (15.5)	1,890 (13.0)	1,740 (12.0)
7	2,730 (18.8)	2,885 (19.9)	2,910 (20.1)	2,900 (20.0)	2,760 (19.0)
28	4,620 (31.9)	5,270 (36.3)	4,600 (31.7)	3,480 (24.0)	4,060 (28.0)

These findings confirm that the Cement-Lock construction-grade cement meets all environmental requirements and is of commercial quality.

A demonstration plant designed to treat from 30,000 to 150,000 cubic yards (22,900 – 114,700 m^3) of sediments is under construction in New Jersey. This project is sponsored through the WRDA and the NJ-DOT Office of Maritime Resources. Another demonstration will be conducted on 2,000 to 5,000 cubic yards (1,530 – 3,820 m^3) of contaminated sediments from the Detroit River. Engineering has been completed on a 100,000 yd^3 (76,500 m^3) per year capacity process module and a 500,000 yd^3 (382,300 m^3) per year capacity plant composed of multiple modules.

Commercial Availability. All equipment required for commercial applications is commercially available. Various vendors willing to provide turnkey production plants have already been identified.

A large-scale (500,000 yd^3/year or 382,300 m^3/year of sediment capacity) Cement-Lock plant would produce in about 200,000 tons (181,400 metric tons) per year of construction-grade cement. This is less than 20 percent of a typical full-scale cement plant. Marketability for this volume of product should be good, since there continues to be an excellent demand for cement. In 1999, about 105 million metric tons of cement was used in the United States of which over 22 million metric tons was imported (Portland Cement Association, 2000).

Beneficial Use Applications. The Cement-Lock product has been tested and shown to be comparable in performance to commercial portland cement. In addition to general commercial use, other markets that are related to environmental and dredged sediment management include:

- general construction for sediment processing stakeholders
- grouting of underground storage tanks at U.S. DOE/U.S. DOD sites
- soil conditioning at landfills
- sediment stabilization processes, and

- construction of retention walls in mines where sediment is used for backfilling.

Logistical and Regulatory Requirements. The Cement-Lock process is only one component of the process train required to move dredged sediment from the harbor or river to its end use. Transportation and storage systems for raw and processed sediment need to be designed and implemented on a site-specific basis. A processing plant requires a constructed site large enough for the equipment and sediment storage with water, fuel, and power utilities. Storage of dredged sediment is required to provide a steady flow of raw sediment to the plant. There is no wastewater generated during processing (excluding wastewater from dewatering included in other processing steps). Debris from the screening process must be transported off-site for disposal. No specific residue requiring special disposal permits has been identified as being generated in the off-gas cleanup process of the Cement-Lock plant.

Variability in the physical and chemical characteristics of the dredged sediment does not to affect the technical performance of the process. The effects of different factors on the technical performance and the economics of the Cement-Lock Technology are presented in Table 4.

TABLE 4. Effect of Various Factors on the Performance and Economics of the Cement-Lock Technology

Technology Performance		Technology Economics	
Factor	Effect	Factor	Effect
Moisture content	No effect	Moisture Content	Affects quantity of cement & revenues
Variations in type and quantity of organic contaminants	No effect	Non-Steady Supply	Larger storage required – affects capital costs
Sediment quality (clayey, loamy, sandy)	No Effect	Use of alternate fuels (sewage sludge and/or refinery tank bottoms	Considerably reduces sediment treatment costs
Sediment particle size	No Effect	Non-continuous batch processing of sediment aliquot	Capital attributed to batch increases processing cost

Favorable costs for the process require a steady and guaranteed supply of sediments or other waste material (i.e., soils, sludges, ash, *etc.*) and a guaranteed buyer and stable price for the cement product for a period of at least 10 to 20 years. Several factors shown in Table 4 are noted that would affect the overall cost. Excessive moisture content would reduce the quantity of cement produced. A non-steady supply of sediments would require larger storage capacity thus increasing capital costs. Use of additional waste materials, particularly high-Btu content materials, would reduce the treatment cost and would supplement and level the feed of raw materials to the process. Non-continuous processing of sediments would increase the product cost.

With respect to regulatory requirements, the Cement-Lock processing plant requires permits similar to those for manufacturing process plants, including an air permit. No wastewater discharge permit is required since none is generated. Specific permitting requirements vary depending on the location in which the system is operated.

Estimated Cost. The processing costs for sediment using the Cement-Lock Technology are shown in Table 5. Operating costs are based on a dedicated 500,000 yd^3 (382,300 m^3) per year capacity plant. Several scenarios are presented including the use of different fuels (natural gas versus high-Btu content waste sludges) and different market prices for the Cement-Lock product. A full market value for cement of about $70 per ton (Cases 1 and 2) and a discounted value of $35 per ton (Case 3) are used for illustrative purposes. The break-even processing costs ranged from -$8.66 (profit) to $32.69 per yd^3 of sediments. The profitable scenario is based on the use of a waste material as a fuel source (such as refinery tank bottom) for which a tipping fee is collected and full market value for the cement product is collected.

TABLE 5. Unit Processing Costs for a 500,000 yd^3 (382,300 m^3) per Year Capacity Cement-Lock Plant

	Case 1	Case 2	Case 3
Credits		x $1000	
Sale of Cement ($70/ton)	$17,440	$23,100	$11,550
Tipping Fees From Other Wastes	--	18,810	18,810
Total Credits	**$17,440**	$41,910	$30,360
Expenses			
Labor	$930	$930	$930
Raw Materials	3,470	4,035	4,035
Utilities	12,110	15,340	15,340
Maintenance/Misc.	5,775	5,775	5,775
Total Expenses	**$22,285**	**$26,080**	**$26,080**
Provision for Lease	$1,500	$1,500	$1,500
Depreciation/Financing Allowance	$10,000	$10,000	$10,000
Net Processing Cost	**$16,345**	**($4,330)**	**$7,220**
Break-Even Tipping Fee, $/yd^3	**$32.69**	**($8.66)**	**$14.44**

Case 1: 100% harbor sediment feed.
Case 2: 75% harbor sediment feed + 25% other waste feed.
Case 3: Same as Case 2 except cement sale price reduced by 50%.

CONCLUSIONS

The Cement-Lock Technology has been successfully demonstrated at the laboratory- and pilot-scale levels with both contaminated harbor and river sediments. Cement-Lock effectively destroys organic contaminants, immobilizes inorganic contaminants, and produces a beneficial use product, namely, construction-grade cement. Because of these favorable environmental attributes, the Cement-Lock Technology offers greater community acceptance. A 30,000 yd^3 (22,900 m^3) per year of contaminated sediments capacity plant is being constructed to demonstrate integrated operation of the technology at a larger scale. The construction-grade cement generated during the demonstration project will be utilized in appropriate projects designated by the NJ-DOT.

Acknowledgments. The work is supported in part under Contract No. 725043 with Brookhaven National Laboratory with funding provided, in part, through the 1996 Water Resources Development Act (Section 226) through Interagency Agreement No. DW89941761-01-1 between the U.S. EPA-Region 2 and the U.S. Department of Energy. Funding from the NJ-DOT Office of Maritime Resources is also gratefully acknowledged.

REFERENCES

Jones, K. W., E. A. Stern, K. R. Donato, and N. L. Clesceri. 1999. "Sediment Decontamination Treatment Train: Commercial-scale Demonstration for the Port of New York/New Jersey," Paper presented at the Nineteenth Western Dredging Association (WEDA XIX) Annual Meeting and Conference and 31st Texas A&M University Dredging Seminar, Louisville, Kentucky, May 15-20.

Portland Cement Association, 2000. *U.S. Cement Industry Fact Sheet (2000 Edition).*

Rehmat, A., A. Lee, M. C. Mensinger, and A. Goyal. 2000. "Co-Production of Electricity and Cement From Coal." Presented at Coal Power Project & Finance Workshop, Porto Alegre, Rio Grande do Sul, Brazil, February 27-March 1.

Rehmat, A., A. Lee, A. Goyal, M. C. Mensinger, and J. I. Bhatty. 1999. "Production of Construction-Grade Cements From Contaminated Wastes," Presented at the Fifth International Conference on Concrete Technology, New Delhi, India, November 17-19.

Rehmat, A., A. Lee, A. Goyal, M. C. Mensinger, J. I. Bhatty, and S. P. Barone. 1998. "Production of Construction-Grade Cements From Wastes Using Cement-Lock™ Technology," Presented at the Fourth Beijing International Symposium on Cement and Concrete, Beijing, China, October 27-30.

TREATMENT OF CONTAMINATED DREDGED MATERIAL IN THE NETHERLANDS

R.H.P. Ringeling (Ministry of Transport, Public Works and Water Management, Dienst Weg-en Waterbouwkunde, Delft, The Netherlands)
J. Rienks (Ministry of Transport, Public Works and Water Management, Institute for Inland Water Management and Waste Water Treatment, Lelystad, The Netherlands)

ABSTRACT

Responsible water quality management employs dredging to improve both the quality of the water as well as the water basin floor. Dredging is used to maintain the accessibility of the waterways and harbours, primarily for depth control and to enhance the influent water flow. Each year enormous quantities of sediment thereby become available. It is, of course, desirable to resourcefully reuse as much of the sediment as possible. Processes and techniques which alter the properties of contaminated dredged material are desirable if they can render the dredged sediment suitable for beneficial use. There are several categories of treatment techniques for contaminated dredged material. One of these categories is physical treatment. Physical treatment techniques are applied with the aim of separating or concentrating the contaminants, and are essentially based on particle separation technology. In this paper the principles and results of two different physical treatment techniques, a sedimentation depot and a hydro cyclone, are compared.

INTRODUCTION

In The Netherlands, a yearly amount of 3.5 million situ m3 of contaminated sediment is dredged. The origin and destination of this amount of dredged material is shown in table 1.

TABLE 1. Supply and destination of contaminated dredged material, The Nederlands 1998

Origin	Supply in 1998 (1,000 situ m3)	Disposal site %	Temporary disposal %	Direct use %	Treatment %
Governmental waters	1,084	93	0	0	7
Port of Rotterdam	623	100	0	0	0
Waterboards	1,374	30	52	12	6
Municipal waters	419	56	38	2	4
SUM	3,500	65	25	5	10

At the moment, the majority is disposed of in dumping sites under water, isolated from the surface water by dikes. Especially the central government and the port authorities of Rotterdam use large dumping sites of tens of millions m3 content. The water boards and the municipal authorities have less and smaller dumping sites to their disposal.

Figure 1 shows the disposal and treatment options for contaminated dredged material according to the Dutch opinion. Particle separation is a widely spread available technology. Cold or hot immobilization are not common practice.

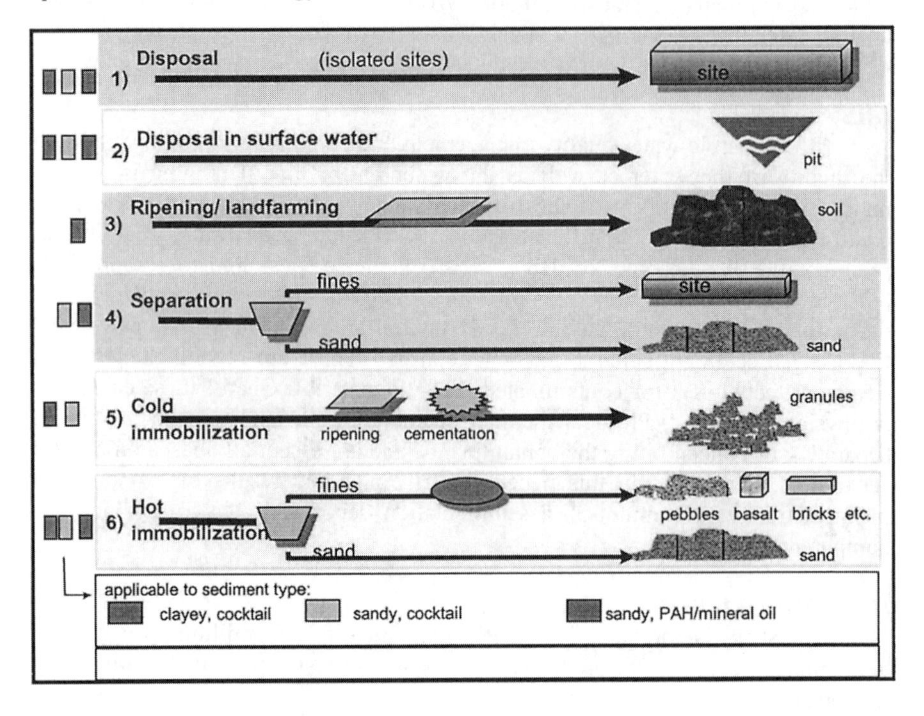

FIGURE 1. Options for finding a destination for contaminated dredging spoil

PARTICLE SEPARATION TECHNOLOGY

The goal of particle separation is to concentrate the pollutants in a small fraction of the material to be treated. This fraction may be stored under controlled conditions or treated by sintering or melting.

The technology is based on differences in grain size, specific gravity and surface properties.

The following technology is applied on sediments:

- sedimentation lagoons;
- hydro cyclones and a fluidised bed up flow column;
- a combination of both options.

A sedimentation lagoon makes use of natural separation during filling: the larger and heavier particles will settle at the first part of the lagoon, whereas the smaller particles will settle later on when the water flow slows down. The result is a spatial particle size gradient in the lagoon from filling point to the discharge. Sedimentation lagoons are used near a dredging site as well as next to a disposal site (figures 2 and 3).

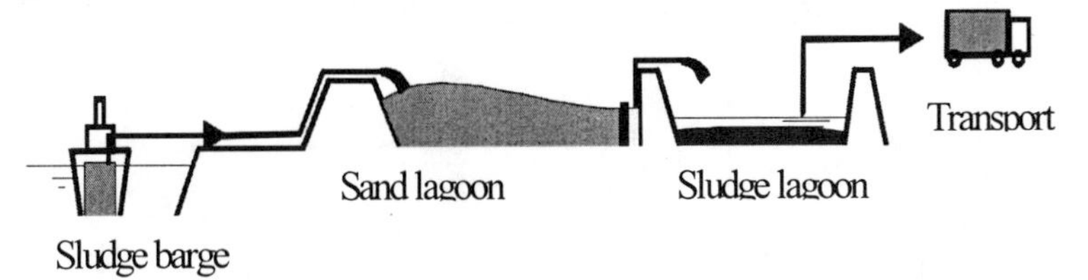

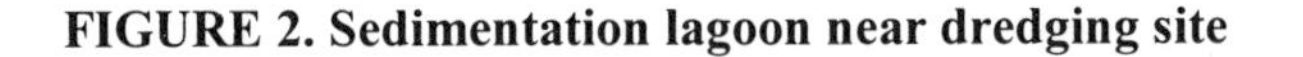

FIGURE 2. Sedimentation lagoon near dredging site

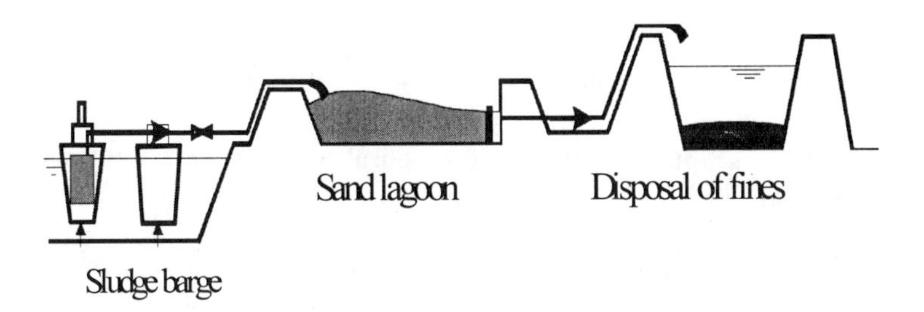

FIGURE 3. Sedimentation lagoon combined with a disposal site

Hydro cyclones are the heart of a particle separation plant. In a cyclone the separation in sand and a fines fraction takes place forced by a combination of tangential and gravity forces. Before entering the cyclone coarse material like rubble is removed by sieves.

BENEFITS OF PARTICLE SEPARATION

The benefits of particle separation are:

- production of re-usable sand;
- saving of limited disposal capacity.

These benefits depend on the sand content and the grain size (fine, medium or coarse) of the contaminated dredged material. The production of re-usable sand is also determined by the type of separation used and the type and speciation of contaminants present in the dredged material. The experience obtained with fine graded sandy sediments shows for a hydro cyclone a recovery of re-usable sand

between 70 to 90% of the sand content while in case of a sedimentation lagoon the recovery amounts usually 60-70%.
The regained fine graded sand is suitable for application as levelling material.

The saving of disposal capacity is subject to research projects. Figure 4 shows a example of the saving of disposal capacity as a function of the sand content for a sample originating from the Zeeland region. The saving is calculated comparing the disposal of separated fines to disposal of the untreated dredged material.

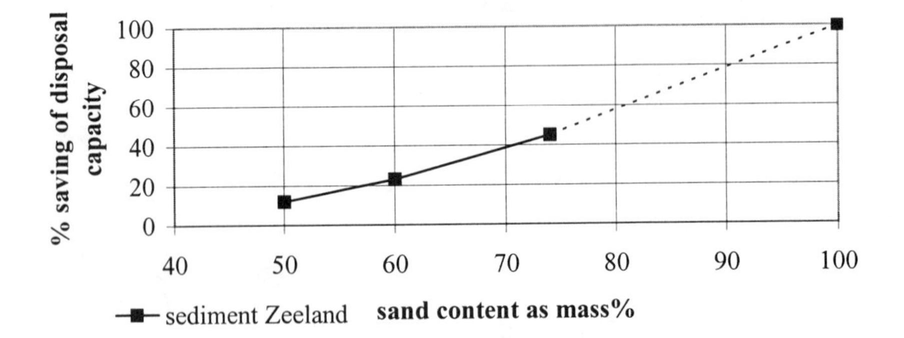

FIGURE 4. Example of saving of disposal capacity for sediment from Zeeland region

Figure 4 indicates that the saving decreases to nil near a sand content of approximately 35-40% by mass for this particular sediment. This implies that nearly half of the yearly supply of contaminated dredged material may be treated by particle separation.

COSTS

Table 2 depicts the costs for particle separation of sediments.

TABLE 2. Costs for particle separation of sediments

Technology	Cost in /situ m3 (disposal of residue excluded)
Sedimentation lagoon at a project	10-20
Sedimentation lagoon at disposal site	5
Mobile Hydro cyclone plant	15-20
Hydro cyclone plant at disposal site	10

Costs can be kept low under condition of sufficient supply during a large enough period.

EFFECTS OF TWEEN 80 AND ELECTRIC FIELD ON BIODEGRADATION OF PHENANTHRENE

Peng She, Zheng Liu, Jiangang Yang, Xiang Liu
(Tsinghua University, Beijing, P.R. China)

ABSTRACT: Effects of nonionic surfactant on the biodegradation of phenanthrene by *pseudomonas aeruginosa* were investigated with emphasis on their impacts on the bioactivity of microorganism. Tween 80 shown the best performance in enhancing the solubilization of phenanthrene in aqueous phase and was thus used as solutizer. Experimental results indicated the existence of a threshold concentration of Tween 80, over which Tween 80 appeared a superior bioavailability over phenanthrene. Below the threshold, the strain's activity in degrading phenanthrene increased gradually. The degradation was mainly accomplished in the stationary phase of bacterium growth. In addition, variation in the initial concentration of phenanthrene gave similar degradation rates. Application of electric field enhanced both the bacterium growth and the degradation of phenanthrene.

INTRODUCTION

Degradation of polycyclic aromatic hydrocarbons (PAHs) has attracted growing attention in recent years due to its highly toxic, carcinogenic and mutagenic potential. The degradation methods available at present include volatilization, photo-oxidation, chemical oxidation, biodegradation and adsorption. Among these methods, microbial degradation has shown to be most promising for the removal of PAHs due to its availability and versatility (Gibson *et al.*, 1975).

The solubility of PAHs is one important factor that affects the efficiency of biodegradation process. Addition of surfactant that functions as solutizer to reaction system has been extensively attempted (Bury and Miller, 1993; Volkering *et al.*, 1995; Kim *et al.*, 2001). While positive effect of surfactant in terms of enhancing the solubility of PAHs was proven (Aronstein and Alexander, 1993, Boonchan *et al.*, 1998), negative impact of surfactant, e.g., the inhibition of microorganism growth, was also identified (Laha and Luthy, 1991, Yuan *et al.*, 2000). These interesting results attracted us to the further understanding of the function of surfactant in the biodegradation of PAHs, especially the impacts of surfactants on the bioactivity of microorganism.

In present study, phenanthrene was selected as a model compound of PAHs. *Pseudomonas aeruginosa,* a phenanthrene-degrading bacterium isolated from the PAHs contaminated sediment taken at the place of Qilu Refinery Company in Shandong province, China, was used in the experiments. The effects of several solutizers including Tween 80, one of the most frequently used nonionic surfactant in biodegradation (Cuny *et al*, 1999; Zheng and Obbard, 2001), β-cyclodextrin (Wang *et al*, 1998), and DOWFAX (Deshpande *et al.*, 2000) on the solubilization of phenanthrene were examined.

In our previous studies on electrokinetics separation, (Liu *et al.*, 2001; Zhou *et al.*, 2001;), we applied an electric field to enhance the mass transport between liquid-solid phase. We found it particular effective in enhancing mass transport inside porous particle. Thus we extended our efforts by applying an electric field into the biodegradation of phenanthrene to enhance phenanthrene transport from aqueous phase to the surface of bacterium. Furthermore, it will be of great interest to examine the effect of electric field on the degradation of phenanthrene.

MATERIALS AND METHODS

Chemicals. Chemicals used in present study are listed in Table 1.

Table 1. Chemicals used in present study.

Chemical	Molecular Formula	Purity	Producer
PAH			
Phenanthrene	$C_{14}H_{10}$	Chemical pure	Beijing Chemicals, China
Solutizers			
DOWFAX Hydrotrope	C6 alpha olefin	Chemical pure	Dow Chemical, U.S.A
DOWFAX Detergent	C16 alpha olefin	Chemical pure	Dow Chemical, U.S.A
DOWFAX C10L	C10 alpha olefin	Chemical pure	Dow Chemical, U.S.A
β-cyclodextrin	$(C_6H_{10}O_5)_7$	Analytical pure	Nankai Chemicals, China
Tween 80	$C_{18}H_{37}S^*_6(OC_2H_4)_{20}OH$[a]	Analytical pure	Beijing Chemicals, China

[a] S^*_6: a sorbitan ring $-C_6H_9O_5-$

Determination of Phenanthrene. Phenanthrene in solution was firstly extracted with hexane (extracted >95%). The upper phase was then subjected to HPLC on a 150X4.6 Shimadzu C18 column and determined spectrophotometrically at 254nm. The volume flow rate of mobile phase, 90% methanol, was 1.0 ml/min.

Determination of Tween 80. 1ml sample was firstly mixed with 2ml chloroform, and 3ml solution of ammonium thiocyanate (20%) and cobaltous nitrate (3%).

After 30 minutes incubation at room temperature, the lower phase was subjected to spectrophotometry at 620nm. The concentration of Tween 80 was interpreted from the calibration curve.

Determination of Biomass. The content of biomass in solution was determined spectrophotometrically at 600nm. For present study, the absorbency was used as an index of biomass content in solution.

Determination of the Phenanthrene Solubility. Excess amount of phenanthrene was loaded into flasks and then incubated at 25°C for 12hours on a rotary shaker (150 rpm). Then the mixture was filtrated using paper filter to remove phenanthrene crystal. Determination of phenanthrene in solution was done by the method described above.

Medium and Culture. Bacterium was cultured in 250ml shake flask containing 100 ml medium, which was prepared by mixing surfactant, phenanthrene (dissolved in hexane) and basal medium containing (in g/l deionized water): K_2HPO_4 (0.8), KH_2PO_4 (0.2), NH_4NO_3 (0.8), $MgSO_4$ (0.25), $FeSO_4.7H_2O$ (0.09), and $CaCl_2$ (0.032). After 1 hour stirring, the mixture was incubated at 40 °C for 1 hour to remove hexane. Solid particles were removed from medium by paper filtration before sterilization.

Batch culture was started by adding 1 ml inoculum acclimatized to phenanthrene into medium. Then the medium was cultivated at 30°C on a rotating shaker for a certain period. Aqueous samples were periodically taken for analysis.

RESULTS AND DISCUSSION

Characterization of *Pseudomonas Aeruginosa.* *Pseudomonas aeruginosa* screened and used in present study is a phenanthrene-degrading bacterial strain. The cells are irregularly rod shaped (0.5 - 2.0 mm) with motility (+) and without capsula and endospores (Figure is not shown here). Our preliminary studies have shown its ability in degrading chain hydrocarbon (liquid paraffin), PAHs (naphthalene and phenanthrene), glucose, and etc. When glucose serves as sole carbon source, *pseudomonas aeruginosa* shows best growth within pH7 to 8.

Solubility Test. The solubility of phenanthrene was investigated as function of the sort and concentration of solutizers. As shown by Figure 1(A), Tween 80 shown the best solubilization function over other solutizers. In Figure 1(B), the solubility of phenanthrene was plotted as a function of the concentration of Tween

80. When the concentration of Tween 80 is over its CMC, 13.4mg/L, the increase of Tween 80 concentration leads to a linear increase in phenanthrene solubility.

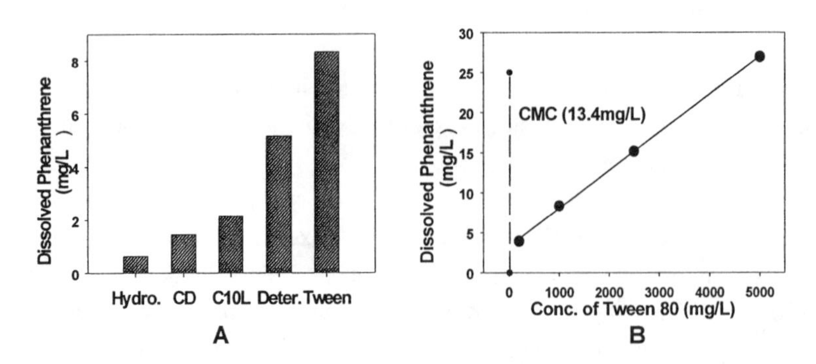

FIGURE 1. Solubilization of phenanthrene by five solutizers(A) and the solubility at the concentration of Tween 80 greater than CMC(B)

Effects of Tween 80 on Biodegradation Process. Figures 2(A), (B), and (C) shown changes of the concentration of bacterium, Tween 80, and phenanthrene in cultivation, respectively. In this set of experiment, the initial concentration of Tween 80 was set at 500, 1000, 2000 and 3000mg/l, respectively, while the initial concentration of phenanthrene was 3mg/l.

As shown by the dashed line in Figure 2(A), the culture process can be divide into two phases according to the growth rate as function of culture time. Phase I is the exponential phase of the bacterial growth, which is accompanied by a rapid decrease of the concentration of Tween 80, as shown by Figure 2(B). Within this phase, about 80% of Tween 80 was consumed while no change in phenanthrene concentration was observed, indicating a superior availability of Tween 80 over phenanthrene.

In phase II, the stationary phase in bacterial growth, the concentration of Tween 80 gradually reduced to a constant level while the phenanthrene concentration decreased at a stead rate, as shown by Figure 2(C). It is interesting to note that at the starting point of phase II, the concentration of Tween 80 were nearly 117, 79, 40, 25 times as much as that of phenanthrene when the starting concentration of Tween 80 was at 3000, 2000, 1000, 500 mg/L, respectively. Compared to phase I, we can conclude that the degradation of phenanthrene can be processed by the bacterium in presence of Tween 80 of low concentration. There seems a threshold value of Tween 80 concentration for the biodegradation, over which the biodegradation of phenanthrene cannot be processed.

As shown by Figure 2(D), the average biodegradation rate of phenanthrene by per unit biomass (denoted by OD_{600}) decreases in respect to the increase of

Tween 80 in medium in the stationary phase. One reason, as described by Guha and Jaffe (1996), is that reduction of the bioavailability of phenanthrene due to the existence of Tween 80. Also it is possible that the existence of Tween 80 inhibits the synthesis of the catabolic enzyme responsible for phenanthrene degradation. This merits further experimental investigation.

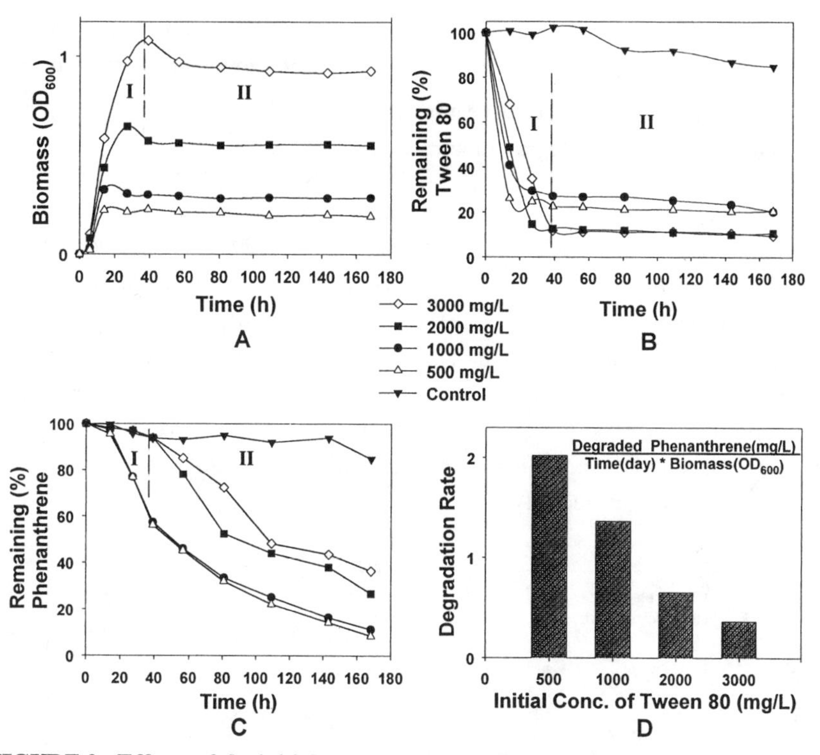

FIGURE 2. Effects of the initial concentration of Tween 80 on bacterial growth (A), degradation of Tween 80 (B), degradation of phenanthrene (C), average calculated biodegradation rate in phase II (D).

Effects of the Initial Concentration of Phenanthrene. The degradation of phenanthrene was operated at starting concentration of 3.0, 2.1 and 1.2mg/l, respectively, while the initial concentration of Tween 80 of 1000mg/l. As shown by Figure 3, the degradation process at different initial concentration of phenanthrene appears similar process characteristics. In all cases, about 80% of phenanthrene were degraded after 120 hours. It is interesting to note the independence of the degradation rate of the initial concentration of phenanthrene.

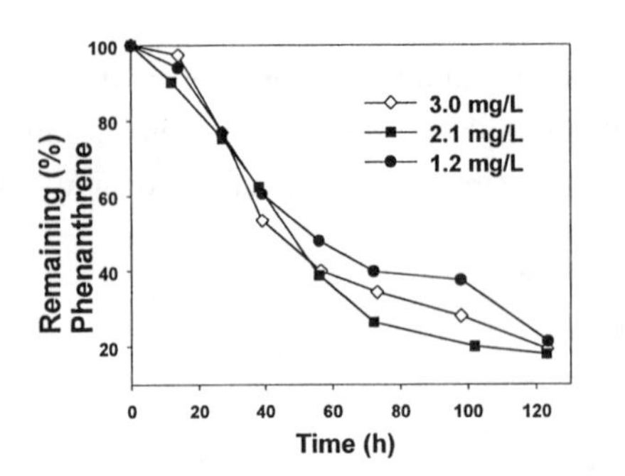

FIGURE 3. Effect of the initial concentration of phenanthrene on biodegradation rate

Application of Electric Field. During the cultivation process, a DC electric field was applied via two Pt electrode mounted in the shake flask. As shown by Figure 4(A) the application of electric field results in an increase in the bacterium concentration in the stationary phase of the bacterium growth. This may be contributed by the enhanced mass transport that improves the bioavailability of phenanthrene. Moreover, as described by Weaver and Chizmadzhev (1996), the application of electric field will generate hydrophilic pores in cell wall, which facilitate the transport of nutrients and thus stimulate the bacterium growth.

As shown by Figure 4(B), the application of electric field lead to a more rapid reduction of phenanthrene, compared to that obtained in the absence of electric field. Moreover, similar result is obtained by using electric field without the presence of strain. This indicates the possibility of oxidative degradation by the oxygen generated by electrolysis.

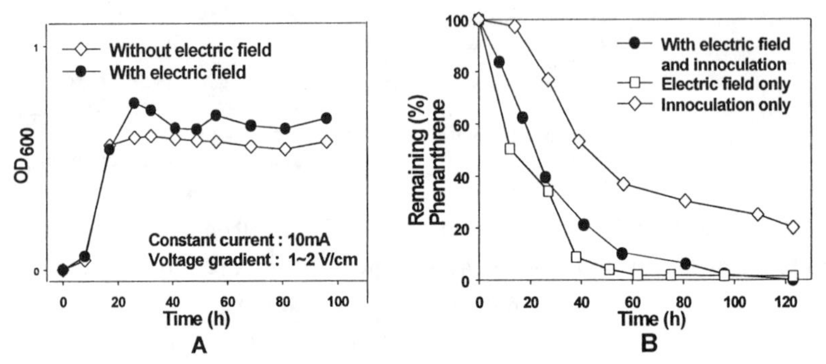

FIGURE 4. Effects of external electric field on bacterial growth (A) and phenanthrene degradation (B).

CONCLUSIONS

The present study is aimed at the influence of non-ionic surfactants on phenanthrene biodegradation process. Preliminary results revealed the competition between phenanthrene and Tween 80 in the bioavailability to *Pseudomonas aeruginosa*. Application of electric field is attempted to enhance the mass transport process, and experimental results reveal it could not only stimulates the growth of the bacterium, but also function in the electro-degradation of phenanthrene. These interesting results will lead us to the further study on the biodegradation kinetics with respect to the utilization of PAHs and surfactants, and the mechanism of electric field enhanced bacterium growth. Also within the scope of our following study is the kinetics of biodegradation of PAHs in the presence of electric field. With its proven effectiveness in phenanthrene degradation and stimulating bacterium growth, this new method is promising for large scale processing.

REFERENCES

Aronstein, B.N. and M. Alesxander. 1993. "Effect of a Non-ionic Surfactant Added to the Soil Surface on the Biodegradation of Aromatic Hydrocarbons within the Soil." *Applied Microbiology and Biotechnology.* 39: 386 –390.

Boonchan, S., M.L. Britz, G.A. Stanley. 1998. "Surfactant-enhanced Biodegradation of High-molecular Weight Polycyclic Aromatic Hydrocarbons by Stenotrophomonas Maltophilia." *Biotechnology and Bioengineering.* 59: 482 –494.

Bury, S.J., and C.A. Miller. 1993. "Effect of Micellar Solubilization on Biodegradation Rates of Hydrocarbons." *Environmental Science & Technology.* 27: 104 –110.

Cuny P., J. Faucet, M. Acquaviva, J.C. Bertrand, M. Gilewicz. 1999. "Enhanced Biodegradation of Phenanthrene by a Marine Bacterium in Presence of a Synthetic Surfactant." *Letters in Applied Microbiology.* 29 (4): 242-245.

Deshpande, S., L. Wesson, D. Wade, D.A. Sabatini, J.H. Harwell. 2000. "Dowfax Surfactant Components for Enhancing Contaminant Solubilization." *Water Research.* 34(3): 1030-1036.

Gibson, D.T., V. Mahadevan, R.M. Jerina, H. Yagi, H.J.C. Yeh. 1975. "Oxidation of the Carcinogens Benzo[a]pyrene and Dibenz[a, h]anthracene to Dihydrodiols by a Bacterium." *Science.* 189: 295-297.

Guha, S., P.R. Jaffe. 1996. "Biodegradation Kinetics of Phenanthrene Partitioned into the Micellar Phase of Nonionic Surfactants." *Environmental Science & Technology.* 30: 605-611.

Kim In S, Jong-Sup Park, Kyoung-Woong Kim. 2001. "Enhanced Biodegradation of Polycyclic Aromatic Hydrocarbons Using Nonionic Surfactants in Soil Slurry." *Applied Geochemistry*. 16:1419 –1428.

Laha, S, and R.G. Luthy. 1991. "Inhibition of Phenanthrene Mineralization by Nonionic Surfactants in Soil–water Systems." *Environmental Science & Technology.* 25:1920 –1930.

Liu Z., G. Yin, S.H. Feng, D.H. Wang, F.X. Ding, N.J. Yuan. 2001. "Oscillatory Electroosmosis-enhanced Intra/Inter-particle Liquid Transport and its Primary Applications in the Preparative Electrochromatography of Proteins." *Journal of Chromatography A.* 921 (1): 93-98

Wang JM, E.M. Marlowe, R.M. Miller-Maier, M.L. Brusseau. 1998. "Cyclodextrin-Enhanced Biodegradation of Phenanthrene." *Environmental Science & Technology.* 32 (13): 1907-1912.

Weaver J.C., Y.A. Chizmadzhev. 1996. "Theory of Electroporation: a Review." *Bioelectrochemistry and Bioenergetics*. 41:135-160.

Yuan S.Y, S.H. Wei, B.V. Chang. 2000. "Biodegradation of Polycyclic Aromatic Hydrocarbons by a Mixed Culture." *Chemosphere*. 41:1463-1468.

Volkering, F., A.M. Breure, J.G.A. Andei, W.H. Rulkens. 1995. "Influence of nonionic surfantants on bioavailability and biodegradation of polycyclic aromatic hydrocarbons." *Applied and Environmental Microbiology.* 6:1699 –1705.

Zheng Z.M., and J.P. Obbard. 2001. "Effect of Non-ionic Surfactants on Elimination of Polycyclic Aromatic Hydrocarbons (PAHs) in Soil-slurry by Phanerochaete Chrysosporium." *Journal of Chemical Technology and Biotechnology*. 76 (4): 423-429.

Zhou J.X., Z. Liu, P. She, F.X. Ding. 2001. "Water Removal from Sludge in a Horizontal Electric Field." *Drying Technology.* 19 (3-4): 627-638.

OZONATION AND BIODEGRADATION OF PERSISTENT BIOACCUMULATIVE TOXINS IN SEDIMENTS

Daniel Cassidy and Duane Hampton (Department of Geosciences, Western Michigan University, Kalamazoo, MI 49008, USA)
Steve Kohler (Department of Environmental Studies, Western Michigan University, Kalamazoo, MI 49008, USA)

ABSTRACT: A laboratory study was conducted to determine the aerobic biodegradability of polychlorinated biphenyls (PCBs) and polycyclic aromatic hydrocarbons (PAHs) treated with ozone. Combined chemical oxidation followed by biodegradation has become an increasingly popular method to biodegrade recalcitrant organics. Ozone is particularly attractive as an oxidant because it works well at a neutral pH. Ozone was sparged into slurries containing the certain PCBs and PAHs. Loss of parent PAHs and PCBs during ozonation was quantified and some ozonation products were identified. Aerobic biodegradation of the ozonation products was quantified using oxygen uptake rate and removal of chemical oxygen demand (COD). The oxidation products were more soluble and more biodegradable than the parent PCBs and PAHs, and were readily degraded aerobically. Chlorine atoms in the PCBs were replaced with hydroxyl groups. Ring cleavage was observed for both PAHs and PCBs. Combined ozonation and biodegradation shows promise for remediation of sediments contaminated with PCBs and PAHs, although some ozone is consumed on native organic material.

INTRODUCTION

Remediation of sediments contaminated with persistent bioaccumulative toxins (PBTs) is one of the more intractable environmental problems. PBTs are contaminants that are very slowly degraded in the environment and accumulate in the food chain because of their lipophilic nature. Two PBTs commonly found in sediments are polychlorinated biphenyls (PCBs) and polycyclic aromatic hydrocarbons (PAHs). For example, PCBs are extremely slowly biodegradable under aerobic and anaerobic conditions (Abramowicz, 1990). Leaving the sediments in place until the PBTs biodegrade (i.e. monitored natural attenuation) is not viable in heavily-contaminated water bodies because, although degradation does occur at low rates, PBTs continue to enter the food chain via benthic organisms and algal growth which are fed upon by zooplankton. Subsequent biomagnification of PBT concentrations in the food chain leads to unacceptable exposures for anglers, birds, and mammals at the top of the food chain. Furthermore, PBT-contaminated sediments are vulnerable to being re-suspended or stripped by flooding, and by the activities of demersal fish.

Dredging is the most widely-used remedial method for PBT-contaminated sediments. Dredging is problematic because it re-suspends contaminated sediments in the stream, thus increasing exposures both downstream and to those handling the dredge spoils, leaves 5 to 25% or more of the contaminants in place, and because there is never a good place to deposit PBT-contaminated sediments. Advocates of dredging draw support from favorable controlled studies of sites remediated via dredging. For example, Lake Jarnsjon, Sweden, was contaminated with paper mill wastes resulting from recycling waste paper containing PCBs. In 1993 and 1994, about 147,000 cubic meters of sediment were dredged, dewatered and landfilled. Roughly 80% of the PCB mass was in the eastern part of the lake. During dredging, this part of the lake was enclosed with a silt curtain to reduce leakage of suspended solids and to control turbidity downstream. About 95% of the PCBs were removed from the lake. Monitoring during and for two years following dredging showed that PCB concentrations in the lake were constant during dredging and decreased thereafter (Bremle, 1997). Two years after dredging, the PCB concentration in one-year old fish in the lake was twice its pre-remediation value. At Waukegan Harbor, Illinois, dredging and subsequent capping of hot spots in 1992 removed or contained most of the PCB mass. Fish flesh sampling showed that 1993 PCB levels in carp were five times higher than those tested in 1983 and 1991 (Clark, 1997).

Subaqueous capping of PBT-contaminated sediments with an impermeable armor has recently been considered as an alternative to dredging. This strategy avoids problems associated with dredging, and isolates the PBTs in sediments from the food chain. However, this approach merely creates an underwater landfill. The impermeable capping material prevents the exchange of pore waters and the nutrients and gasses dissolved therein, which are essential to microbial degradation. Responsible management of PBT-contaminated sediments should minimize PBT mobility via sediment suspension and entry into the food chain, while simultaneously promoting *in situ* biodegradation.

Two novel remediation methods are being tested by our research team for the *in situ* treatment of PBT-contaminated sediments in "hot spots" (i.e., heavily PBT-contaminated areas): (1) the use of permeable geotextile materials anchored with sand and gravel to serve as a "breathing cap" for PBT-laden sediment; and (2) ozonation of PBTs to enhance their biodegradability and reduce their residence time in sediments. The geotextile barrier will reduce the suspension of PBT-contaminated sediments caused by fluvial erosion and the activity of demersal fish and reduce or prevent the entry of PBTs into the food chain by isolating benthic organisms from the water column. The blanketed sediments can then be sparged with ozone.

Combined ozonation followed by biodegradation has become an increasingly popular method to biodegrade recalcitrant organics. The purpose is not to completely oxidize PCBs, but only to partially oxidize them so that the products can be biodegraded. While many chemical oxidation methods exist, ozone is particularly attractive because it works well at a neutral pH. In contrast, other oxidants such as Fenton's Reagent work best at a pH less than 3 (Carberry and Yang, 1994), which would require subsequent pH adjustment to encourage microbial activity. Furthermore, Fenton's Reagent releases considerable heat upon reaction (O'Brien & Gere Engineers, 1995), which can volatilize contaminants and kill microorganisms. Ozone produces hydroxyl free radicals, which are known to oxidize PAHs (Brown et al., 1997; Nelson et al., 1997; Marvin et al., 1998; Clayton, 1998). Sparged ozone dissolves in water (ozone is 13 times more soluble than oxygen) and readily diffuses away from the source (Clayton, 1998). However, reaction of ozone with native organic material increases ozone doses.

MATERIALS AND METHODS

Two PAHs (anthracene and fluoranthene) and two PCBs (2-,2'-dichlorobiphenyl (DCB) and 2-,3-,4-,2'-,3'-,4'-hexachlorobiphenyl (HCB)) were added to kaolinite slurries. Separate slurries were maintained for the PAHs and PCBs. All four PBTs were added to achieve a concentration in the slurry of 1 g/kg. After dosing, the slurries were allowed 4 months of contact time before beginning the ozonation experiments to allow sorption of the PCBs to the solids. The slurries were then allowed to settle overnight, and thickened slurry was placed in 1 L glass columns with fritted-glass openings at the bottom to allow gas to be sparged upward through the sediment. The solids content of the thickened slurry was approximately 80% w/v (i.e., 0.8 kg kaolinite/L slurry). The reactors were sparged with ozone using a laboratory ozone generator (OL-100, Ozone Services, Burton, British Columbia). The ozone generator provided known and fixed O_3 concentrations in the influent gas stream. An on-line O_3 monitor was used to measure ozone concentrations exiting the reactor. Effluent air was passed through an activated carbon trap to quantify volatile losses of organic material. Two different O_3 concentrations were tested, 0.5% and 5%, to investigate how ozone concentration affected rates of oxidation. Control reactors were sparged with N_2 gas. Sparging provided the only mixing, in order to simulate conditions encountered in *in situ* sediment sparging. Each reactor type was run in triplicate.

During 20 days of sparging, samples were taken of the slurry (i.e., solids and liquid) to measure the parent PBT concentration. After 20 days, kaolinite solids were separated from the liquid by centrifuging and were extracted with petroleum ether to quantify the remaining PBT using gas chromatography with electron capture detection (GC/ECD) for the PCBs and flame ionization detection (GC/FID) for the PAHs. GC/mass spectroscopy (GC/MS) was used to tentatively identify ozonation products of the PCBs. The liquid fraction of the reactors was sampled for Chemical Oxygen Demand (COD) using the Hach COD test and a Hach DR-4000 spectrophotometer. Chlorine (Cl^-) concentration was also measured (using ion chromatography, IC) in the reactors containing PCBs to quantify the release of chlorine atoms during ozonation and biodegradation. The remainder of the liquid was placed in closed, 500 mL respirometer (BOD) bottles with added nutrients and inoculum from the Hamilton, Ontario municipal wastewater treatment plant. Oxygen consumption, COD, and Cl^- were measured with time to determine the aerobic biodegradability of the ozonation products of the PBTs. Oxygen uptake in the bottles was measured automatically with a YSI BOD Monitor (Yellow Springs, Ohio).

RESULTS AND DISCUSSION

PCBs. The results from the ozonation and biodegradation of DCB and HCB are summarized in Table 1. The removal of DCB was 93% after 20 days, compared with 81% for HCB. The lower removal of HCB may be due to its lower aqueous solubility than DCB. The concentration of O_3 in the reactors had no impact on the removal of DCB and HCB, indicating that rates of reaction with O_3 were limited by desorption of PCBs from the kaolinite. Removal of approximately 2 and 6 chlorine atoms from dichlorobiphenyl and hexachlorobiphenyl, respectively, shows that chlorine atoms were stoichiometrically removed from the PCBs during ozonation. Close agreement between percent removal of the two PCBs and percent Cl^- released also indicates stoichiometric removal of chlorine atoms from the PCBs by ozone. This means that ozonation achieved chlorine removal from the PCBs, but does not necessarily mean that further oxidation of the PCBs products ceased after dechlorination. GC/mass spectroscopy (GC/MS) results identified hydroxylated benzoates, oxalate, and formate. These products are known to be biodegradable. These results suggest that ozonation replaces chlorine atoms in the PCBs with hydroxyl (OH) groups, and that ring cleavage also occurs. A COD reduction of over 85% in the respirometers for both PCBs, accompanied by elevated oxygen uptake rate compared with control reactors without added nutrients, shows that the ozonation products of DCB and HCB are readily biodegradable. The soluble COD of the slurries before ozonation was zero. After ozone treatment, all the residual organic carbon (measured as COD) was present in the aqueous phase and not sorbed to the kaolinite solids.

Table 1. Summary of results for ozonation of dichlorobiphenyl (DCB) and hexachlorobiphenyl (HCB) followed by biodegradation.

DCB	
DCB Removed by Ozonation (%)	93 ± 6 (3)[a]
Cl^- Released by Ozonation (%)	90 ± 4 (3)
Cl^- Released/DCB Removed (mol/mol)	1.8 ± 0.4 (3)
O_3 Consumed/DCB Consumed (mol/mol)	4.3 ± 0.6 (3)
COD Removed by Biodegradation (%)	86 ± 7 (3)
HCB	
HCB Removed by Ozonation (%)	81 ± 8 (3)
Cl^- Released by Ozonation (%)	83 ± 3 (3)
Cl^- Released/HCB Removed (mol/mol)	6.1 ± 0.7 (3)
O_3 Consumed/HCB Consumed (mol/mol)	13.3 ± 1.1 (3)
COD Removed by Biodegradation (%)	89 ± 5 (3)

[a] average ± standard deviation (number of measurements).

The results from Table 1 clearly show that; (1) ozonation decreased DCB and HCB concentrations in the slurry, (2) stoichiometric removal of chlorine atoms accompanied removal of DCB and HCB during ozonation, (3) the residual organic carbon after ozonation was readily biodegradable, and (4) ozonation increased the aqueous solubility of the residual organic carbon relative to the parent compounds. Aronstein and Rice (1995) obtained similar results using Fenton's reagent to treat PCBs. In this study, Fenton's reagent increased the overall amount of PCBs degraded by 4 times relative to sediment samples not treated with Fenton's reagent. Their study also showed that the partial oxidation products were highly soluble compared with the parent PCBs. This study showed that chlorine atoms on the PCBs were replaced by hydroxyl groups during ozonation. Heinzle et al. (1995) also observed stoichiometric chlorine replacement with OH groups during ozonation of chloroguaiacols (i.e., chlorinated methoxy phenols). Heinzle et al (1995) also observed ring cleavage, and a 10-fold increase in biodegradation of the partial oxidation products compared with the original chloroguaiacols. The stiochiometric replacement of chlorine with hydroxyl groups means that ozone dose increases with increasing degree of chlorination on the PCBs (Figure 1). The fact that hexachlorobiphenyl used approximately three times more ozone than dichlorobiphenyl, and that the molar ratio of O_3 consumed/PCB consumed was also about 3 times greater for HCB than DCB suggests that ozone is used to achieve complete dechlorination before ring cleavage occurs. Heavily chlorinated PCBs would then exert a high ozone demand. Marvin et al. (1998) also reported preferential oxidation of chlorinated compounds (pentachlorophenol) to non-chlorinated organics (PAHs) with ozone. Non-chlorinated organic wastes exert ozone demands ranging from approximately 4-5 g O_3/g COD (Narkis and Schneider-Rotel, 1980; Jones et al., 1985). The identification of hydroxylated benzoates in the residual COD in these studies is also consistent with advanced oxidation of PCBs reported by Brubaker and Hites (1998). The oxalate and formate probably formed from further oxidation of benzoate.

PAHs. The results from the ozonation and biodegradation of anthracene (a 3-ring PAH) and fluoranthene (a 4-ring PAH) are summarized in Table 2. The removal of anthracene and fluoranthene by ozone treatment was 99% and 94%, respectively. The slightly lower removal of fluoranthene may have been a result of its lower aqueous solubility. The molar ratio of O_3 consumed/PAH Consumed was slightly higher for fluoranthene (4.1) compared with anthracene (3.4). These values are consistent with those reported for ozonation of oil shale wastewaters (Jones et al., 1985). In excess of 95% of the residual COD from ozone treatment of both PAHs was biodegraded under aerobic conditions. These results are similar to those obtained by several other studies (Brown et al., 1997; Nelson et al., 1997; Marvin et al., 1998). Ozonation products were not identified for the PAHs.

CONCLUSIONS

The results from the ozonation of PCBs and PAHs indicate that ozone sparging combined with biodegradation can be used in conjunction with a geosynthetic liner system to treat PCB-contaminated sediments in place. The high cost, relative to dredging, will likely make such a remediation strategy feasible only in areas with very high contamination (i.e., "hot spots"). Ozonation could also be used to treat heavily contaminated sediments that have already been dredged prior to disposal. The ozone doses in sediments with high concentrations of native organic material are likely to be higher than those in this study. Furthermore, the ozone dose for PCBs increases with increasing substitution in the biphenyl molecule.

Table 2. Summary of results for ozonation of anthracene and fluoranthene followed by biodegradation.

Anthracene	
Anthracene Removed by Ozonation (%)	99 ± 1 (3)[a]
O_3 Consumed/Anthracene Consumed (mol/mol)	3.4 ± 0.5 (3)
COD Removed by Biodegradation (%)	96 ± 6 (3)
Fluoranthene	
Fluoranthene Removed by Ozonation (%)	94 ± 7 (3)
O_3 Consumed/Fluoranthene Consumed (mol/mol)	4.1 ± 0.4 (3)
COD Removed by Biodegradation (%)	98 ± 5 (3)

[a] average ± standard deviation (number of measurements).

ACKNOWLEDGEMENTS
The research presented in this paper was funded by the Michigan Department of Environmental Quality-Michigan Great Lakes Protection Fund.

REFERENCES

Abramowicz, D. A. 1990. Aerobic and anaerobic biodegradation of PCBs--a review. *Crit. Rev. Biotechnol.* 10: 241-249.

Aronstein, B. N., Rice, L. E. 1995. Biological and Integrated Chemical-Biological Treatment of PCB Congeners in Soil/Sediment-Containing Systems, *J. Chem. Tech. Biotechnol.*, 63, 321-328.

Bremle, G. 1997. Polychlorinated biphenyls (PCB) in a river ecosystem, Ph.D. Thesis, Ecotoxicology, Lund Univ., Lund, Sweden, 1997, URL: http://www.ecotox.lu.se/staff/gbr/abstract.html, http://www.hultsfred.se/miljo/ironeng.htm.

Brown, R. A., Nelson, C. H., and M. C. Leahy. 1997. "Combining Oxidation and Bioremediation for the Treatment of Recalcitrant Organics." In *In-Situ and On-Site Bioremediation: Vol. 4 (4)*, pp. 457-462. Battelle Press. Columbus, Ohio.

Brubaker, W. W., Jr., and R. A. Hites. 1998. "Gas-Phase Oxidation Products of Biphenyl and Polychlorinated Biphenyls." Environ. Sci. & Technol., vol. 32, 3913-3918.

Carberry, J. B., Yang, S. Y. 1994. Enhancement of PCB Congener Biodegradation by Pre-Oxidation with Fenton's Reagent, *Wat. Sci. Tech.*, 30:7, 105-133.

Clark, J. M. 1997. Policy Implications—a presentation to the Great Lakes Roundtable—Achieving the Future, Niagara-on-the-Lake, Oct. 31, 1997, cited with fish flesh PCB data tables in URL: http://www.ijc.org/boards/wqb/cases/waukegan/waukegan.html

Clayton, W. S. 1998. "Ozone and Contaminant Transport During In situ Ozonation." In *Remediation of Chlorinated and Recalcitrant Compounds-Physical, Chemical, and Thermal Technologies* (eds. G. B. Wickramanayake, and R. E. Hinchee): pp. 389-395. Battelle Press. Columbus, Ohio.

Heinzle, E. Stockinger, H., Stern, M., Fahmy, M., and O. M. Kut. 1995. "Combined Biological-Chemical (Ozone) Treatment of Wastewaters Containing Chloguaiacols." *Journ. Chem. Technol. Biotechnol., vol 62*, 241-252.

Jones, B. N., Sakaji, R. H., and C. G. Daughton. 1985. "Effects of Ozonation and Ultraviolet Irradtion on Biodegradability of Oil Shale Wastewater Organic Solutes." *Water Research, vol. 21*, 1421-1428.

Marvin, B. K., Nelson, C. H., Clayton, W., Sullivan, K. M., and G. Skladany. 1998. "In-Situ Chemical Oxidation of Pentachlorophenol and Polycyclic Aromatic Hydrocarbons: From Laboratory Tests to Field Demonstration." In *Remediation of Chlorinated and Recalcitrant Compounds-Physical, Chemical, and Thermal Technologies* (eds. G. B. Wickramanayake, and R. E. Hinchee): pp. 383-388. Battelle Press. Columbus, Ohio.

Nelson, C. H., Seaman, M., Peterson, D., Nelson, S., and R. Buschorn. 1997. "Ozone Sparging for the Remediation of MGP Contaminants." In *In-Situ and On-Site Bioremediation: Vol. 4 (3)*, pp. 468-473. Battelle Press. Columbus, Ohio.

Narkis, N. and M. Schneider-Rotel. 1980. "Evaluation of Ozone Induced Biodegradability of Wastewater Treatment Effluent." *Water Research, vol. 14*, 929-939.

O'Brien & Gere Engineers. 1995. *Innovative Engineering Technologies for Hazardous Waste Remediation*. R. Bellandi, ed., Van Nostrand Reinhold publishers, New York, New York.

SULPHATE-REDUCING AND *LACTOBACILLUS* BACTERIA FOR THE TREATMENT OF CONTAMINATED SEDIMENTS

E. D'Addario, G. Lucchese, G. Scolla, R. Sisto
(EniTecnologie S.p.A., Monterotondo, Rome, Italy)
P. Carrera (Ambiente S.p.A., San Donato Milanese, Milan, Italy)

ABSTRACT: A new process based on sequential anaerobic and aerobic treatments for the decontamination of polluted sediments has been tested at laboratory and bench scale. The process lasts around 300 days (200 anaerobiosis and 100 aerobiosis) and is able to give more than 90% total aliphatic and aromatic chlorinated molecules removal (initial concentration 3260 $mg \cdot kg^{-1}$sediment dwb), around 70% PCDD/PCDF removal (initial concentration $2 \cdot 10^{-4}$ mg $TE \cdot kg^{-1}$ sediment dwb) and to stabilise the leacheable metals. Because of the: i) low chemicals consumption, ii) rather low utilities requirements and, iii) technological simplicity of the tested slurry bioreactor, the process appears competitive with sediment washing and thermal processes such those for building materials production or thermal desorption.

INTRODUCTION

Contaminated sediments are generally discharged into confined disposal facilities (CDFs) which in many cases are filled or tends to reach their maximum capacity. Many approaches are currently proposed to modify CDFs facilities or to develop systems producing clean materials for reuse. Bioremediation, either in-situ or in bioreactors, is certainly one of the most promising and extensively studied approach. Most of the published works address different aspects of biodegradation of polycyclic aromatics hydrocarbons usually present in contaminated sediments.

Enrichment cultures from native microbial communities of sediments have been also investigated for the dehalogenation of polychlorodibenzodioxins (PCDD), tetrachlorobisphenol A and brominated phenols (Fennel et al., 2001). The attention has been payed to: i) the stimulation of dehalogenation with different electron-accepting processes (sulfate reducing, iron(III) reducing and methanogenesis) and, ii) the molecular and physiological characterisation of dehalogenating consortia.

In the past we have developed a number of proprietary processes based on the use of sulfate reducing bacteria (SRB) as agents capable to concentrate and to immobilise heavy metals ions contained in sludge, liquid and solid waste (e.g.. Lucchese et al., 1995). In these processes SRB were cultivated simultaneously with *Lactobacillus sp.* which were allowed to proliferate on a proper and cheap carbon source such as cheese whey. This in order to provide SRB with a preferred substrate such as lactate originated from transformation of the lactose contained in cheese whey.

Objective. Aim of this work was to check the capabilities of SRB / *Lactobacillus* mixed cultures to: i) accomplish heavy metals immobilisation and, ii) create redox conditions for the dehalogenation of chlororganics, including PCDD and polychlorodibenzofurans (PCDF), so that an aerobic oxidation can result in a clean solid. The goal was to perform some technical economical projections based on experiments at bench level with a bioreactor potentially usable at full scale.

MATERIAL AND METHODS

Sediments. Two samples have been used in this work, dredged from hot spots in Venice lagoon, both contaminated by heavy metals and varying chlororganic content: sediment A (ca. 13000 $mg \cdot kg^{-1}$dwb) for laboratory activities, sediment B (ca. 3000 $mg \cdot kg^{-1}$dwb) for the bench phase. Both samples had 48% w/v dry matter. Analytical details are given below (see RESULTS AND DISCUSSION).

Sediment bioreactors. Three anaerobic glass reactors (R-1, R-2 and R-3) were utilised at laboratory scale. Each of the 2.5 L reactors was mixed with a turbine stirrer at 300 rpm. The reactors were operated at room temperature (approx. 20°C) and equipped with pH and redox potential monitoring systems. The reactors were also equipped with air feeding systems and flow meters.

At bench scale the 120 L slurry reactor schematically shown in Figure 1 was used. Sediments were mixed with 3 vomer stirrers mounted on a system placed through the width of the reactor. The mixing system was moved from one side to the other of the length of the reactor at 0.5–1 $stroke \cdot min^{-1}$ by means of a translating screw equipped with an electric motor. The reactor was closed with a transparent plastic cover and equipped with pH, temperature, redox potential probes installed in different places to check the homogeneity of the material and the efficiency of the mixing system. The reactor was operated at room temperature (15-20°C).

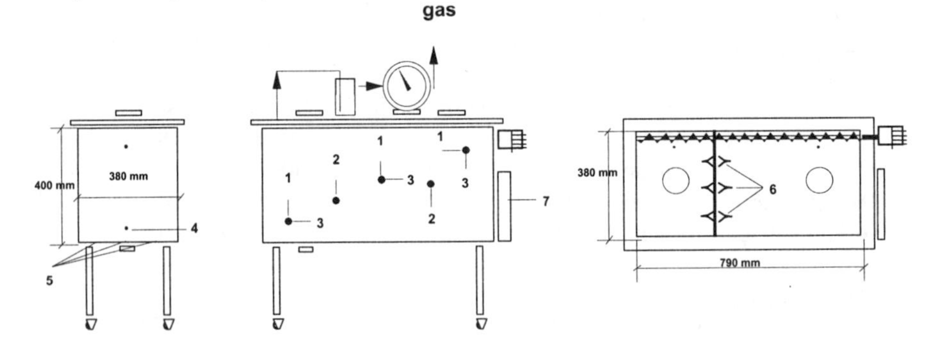

FIGURE 1. Schematic of the bench scale bioreactor (1 redox probes, 2 thermometers, 3 pH probes, 4 sample ports, 5 air feeding, 6 vomers)

Experimental. Commercial powder cheese whey as hydrogen release compound (HRC) was supplied discontinuously on a weekly basis to the laboratory reactors, each filled with 2.2 L of sediment A. pH was monitored and kept in the range 7.2-7.7 by adding on a daily basis a 20% w/v $Ca(OH)_2$ suspension. After complete depletion of sulfate originally contained in sediment, cultures were supplemented with Na_2SO_4 fed on a weekly basis. Three different HRC/ Na_2SO_4 w/w ratios were used: 1/1 (R-1), 2/1 (R-2) and 4/1 (R-3). The reactors were operated for 200 days under anaerobic conditions. At the beginning of the experiments 7.5 g HRC per reactor were weekly added. The addition of HRC was progressively reduced throughout the experiment till 2.6 g per reactor final value (0.125 $g \cdot d^{-1}$ kg^{-1} sediment).

After the anaerobic step the mineralisation of residual organics was attempted via aerobiosis. At this purpose the reactors were fed with 0.5 v/v,min air under mixing at 800 rpm for 12 days to raise the redox potential from –400 to +50 mV. After an initially supplement of 0.5 g ammonium nitrate kg^{-1} sediment as N source, air feeding was continued till 100 days under mixing keeping the pH at around 7 by adding concentrated phosphoric acid. During the experiments slurry samples were drawn out and analysed for aliphatic and aromatic chlorocarbons, sulfate, PCDD/PCDF and leachable heavy metals. The bench reactor was filled with 100 kg of sediment B and operated under the most favourable conditions resulted from laboratory experiments, namely: 0.125 g HRC $d^{-1} \cdot kg^{-1}$ sediment and 0.03 g Na_2SO_4 $d^{-1} \cdot kg^{-1}$ sediment.

Analytical Methods. Chlorocarbons in sediments were determined by a HP 5890 Gas Chromatograph equipped with a HP 5921A atomic emission detector, according to Szelewski (1989), on CH_2Cl_2 extracts prepared according to the IRSA/CNR method (1985). PCDD/PCDF analysis was performed by Scientific Analysis Laboratories Ltd., Manchester (UK). Heavy metal content was determined by a Spectra A-10 Varian atomic absorption spectrometer on either sediments, after mineralisation with a HNO_3/HCl mixture, or acetic acid leachates prepared according to the IRSA/CNR method (1985). The most probable number (MPN) of SRB was determined according to the procedure reported by Koch (1981). The toxicity was measured by Microtox™ equipment (AZUR Environmental, Carlsbad, CA, USA).

RESULTS AND DISCUSSION

Laboratory Experiments. Microbial counts in laboratory reactors during the anaerobic process are reported in Table 1.

TABLE 1. Microbial counts in laboratory experiments (MPN/g sediment)

Time (days)	*Reactors*					
	R1		R2		R3	
	TA	SRB	TA	SRB	TA	SRB
0	$6 \cdot 10^2$	$3 \cdot 10^2$	$6 \cdot 10^2$	$3 \cdot 10^2$	$6 \cdot 10^2$	$3 \cdot 10^2$
20	$7 \cdot 10^3$	$5 \cdot 10^3$	$9 \cdot 10^4$	$8 \cdot 10^4$	$9 \cdot 10^4$	$6 \cdot 10^4$
40	$2 \cdot 10^6$	$1 \cdot 10^6$	$8 \cdot 10^6$	$2 \cdot 10^6$	$7 \cdot 10^6$	$2 \cdot 10^6$
60	$7 \cdot 10^5$	$2 \cdot 10^5$	$4 \cdot 10^6$	$1 \cdot 10^6$	$7 \cdot 10^6$	$2 \cdot 10^6$
200	$8 \cdot 10^5$	$7 \cdot 10^5$	$9 \cdot 10^6$	$7 \cdot 10^5$	$3 \cdot 10^6$	$9 \cdot 10^5$

HRC/ Na_2SO_4 ratio: R-1 1:1; R-2 2:1; R-3 4:1. TA total anaerobics

The table shows that: i) in spite of the toxicity of the sediment, due to high concentration of chlororganics and heavy metals, total anaerobic micro-organisms and autochthonous SRB rapidly increase, ii) the SRB population is slightly lower than total anaerobics since the beginning and this difference tends to remain almost unchanged along all the experiments, iii) in the reactors R-2 and R-3 operating at higher HRC/sulfate ratio the initial lag phase appears shorter than in reactor R-1, iv) after 40 days the bacterial proliferation reached a plateau phase in all the reactors and, v) quite higher bacterial densities are observed in reactors R-2 and R-3 operating under more favourable HRC conditions.

Table 2 shows the initial content of heavy metals in sediment A as well as the acidic leachates from samples drawn from reactor R-3 through the experiments. Similar results have been obtained in bioreactors R-1 and R-2.

TABLE 2. Heavy metals in the sediment A (mg Kg^{-1} dwb) and in the acidic leaching solutions (reactor R3, values mg L^{-1})

	Sediment A	Acidic leaching solutions			
		t = 0	t = 22 d	t = 40 d	2B Landfill (1)
Arsenic	165	< 0.5	<0.5	<0.5	0.5
Cadmium	0.6	0.08	0.06	0.01	0.02
Chromium VI	46	<0.2	<0.2	<0.2	0.2
Copper	126	<0.1	<0.1	<0.1	0.1
Mercury	127	0.03	0.09	0.004	0.005
Lead	627	0.5	<0.2	<0.2	0.2
Iron	12230				
Nickel	40				
Zinc	627				

(1) Interministerial commission on waste management, deliberation 27/7/1984.

On the basis of the Italian laws, the sediment has to be considered a toxic waste and disposed of in proper landfills (2C group). In fact, Cd and Hg leach at concentrations higher than the limits for the disposal in landfills for special waste (2B group). Data reported in Table 2 indicate that after around 40 days the SRB activity results in the immobilisation of Cd and Hg at the extent required (2B landfills). It is worth emphasising that this result has been obtained in the presence of rather high content of Fe, Ni and Zn which compete for sulfide produced by the anaerobic reductive pathway of sulfates.

The time course of chlororganic removal in the three systems is shown in Figure 2, and the concentrations of single compounds in the sediment as well as in the material after the anaerobic and aerobic procession is reported in Table 3.

Although the three cultures showed basically similar dehalogenating profiles, the reactors R-2 and R-3 operated at higher HRC levels, tend to perform better. This seems to be in agreement with the inhibiting phenomena of dehalogenating microorganisms caused by sulfate under limiting concentrations of hydrogen donors described by Yang and McCarty (2001).

A quite good removal of total chlorinated molecules (c.a. 60%) was observed after the anaerobic step. After the aerobic treatment, the Italian legal limits for the use of the treated sediments in the restoration of commercial or

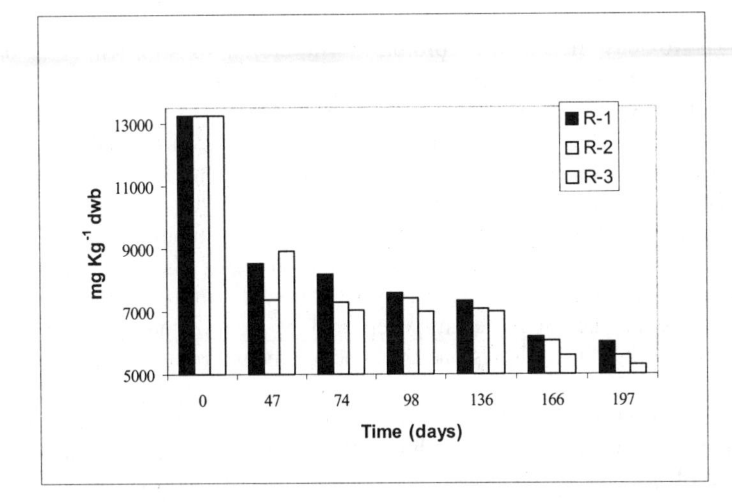

FIGURE 2. Time course of chlororganic concentrations in the three reactors (HRC/Na_2SO_4 ratios: R1=1:1, R2=1:2, R3=4:1)

industrial areas was reached for all the compounds except pentachlorobutadiene and perchlorinated molecules.

TABLE 3. Concentrations of chlororganics in sediment A before and after the anaerobic and aerobic treatment (Reactor R3, Values, mg Kg^{-1} dwb)

Chlorinated Molecules	Sediment A	Anaerobic Treatment (200 d)	Anaerobic and Aerobic Treatment (200+100 d)	Legal Limit (1)
Tetrachloroethylene	1089	27	ND	20
Tetrachloroetane	901	167	ND	10
Hexachloroetane	57	10	4	10
Tetrachlorobutadiene	825	189	ND	10
Pentachlorobutadiene	1225	322	41	10
Hexachlorobutadiene	2679	540	237	10
Pentachlorobutane	374	23	ND	10
2,4,5-Trichlorotoluene	191	4	ND	10
Hexachlorobutene	1044	600	61	10
Hexachlorobutane	4643	3106	512	10
Hexachlorobenzene	337	198	102	5
Total	13264	5186	957	-

(1) Decree of the Italian Environmental Ministry 471/99. Material to be used for soil restoration in commercial or industrial areas. ND Not Detectable

It has to be pointed out that, results presented in Table 3, were reached in the presence of noticeable amount of PCDD/PCDF ($6.7 \cdot 10^{-4}$ mg Toxicity Equivalent, TE, kg^{-1} dwb) which were strongly reduced (around 60% removal) after the anaerobic treatment.

The toxicity measured according to Microtox™ of sediment A before the treatments is equivalent to 0.0062 (percent of sediment necessary to get a 50% decrease of luminosity expressed by a standard culture of *Vibrio Fischeri*). After the anaerobic step that value became 0.044 (7 times reduction) and at the end of aerobic treatment was further increased to 0.38 (62 times toxicity reduction).

The above results can be explained in terms of the presence of a mixed culture formed of *Lactobacillus sp.*, SRB and dehalogenating micro-organisms, and can be regarded as a good starting point towards the elucidation of the mechanism of the reactions catalysed by such mixed cultures. Different species, similar to Dehalococcoides recently reported as heterogeneous dechlorinators (Ritalahti et al., 2001), could also be contained in the microbial population.

Bench experiments. With the aim of collecting data useful for a preliminary evaluation of the economics, a bench experiment was carried out using sediment B, having a lower content of chlorinated compounds than sediment A.

According to laboratory results the test was performed using 4:1 HRC/sulfate ratio. In this case, considering the lower polluting load, HRC was supplied at 0.125 $g \cdot kg^{-1}$ sediment since the beginning. Further details concerning the experiments such as substrate supplying procedures, mixing, control of the cultures, etc., are part of technological know how under patent application. The results obtained are summarised in Table 4.

TABLE 4. Concentrations of chlororganics in sediment B before and after the anaerobic - aerobic treatment at bench scale level (Values, $mg \cdot Kg^{-1}$ dwb)

Chlorinated Molecules	Sediment B	Anaerobic Treatment (200 d)	Anaerobic and Aerobic Treatment (200+100 d)
Tetrachloroethylene	186	1	ND
Tetrachloroetane	161	2	ND
Hexachloroetane	100	13	ND
Tetrachlorobutadiene	185	45	ND
Pentachlorobutadiene	308	15	ND
Hexachlorobutadiene	712	70	ND
Pentachlorobutane	46	1	ND
2,4,5-Trichlorotoluene	10	2	ND
Hexachlorobutene	273	13	ND
Hexachlorobutane	1214	55	ND
Hexachlorobenzene	67	23	11
Total	3262	239	-

All the chlororganics were removed under detectable values except hexachlorobenzene (HCB). HCB confirms to be the most recalcitrant compound and, whether the previously mentioned legal limits (see Table 3) should be set as target, only sediments containing less than approx. 30 mgHCB Kg^{-1} dwb can be considered for the treatment. That value has been calculated on a 170 days half time basis which was observed after the whole biological process.

PCDD/PCDF initially present in sediment B at $2 \cdot 10^{-4}$ mg TE Kg^{-1}dwb were removed at a quite rather high extent (around 70 %), reaching a final concentration below the limit of the Italian law for the use in commercial or industrial areas ($1 \cdot 10^{-4}$ mg TE Kg^{-1} dwb).

Table 5 shows the initial content of heavy metals in sediment B as well as the acidic leachates from samples drawn from bench scale through the experiments.

TABLE 5. Heavy metals in the sediment B (mg Kg^{-1} dwb) and in the acidic leaching solutions values mg L^{-1}

	Sediment B	Acidic leaching solutions			
		t = 0	After Anaer.	After Aerob.	2B Landfill (1)
Arsenic	120	< 0.5	<0.5	<0.5	0.5
Cadmium	2,0	0.03	0.003	0.02	0.02
Chromium VI	48	<0.2	<0.2	<0.2	0.2
Copper	143	<0.1	<0.1	<0.1	0.1
Mercury	485	0.03	0.002	0.002	0.005
Lead	300	0.5	0.005	0.02	0.2
Iron	9750				
Nickel	51				
Zinc	762				

The results indicate that the anaerobic step allows the immobilisation of Cd and Hg at levels required by the above mentioned Italian law (2B landfills). It is important to notice that, in this case, the subsequent aerobic treatment does not result in a significant metal leaching.

Economical estimations and concluding remarks. Table 6 shows the consumption of utilities and chemicals for the bench reactor. Electricity and cheese whey give the major contributes to total costs (53 and 21 % respectively). About cheese whey it has to be pointed out that powder currently marketed as animal feed has been preliminarily considered. On large scale application cheap liquid whey could be used. Considering the quite low operating cost and the simplicity of the process which implies low investments, total costs lower than those currently indicated for sediment washing (60-80 Euro t^{-1}) or thermal processes for building materials production (55-220 Euro t^{-1}) are expected.

TABLE 6. Chemicals and utilities consumption and related operating for the slurry bioreactor.

	Unit	Unit Costs (Euro)	Consumption (Unit/t sediment)	Costs (Euro/t sediment)
ANAEROBIOSIS (200 d)				
Cheese whey	kg	0.25	25.0	6.3
Na_2SO_4	kg	0.12	6.0	0.7
Calcium hydrate	kg	0.12	11.2	1.3
Electricity	kWh	0.05	137.0	6.9
Total				***15.2***
AEROBIOSIS (100 d)				
Ammonium nitrate (34 %)	kg	0.29	9.3	2.7
Phosphoric acid (98 %)	kg	0.51	5.5	2.8
Electricity (compressed air)	kWh	0.05	110.0	5.5
Electricity (mixing)	kWh	0.05	68.0	3.4
Total				***14.4***
TOTAL				**29.6**

REFERENCES

Fennel, D.E., Rhee, S.K., Voordeckers, J.M., Kerkhof, L.J., and Häggblom, M.M., 2001. "Molecular and physiological characterization of dehalogenating consortia in marine sediments". *In Situ and On-Site Bioremendiation, The Sixth Int. Symposium, San Diego, Ca, Platform Abstract.* p. E-6.

IRSA-CNR, 1985. "Metodi analitici per i fanghi. Parametri chimico-fisici". Quaderni IRSA-CNR n.64

Koch, A. 1981. "Most probable number". In Gerhardt, Murray and Pillips (Eds). Manual of methods for general bacteriology. American Society for Microbiology, Washington D C, USA.

Lucchese, G., Scolla, G., and Robertiello, A. 1995. Inertization of liquid waste mud and solid waste containing heavy metals by sulfate-reducing bacteria". EP 95110068.4-2104, US patent 5660730

Ritalahti, K. M., Krajmalnik-Brown, R., Tiedje, J.M., and Löffler, F.E., 2001. "*Dehalococcides* species are the dominant vinyl chloride dechlorinating bacteria in anaerobic environment". *In Situ and On-Site Bioremendiation, The Sixth Int. Symposium, San Diego, Ca, Platform Abstract.* p. C-1.

Szelewski, M. 1989. "Analysis of an environmental sediment extract using the HP 5921 A to emission detector" HP Application note 228-71

Yang, Y., and McCarty, P. L. 2001. "Sulfate impact on anaerobic reductive dehalogenation". *In Situ and On-Site Bioremendiation, 6th Int. Symposium, San Diego, Ca, Platform Abstract.* p. C-6.

HYDROGEN-ENHANCED DECHLORINATION IN CONTAMINATED COASTAL SEDIMENTS

Cyndee L. Gruden, Iris D. Albrecht, and Peter Adriaens (Environmental Water Resources Engineering, The University of Michigan, Ann Arbor, MI)

ABSTRACT: Contaminated freshwater, estuarine and marine sediments are a priority of national concern, due to impacts on ecosystems and marine resources, public health and economic development. Contaminants such as chlorinated aromatic compounds tend to be strongly sorbing and bioaccumulative, thus impacting ecosystem degradation, public health, and economic development. Hydrogen amendments in sediments contaminated with dioxins resulted in extensive dechlorination of Cl_8DD to Cl_1DD at the expense of a decrease in 2,3,7,8-Cl_4DD production. In addition, dechlorination occurred at accelerated (200-fold) rates. An increase in hydrogen flux (dissolved hydrogen concentration) resulted in a statistically significant increase in percent active sediment-eluted microorganisms. Evidence from groundwater literature and our previous sediment research indicates a role for hydrogen as a driving force for dechlorination reactions. The purpose of this work is to determine the efficiency of a microbial enhancement technology through hydrogen amendments in sediments of varying geochemistry and contaminant history.

INTRODUCTION

Sediments pose some of the most challenging contaminated site issues facing regional and national freshwater and estuarine coastal environments, due to point-source releases and diffuse contaminant sources. An estimated 14-28 million cubic meters of contaminated sediment must be managed annually (NRC, 1997). Whereas volumes of contaminated sediments tend to be large, contaminant concentrations are low such that application of expensive conventional control technologies such as incineration or disposal in a secure landfill cannot be justified. Under such conditions, natural or enhanced recovery may be ideal approaches for reduction of human and ecological risks from sediment contamination. Despite the significance of the problem, remedial action is limited due to inadequate understanding of the natural processes governing sediment dispersion and contaminant bioavailability, and the high costs and technical challenges involved in sediment characterization, removal, containment and treatment (NRC, 1997).

The experience with in situ bioremediation technologies in land-based soils or sediments and groundwater systems (e.g. Adriaens and Vogel, 1995; Adriaens et al., 1999) have limited potential for transfer to in situ (freshwater or marine) sediments, due to: (i) the dearth of knowledge on the degradative potential of (particularly) marine microbial consortia, (ii) difficulties with nutrient, oxidant or reductant delivery to in situ sediments, and (iii) marine sediment biota are intimately linked into the food chain, and hence, strategies aimed at increasing contaminant bioavailability to microorganisms may negatively impact the food chain community (NRC, 1997).

Considering that anaerobic inorganic electron acceptors are generally found in excess in sediments and respiratory activities coexist, Kerner (1993) and Postma and Jakobsen (1996) have argued that key-metabolites such as hydrogen and acetate produced from fermentation may be rate limiting for microbial reduction in sediments. Growing evidence indicates that hydrogen is a key electron donor used in the dehalogenation of lesser chlorinated organics (e.g. cis-DCE and VC to ethene), and organic electron donors appear to serve mainly as primary precursors to supply the needed hydrogen via fermentation (DiStefano et al., 1992; Fennell et al., 1997). The number of reports describing the use of hydrogen to stimulate anaerobic dechlorination processes is limited. Newell et al. (1997) reported on amendments of aquifer material contaminated with chlorinated solvents with hydrogen gas. Complete dechlorination to ethene was observed, but no mechanistic information or correlations to respiratory activity was inferred. Recently, Yang and McCarty (1998) reported on the use of different organic acids to stimulate reductive dechlorination in a benzoate-acclimated dehalogenating mixed culture. This suggests that the dehalogenating organisms occupy a niche in the anaerobic system similar to that occupied by hydrogen-utilizing methanogens, homoacetogens and sulfidogens, all of which compete for hydrogen in mixed microbial communities.

Objectives. The overarching goal of this work is to develop a cost effective sediment management technology for contaminated sediments, based on the application of hydrogen-based microbial activity enhancement processes. Specific objectives are: (i) verification of hydrogen-enhanced dechlorination activity in sediment systems and (ii) evaluation of hydrogen uptake capacity and microbial respiration (hydrogenase activity) in sediments of varying geochemistry and contaminant history under mixed conditions, and as a function of hydrogen input fluxes. Emphasis will be placed on the remediation of halogenated aromatic contaminants such as polychlorinated biphenyls (PCBs), polychlorinated dibenzo-p-dioxins (PCDD), and polychlorinated phenols.

MATERIALS AND METHODS

Objective 1: Verification of hydrogen-enhanced dechlorination activity in sediment systems. Historically-contaminated sediment cores collected from the Passaic River, NJ were extruded, and core material was preincubated in the presence of an organic acid cocktail representative of concentrations in porewater to stimulate microbial activity (Adriaens et al., 1995; Barkovskii and Adriaens, 1996). Following a 3-month anaerobic incubation period, 200 g (wet weight) subsamples were placed in 250 mL wide-mouthed Mason jars. The sediment was overlaid with 50 mL of river water, resulting in a 50 mL nitrogen headspace. Jars were sealed tightly with Teflon tape. Triplicate microcosms were sacrificed at time zero to provide a baseline congener profile. Microcosms were amended with organic acid (OA) cocktail (100 mg/L), except for one triplicate set of the autoclaved controls, and incubated for 3 months. After incubation, one triplicate set of microcosms was subjected to a daily replacement of headspace with pure hydrogen gas for 10 days. During this time, microcosms were manually inverted on a daily basis. All samples were sacrificed for dioxin analysis at the end of the experiment.

Objective 2: Evaluation of hydrogen uptake capacity and microbial respiration (hydrogenase activity) under mixed conditions, and as a function of hydrogen input fluxes. Historically-contaminated sediments were preincubated with estuarine media, in the presence of organic acid cocktail. Sediment-eluted microorganisms (Fu et al., 1999, Barkovskii and Adriaens, 1996) were dispensed in a SIXFORS Bacteria/Yeast fermentation System (300 mL). This 6-reactor system is equipped with a H_2/N_2 gas mixing/delivery system, temperature, and pH control. The sediments were amended with varying H_2 fluxes to achieve optimum dissolved hydrogen concentrations as defined by microbial respiratory activity. Subsamples were collected, amended with 5-cyano-2, 3-ditolyl tetrazolium chloride (CTC), a redox dye which is reduced to a fluorescent red intracellular precipitate by metabolically active (respiring) bacteria. In this preliminary experiment, 10%H_2 balanced with N_2 was sparged through the headspace or a ring sparger in the estuarine media of duplicate microscosms at a rate of 167 mL/min for a 36-hour period. After 28 hours of incubation, the microcosms were spiked with OA to eliminate the possibility of substrate limitation. Due to a significant increase in transfer efficiency, dissolved hydrogen concentration is expected to be significantly greater in the microcosms receiving H_2 through the ring sparger.

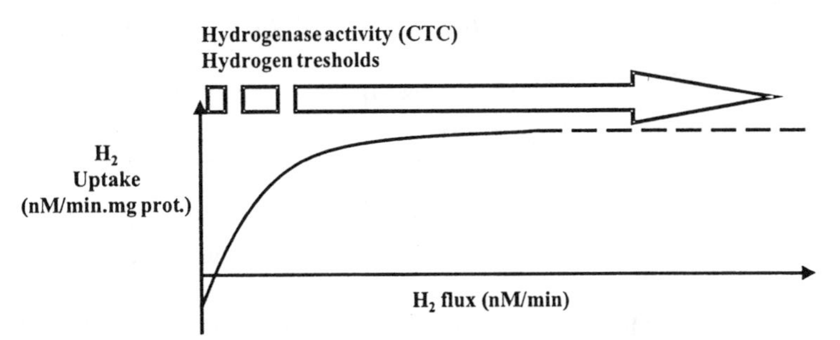

Figure 1: Schematic representation of the methodology followed to achieve objective 2

RESULTS AND DISCUSSION

Hydrogen-Enhanced Microbial Dechlorination Activity. The baseline congener profile (in mol%) was comprised of approximately 50% Cl_8DD, 20% of both 1,2,3,4,6,7,9-Cl_7DD and 2,3,7,8-Cl_4DD, and 5% 1,2,3,4,6,7,8-Cl_7DD. The abiotic (autoclaved) samples resulted in a decrease in mol% of Cl_7DD accompanied by an increase in mol% Cl_4DD as compared to the baseline congener profile. Contrary to thermodynamic predictions, the congeners resulting from biotic reduction indicated the predominance of a single *peri*-dechlorination step from Cl_8DD to 1,2,3,4,6,7,8-Cl_7DD. Hydrogen amendments in sediments contaminated with dioxins resulted in the production of 23 mol% of 2-Cl_1DD without significant accumulation of Cl_7DD (8.4%) or Cl_4DD (~13%) (Figure 2). No other amendment demonstrated a measurable amount of Cl_1-DD production. In comparison to the biotic system (amended with OAs only), the Cl_8DD fraction of the hydrogen-amended sample remained the same while the 1,2,3,4,6,7,8-Cl_7DD and 2,3,7,8-Cl_4DD fractions decreased by 76% and 25%, respectively. The observed reduction in 2,3,7,8-Cl_4DD and concurrent 2-Cl_1DD production indicates a predominance of lateral dechlorination reactions, which have been previously observed in historically-contaminated sediment-derived cell suspensions (Barkovskii and Adriaens, 1996).

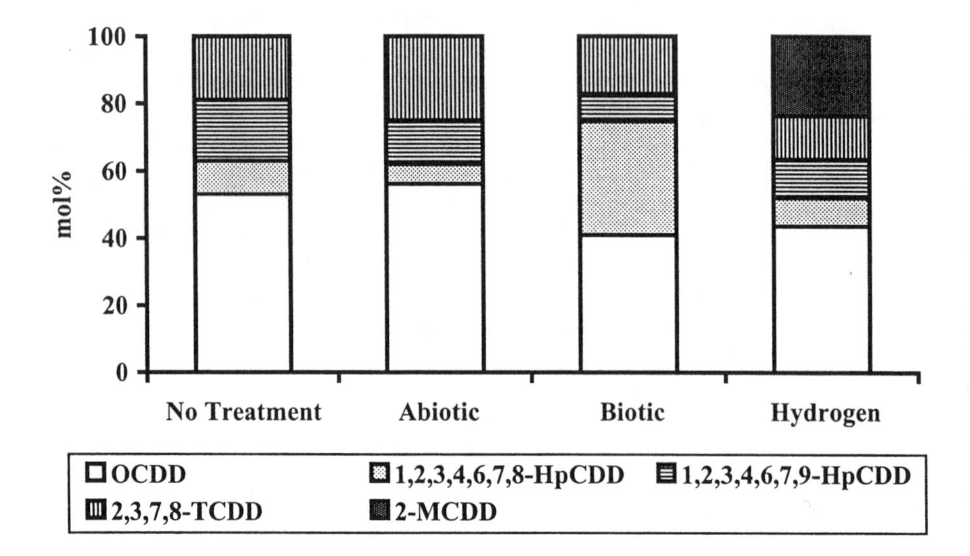

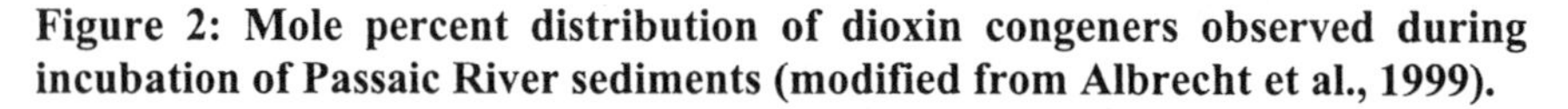

Figure 2: Mole percent distribution of dioxin congeners observed during incubation of Passaic River sediments (modified from Albrecht et al., 1999).

Dechlorination Kinetics in Biotic, Abiotic, and Sediment Systems. Since disappearance of dioxin congeners is not limited to dechlorination reactions, dechlorination kinetics were based on product appearance. Considering only two time points were evaluated, absolute values and magnitudes of dechlorination rates are primarily of interest. Rates for production of the various dioxin homologue groups in model systems generally decrease with subsequent dechlorination, corresponding to thermodynamic predictions: both the Gibbs free energy for dechlorination and the HOMO-LUMO gap (Lynam et al., 1998). Production rates of lesser-chlorinated congeners range from 0.05 to 0.90 ng/d in sediment-free systems (Fu et al., 1999, Barkovskii and Adriaens, 1996) and 0.01 to 17.29 ng/d in sediment systems with varied amendments (Table 1) (Albrecht et al., 1999).

Table 1: Dioxin dechlorination rates (pmol/d) in historically-contaminated sediments.

Sample preparation	Cl_8DD	1,2,3,4,6,7,8-Cl_7DD	2,3,7,8-Cl_4DD	2-Cl_1DD
Abiotic (autoclaved)	0.75	-0.28	0.36	0
Biotic (OA added)	-5.66	3.97	0.08	<0.005
H_2-amended	-6.76	-8.94	-28.59	79.32

*Rates are based on 25 g dry weight of historically-contaminated sediments.

In abiotic reactions, the time-dependent appearance of lesser-chlorinated products as a function of incubation conditions shows an increase in all congener groups with time, whereas dechlorination in cell-mediated incubations resulted in sequential production of lesser-chlorinated congeners (Fu et al., 1999). Production rates of Cl_7DD far exceed 2,3,7,8-Cl_4DD in biotic (OA amended) systems suggesting that Cl_7DD dechlorination may be rate-limiting. Contrary, the rate of 2-Cl_1DD production in the H_2-amended system is two orders of magnitude higher than the combined disappearance of other measured congeners. The relative rates of 2,3,7,8-Cl_4DD production from Cl_7DD and Cl_8DD and 2,3,7,8-Cl_4DD dechlorination to Cl_1DD are of particular interest due to toxicity concerns.

Evaluation of hydrogen uptake capacity and microbial activity as a function of hydrogen input fluxes. Preliminary results indicate that the population composition and respiratory counts (as fraction of total) increased five-fold (50%) when sparged with H_2/N_2 relative to headspace H_2/N_2 (10%) after 28 hours of incubation and an OA amendment (Figures 3a and 3b). Microscopic observation showed a decrease in community diversity (predominance of ~1um cocci bacteria) and total counts indicating that a selection process has taken place in microcosms sparged with H_2/N_2 (Figure 3a). The original microbial community (1:1 rods to cocci ratio) indicated CTC activity in < 5% of all bacteria.

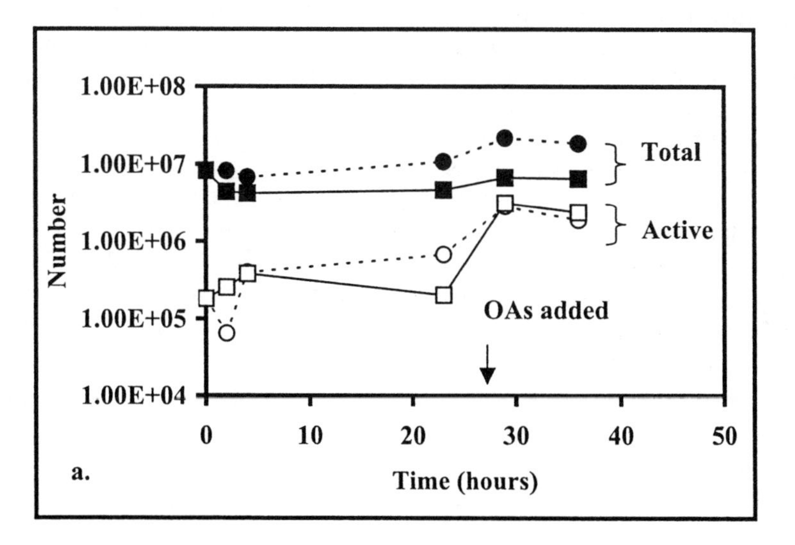

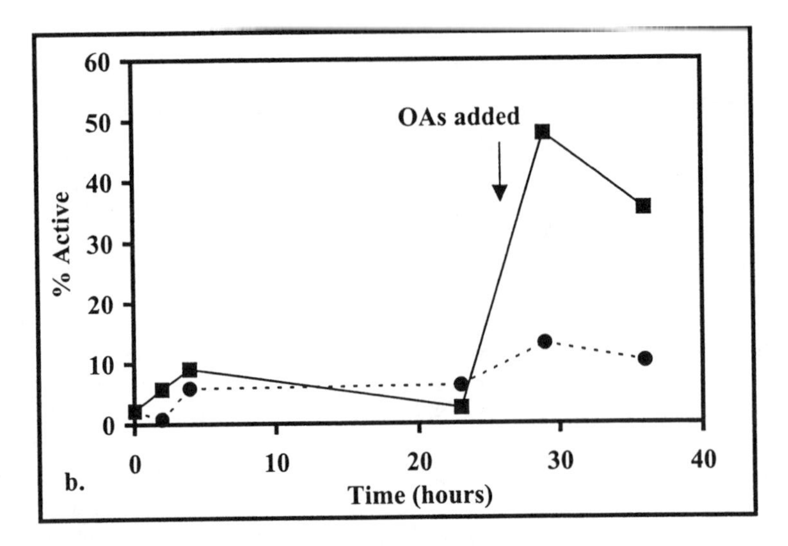

Figure 3: (a) Total (DAPI) and CTC active bacteria and (b) percent active bacteria for both headspace (•) and sparged (▪) H_2/N_2 samples. Error bars are smaller than symbols.

CONCLUSIONS

Hydrogen amendments in freshwater sediments resulted in extensive dechlorination of Cl_8DD to Cl_1DD at 200-fold accelerated rates without accumulation of 2,3,7,8-Cl_4DD. Further, fatty acid turnover was complete and residual hydrogen concentrations were significantly lower than those noted in the literature, implicating the role of hydrogen in dioxin dechlorination. Differences in equilibrium hydrogen concentrations resulting from fermentation and direct additions affect the extent and kinetics of dechlorination. Preliminary experiments involving variation of hydrogen fluxes demonstrated an increase in percent CTC activity in response to increased dissolved hydrogen concentration. Microscopic observation suggested a shift in population dynamics in response to sparging with H_2/N_2. Results from this work indicate that hydrogen has the potential to stimulate microbial activity resulting in enhanced dechlorination of halogenated aromatic compounds. This ongoing research endeavors to determine the efficiency of a microbial enhancement technology through hydrogen amendments in sediments varying geochemistry and contaminant history. If successful, the technology would be widely applicable to sediments of diverging origin, and would help decision-makers implement the most effective in situ treatment strategy.

REFERENCES

Adriaens, P. A.L. Barkovskii, and I.D. Albrecht. 1999. "Fate of Chlorinated Aromatic Compounds in Soils and Sediments." In D.C. Adriano, J.-M. Bollag, W.T. Frankenberger, and R. Sims (Eds), *Bioremediation of Contaminated Soils*, Soil Science Society of America/American Society of Agronomy Monograph, Soil Science Society of America Press, Madison WI. Pp. 175-212.

Adriaens, P. and T.M. Vogel. 1995. "Treatment Processes for Chlorinated Organics." In L.Y. Young and C. Cerniglia (eds.), *Microbiological Transformation and Degradation of Toxic Organic Chemicals*, pp. 427-476. John Wiley and Sons, Inc., New York, N.Y.

Adriaens, P., Q.S. Fu, and D. Grbic'-Galic. 1995. "Bioavailability and transformation of highly chlorinated dibenzo-p-dioxins and dibenzofurans in anaerobic soils and sediments." *Environ. Sci. Technol.* 29: 2252-2261.

Albrecht, I.D., A.L. Barkovskii, and P. Adriaens. 1999. "Production and Dechlorination of 2,3,7,8-Tetrachloro-Dibenzo-p-Dioxin (TCDD) in Historically-Contaminated Estuarine Sediments." *Environ. Sci. Technol.* 33: 737-744.

Barkovskii, A.L., and P. Adriaens. 1996. "Microbial Dechlorination of Historically Present and Freshly Spiked Chlorinated Dioxins and the Diversity of Dioxin-Dechlorinating Populations." *Appl. Environ. Microbiol.* 62: 4556-4562.

DiStefano, T.D., J.M. Gossett, and S.H. Zinder. 1992. "Hydrogen as an Electron Donor for Dechlorination of Tetrachloroethene by an Anaerobic Mixed Culture." *Appl. Environ. Microbiol.* 58:3622-3629.

Fennell, D.E., Gossett, J.M., and S.H. Zinder. 1997. "Comparison of Butyric Acid, Ethanol, Lactic Acid, and Propionic Acid as Hydrogen Donors for the Reductive Dechlorination of Tetrachloroethene." *Environ. Sci. Technol.* 1997, 31:918-926.

Fu, Q.S., A.L. Barkovskii, and P. Adriaens. 1999. "Reductive Transformation of Dioxins: An Assessment of the Contribution of Dissolved Organic Matter to Dechlorination Reactions." *Environ. Sci. Technol.* 33: 3837-3842.

Kerner, M. 1993. "Coupling of Microbial Fermentation and Respiration Processes in an Intertidal Mudflat of the Elbe Estuary." *Chemosphere* 38:314-330.

Lynam, M.M., M. Kuty, J. Damborsky, J. Koca, and P. Adriaens. 1998. "Molecular orbital calculations to describe Microbial reductive Dechlorination of Polychlorinated Dioxins." *Environ. Toxicol. Chem.* 17(6):988-997.

National Research Council. 1997. *Contaminated Sediments in Ports and Waterways: Cleanup Strategies and Technologies*. National Academy Press, Washington DC.

Newell, C.J., R.T. Fisher, and J. Hughes. 1997. "Direct Hydrogen Addition for the in-situ Biodegradation of Chlorinated Solvents." In Proceedings of the Petroleum Hydrocarbons and Organic Chemicals in Ground Water. *Prevention, Detention, and Remediation Conference*, pp. 791-801. Nov 12-14, 1997. Houston, Texas. The National Ground Water Association, Ground Water Publishing Company, Westerville, Ohio.

Postma, D. and R. Jakobsen. 1996. "Redox Zonation: Equilibrium Constraints on the Fe(III)/SO4- Reduction Interface." *Geojchim. Cosmochim. Acta* 60: 3169-3175.

BIOSURFACTANTS AND ENVIRONMENTAL IMPROVEMENT IN THE OIL AND PETROCHEMICAL INDUSTRY AND THE ECOSYSTEM

I.M. Banat (University of Ulster, Northern Ireland, UK)
Ivo Rancich and Piera Casarino (Idrabel Italia, Arenzano, Genoa, Italy)

ABSTRACT: Biosurfactants are natural biodegradable complex polymers that are mainly produced by microorganisms through a bacterial fermentation process. Interest in this kind of product is steadily increasing due to the fact that they are none toxic and are biodegradable. Many studies have been directed in recent years to develop the possibility of using biosurfactants in various promising industrial fields, such as enhanced oil recovery, hydrocarbon bioremediation, crude oil drilling, lubricants and surfactant-aided bioremediation of water-insoluble pollutants. Both laboratory and filed applications have demonstrated significant potential market for biosurfactants in the oil and petrochemical industry applications directed to the removal/mobilisation of oil sludge from storage tanks, to the oil spill bioremediation/dispersion both inland and at sea, oil contaminated sediment treatment and enhanced oil recovery. In this paper we outline our experience in biosurfactants' use in the oil and petrochemical field including crude oil storage tanks cleaning, sludge reduction in thickeners and decontamination of sites and sediments polluted by crude oil spills or seepage. Future potential use of these compounds for the benefit of reducing oil sediment sludge accumulation and reduction of environmental pollution is also discussed.

INTRODUCTION

Concern about environmental protection has increased recently from a global viewpoint. Accordingly, chemical products that can be used in harmony with environments are being sought and in this respect natural products are advantageous as substitutes to synthetic chemical products. Biosurfactants are natural extracellular amphiphilic compounds produced by microorganisms. They contain both a hydrophobic and a hydrophilic moiety. The hydrophobic domain is usually a hydrocarbon whereas the hydrophilic domain can be non-ionic, positively or negatively charged or amphoteric. The presence of the hydrophobic and hydrophilic moieties within the same molecule renders them capable of partitioning preferentially at the interfaces between fluid phases, which has different degrees of polarity or hydrogen bonding. This leads to the formation of a film at the interfaces leading to a reduction in the surface tension and interfacial tensions between phases at the surfaces and interfaces and the formation of micro-emulsions of oil in water or water in oil. These properties make surfactants suitable for an extremely wide variety of industrial applications involving emulsification, foaming, detergency, wetting and phase dispersion or solubilisation.

Rapid advances in biotechnology and increased environmental awareness combined with expected new legislation has provided further impetus for consideration of biological surfactants as possible alternatives to the existing products. This is because most biosurfactants have lower possible toxicity and persistence in the environment in comparison to chemical surfactants (Georgiou *et al.,* 1994). Low toxicity, biodegradable nature and diversity have gained them considerable interest in recent years. The range of potential industrial applications includes enhanced oil recovery (EOR); surfactant aided bioremediation and oils sludge storage tank cleaning. Other developing areas of biosurfactants use include health care, cosmetic and foods industries (Banat *et al.,* 2000).

Several types of biosurfactants have been isolated and characterised including glycolipids, phospholipids, neutral lipids, fatty acids, peptidolipids, lipopolysaccarides and others not fully characterised (Desai and Banat, 1997). Most biosurfactants are produced by bacteria and are either neutral or negatively charged. Certain microorganisms are likely to be found better adapted to particular environments such as oil reservoirs, soil or the ocean environment.

BIOSURFACTANTS APPLICATIONS

Oil industry is the largest market expected for biosurfactants use, both in petroleum production and incorporation into oil formulations. Other applications related to the oil industries includes oil spill bioremediation/dispersion, both inland and at sea, removal/ mobilisation of oil sludge from storage tanks and enhanced oil recovery (Banat 1995a). The second largest market for biosurfactants is emulsion polymerisation for paints, paper and industrial coatings. Surfactants can also used in food and cosmetic industries, industrial cleaning of products as well as in agricultural-chemicals as pesticides or to dilute and disperse fertilizers and enhance penetration of active compounds into plants. A summary of potential applications are shown in Table 1 and some are discussed bellow.

BIOSURFACTANT USE IN HYDROCARBON BIOREMEDIATION

Bioremediation options range from simply monitoring a site to aggressive treatment. Accelerated natural attenuation is an option involving the addition of a nutrient mainly a terminal electron acceptor or others limiting agents can enhance remediation. The ability of biosurfactants particularly rhamnolipids to emulsify hydrocarbon-water mixtures and consequently enhance the degradation of hydrocarbons in the environment has been explored. The ability of a surfactant to enhance biodegradation of slightly soluble organic compound depends on the extent to which it increases the bioavailability of the compound.

Harvey *et al.,* (1990) tested a biosurfactant from *P. aeruginosa* for its ability to remove oil from contaminated Alaskan gravel samples under various conditions including concentration of surfactant, time of contact, temperature of the wash and presence or absence of gum. They reported increased oil displacement (about 2-3 folds) in comparison to water alone. Necessary contact time for the maximum effect was also reduced from 1.5-2 min. for water to 1 min. In addition, the Environmental Technology Laboratory at University of Alaska, Fairbanks reported complete removal of diesel range petroleum hydrocarbons (to the limit of 0.5mg/kg) while semi volatile petroleum hydrocarbons were reduced to 70% level, a removal of 30% (Tumeo *et al.,* 1994).

TABLE 1. Areas of possible potential applications for biosurfactants in industry. (Adapted form Banat *et al.*, 2000).

Property	Potential field of application
Emulsification and disperdant	Cosmetics, paints, bioremediation, oil tanks cleaning.
Solubilizers and microemulsions	Toiletries and pharmaceuticals
Wetting and penetrating agents	Pharmaceuticals, textile industry and paints
Detergents	Household and agriculture products
Foaming agents	Toiletries, cosmetics and ore floatation
Thickening agents	Paints
Metal sequestering agents	Mining
Vesicle forming materials	Cosmetics, drug delivery system
Microbial growth enhancers	Sewage sludge treatments for oily wastes
Demulsifiers	Waste treatment and oil recovery/separation
Fungicide	Biological control of some plant pathogens
Viscosity reducing agents	Pipeline transportation
Dispersants	Coal –oil and coal-water slurry mixing

Biodetox (Germany) described a process to decontaminate soils, industrial sludge and wastewater (Van Dyke *et al.,* 1991). The procedure involves an *in situ* bioreclamation using "Biodetox foam", which contains bacteria, nutrients and biosurfactants and can be biodegraded. Jain *et al.,* (1991) found that the addition of *Pseudomonas* biosurfactant enhanced the biodegradation of tetradecane, pristane, and hexadecane in a slit loam with 2.1% organic matter. Similarly Zhang & Miller (1995) reported the enhanced octadecane dispersion and biodegradation by a *Pseudomonas* rhamnolipids surfactant. Falatko and Novak (1992) studied biosurfactant-facilitated removal of gasoline overlaid on the top of coarse grain sand packed column. Up to 15-fold increase in the effluent concentration of four gasoline constituents; toluene, m-xylene, 1,2,4-trimethylbenzene and naphthalene was observed upon adding a biosurfactant solution (600mg/l). These results demonstrated the capacity of biosurfactants to remove oil from naturally occurring substrate which lead to increased interest in biosurfactants applications in treating hydrocarbon-contaminated soils (Banat *et al* 1995b). Partially purified biosurfactants can be either used in bioreactors or *in situ* to emulsify and increase the solubility of hydrophobic contaminants. Alternatively, either biosurfactant producing microorganisms or growth limiting factors may be added to the soil to enhance growth of added or indigenous microorganisms capable of producing biosurfactants.

Herman *et al.* (1997) investigated the effects of rhamnolipids biosurfactants on *in situ* biodegradation of hydrocarbon entrapped in porous matrix and reported a mobilisation of hydrocarbon entrapped within the soil matrix at biosurfactants concentration higher than the critical micelle concentration (CMC). At concentrations lower than CMC enhanced in-situ mineralization of entrapped hydrocarbon occurred. One of the methods of removing oil contaminants is to add biosurfactants into soil to increase hydrocarbon mobility. The emulsified hydrocarbon can then be recovered by a production well and degraded above ground in a bioreactor. Bai *et al.* (1997) used an anionic mono rhamnolipid biosurfactant from *P. aeruginosa* to remove residual hydrocarbons from sand columns. They recovered approximately 84% of residual hydrocarbon (hexadecane) from sand column packed with 20/30-mesh sand and 22% hydrocarbon from 40/50-mesh sand, primarily because of increased mobilisation. They reported the optimal concentration of rhamnolipid of 500 mg/l and a range of possible use of 40-1500 mg/l. Recently Shulga *et al.* (2000) reported 92-95% efficiency for oil removal form sand polluted oil using rhamnolipid biosurfactant extracted form a *Pseudomonas sp.* PS-17. They also reported efficient removal of oil from oil-contaminated birds' feathers and animals' fur, with no subsequent harmful effects to the animals involved.

BIOSURFACTANTS AND PAH & METAL BIOREMEDIATION

Applying surfactants as immobilising agents might be one way to enhance the solubility of PAHs as they may increase their solubilisation or emulsification, to release hydrocarbons sorbed to soil organic matter and increase the aqueous concentrations of hydrophobic compounds resulting in higher mass transfer rates. In an investigation of the capacity of PAH utilising bacteria to produce biosurfactants using naphthalene and phenanthrene Daziel *et al.* (1996) concluded that biosurfactant production was responsible for an increase in the aqueous concentration of naphthalene. This indicates a potential role for biosurfactant in increasing the solubility of such compounds. Similarly Zhang *et al.* (1997) tested two rhamnolipid biosurfactants' effects on dissolution and bioavailability of phenanthrene and reported increased solubility and degradation rates. Noordman *et al.*(1998) also tested rhamnolipid (500 mg/l) biosurfactant solution's ability to enhanced removal of phenanthrene from contaminated soil in a laboratory columns' study and detected significant enhanced removal of phenanthrene compared to controls.

Recently Page *et al.* (1999) investigated the effects of biosurfactants and chemical surfactants on the solubilisation and mobilisation of phenanthrenes, fluorenes, pyrenes and chrysenes and reported 2.5 to 35 times increase in solubility in the presence of biosurfactants compared to chemical surfactants. Similarly Vipulanandan *et al.* (2000) reported more than 30 fold enhancement of solubilisation of naphthalene when using 5g/l biosurfactant concentration. Such ability to solubilise or mobilise nonaqueous phases is important in instances in which biodegradation is too slow or infeasible and *ex situ* washing or *in situ* soil flushing is the more logical process to use.

Biosurfactant have also been reported to promote heavy metals de-sorption from soils in two ways (Miller, 1995). The first is through complexation of the free form of the metal residing in solution which decreases the solution–phase activity of the metal and therefore promotes de-sorption. The second occurs under reduced interfacial tension conditions; the biosurfactants will accumulate at the solid-solution interface that may allow the direct contact between the biosurfactant and the sorbed metal (Mulligan *et al.,* 1999).

MICROBIAL ENHANCED OIL RECOVERY

An area of considerable potential for biosurfactant application is in the field of microbial enhanced oil recovery (MEOR). Biosurfactants can aid in oil emulsification and assist in the detachment of oil films from rocks (Banat 1995a, Banat 1995b). *In situ* removal of oil is due to multiple effects of the microorganisms on environment and oil. These effects include gas and acid production, reduction in oil viscosity, plugging by biomass accumulation, reduction in interfacial tension by biosurfactants and degradation of large organic molecules. These are all factors responsible for decreasing the oil viscosity and making its recovery easier.

The strategies involved in the MEOR depend on the oil reservoir prevalent conditions including temperature, pressure, pH, porosity salinity, geologic make up of the reservoir, available nutrients and the presence of indigenous microorganisms. These factors should be considered before devising a strategy for use in an oil well. There are three main strategies for use of biosurfactants in Enhanced Oil Recovery (EOR) or mobilisation of heavy oils.

a- Production in batch or continuous culture under industrial conditions followed by addition to reservoir through water flooding (*ex situ* MEOR).

b- Production of microbial biosurfactants at the cell-oil interface within the reservoir implying penetration of active cells into the reservoir.

c- Injection of selected nutrients into a reservoir, thus stimulating growth of indigenous biosurfactant producing microorganisms.

The first strategy is expensive due to capital required for bioreactors operation, product purification and introduction into oil containing rocks. The second and third strategy requires that the reservoir contain bacteria capable of producing sufficient amounts of surfactants. For production of biosurfactants, microorganisms are usually provided with low cost substrates such as molasses and inorganic nutrients, which promote growth and surfactant production. Alternatively surfactant-producing strains may be introduced into the well. The introduced organism faces competition from the indigenous population of microbes for the binding sites on rocks and for the added nutrients, which may make the results unpredictable.

Another application of biosurfactants is oil storage tank cleaning. Surfactants have been studied for use in reducing the viscosity of heavy oils, thereby facilitating recovery, transportation and pipelining. In a full-scale field investigation carried out in Kuwait (Banat *et al.,* 1991) tested the ability of biosurfactant to clean oil storage tanks and to recover hydrocarbon from the emulsified sludge. Two tones of biosurfactant-containing whole cell culture were used to mobilise and clean 850 m^3 oil sludge. Approximately 91% (774 m^3) of this sludge was recovered as re-sellable crude oil and 76 m^3 non-hydrocarbon materials remained as impurities to be manually cleaned. The value of the recovered crude covered the cost of the cleaning operation ($100,000-150.000 per tank). Similar results were obtained when carrying other clean up operation on fuel storage tanks and thickener storage tanks in Italy (Table 2). Such a clean up processes is therefore economically rewarding and less hazardous to persons involved in the process compared to conventional process. It is also an environmentally sound technology leading to less disposal of oily sludge in the natural environment. To our knowledge however, further commercial applications of this technology has not been carried out.

TABLE 2. The results of oils storage tank cleaning and oil recovery for four different types of oil storage tanks carried out in oil storage refineries in Kuwait and Italy.

No.	Storage tank type	Capacity (m^3)	Oil content in sludge (m^3)	Recovered oil (m^3)	% Recovered oil	% hydrocarbon in recovered oil
1*	Kuwaiti Crude oil	26000	850	774	91.00	> 99 %
2**	Heavy fuel oil	3000	100	90	90.00	96 %
3**	Sludge Thickener	830	170	170	100	90%
4**	Heavy fuel oil	50000	913	762	83.50	99%

* Carried out in Kuwait (Banat *et al.,* 1991).
** Carried out in Italy (Idrabel Italia, srl).

CONCLUSION

The usefulness of biosurfactants in bioremediation is expected to gain increasing importance in the future. To date, biosurfactants are unable to compete economically with the chemically synthesised compounds in the market mainly due to their high production costs and lack of comprehensive toxicity testing. Their success in bioremediation will require precise targeting to the physical conditions and chemical nature of the pollutant affected areas. Encouraging results have been obtained for use of biosurfactants in hydrocarbon pollution control in marine biotypes, in closed systems (oil storage tanks) and although many laboratory studies indicate potentials for use in open environment a lot remains undemonstrated in pollution treatment in marine environments or coastal areas. The possible use of biosurfactants in MEOR has many advantages, yet more information about structures and factors such as interaction with soil, structure function analysis of surfactant solubilisation, scale up and cost analysis for *ex-situ* production are required.

ACKNOWLEDGEMENT

I.M. Banat would like to thank the Environment & Heritage Service, DOE, for FRDF financial support under N. Ireland Single Programme (Ref. WM 47/99).

REFERENCES

Bai, G.Y., Brusseau, M.L. and Miller, R.M. 1997."Biosurfactant enhanced removal of residual hydrocarbons from soil". *J. Contam Hydrol.* 25: 157-170.

Banat, I.M. 1995a."Biosurfactants production and possible uses in microbial enhanced oil recovery and oil pollution remediation: A review". *Bioresource Technol.* 51: 1-12.

Banat, I.M. 1995b."Characterisation of biosurfactants and their use in pollution removal, State of the art review". *ACTA Biotechologica.* 15: 251-267.

Banat, I.M., Makkar R.S. and Cameotra, S.S. 2000."Potential commercial application of microbial surfactants". *Appl. Microbiol. Biotechnol.* 53: 495-508.

Banat, I.M., Samarah, N., Murad, M., Horne, R. and Benerjee, S. 1991. "Biosurfactant production and use in oil tank clean-up". *World J. Microbiol. & Biotechnol.* 7: 80-84.

Daziel, E., Paquette, G. Vellemur, R., Lepins, F. and Bisaillnon, J.G. 1996. "Biosurfactant production by a soil *Pseudomonas* strain growing on PAH's". *Appl. Environ. Microbiol.,* 62: 1908-1912.

Desai J.D. and Banat I.M. 1997."Microbial production of surfactants and their commercial potential". *Microbiol. & Molec. Biol. Reviews* 61: 47-64.

Falatko, D.M. and Novak, J.T. 1992."Effects of biologically produced surfactants on mobility and biodegradation of petroleum hydrocarbons". *Wat. Environ. Res.* 64: 163-169.

Georgiou, G., Lin, S.C. and Sharma, M.M. 1990."Surface active compounds from microorganisms". *Bio/Technol.* 10: 60-65.

Harvey, S., Elashvili, I., Valdes, J.J., Kamely, D. and Chakrabarty, A.M. 1990. "Enhanced removal of Exxon Valdez spilled oil from Alaskan gravel by a microbial surfactant. *Bio/ Technol.* 8: 228-230.

Herman, D.C., Zhang, Y.M. and Miller, R.M. 1997."Rhamnolipids (biosurfactant) effects on cell aggregation and biodegradation of residual hexadecane under saturated flow conditions". *Appl. Environ. Microbiol.,* 63: 3622-3627.

Jain, D.K., Thompson, D.L.C., Lee, H. and Trevors, J.T. 1991."A drop- collapsing test for screening surfactant producing microorganisms". *J. Microbiol. Methods,* 13: 271-279.

Miller. R.M. 1995."Biosurfactant-facilitated remediation of metal-contaminated soils Source". *Environ. Health Perspect*. 103: 59-62.

Mulligan, C.N., Young R.N. and Gibbs, B.F. 1999."On the use of biosurfactants for the removal of heavy metals from oil-contaminated soil". *Environ. Prog.* 18: 50-54.

Noordman, W.H., Ji W., Brusseau, M.L. and Jamssen. D.B. 1998."Effects of rhamnolipids biosurfactants on removal of phenanthrene from soil". *Environ. Sci. Technol.,* 32: 1806-1812.

Page, C.A., Bonner, J.S., Kanga, S.A., Mills, M.A. and Autenrieth, R.L. 1999. "Biosurfactant solubilisation of PAHS". *Enviorn. Eng. Sci.* 16, 465-474.

Shulga, A., Karpenko, E., Vildanova-Martsishin, R., Turovsky, A. and Soltys, M. 2000."Biosurfactant-enhanced remediation of oil-contaminated environments. *Adsorp. Sci & Technol.* 18: 171-176.

Tumeo, M., Bradock, J., Venator, T., Rog, S. and Owens, D. 1994."Effectiveness of a biosurfactant in removing weathered crude oil from subsurface beach material". *Spill Sci. Technol. Bull.* 1: 53-59.

Van Dyke, M.I., Lee, H. and Trevors, J.T. 1991."Application of microbial surfactants". *Biotechnol. Adv.* 9: 241-252.

Vipulanandan, C. and Ren, X. 2000."Enhanced solubility and biodegradability of naphthalene with biosurfactant". *J. Environ. Eng.* 126: 629-634.

Zhang, Y., Maier, W.J. and Miller, R.M. 1997."Effect of rhamnolipids on the dissolution, bioavailability and biodegradation of phenanthrene". *Environ. Sci. Technol.* 31: 2211-2217.

Zhang,Y. and Miller, R.M. 1995." Effect of rhamnolipid (biosurfactant) structure on solubilisation and biodegradation of n-alkanes". *Appl. Environ. Microbiol.* 61: 2247-2251.

MOLECULAR ENGINEERING OF SOLUBLE BACTERIAL PROTEINS WITH CHROMATE REDUCTASE ACTIVITY

C-H. Park, C. Gonzalez, D. Ackerley, M. Keyhan, and ***A. Matin***
Stanford University, Stanford, CA 94305, USA

ABSTRACT: Cr(VI) (chromate), which is a serious environmental pollutant, is amenable to bacterial bioremediation. Molecular and genetic engineering can enhance this capacity. Such approaches can, for example, decrease the sensitivity of these enzymes to other toxic waste present in chromate-contaminated sites; increase their substrate affinity and V_{max}; and broaden their substrate range, enabling an individual enzyme to remediate multiple pollutants. To initiate such improvements, it is necessary to learn more about the properties and biological role of the wild type enzymes. We have cloned six bacterial genes, which fall into two homology classes, and have studied, using pure proteins, the kinetics of reduction of chromate and other compounds. All can reduce quinone; the three class I and two of the three class II proteins can also reduce chromate; and all the class II, but not class I proteins can in addition reduce nitrocompounds, which are also serious environmental pollutants. The K_m of class II enzymes for chromate reduction is lower than that of Class I. Our ongoing studies indicate that the biological role of these proteins may be to prevent channeling of reducing equivalents into reactive oxygen species-generating pathways. Thus, strengthening the activity of these gene products may also bolster cell survival.

INTRODUCTION

Chromate, whose toxicity is well established, is a widespread environmental pollutant due to the many industrial uses of chromium compounds. At the U.S. Department of Energy (DOE) sites, for example, which occupy approximately 2,800 sq. miles, it is the second most common heavy metal contaminant. Its concentration at these sites ranges between 0.008 to 173 μM in groundwater, and 98 nM to 76 mM in soil and sediments – the latter may represent even higher concentrations because soil water is stored in small capillaries (Riley et al., 1992). Estimates for the cost of cleanup of DOE sites run into hundreds of billions of dollars (McIlwain, 1996).

Bacteria can reduce chromate to the relatively insoluble and immobile Cr(III) species (Suzuki et al., 1992; Park et al., 2000). However, bioremediation of chromate and other pollutants faces two major challenges. One arises from the fact that most polluted environments contain mixtures of toxic compounds. For example, the waste at DOE sites includes heavy metals, radionuclides, chlorinated and fuel hydrocarbons, explosives and many other organic pollutants (Riley et al., 1992). Such conditions inhibit individual enzymes and bacteria, precluding effective bioremediation. The other problem is the inherent sluggishness of bioremediation. Addition of nutrients to the contaminated environments to stimulate the growth of indigenous bacteria can ameliorate this problem. But it leads to the production of a large biomass, which can cause clogging, confining remediation to a narrow zone (McCarty, 1994).

One way to address these problems is to use genetic- and protein-engineering approaches. These can help engineer enzymes with decreased sensitivity to other pollutants, improved K_m and V_{max}, and a broad substrate range, permitting a given enzyme to remediate several pollutants. Also, their expression can be controlled by regulatory elements that permit abundant synthesis under the harsh in-situ conditions that restrict growth. Such elements, called the starvation promoters, are maximally activated under slow-growth conditions. Their use not only permits effective expression under in-situ conditions, but can also minimize biomass formation, and mitigate the clogging problem (Matin et al., 1995).

Before attempting such improvements in the chromate-reducing enzymes, it is necessary to obtain a fuller understanding of their kinetics, inhibition characteristics, regulation and biological role. Toward this end, we have cloned six bacterial genes, the function of most of which is as yet unknown, and describe here the substrate range and kinetics of their pure products. These genes are listed as annotated opening reading frames (ORFs) in various data bases, and are identified here by the number they have been assigned.

MATERIALS AND METHODS

Enzyme Assays. Chromate reductase activity was measured as described (Park et al., 2000). Quinone reductase activity was measured spectrophotometrically by monitoring NAD(P)H disappearance at 340 nm, and that of nitrofurantoin reductase activity by measuring its disappearance at 373 nm. One ml reaction mixtures contained 50 mM citrate or phosphate buffer, 250 μM coenzyme Q_1 [1 mM for YieF protein, which has a high K_m for quinone (Table 3)], or 50 μM of nitrofurantoin, and the enzyme protein (10–20 μg). Reactions were started by adding 250 μM NAD(P)H. Stock coenzyme Q_1 solution was prepared in 10 mM N,N-dimethylformamide. Controls without enzyme gave no activity.

Gene Cloning. The genes were cloned using the polymerase chain reaction (PCR) and the primers given in Table 1. The cloned products were sequenced and checked against the databases: [The Institute of Genomic Research (TIGR: http://www.tigr.org/tdb/mdb/mdbinprogress.html); Blattner et al. (1997); and Kunst et al. (1997)].

Protein Overproduction and Purification. The coding regions of the genes were cloned in pET-28 vectors. The recombinant plasmids were transformed in *Escherichia coli* BL21 (DE3) strain. Following induction with 0.5–1.0 mM IPTG, recombinant protein production was detected by SDS gel electrophoresis. Inclusion bodies were renatured, where necessary using the Novagen Protein Refolding Kit; and thrombin was cleaved as per Novagen protocol. The resulting proteins were electrophoretically pure (not shown).

Computer programs. The ones used in data analysis are listed in Table 2.

TABLE 1. Primers used for PCR cloning

STRAIN	TARGET GENE	PRIMERS [a]
P. putida MK1	Park et al. (2000) enzyme	Forward: 5'-TGTCGACTCATATGAGCCAGGTGTATTCGGTAGCAGTCG-3' Reverse: 5'-TTAGGAATTCTCAGACCGCCCTGTTCAACTTCACCC-3'
P. putida KT2440	10732 [c]	Forward: 5'-TGTCGACTCATATGAGCCAGGTGTATTCGGTAGCAGTCG-3' Reverse: 5'-TTAGGAATTCTCAGACCGCCCTGTTCAACTTCACCC-3'
E. coli	401621 [b]	Forward: 5'-AGTATATCCATATGATGTCTGAAAAATTGCAG-3' Reverse: 5'-GAATTCTTAGATCTTAACTCGCTGA-3'
P. putida KT2440	10712 [c]	Forward: 5'-TGTCGACTCATATGAGCCTTCAAGACGAAGC-3' Reverse: 5'-ATCGAATTCTCAGCGCAGGCCGAAACCTAG -3'
B. subtilis	3225092 [b]	Forward: 5'-CTATGGCTAGCGGTGTCTCATATGAATGAAGT-3' Reverse: 5'-ATCGAATTCTAACCGGGTAAATGAAAG-3'
E. coli	*NfsA* [d] 730007 [b]	Forward: 5'-ATACGTCGCATATGACGCCAACCATT-3' Reverse: 5'-AGTGAATTCATTAGCGCGTCGCCCA-3'

[a] Underlined sequences denote the restriction sites used to clone the genes into plasmids.
[b] Protein accession number in PubMed data base.
[c] Contig number in TIGR data base.
[d] The only gene previously studied.

TABLE 2. Computer programs

DATA ANALYSIS	ELECTRONIC TOOL	SOURCE
Homology search	BLAST	http://www.ncbi.nlm.nih.gov/BLAST/
	TIGR-BLAST	http://www.tigr.org/cgi-bin/BlastSearch/blast.cgi?
Primer design	Oligo-Primer Analysis program	Version 6.13, Mol. Biol. Insights
Sequence alignment	Clustal W, Clustal X	http://www2.ebi.ac.uk/clustalw/
Structure analysis	J-PRED *	http://jura.ebi.ac.uk:8888/submit.html
	PHD *	http://tw.expasy.org/tools/

* Conserved secondary structures in each protein with > 82% expected average accuracy were used.

RESULTS

Cloning of the *Pseudomona putida* gene encoding chromate reductase activity. This investigation began with the cloning of the gene that encodes the novel soluble chromate reductase activity we purified previously from *P. putida* (Park et al., 2000). This was done by "reversed genetics." The N-terminal and internal amino acid composition of the pure enzyme was determined (Lomovskaya et al., 1994), and based on the derived nucleotide sequence, the gene was identified on contig 10732 of the TIGR genomic sequence of *P. putida*. Using the primers given in Table 1, the gene was cloned by PCR from *P. putida* strain KT2440 and also from strain MK1 (Park et al., 2000); the two genes are 90% homologous.

Homologs of Genes Encoding Chromate Reductase Activity. In 1996, Suzuki et al. reported cloning of a gene (Genbank protein accession number D83142) from *Pseudomonas ambigua*, which encodes a soluble chromate reductase activity (Suzuki et al., 1992). The Suzuki et al. enzyme is 243-amino acid long, whereas ours has 183 amino acids, and the two genes bear no homology. Neither gene has been previously studied.

To gain an insight into the potential substrate range of the proteins encoded by these genes, we conducted a blast search, using the programs listed in Table 2, based on their derived amino acid sequence. Some 16 bacteria and a plant possessed homologues of each gene, with 30% or more amino acid identity. For convenience, we refer to the homologues of our enzyme as class I, and those of the Suzuki et al. enzyme, as class II proteins. While most of the proteins in both classes are identified as ORFs, one protein in each class has been examined in detail. Thus, the class I *Arabidopsis thalianus* protein (accession #5002232) has been identified as a flavoprotein that catalyzes a two-electron reduction of quinones, using NAD(P)H as electron donor (Sparla et al., 1999); and the class II protein, the *E. coli* NfsA, has been extensively studied for its role in reducing nitrocompounds [which are important environmental pollutants (Zenno et al., 1998)], and quinones.

The amino acid identity might signify functional similarity, a possibility that would be reinforced if the proteins also bore structural similarity. We therefore compared the predicted secondary structure of several proteins of class I (see Table 2 for the computer program used) with that of the *Arabidopsis* protein; and that of class II with the *E. coli* NfsA protein – proteins, as mentioned above, whose activity has been partially elucidated. Strong similarities were found. This is illustrated for selected class I proteins in Fig. 1; a similar situation was found for class II proteins and NfsA (not shown).

Substrate Range and Kinetic Characteristics of Selected Class I and II proteins. The predicted secondary structural similarity suggests that the class I proteins should have both chromate and quinone reductase activity, and the class II proteins chromate, quinone and nitrocompound reductase activity. This prediction was borne out, by and large, by the proteins studied so far. Thus, all the class I proteins examined possessed chromate as well as quinone reductase activities, with NAD(P)H as electron donor (Table 3). Similarly, all class II proteins, except *P. putida* 10712, possessed the predicted three reductase activities; 10712, however, failed to show chromate reductase activity through an extensive range of pH (4.6 to 8.5), and temperature (20-80°C). This protein differed from the others also in that, while the others utilized both NADH and NADPH equally efficiently as electron donor, it was inactive with NADH.

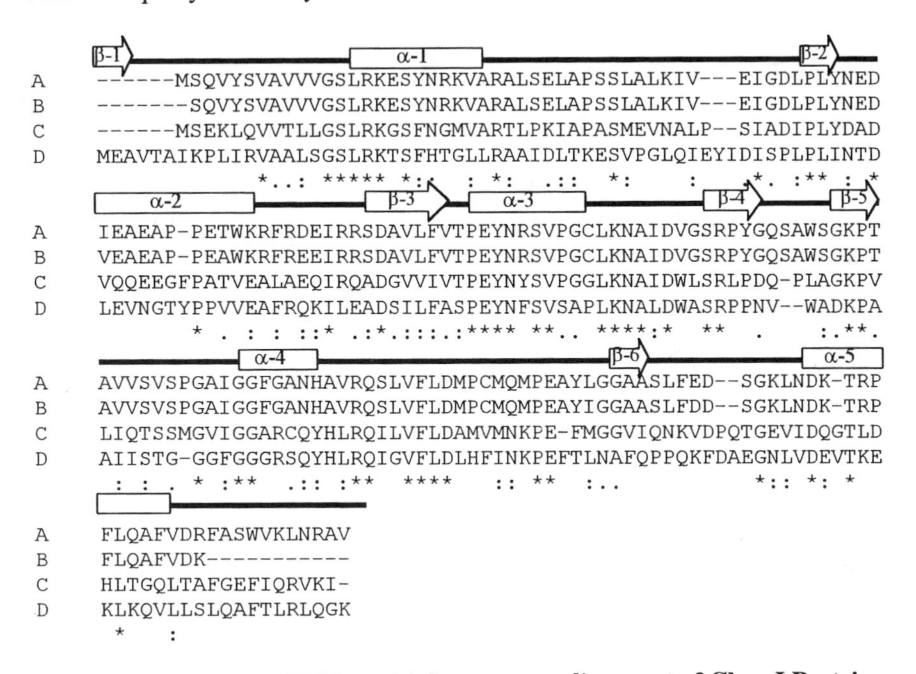

FIGURE 1. CLUSTAL multiple sequence alignment of Class I Proteins.
[A, AF5682; B, AF375642; C, AF385329; D, *A. thaliana* NAD(P)H quinone oxidoreductase (5002232).] Arrows indicate β-sheets, and boxes represent α-helices.

The class I proteins had similar K_m (or $K_{0.5}$) and V_{max} for chromate, but they differed in their kinetics for quinone reduction. The *E. coli* protein 40162, belonging to this class, exhibited sigmoidal kinetics with a Hill's coefficient of 1.2 and 1.1 for chromate and quinone, respectively. None of this class of proteins reduced the nitrocompound, nitrofurantoin. The class II proteins had similar K_m for quinone, nitrofurantoin, and chromate reduction; their K_m for chromate reduction was lower than that of class I.

All six proteins have a dialysis-resistant yellow color, suggesting that they are flavoproteins. The substrate range for these proteins shown here has been deposited in Genbank under the following accession numbers: AF56852 for the chromate reductase enzyme of Park et al. (2000); AF375642 for gene on contig #10732 in *P. putida* genomic sequence (TIGR); AF385329 for gene on contig #401621 in *E. coli* genomic sequence (Blattner et al., 1997); AF417208 for

3225092 (PubMed); AF417209 for gene on contig #10712 (TIGR). The updated information on NfsA retains the previous number for this protein (730007). See Table 1.

TABLE 3. Substrate range and kinetics of selected Class I (first three columns) and II proteins (next three columns).

1	2	3	4	5	6	7
Park et al (2000) enzyme	*P. putida* MK1	Cr (VI)	208	1250	70	5.0
		Quinone	N.D[c]	N.D.	60	5.0
		NF[b]		NO ACTIVITY		
AF375642	*P. putida* KT2440	Cr (VI)	193	1715	70	5.0
		Quinone	20	8200	60	8.5
		NF		NO ACTIVITY		
AF385329	*E. coli*	Cr (VI)	152	1840	35	5.0
		Quinone	541	43000	40	5.0
		NF		NO ACTIVITY		
10712	*P. putida* KT2440	Cr (VI)		NO ACTIVITY		
		Quinone	16	560	40	7.0
		NF	39	3800	40	7.0
730007 (*nfsA*)	*E. coli*	Cr (VI)	3	151	50	5.0
		Quinone	4	860	30	9.5
		NF	3	158	50	9.5
3225092	*B. subtilis*	Cr (VI)	13	93	20	5.0
		Quinone	4	3240	40	5.5
		NF	6	7250	40	5.5

[a] Arabic numerals in the top column refer to: 1. protein; 2. bacterium; 3. substrate; 4. $K_m/K_{0.5}$ (μM); 5. V_{max} (nmol min^{-1} mg $protein^{-1}$); 6. optimal temperature (°C); and 7. optimal pH.
[b] Nirofurantoin
[c] Not determined

X-ray Absorption Near-edge Spectroscopic (XANES) Analysis of Chromate Reduction by NfsA. This is the first time that the *E. coli* protein, NfsA, has been shown to possess chromate reductase activity.

To be useful in chromate bioremediation, it is important that an enzyme be able to convert Cr(VI) to Cr(III), since partially reduced Cr species are harmful. We previously showed, using XANES, that the class I *P. putida* protein (AF56852) quantitatively converted Cr(VI) to Cr(III) (Park et al., 2000). Similar measurements showed that NfsA also converts Cr(VI) quantitatively to Cr(III).

DISCUSSION

Suzuki et al. (1992) and Park et al. (2000) purified different proteins from *Pseudomonas* species that possessed chromate reductase activity. These proteins bear no homology to each other. However, each bears close structural homologues in several other bacteria; these are referred to here as class I and class II proteins. While most proteins in these classes are identified as putative ORFs, one in each class has been studied in considerable detail: the *A. thaliana* protein belonging to class I, and the *E. coli* NfsA protein belonging to class II. The similarities suggested that the class I proteins possess chromate and quinone reductase activities, and class II proteins possess, in addition, nitrocompound reducing activity. This prediction is borne out, by and large, by the data presented here, involving the study of pure proteins generated from cloned genes. The only exception is the class II *P. putida* protein (Table 3) which while possessing the other predicted activities, fails to reduce chromate. Thus, six genes possessing important bioremediation activity were cloned in this study and used to generate pure proteins all of which reduced quinone, five reduced chromate and three also nitrocompounds. All but one could use both NADH or NADPH as electron donor equally effectively.

Representative members of both the class I and II enzymes can quantitatively reduce chromate to Cr(III). Whether other proteins studied here can also do so is under investigation. Given the strong predicted secondary structure similarity among the class members, and the fact that another class II enzyme, the *P. ambigua* enzyme (D83142) purified by Suzuki et al. (1992), can also fully reduce Cr(VI), it is likely that they can. Thus, these enzymes can be effective agents of chromate bioremediation, as opposed to several others, such as glutathione reductase, lipoyl dehydrogenase, and ferredoxin-NADP oxidoreductase, that reduce chromate only to Cr(V). Cr(V) generates reactive oxygen species (ROS), and is centrally involved in chromate toxicity (Shi and Dalal, 1990; Singh et al., 1998).

The availability of these genes and proteins can benefit the current chromate bioremediation technology. For example, the genes can be brought under the control of strong promoters and cloned in other organisms such as *Deinococcus radiodurans*. This bacterium has a low intrinsic chromate reducing activity (Keyhan and Matin, unpublished) and its chromate reductase-encoding gene has not been cloned. It is strongly resistant to radiation and is therefore suitable for bioremediating DOE-type sites. Similarly, the proteins generated in this study can be used in bioreactors.

Moreover, this availability opens up ways for bioengineering superior chromate reductase activities. Established protocols for generating large quantities of pure proteins from cloned genes now make the detailed studies of these proteins feasible. These reductases appear to have a broad substrate range; thus, it may be possible to extend their decontaminating activity to other pollutants, such as U(VI). This, and other improvements can be attempted through structural studies as well as by the use of the DNA shuffling technology. The latter has proven to be extremely powerful for generating enzymes with desired characteristics, and is particularly effective in situations, such as the present one, where gene families from diverse species can be used (Crameri et al., 1998).

In proposing protein engineering, it is important to consider the biological role of the protein concerned. Studies to be reported elsewhere show that mutations in the *chrR* and *nfsA* genes affect not only *P. putida*, and *E. coli* chromate and nitrofurantoin-reducing ability, but also make them more sensitive to oxidative stress in general. The regulatory pattern of these genes and the manner of their involvement in oxidative stress protection suggest that their biological role is to prevent channeling of reducing equivalents into pathways that lead to the generation of ROS (in preparation). Thus, strengthening the activity of these gene products may also bolster cell survival.

CONCLUSIONS

1. Six genes and their pure protein products containing important bioremediation potential were cloned and generated.
2. Their availability makes it possible to conduct structural studies on the proteins.
3. It also makes it possible to conduct DNA shuffling to evolve more efficient bioremediating enzymes.

ACKNOWLEDGMENTS

This research was funded by the Natural and Accelerated Bioremediation (NABIR) program, Biological and Environmental Research (EBER), U.S. Department of Energy (grant# DE-FG03-97ER-62494 to A.M). CG was supported by a grant from Conicet, and DA by a grant from FRST, New Zealand. We thank Scott Fendorf for carrying out the XANES analysis.

REFERENCES

Blattner F. R., G. Plunkett, et al. 1997. The complete genome sequence of *Escherichia coli* K-12. Science 277:1453-1474.

Crameri, A., S. Raillard, E. Bermudez, and W. P. C. Stemmer. 1998. DNA shuffling of a family of genes from diverse species accelerates directed evolution. Nature 391:288-291.

Kunst F, Ogasawara N, et al. 1997. The complete genome sequence of the gram-positive bacterium *Bacillus subtilis*. Nature. 390:237-238.

Lomovskaya, O.L., J.P. Kidwell, A. Matin. 1994. Characterization of the sigma 38-dependent expression of a core *Escherichia coli* starvation gene, *pexB*. J. Bacteriol.176:3928-3935.

Matin, A., C. D. Little, C. D. Fraley, and M. Keyhan. 1995. Use of starvation promoters to limit growth and selectively express trichloroethylene and phenol transformation activity in recombinant *Escherichia coli*. Appl. Environ. Microbiol. 61:3323-3328.

McCarty, P. L. 1994. Ground-water treatment for chlorinated solvents, p. 87-116. *In* J. E. Matthews (ed.), Handbook of bioremediation. Lewis Publishers, Ann Arbor, Mich.

McIlwain, C. 1996. Science seeks weapons clean-up role. Nature 383:375-379.

Park, C. H., M. Keyhan, B. Wielinga, S. Fendorf, and A. Matin. 2000. Purification to homogeneity and characterization of a novel *Pseudomonas putida* chromate reductase. Appl. Environ. Microbiol. 66:1788-1795.

Riley, R. G., J. M. Zachara, and F. J. Wobber. 1992. Chemical contaminants on DOE lands and selection of contaminant mixtures for subsurface science research. U.S. Department of Energy DOE/ER-0547T.

Shi, X. and N.S. Dalal. 1990. NADPH-dependent flavoenzymes catalyze one electron reduction of metal ions and molecular oxygen and generate hydroxyl radicals. FEBS Lett. 276:189-191.

Singh, J., D. L. Carlisle, D. E. Pritchard, and S. R. Patierno. 1998. Chromium-induced genotoxicity and apoptosis: relationship to chromium carcinogenesis. Oncology Reports 5L1307-1318.

Sparla, F., G. Tedeschi, P. Pupillo, and P. Trost. 1999. Cloning and heterologous expression of NAD(P)H:quinone reductase of *Arabidopsis thaliana*, a functional homologue of animal DT-diaphorase. FEBS Lett. 436:382-386.

Suzuki, T., N. Miyata, H. Horitsu, K. Kawai, K. Takamizawa, Y. Tai, and M. Okazaki. 1992. NAD(P)H-dependent chromium(VI) reductase of *Pseudomonas ambigua* G-1: a Cr(V) intermediate is formed during the reduction of Cr(VI) to Cr(III). J. Bacteriol. 174:5340-5345.

Zenno, S., T. Kobori, M. Tanokura, K. Saigo. 1998. Conversion of NfsA, the major *Escherichia coli* nitroreductase, to a flavin reductase with an activity similar to that of Frp, a flavin reductase in *Vibrio harveyi*, by a single amino acid substitution. J. Bacteriol. 180:422-425.

BIOREMEDIATION OF POLLUTED SEDIMENT: A MATTER OF TIME OR EFFORT? — PART II*

Joop Harmsen and ***Lucas Bouwman*** (Alterra, Wageningen-UR, Wageningen, The Netherlands)

ABSTRACT: PAH- and mineral oil contaminated sediments were bioremediated for 5 years in drained open-air basins in layers 90 cm deep. During this period the effects of optimisation on the remediation process were studied in the field situation as well as in separate experiments. Physical impacts such as oxygenation, raising temperature, tillage and sieving are in particular effective in the early stage of landfarming but the effects decrease in the course of time, parallel to the decrease of the bioavailable fraction. A pot-experiment with sediments from the passive stage of landfarming and various crops is thought to indicate if, and by means of what mechanism(s), vegetation and its accompanying soil ecosystems affects bioavailability and degradation of the organic contaminants. Development of the micro- and mesobiological soil system was investigated in relation to the chemical and physical improvement of the quality of the created soil.

INTRODUCTION

In the Netherlands annually 5.10^6 cubic meters dredged sludge should be remediated, in particular the 60% that is seriously contaminated. For landfarming this amount, at least 500 ha is required for a number of years, which is a relatively large area and expensive investment. Remediation of sediment should lead to a safe product, useful for various purposes (multifunctional) but as the time for remediation has to be shortened, nowadays higher amounts of contaminant residue in the final product are accepted; consequently the remediated sediment is useful only for a restricted number of destinations.

Landfarming is an effective, though time consuming, biological method to reduce amounts of organic contaminants such as mineral oil and polycyclic aromatic hydrocarbons (PAH) (Harmsen et al., 1997). In the practice of landfarming of dredged sediments the first treatment is dewatering and, if necessary, desalination. This may be shortened with technical impacts, but if stored in an open-air basin and equipped with draining devices, it will take more than one year to get aerobic conditions in a sediment layer of ± 1 m depth, under Dutch climatic conditions. Aerobic conditions are a prerequisite for microbiological degradation of organic contaminants. As soon as aerobic circumstances have been achieved, actual degradation starts. Its rate can be enhanced in various ways: temperature and oxygenation can be improved, soil tillage and addition of mineral nutrients (NPK) and of specific microbes (for example fungi) can be applied and development of a spontaneous or specific vegetation can be stimulated. It must be realised that degradation already starts in a partly aerobic sediment, which occurs parallel to the

* This article is a update of a paper published in the proceedings of the Sixth In Situ and On-Site Bioremediation Symposium (Harmsen, 2001).

oxygenation of the sediment. The first stage of landfarming is the intensive one during which impacts may be effective with respect to the reduction of the remedial period. This is normally possible during oxygenation and the following year for a 1 m sediment layer in an open-air basin. During this stage the (potential) bioavailable fraction is mineralised if the process conditions allow this.

After the intensive stage a strongly adsorbed recalcitrant fraction remains and is degraded slowly during the second, the passive stage of landfarming. Although the contaminant in the passive stage of landfarming is more recalcitrant to degradation than during the previous stage, the rate of mineralisation of this fraction is probably not the limiting factor for the remedial process as is in fact availability or the desorption of this fraction from the soil matrix (Harmsen and Ferdinandy, 1999). The least thing to do during the passive phase is the prevention of anaerobic conditions, which certainly decrease the rate of degradation of contaminant residues; maintaining vegetation is a simple method in this context, and probably the best way to let the soil pass through the passive stage.

In the first part of this research it was investigated if intensivation of treatment could decrease the time necessary for physical and chemical quality improvement (Harmsen, 2001). The presented research also describes the successive stages of several landfarmed dredged harbour sediments in relation to their biological ripening. A pot-experiment with passive landfarmed sediments to research (driving mechanisms behind) the effects of growing crops and soil biota in relation to the degradation of the contaminants, is described, and preliminary results are presented.

MATERIALS AND METHODS

In 1994, basins sealed with HDS-foil, 20x7x1,5 m, were filled with 90 cm PAH/oil-contaminated dredged harbour sediment from Petroleum H(arbour) Amsterdam and Wemeldinge H. Sealand, on top of a layer of 50 cm river sand; the basins were provided with devices for draining and dewatering. The two sediments differed widely: Petroleum H. contained 26% clay and 9% organic matter and Wemeldinge H. 8.3% and 7.5% respectively, and while Petroleum H. contained 527 mg PAH kg^{-1} dry wt, and 13.400 mg oil, these figures were 53 mg and 2015 mg respectively for Wemeldinge H. The various basins got different treatments with respect to tillage and active introduction of crops, including also a zero treatment. Besides, with the same sediments parallel investigations were carried out in a bioreactor, by means of greenhouse farming, and after application of fungal compost. Total amounts and species spectra of PAH and of mineral oil were monitored as well as the bioavailable fraction of PAH; the latter fraction was measured by adsorption with the tenax procedure (Cornelissen et al., 1997)

Bacterial biomass and –growth and the numbers of nematodes were measured to get a picture of the biological maturation of the sediment ecosystems (Bouwman et al., 2001); therefore biota were also measured in two other sediments, landfarmed since 1990 (Zierikzee H. and Geul H.).

Vegetation has several impacts on the sediment and the degradation of contaminating residues. In the early phase of landfarming vegetation stimulates dewatering and structuring of the soil and consequently growing roots create a pore-canals system and affect formation and accessibility of soil aggregates; as a consequence of the effects, gas (O_2) diffusion is influenced and so the degradational activity of the microbes. Besides physical effects, growing crops influence microbial activity as suppliers of substrate: fast turn-over roots and root exudates; they also facilitate the formation of mycorrhizal fungi. A separate experiment was established with grass and lucerne growing in pots on second stage landfarmed sediments, Petroleum H. and Wemeldinge H., including also a bare soil treatment with and without organic substrate added (glucose, glutamic acid). In the experiment the effect of the treatment on PAH/oil degradation and on PAH bioavailability is measured and, if possible also related to soil biota.

RESULTS AND DISCUSSION

As a consequence of the textural differences between the two sediments, in the untilled Petroleum H. dewatering and desalination was much faster than in untilled Wemeldinge H. sediment. After two years 75% of the first and only 40% of the second species was (partially) aerobic. In the tilled treatment both species were (partially) aerobic over the complete depth in about one-year. Figure 1 presents the relative decrease of PAH in both sediments in the tilled treatment. After two years the intensive phase ended in both sediments and the residual PAH concentration slowly degraded in the following three years of extensive treatment while also spontaneous vegetation had developed. During the first two years of landfarming decrease in PAH content was much faster in Petroleum H. as compared to Wemeldinge H.

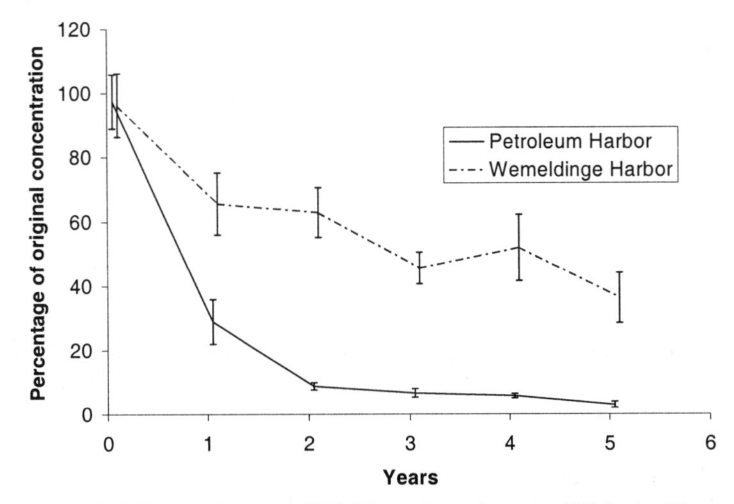

FIGURE 1. Degradation of PAH on the reference fields (cultivated). PAH as percentage of the original concentration, including 95% confidence values

In separate experiments with the same species (sediments) it was demonstrated that various methods of oxygenation of slurry and of already dewatered sediment considerably enhanced the degradation of PAH and oil; in treatments ranging from 5 to 70 days decreases in PAH/oil contents of 65% to 92% were achieved. In greenhouse experiments with sieved sediments and with enhanced temperatures (± 23 °C, completely aerobic soil), degradation of PAH and oil was considerably enhanced in Petroleum H. sediment, but PAH degradation was not affected in Wemeldinge H. sediment, while degradation of mineral oil was also enhanced in this sediment. Addition of fungi (Pleurotus sp. and Agaricus bisporus) to the sediments did not affect the rate of degradation of PAH and oil. Figure 2 shows the effect of several treatments on the bioremediation proces in Petroleum H. sediment. In this degradation curve the chemical measured bioavailable part (TENAX-procedure) is also given. On the landfarm, the bioavailble fraction was degraded in the following one to two years. Arrows are results of optimization. Only the available part could be removed. Stimulation of this biodegradation is in particular effective during the early stage of dewatering/oxygenation of the sludge and the size of the effects decreases in the course of time.

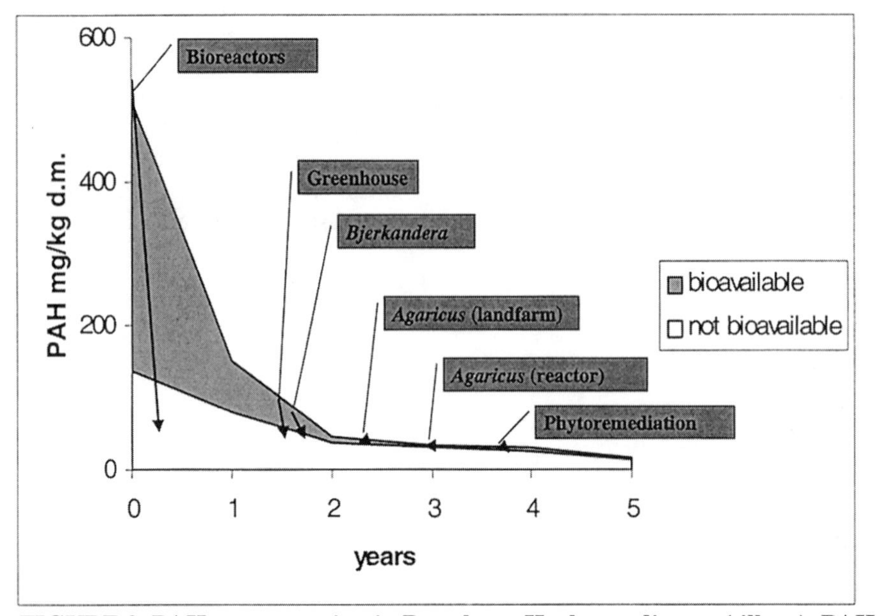

FIGURE 2. PAH concentration in Petroleum Harbor sediment (tillage). PAH concentration divided in a bioavailable part and a not bioavailable part. Arrows are results of optimization

Not only the increase in chemical quality as presented in Figure 2 is important, but also the biological quality. So the treated sediments have to develop into a healthy soil. In both sediments and in two additional sediments from landfarming plots established four years earlier (1990), Zierikzee H. and Geul H., bacterial biomass and –activity were measured, as well as numbers of nematodes (Table 1a, b). From the table it becomes evident that time and active impacts (tillage) positively affect bacterial biomass and numbers and diversity of nematodes. Bacterial activity is enhanced by tillage. The ecosystems in Petroleum H. and

Wemeldinge H. still have to develop in the direction of the more stable situation in Zierikzee H. and Geul H.

TABLE 1a Biological parameters as measured in 1999 in landfarmed sediments from 4 locations, deposited in 1991 and 1994. Bacterial biomass in μgc.g^{-1} soil, -activity as rate of incorporation of ^{3}H-labelled thymidine and ^{14}C-labelled leucine in pmol.g^{-1}.h^{-1}, -diversity as numbers of DNA-bands obtained with the DGGE/PCR method, nematodes in numbers per 100 g dry soil, nematode diversity in numbers of distinguished taxa.

Sediment (location)	Bacteria				Nematodes	
	Bio-mass	Activity		Diver-sity	Numbers	Diver-sity
		Thymidine	Leucine			
Zierikzee H. (91)	144*	5	84	82	4000	13
Geul H. (91)	103	5	81	85	16000	12
Petroleum H. (94)	51 (6)**	11 (22)	285 (450)	66 (89)	1100 (13600)	11 (4)
Wemeldinge H. (94)	15 (37)	3 (30)	112 (457)	68 (86)	2000 (3300)	6 (7)

*figures from untilled fields
**figures from tilled fields ( )

TABLE 1b Presence (+) of nematode taxa per nematode feeding guild in sediments land farmed since 1991 and 1994.

Feeding guild Taxa	1991 Zierikzee H. Geul H.	1994 Petroleum H. Wemeldinge
Bacterivores		
Cephalobids	+	+
Rhabditids	+	+
Monhysterids	+	+
Plectids		+
Fungivores		
Aphelenchids	+	+
Aphelenchoids	+	+
Herbivores		
Tylenchids	+	+
Paratylenchids	+	+
Pratylenchids	+	
Rotylenchids	+	
Tylenchorhynchids	+	
Omnivores/Predators		
Diphtherophorids	+	+
Dorylaimids	+	
Mononchids	+	

Short-term observations (2,5 month) in the pot experiment with bare soil, bare soil plus substrate, grass and lucerne did not yet indicate effects of the treatment on PAH/oil degradation or bioavailability, as incubation time was too short for this. Treatments, however, had a great impact on the numbers of bacteria and nematodes (Table 2a, b). Again a stable situation is not yet obtained.

TABLE 2a Bacterial numbers and –activity in soil from a pot experiment with bare soil, bare soil plus substrate and with grass and lucerne growing on third stage landfarmed sediment from Petroleum H. and Wemeldinge H. Bacteria as numbers.g soil x10^{9}, bacterial activity as incorporation rate of ^{3}H-thymidine en ^{14}C-leucine in pmol.g^{-1}.h^{-1}.

Sediment (location)	Bare soil	B s plus s	Grass	Lucerne
Petroleum H				
Bact. nb.	0.20	1.47	0.95	0.76
Bact.act.t.	0.39	0.25	0.32	0.24
Bact.act.l.	0.56	0.75	0.47	0.49
Wemeldinge H.				
Bact. nb.	0.16	0.36	0.20	0.18
Bact.act. t.	0.18	0.28	0.34	0.47
Bact.act.l	0.37	1.32	0.77	0.43

TABLE 2b Numbers of nematodes belonging to different feeding guilds in sediment from the potexperiment (2a).Numbers per 100g fresh soil.

Sediment (location)	Bare soil	B s plus s	Grass	Lucerne
Petroleum H.				
Fungivores	89	493	49	134
Bacterivores	288	1919	222	404
Herbivores	36	70	1268	51
Totals	413	2614	1539	589
Wemeldinge H.				
Fungivores	26	61	58	145
Bacterivores	644	4980	1538	1822
Herbivores	56	60	167	71
Totals	726	5101	1763	2038

However, if bioavailability of PAH in the pot experiment is compared to availability under field conditions, great differences become visible (results not presented). Obviously the various treatments of pot soil as there are sieving, fertilisation, increased temperature, crop growth and substrate addition stimulated the microbes to reduce the available fraction faster than in the field basins. The fast reduction of the bioavailable fraction will probably have a positive effect on effects as measured with bioassays. As supplementation of the available PAH-fraction was not enhanced in the pot experiment, on this short term no effect of the treatments on total PAH/oil could be observed. The effects of the various treatments on bacteria and nematodes depended on the sediments origin and will be taken into account if on the long term (2,5 year) differences in PAH/oil degradation are observed. For mineralisation of the recalcitrant fraction of

PAH/mineral oil during the passive stage of landfarming a more sophisticated ecosystem is probably needed with a larger variety of participants (microbes and microbivores) than for mineralisation of the less complicated compounds during the first stage of landfarming by the then present pioneer ecosystem. The development of the appropriate ecosystem simply needs time.

CONCLUSIONS

Passive landfarming renders the same results as intensive landfarming as in both approaches the core-process is the same: the degradation of the bioavailable part of the PAH/oil contamination. Soils with the same physico-chemical properties are created with both methods. Biological measurements also show that several years will be necessary to create a soil ecosystem that indicates a healthy stable soil. Such an ecosystem needs more then 5 years to develop. This is possible within the time needed for successful passive landfarming.

If vegetation and/or soil biota have a stimulating effect on the development of the microbiological population and the PAH/mineral oil degradation during the passive stage, this will be observed in the course of the pot experiment. Thus the ripening of the soil can also be described as the ripening of the soil ecosystem. It is not clear yet if the development of a more healthy (complex) soil eco-system has a positive effect on degradation of PAH/mineral oil or on the magnitude of the biological available fraction. If the latter is true vegetation will have an extra remediation effect on the longer term.

REFERENCES

Bouwman, Lucas A., Jaap Bloem, Paul F.A.M. Rőmkens, Gerben T. Boon and Jaco Vangronsveld. 2001. "Beneficial effects on the growth of metal tolerant grass on biological and chemical parameters in copper- and zinc- contaminated sandy soils." *Minerva Biotecnologica 13:* 19-26

Cornelissen, G., P.C.M. van Noort en A.J. Govers, 1997. "Desorption Kinetics of Chlorobenzenes, Polycyclic Aromatic Hydrocarbons, and Polychlorinated Biphenyls: Sediment Extraction with TENAX and Effects on Contact Time and Solute Hydophobicity." *Environmental Toxicology and Chemistry, 16*(7): 1351-1357.

Harmsen, J. 2001. "Bioremediation of Polluted Sediment: A matter of Time or Effort." In: A. Leeson et al. (eds.), *Phytoremediation, Wetlands and Sediments*, pp. 279-287. Battelle Press, Columbus.

Harmsen, J. and M. Ferdinandy, 1999. "Measured Bioavailability as a Tool for Managing Clean-up and Risks on Landfarms." In: A.Leeson and B.C. Alleman (eds), *Bioremediation Technologies for Polycyclic Aromatic Hydrocarbon Compounds,* 57-62. Battelle Press, Columbus.

Harmsen, J., H.J.J. Wieggers, J.J.H. van den Akker, O.M. van Dijk-Hooyer, A. van den Toorn and A.J. Zweers. 1997. "Intensive and Extensive Treatment of Dredged Sediments on Landfarms." In: *In Situ and On-Site Bioremediation:* Volume 2, pp 153-158. Battelle Press, Columbus.

REMEDIAL ACTIONS IN RESPONSE TO A KEROSENE RELEASE IMPACTING A WETLAND

S.W. Rumba and W.E. Baird (Web Engineering Associates, Inc., Norwell, Massachusetts, USA)

ABSTRACT: Web Engineering Associates, Inc. (Web Engineering) has conducted remedial actions in response to a release of an estimated 50,000 gallons (189,250 liters) of kerosene that had migrated to a wetland area. The remedial actions were designed with the following objectives: to recover separate phase kerosene in order to prevent any impact on down gradient receptors and any further impact on the wetland area; reduce the levels of dissolved groundwater contamination; and to degrade the petroleum hydrocarbons in the wetland sediments in a manner that would not cause significant damage to the wetland ecology. Implementation of the remedial actions, which included the operation of a groundwater treatment / product recovery system and in situ bioremediation, have resulted in the recovery of nearly all non-aqueous phase liquids (NAPL) and a full recovery of the wetland vegetation. This was accomplished with minimal impact on the wetland and required no excavation of wetland sediments or wetland replication.

GENERAL SITE INFORMATION

The subject site is located within a large mobile home park located on the northern edge of Turkey Swamp in Halifax, Massachusetts. Each mobile home is serviced by municipal water and shared on-site septic systems. The nearest surface water body is an unnamed intermittent brook located south of the site. The brook drains to the west toward commercial cranberry bogs located approximately 1/2 mile (0.85 km) away.

An underground pipeline was installed when the mobile home park was developed in the early 1970's in order to supply kerosene to heat the mobile homes. Kerosene was stored in an underground storage tank and pumped through the pipeline to the mobile homes.

RELEASE HISTORY AND SITE ASSESSMENT

The presence of kerosene within Turkey Swamp was fist observed by a local resident in the summer of 1991. Initial investigations determined that separate phase kerosene was present in the shallow groundwater within the wetland and in the wetland sediments. In addition, severely stressed vegetation and numerous dead trees were observed in the portion of the wetland located hydraulically down gradient of the mobile home park, and an odor of kerosene permeated the impacted portion of the wetland. Web Engineering was retained by the operator of the kerosene distribution system in August of 1991 to assess the site and to design and implement remedial actions in response to the release.

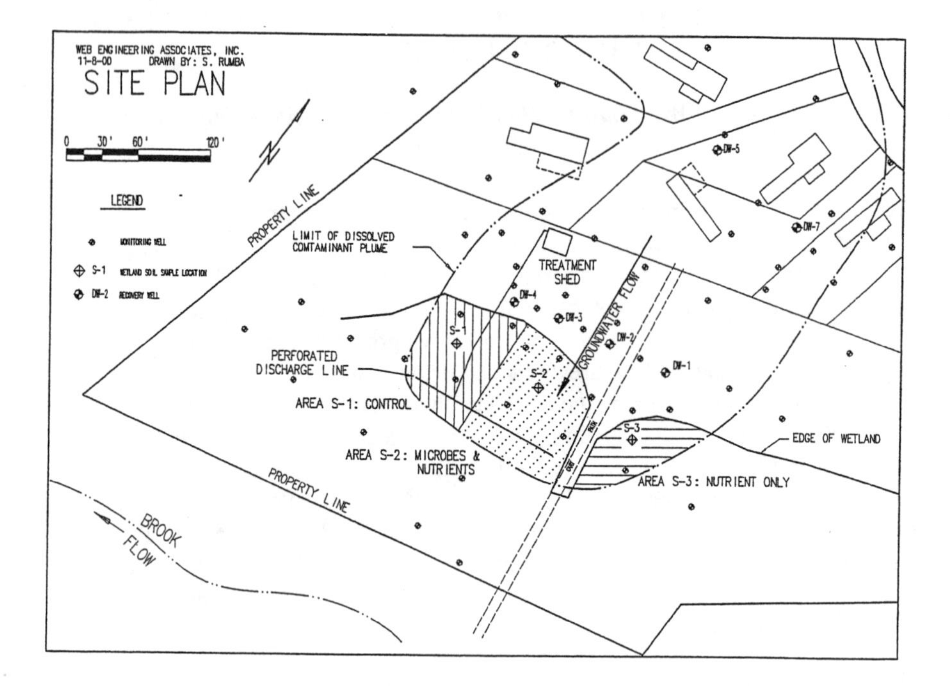

FIGURE 1. Site Plan

The subsurface conditions within the wetland and along the immediate upland boundary were initially assessed via hand augered soil borings and monitoring wells. The borings within the wetland encountered up to ten feet (3 meters) of wetland sediments (peat) overlying silty sand with layers of dense silt (Figure 2). Sampling from down gradient borings determined that the kerosene was being absorbed by the organic wetland sediments along the edge of the swamp and neither the separate phase product nor the dissolved contaminants had reached the nearby brook. Accordingly, there had been no impact to the brook or to the downstream cranberry bogs.

Soil borings and monitoring wells in the upland (hydraulically up gradient) portion of the site were installed using a truck-mounted drill rig. Soils in this portion of the site were glacial outwash deposits consisting of inter-layered fine to medium sand and sandy silt.

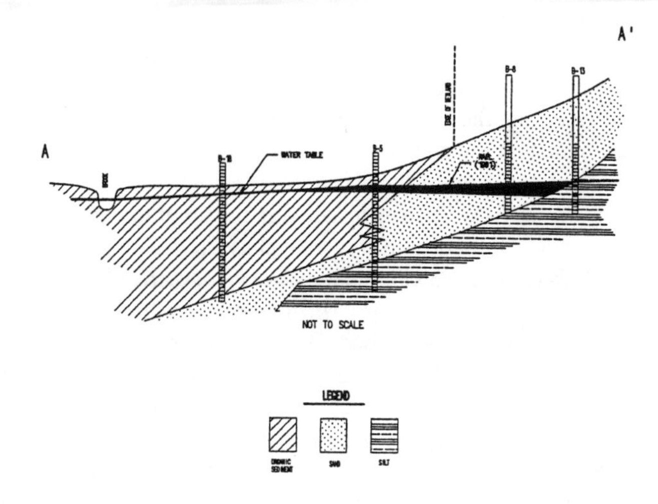

FIGURE 2. Geologic Cross Section of Impacted Wetland Area

The depth to groundwater at the site ranges from approximately 30 feet (9.1 meters) below grade in the up gradient portion of the site to less than one foot (0.3 meters) within the wetland. Initial well gauging results indicated that up to 1.0 feet (0.3 meters) of separate phase kerosene was present in the wells within the wetland and up to 4.0 feet (1.2 meters) was present in the up gradient wells. The volume of separate phase kerosene released to the subsurface environment was calculated to be at least 50,000 gallons (189,250 liters).

The source of the release was determined to be a leak in the underground kerosene distribution system at the mobile home park. Given its relatively large volume, the release was probably the result of a small leak that persisted over a long period of time and, therefore, was not detected through inventory records. The distribution system was immediately abandoned upon discovery of the release.

REMEDIAL ACTIONS

The objectives of the remedial actions were to recover separate phase kerosene in order to prevent any impact on down gradient receptors and any further impact on the wetland area; reduce the levels of dissolved groundwater contamination; and to degrade the petroleum hydrocarbons in the wetland sediments in a manner that would not cause significant damage to the wetland ecology. In order to meet these objectives, a treatment system was designed to recover NAPL and to remove dissolved contaminants from the groundwater. Once the migration of NAPL and contaminated groundwater to the wetland was controlled, a bio-treatment program was implemented to degrade the residual petroleum hydrocarbons in the wetland sediments and in the subsurface soils and groundwater in the up gradient portion of the site.

Groundwater Treatment / NAPL Recovery System. A series of four (4) recovery wells were initially installed along the edge of the wetland to recover the separate phase product. Although a recovery trench may have been more effective in intercepting the NAPL plume, the recovery wells could be installed with minimal impact on the wetland. The spacing of the recovery wells was based on the calculated capture zones determined from pump tests. Three (3) additional recovery wells were later installed in the up gradient portion of the site, where the greatest accumulations of separate phase kerosene had been observed.

Total fluids (groundwater and kerosene) were pumped from each recovery well using electric diaphragm pumps. The total fluids were pumped to a separator tank where free kerosene was skimmed via gravity from the influent stream. Residual kerosene was further removed from the groundwater stream via an oil/water coalescing separator. The recovered kerosene was fed to an oil storage tank. The effluent stream from the separator was pumped through an activated carbon treatment system to remove dissolved kerosene constituents. The treated water was discharged into the wetland through a 100 foot (30.5 meters) long perforated pipe, located at the down gradient extent of the plume. The objective of the water distribution system was to prevent further migration of the plume into the swamp and to maintain saturated conditions within the wetland.

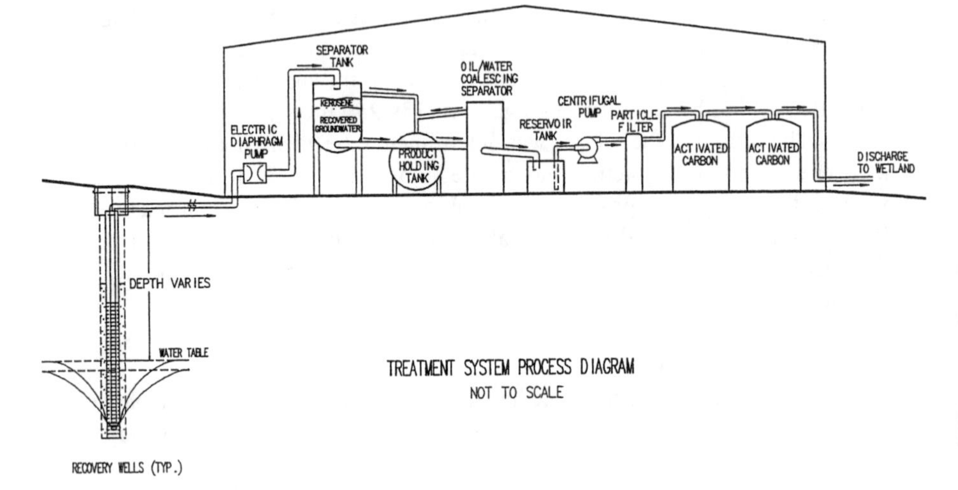

FIGURE 3. Treatment System Process Diagram

The treatment system operated from 1991 until 1998, when the system was shut down due to diminishing recovery rates. The total quantity of groundwater treated by the system was approximately 42 million gallons (1.6 x 10^8 liters). The total volume of separate phase kerosene recovered was approximately 38,000 gallons (143,830 liters).

Bio-Treatment Of Wetland Sediments. Bio-treatment of the impacted organic wetland sediments was conducted in 1993. The bulk of the contamination was determined to be in the upper three feet of the sediments. Bio-treatment was selected over other remedial alternatives such as excavation of the contaminated sediments in order to minimize the impact on the wetland ecology and to reduce cleanup costs.

The impacted wetland was divided into three areas of equal size (S-1, S-2, and S-3). A mixture of hydrocarbon degrading microbes and nutrients were applied to one area, nutrients only were added to another, and the third was left untreated as a control area. The microbes were selected based on the results of treatability studies conducted on sediment samples. The biological products were applied to the shallow wetland soils using backpack type spraying equipment.

Prior to treatment, the indigenous microbes and the exogenous microbes were typed by an independent treatability laboratory. This procedure was instituted in order to determine if the applied exogenous microbes could be distinguished from the indigenous microbes present in the swamp.

Soil samples were collected 45 days after application and analyzed to identify the microbe species present in each area. The results of the analyses indicated that the bacteria species in the bio-treatment product were present in all three areas of the impacted wetland. Analysis of the soils also indicated that a significant reduction in the levels of petroleum hydrocarbons had occurred in all three areas.

A Phospholipid fatty acid (PLFA) analysis was conducted on soil samples collected one year after treatment in order to determine the structure of the microbial community. The sample from area S-2 (treated with microbes and nutrients) was found to contain the greatest biomass and the highest concentration of Gram negative bacteria (hydrocarbon degraders).

Bio-Treatment Of Subsurface Soils And Groundwater. Bio-treatment of the subsurface soils and groundwater at the site began in 1997 when most of the NAPL had been recovered and a decline in the efficacy of the treatment system was first noted. The bio-treatment consists of the injection of a mixture of hydrocarbon degrading microbes, biocatalyst, and nutrients into the subsurface through selected monitoring wells within the plume.

The initial injections took place while the treatment system was still operating, and immediate effects on the oil/water separator and activated carbon filtration system were observed. The amount of biomass in the initial separation tanks was noticeably reduced. The biomass had historically been the cause of a persistent problem with fouling of the oil/water separator and the activated carbon canisters and was assumed to be the result of nutrient loading from nearby septic systems. The biomass would cause excessive pressure buildup and short-circuiting in the canisters, requiring carbon changes every 2-3 months. Once the bio-treatment began, the biomass was reduced and the activated carbon was observed to last up to six months without any pressure build-up or break through.

A significant reduction in the dissolved contaminant concentrations in the influent to the carbon treatment system was also noted. Prior to treatment, the average total BTEX concentration in the influent samples was 390 ppb and the average TPH concentration was 22.4 ppm. Immediately after treatment the average concentration of total BTEX was 113 ppb and the average concentration of TPH was 12.1 ppm. These results represent a 71% reduction in total BTEX and a 46% reduction in TPH.

Recent sampling indicates that a significant reduction in residual separate phase product, dissolved groundwater contaminant levels, and soil contaminant levels has occurred throughout the site as a result of the bio-treatment.

PRESENT WETLAND CONDITIONS

As a result of the remedial actions conducted by Web Engineering, all separate phase kerosene has since been recovered or biologically degraded in the wetland area. Figure 4 shows the average annual separate phase product thickness in a monitoring well located within the wetland (MW-65) between 1993 and 1997. The well, which had originally contained over 12 inches (0.3 meters) of separate phase kerosene, has not exhibited any product since 1997. A significant reduction of contaminant levels has also occurred in the wetland soils. Prior to treatment in 1993, the average TPH level in the shallow wetland soils was 114,700 ppm. In 2000, the average TPH level in the same soils was 6,300 ppm. Figures 5-7 depict the TPH levels in the wetland soils in each of the three (3) wetland areas between 1993 and 2000. The dissolved petroleum hydrocarbon levels in the groundwater within the wetland presently meet applicable state cleanup standards and no contaminants have been detected in the surface water of the nearby brook.

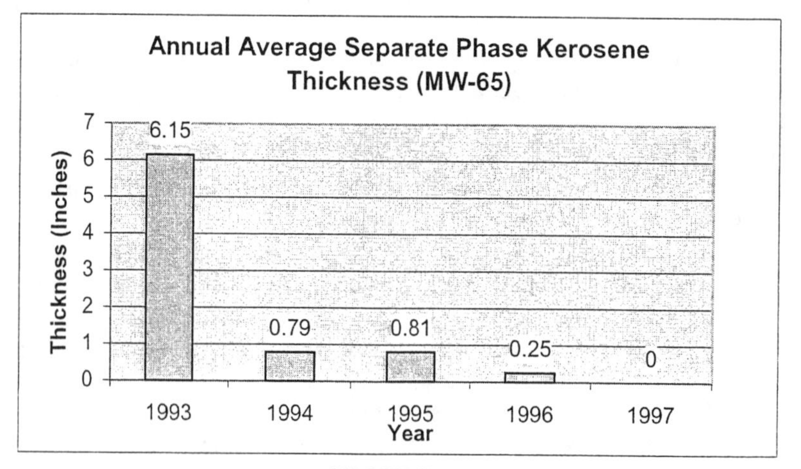

FIGURE 4.

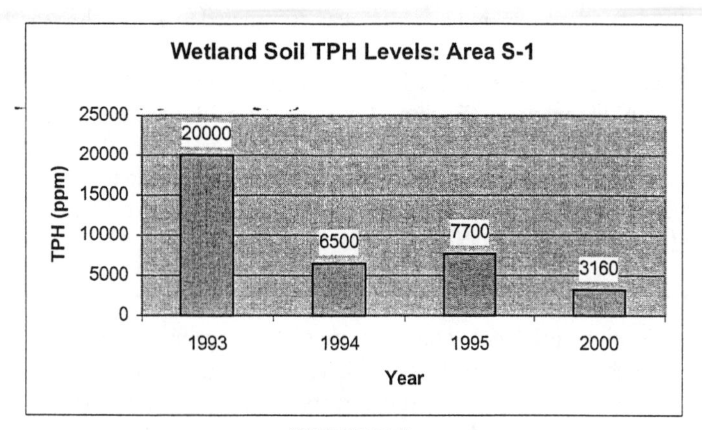

FIGURE 5.

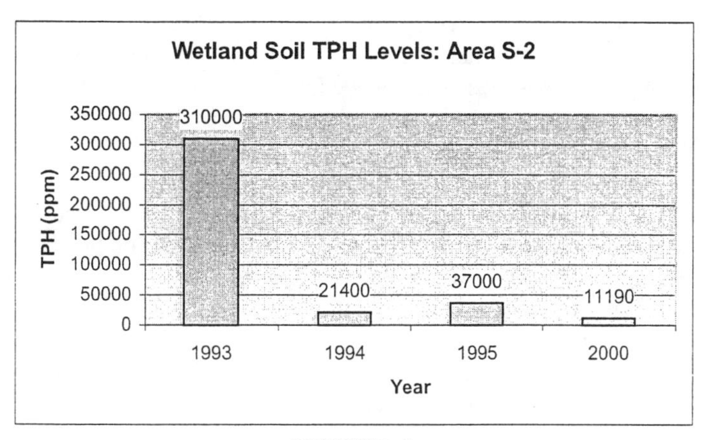

FIGURE 6.

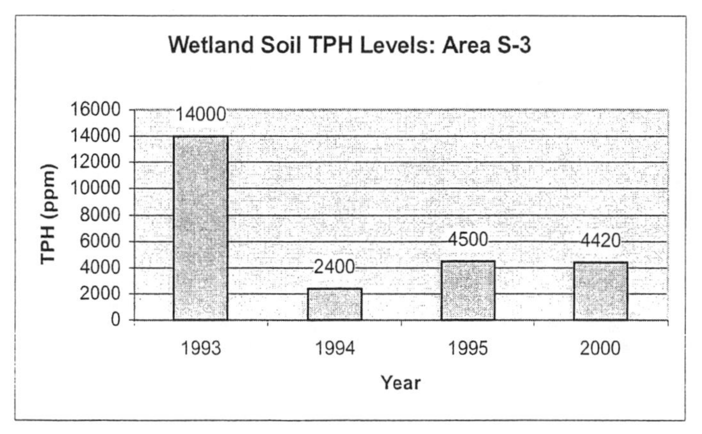

FIGURE 7.

An independent wetland scientist was contracted to assess the recovery of the wetlands in 1995. Two grids were established within the wetland area. One grid was located in the center of the impacted area and one in a similar non-impacted area of the wetland. Each grid was periodically assessed over a two-year period. The wetlands vegetation in the impacted area exhibited continuous recovery throughout the study period. In October of 1997, a final report was prepared which stated that the remedial actions at the site had resulted in a full recovery of the wetland vegetation.

SUMMARY

Web Engineering has conducted remedial actions in response to a release of over 50,000 gallons (189,250 liters) of kerosene from an underground distribution line, which had migrated to a wetland area. The response actions include the operation of a product recovery / groundwater treatment system and the bio-treatment of the wetland soils, the subsurface soils, and the groundwater at the site. The results of recent assessment work indicated that most of the separate phase kerosene in the up gradient portion of the site has been recovered and no separate phase kerosene is present in the wetlands. In addition, contaminant levels in the wetland soils, subsurface soils, and groundwater have been significantly reduced. A two-year study of the wetland vegetation by an independent Professional Wetland Scientist has concluded that the remedial actions at the site had resulted in a full recovery of the wetland vegetation.

These results have been achieved with minimal impact on the wetland itself. The full recovery of the wetland vegetation required no excavation or removal of wetland soils and therefore required no wetland replication.

Continuation of the bio-treatment is on going with a goal of degrading the residual contaminants in the wetland soils, subsurface soils, and groundwater to levels that no longer pose any significant risk.

INNOVATIVE TREATMENT STUDIED ON SEDIMENT SAMPLES FROM VENICE

Dr. Volker Friehmelt and Dr. Evangelos Gidarakos,
Battelle Ingenieurtechnik GmbH, Düsseldorfer Str. 9, 65760 Eschborn, Germany

ABSTRACT: In order to investigate sediment phenomena and separation techniques a detailed research project about integrated coastal management has been performed by Battelle in Sardinia. This includes selection of instruments and well-accepted methods according to standards of quality management as well as training of staff. Innovative methods for the treatment of sediments have been compiled and tested. Much emphasis was placed on achieving a maximum yield of data at low costs of instruments. As a starting basis samples from Alghero, Porto Torres and Venice have been analysed, characterised and examined as concerns Beneficial Use Options, e.g. as building material. During this examination especially frames given by law, regional aspects and technical conditions have been considered.

INTRODUCTION

The contaminated sediments project was established to develop expertise in Sardinia and to provide services to solve problems related to the management and treatment of sediments (Friehmelt, 1999).

Sedimentation is a well known and well studied phenomenon. The reason for sedimentation is that particles have less buoyancy than the surrounding medium. Therefore, sediments mostly are of inorganic composition. Organic compounds are present usually in form of microscopic particles which are coagulated together with inorganic tiny particles to a flocculated unit like tar adhering to a sand grain. Contaminated sediments can pose threats if the material is containing hazardous compounds, e.g. heavy metals or toxic organic compounds that are harmful for the environment.

New international regulations relating to dredging and disposal have been implemented in European law and have to be applied requiring significant changes in procedures being used in former times. Innovative concepts for the management of sediment have been implemented to prepare for dredging according to the new regulations. These concepts consider techniques for cleaning the coarse material so that utilisation of gravel and sand will be possible as well as techniques for waste management including waste water treatment, conditioning of fines, management of contamination (organic and heavy metals) and last not least final disposal. The experience gained and the results obtained by this project are not strictly limited to dredged sediments but can be transferred to other problems concerning contaminated soil or industrial wastes.

Special emphasis was placed on the overall waste management process so that finally the implemented techniques and concepts fit into a feasible path. The engineering has been built up upon world-wide proven technologies. Certain demands of sediment composition and contamination, however, have to be taken into account as well as official regulations and regional aspects and demands so that the integration of innovative techniques into the entire path will be possible.

So, one main task of the sediment project was the transfer of know-how to Sardinia and its establishment there. This task includes selection of equipment, set-up of standard operating procedures, training of staff, and finally, application of instruments and practice of techniques using actual sample material.

Samples from the lagoon of Venice have been investigated which are contaminated by hazardous compounds. These sediments need to be cleaned. A lot of money, to be paid for final disposal of the total material, can be saved, however, if clean inorganic material can be separated and can be put to a cheap disposal place as a Beneficial Use Option. Therefore, samples have to be characterised by physical and chemical analyses prior to the test and application of the treatment procedure.

The expected benefit of the project is that contaminated sediments will be cleaned instead of dumping. The resulting products, especially clean sand can be used for Beneficial Use Options. The concentrated contamination will be transferred to an environmental safe form by solidification. Some of the organic contamination can be degraded by biological technologies.

Through experiences gained on this project a pool of knowledge has been created containing analytical characterisation, material data, and innovative treatment techniques obeying the demands and regulations. This can be used, finally, to offer a service to clients to resolve similar environmental problems such as the cleaning of contaminated sand and soil, treating of waste water resulting from that process, and conditioning and final disposal of the waste.

OBJECTIVES

Several tasks have been designed for the Contaminated Sediments (CS) project to demonstrate the feasibility of innovative treatment technologies and to serve as a model for future sediment management options. The overall objectives of the project are to provide:

- an experimental compilation of the technical, physical, and chemical data of the sediments
- a compilation of regional demands and official regulations
- an establishment of the required equipment, procedures, and quality management
- an assessment of the concepts for the treatment of sediments
- implemention of different innovative techniques (separation, treatment, conditioning)
- a comparison of economical and ecological aspects of sediment treatment
- an assessment of the entire path (treatment, utilisation, disposal)

TASKS

The first task was to create a pool of knowledge by a study of the site, the regional aspects, laws, and techniques already applied for similar tasks. In the second task the technical description of the available techniques has to be compared with the demands of material composition in sediment samples and the demands resulting from laws and other aspects. The material will be cleaned in the third task by the technique with the best ecological and economical efficiency. This means also that for the cleaned material as well as for the waste Beneficial Use Options will be investigated. Finally, the progress achieved by the project has to be documented. So, the (CS) project shall serve as a demonstration model to showcase innovative technology being applicable to treat a variety of locally obtained waste streams and problems which can be transferred to other contaminated areas of the Mediterranean Sea or other European regions.

A main task of the project, furthermore, was the know-how transfer. Therefore, a benefit was achieved besides the environmental progress as well by the increase of knowledge of the staff engaged in this project. This includes the establishment of a laboratory with equipment including legal aspects, instruments, standard operating procedures, and quality management. The management of this project was performed according to the Quality Rules. In detail this means that standard operating procedures were established according to rules of engineers.

EXPERIENCE

During more than 15 years, Battelle has gained experience in management of dredged material from German rivers and harbours by on-shore treatment. Because in Germany those sediments may not be dumped, there were great efforts in developing technologies. Technical, economical and ecological assessments have been performed in a lot of projects as well as experimental investigations and development of technologies ending up in co-ordination, instrumentation and maintenance of the sediment treatment plant METHA II (MEchanical Treatment of HArbour Sludge). In this plant, which has been built for the harbour of Hamburg, a mechanical classification and separation of sand is performed, and the water content of the fines is extremely reduced before disposal (Koehling and Schneider-Fresenius, 1990). The material balance and the separation efficiency have been calculated from the measured data and compared with results from modelling. Several disposal techniques, e.g. direct deposit on hills, solidification using additives and thermal treatment yielding ceramic pellets, were tested. Innovative technologies developed in other Battelle projects have been used to create solutions for the treatment of sediments and for the utilisation of dredged material.

Similar problems with dredged materials have become obvious at many harbours all over the world. The treatment procedures optimised for these harbours, however, can not be directly transferred to other harbour situations because of different composition and contamination of sediments or regional ecological and economical aspects and demands. However, the experience gained by Battelle in these projects will form a good basis for identifying the specific problems and for finding suitable tools and solutions. Therefore, some summaries are presented in the following project descriptions.

CONVENTIONS, LAWS, AND REGIONAL ASPECTS

By international regulations, e.g. London Convention (LC72) dumping will be restricted more and more in future. It is to be expected that these regulations will be implemented also in the national laws. International regulations give some guidance which is accepted by several nations. The regulations derived from that conventions and put into legislation are different, however, for several countries or even different regulations or performance are to be found in regions of one country(Burt and Fletcher, 1997). Therefore, harmonisation of regulations is urgently required.

The Conventions include either general lists of substances which are prohibited to be dumped ("black list" given in Annex of LC72), or substances for which a special permit is required ("grey list" of LC72) or exemptions or a list of materials that may be dumped (OSPAR Convention). Details (limiting values, action levels, measures, etc.) are described, however, in the laws of each nation.

Working groups are continuously reviewing the conventions (e.g. HELCOM, OSPARCOM). As a result the LC72 has adopted a new method of assessment of the suitability of material for disposal (WAF Waste Assessment Framework) to replace the former annex structure of the convention. The DMAF (Dredged Material Assessment Framework) has been accepted and it is advised that contracting parties adopt it accordingly. The DMAF is a guideline for: reducing dredging activities, physical, chemical and biological analyses, evaluation of disposal options, Beneficial Use, sea disposal site selection, impact assessment, permit issue, and monitoring. One important item is that permission for sea disposal shall be given only after presenting an assessment of the potential of Beneficial Use. Furthermore, it is intended to bring most of the dredging activities (not just disposal) under the control of the convention.

It is to be expected that the Italian regulations will be modified in the near future in respect to the modification of LC72 and the EEC directives. Especially, it is to be hoped that the limiting values will be indicated more clearly. There is a continuous change of regulations and the relevant regulations are difficult to find or even still not published. The situation in Italy is that in general (there are some exceptions) two cases (A and B) have to be considered:

A. The Italian law does not allow dumping if the dredged material is contaminated. Contamination is defined by limiting values for several compounds which may not be exceeded for one compound or in a sum of compounds to be calculated by a defined formula.
Contaminated material must be brought to a deposit. Depending on the degree of contamination and other demands the class of deposit (2a, 2b, 2c, or 3) has to be selected.
Therefore, economic benefit is to be expected by saving transport and disposal costs if a high amount of material can be placed cheaper or even be used in a nearby location after separation, using a treatment technology. This is the only chance to find a real market for Beneficial Use Options.

B. For material contaminated below the limiting level permission for dumping can be obtained by a rather complicated procedure. Samples have to be taken according to the rules and analyses may not be older than 6 months, so rapid assessment techniques are not required. In order to get the permis-

sion, it has to be presented by the dredging company to the Ministry that Beneficial Use Options for the material have been considered.
This more or less theoretical consideration usually from economic reasons will lead to the conclusion that Beneficial Use is not possible, because transport and treatment costs are higher than the money earned by selling the products. A tiny chance for Beneficial Use is given only if the dumping place is far away and/or material can be used nearby for coastal protection.

Attention has to be paid to the fact that the Italian limiting values for "Dumping not allowed" are extremely high compared to other European standards. Some limiting values are, e.g.: 100 ppm for Cd, Hg, As, Se, Te, and Cr; 500 ppm for Be; and 5,000 ppm for Pb and Cu related to the dry sediment. The sum formula says that the sum of the concentration for each element divided by its limiting value may not exceed 1. That means material contamination below that limiting values still can be very high and so that material, depending on the degree of contamination, may only be brought to a deposit, because its use in agriculture is restricted. This restriction is given by limiting values defined by EEC-regulations for clean soil and use in agriculture which are binding also for Italy. Material being considered toxic, so that dumping is not allowed, has to be carried to a deposit class 2 b if the contamination is in the range up to the toxic level plus 1 %. If it is higher but lees than 10 times that of the toxic level it has to be brought to class 2c or even to a deposit class 3 if the contamination is higher that 10 times the toxic level. Furthermore, other contaminants have to be considered.

A lot of details have been compiled and stored, including detailed bathimetric maps for harbours, geological maps for surrounding areas, maps of the watershed, information about nearby industries and agriculture, harbour traffic regarding passengers and goods, etc. All around Sardinia there are more than 160 small harbours and 6 larger ones. Dredging has been performed e.g. at the inner harbour of Porto Torres. The material was carried by truck to a nearby sand mine and deposited there by special permission. So, by regional aspects in this case disposal on shore was easier than dumping. Large amounts are dredged at the harbour of Cagliari and dumped. A high amount of material dredged from channels near to Cagliari is used in agriculture by special permission in a research project.

There are a lot of sand mines next to all towns and sand is rather cheap. The cost for different qualities are in the range of 5 to 7.5 Euro/t, usually including even the transport costs if the place of utilisation is within a radius of 20 km. Therefore, it cannot be expected that much money will be won by separation of dredged material and selling part of it as building material. Nevertheless, money can be saved by reducing the disposal costs.

In Sardinia there are only deposits up to class 2b. The costs to bring material to a class 2a deposit are between 40 and 55 Euro/t , whereas the cost are already between 65 and 100 Euro/t for class 2b. Usually, a class 2a deposit may accept only solids, but sludge will only be accepted by class 2b and 3. Therefore, money can be saved by solidification of the sediment, because solids can be disposed at class 2a.

At the lagoon of Venice treatment of sediments is urgently required. There is a rather high contamination by organic compounds in addition to the heavy metal contamination. This problem is well known and some activities have already been started. There are deposits class 2c in Italy and the costs are between 150 and 190 Euro/t. The exact costs are depending on the amount of material and its consistence and most of all whether the delivering company is affiliated to the local association of industries. It becomes obvious that a lot of money can be saved if the disposal to a class 2c can be avoided especially since additional transport costs might be high, because the location of that deposit is far away from dredging.

All the details can be summarised by the statement that from the economic point of view it is not to be expected that money can be earned by the utilisation of material, but it might be possible to save a lot of costs as concerns the disposal if dumping of dredged material is not allowed.

BENEFICIAL USE OPTIONS AND TREATMENT TECHNOLOGIES

Opportunities for Beneficial Use of sediments appear to be unlimited, and over 1,300 cases of Beneficial Use have been documented in North America alone (Burt, 1996). This list and another one with 200 ideas have been reduced after a first assessment of sediment samples from Venice and Sardinian harbours, however, and it was decided to concentrate the investigations on the items:

- use of sand and gravel for building material
- use of material for coastal protection (refilling of eroded beaches)
- use of material in agriculture
- use of material for industrial filling (recultivation of mines)
- deposit (for comparison)

From the economic point of view it is not to be expected that money can be earned by the separation and utilisation of material but compared to bringing the whole amount of dredged material to a class 2c or class 3 deposit it might be possible to save a lot of costs.

Different paths for management of dredged material have been established in a lot of harbours all over the world. The volume of material that is dredged in the maintenance of waterways and harbours amounts to some million m^3 each year. Detailed technical descriptions of the techniques (material separation, flocculation of fines, dewatering, wastewater treatment and disposal) as well as economical data are present, and there seems to be no contamination that can not be treated. For treatment on shore the easiest way is to pump the material to spoil fields where the coarse material is depleted whereas the fines having a lower sedimentation velocity are transported to the back end. In small scale it can be performed using containers. The advantage is that these containers are leak-proof and can be easily removed to another site or place of utilisation by truck without reloading the sediment in it.

The more sophisticated techniques can be subdivided in physical, chemical and biological treatment. For any kind of contamination at least one procedure can be presented as an example. and the most suitable has to be selected with special respect to the material properties. Many of these techniques have been already commercially available since a long time. There are companies offering complete service and companies only offering large scale equipment, e.g. hydro cyclones, upstream-classifiers, centrifuges, settling-tanks, etc. A lot of data about techniques and offers about large scale equipment have been compiled to perform the economical assessment.

EXAMPLES

Several samples of Sardinian sediments, reference material from beaches, and dredged material were analysed. Most effort was spent for the analyses of samples from the lagoon of Venice which were contaminated by hydrocarbons and heavy metals. Investigations of procedures to clean or solidify the material have been performed, including physical separation by wet sieving, dry sieving, centrifugation, sedimentation, washing, thermal treatment, solidification and electrochemical treatment. Analyses of the so treated samples have already shown, that separation of clean sand is possible by these easy to perform techniques. Depending on the composition of samples from Venice up to 50 % of clean sand could be separated from a contaminated fines fraction, e.g. by use of washing and a centrifuge. The sand was clean enough to be used for building material. Contamination was concentrated in the extracted mud. The mud was dewatered and solidified.

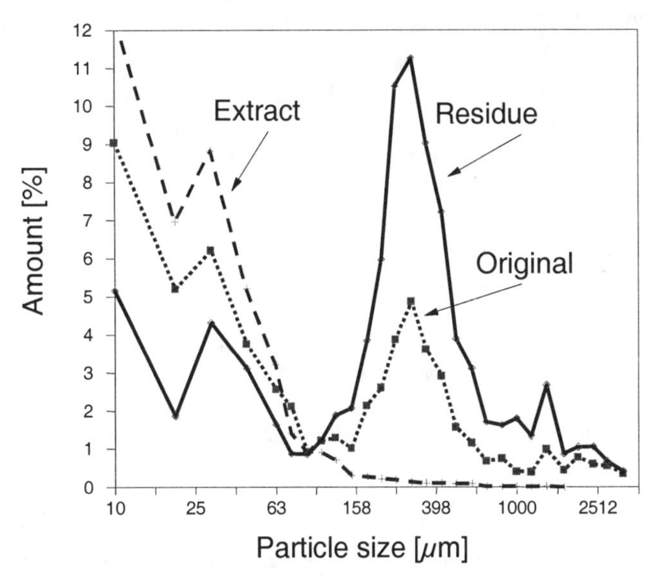

FIGURE 1. Particle size of an original sample from Venice and of residue of the sample after washing and of extracted sludge.

ASSESSMENT

It was confirmed that basic laws of nature can be utilised as well in the laboratory scale as for large scale technical application and the following statements can be given:

- innovative cost-effective treatment techniques are available
- analyses of samples indicate that a high quantity of material can be separated (up to 50 %)

utilisation potential has been detected (filling of eroded beaches, building material)

- ecological benefit is resulting from reducing pollution of the sea caused by dumping
- economic benefit is resulting from saving disposal costs

Potential business fallout/business opportunities:

- knowledge has been compiled that can be utilised to elaborate feasible paths for any case of handling, treatment and disposal of contaminated solids and waste water
- instrumentation and technologies have been installed and can be applied to determine properties of hazardous mixtures of inorganic/organic waste streams and to select optimal treatment procedures
- the method of ecological and economical assessment can be transferred to any problem of waste water or solid material handling taking disposal and utilisation into account
- Compared to sediments similar tasks and problems are to be found at mine tailings, sewage sludge, residues and waste from building and construction, residues from incineration, paint sludge and other industrial waste

REFERENCES

Burt, T. N. 1996. "Guidelines for the Beneficial Use of Dredged Material." HR Wallingford Report SR 488.

Burt, T. N., Fletcher, C.A.1997. "Environmental Aspects of Dredging; Conventions, Codes and Conditions." IADC/CEDA Vol. 2a, ISBN 90-75254-05-9.

Friehmelt, V. 1999. "Contaminated Sediments". Final Report, WBS No. 1.3, Battelle 68392.

Koehling, A., Scheider-Fresenius,. W. 1990. "Planung und Koordination des Versuchsbetriebes sowie Auswertung der Versuchsergebnisse der METHA II im Hinblick auf eine Ablagerung des Schlicks." Report for Hansestadt Hamburg, Battelle R-67.176.

ADVANCED BIOTECHNOLOGICAL PROCESSES FOR ENVIRONMENTAL REMEDIATION OF HARBOUR, COASTAL, AND CHANNEL SEDIMENTS

Ivo Rancich and Piera Casarino (IDRABEL Italia S.r.l., Arenzano, Genoa, Italy)
Ibrahim M. Banat (University of Ulster, Northern Ireland, UK)

ABSTRACT : Bioremediation technologies have been steadily increasing in recent years as an environmental friendly application for the treatment of polluted waters and organic sludge removal in ditches, ponds, channels, harbours and coasts. Closed water basins, both marine and internal, are complex ecosystems, where an increase in the organic load produced by different pollutants such as hydrocarbons and other organic compounds causes an imbalance that results in high turbidity and eutrophication of these water bodies. This may lead to septisation in the sediments and consequent unpleasant odours release. In this paper we report on successful applications of biological remediation by means of fixed microorganisms inoculation to reduce organic sediments accumulation in sewage collector tanks in the Commune of Pisa and in the marine sediments in the area of Genoa Ancient Harbour. An overall average reduction of 60% of the sediment sludge volume in the sewage Commune of Pisa was achieved within 6 months of the treatment. Successful bioremediation in the Genoa Ancient Harbour was also achieved with an overall average reduction equivalent to $6000 m^3$ volume of the sediment layer within the harbour area. This technology had positive effects on the ecosystems, without causing any danger or pathogenic diseases to the inhabitants including humans, aquatic organisms or the environment in general.

INTRODUCTION

The bioremediation process is a biological treatment, based on microbial biodegradation capability to decompose the pollutant substances (organic compounds in general) and to reduce their toxicity (Rocchiccioli, 2000). It is generally the result of the synergic action of several bacterial species and develops through different metabolic processes. Bioremediation is currently receiving a remarkable consent due to the possibility of decontaminating particularly polluted environments. This kind of treatments is directed to favour and enhance the natural biochemical mechanisms, which lead to the toxic organic compound decomposition and mineralisation. The efficacy of the biological degradation processes depends on several parameters, such as :

- the chemical composition and the amount of pollutants to be removed;
- the amount of microbial species to be inoculated;
- the chemical-physical interactions that occur inside the matrix among the different parameters (pH, temperature, solved oxygen, amount of pollutants, presence of nutrients and trace elements, etc.).

Bioremediation is possible if the pollutants to be removed are degradable by the micro-organisms. Use of specific bacteria can allow degradation of complex organic compounds (such as hydrocarbons, phtalates, IPA) deriving from anthropic activities.

Microbial systems therefore, can be composed both of autochthonous micro-organisms (used in biofarming or landfarming) or by selected inoculations with specific strains in the site to be remediated (Rancich and Munizzi, 2000). Applying the biofarming technique, the degradation capability of the biomass present inside the polluted matrix is used. In an effort to optimise the conditions in which the indigenous micro-organisms grow: the matrix to be treated is conditioned, in order to favour an increase of the microbial growth through activating the degradation processes of the organic matter and reinstating the biological balance.

Due to the irreversible biological imbalance in heavily polluted sites, it is necessary to have an external intervention that reactivates the degradation in an environment containing low microbial counts and mixed pollutants. To be able to remove these substrates, it is necessary to use microbial strains and suitable compounds that are applied in the site to be treated – *inoculation.* The use of products containing high amounts of micro-organisms, properly selected allows the balance of natural degradation upon reactivating the biological processes.

Both treatments facilitate use *in situ*, and eliminates the following:

- moving the matrix to be treated,
- employing particular mechanical means,
- needing wide areas,
- disposing refuse products,
- creating "environmental troubles" (unpleasant odours, environmental impact and unbalance, difficulty for the personnel, etc.).

Applying this kind of technologies, it is possible to obtain the following:

- an increase of the microbial population, strictly necessary to degrade the organic substances, including also hydrocarbons;
- the reinstatement of the correct biological balance, with an increase of bentonic species, development of the meiofauna, reappearance of fishes mussels, etc.;
- the reduction of the stratified organic sludge (biodredging) without emptying the water bodies and disposing the refuse products;
- the reduction of unpleasant odours, deriving from the sediment putrefaction;
- the clarification of the water column and the reduction of the pollution indexes;
- the management of activities that don't involve the use of means/interventions producing a strong environmental impact;
- the satisfaction of the local users and of the population in general.

MATERIALS AND METHODS

The biotechnological products, that have been applied in order to carry out the bioremediation treatments, contain micro-organisms that have been inserted or biofixed, in extremely high concentrations, into the capillaries of a porous natural material by an active ion-exchange process.

Biofixation processing begins with carefully selected groups of live, natural micro-organisms (Lantero and Rancich, 2001). More than a dozen kinds of natural occurring micro-organisms in each group are chosen for their ability to biodegrade targeted pollutants into water and gases. They are organized into structured clusters to optimize their collective power. The resulting compound protects and preserves the clustered micro-organisms while also adsorbing targeted pollutants for biodegradation.

Because their food supply is organic pollutants, micro-organisms can biodegrade organic wastes into water and gases with exponential speed. In their free living natural condition these micro-organisms would be vulnerable to destruction and competition by other micro-organisms in highly polluted environments. The biofixation process technologies make, in practice, these micro-organisms less "vulnerable" to such destruction.

The bioremediation treatment (Figure 1) can be carried out both by autochthonous micro-organisms (biofarming or landfarming*)* and by selected inoculations of properly selected micro-organisms (Munizzi and Rancich, 2000) :

- BIOFARMING *:* the conditioning of the matrix to be treated is carried out by means of a fossil calcareous product, extracted from properly selected millenarian deposits, composed by algae and fossilised shells, whose only manufacture is the milling. The granules of fossil alga have a very high porosity and wide adsorption surface, so that they are able to link the microscopic suspended particles bringing them slowly to the bottom (2 or 3 days), through a physical-mechanical action. The main results are : the re-oxygenation of the bottoms and the water clarification. Once settled on the bottom, the granules create the ideal substrate for the growth of autochthonous micro-organisms that, protected inside the capillaries, are able to attack and to degrade the exceeding organic substances in a more effective way.
  The macroscopic resulting effects are : the improvement of the organolectic parameters, the reduction of the sediment thickness and the reinstatement of the balance in the ecosystem, with an increase of the microbial species, development of the meiofauna and of the bentonic species, reappearance of different aquatic animal species.
- INOCULATION : In the case of strongly polluted sites, mineral inert supports are enriched, through the above mentioned "biofixation" process, with properly selected micro-organisms that significantly increase the degradation capacity of the organic sediments (also of complex molecules such as hydrocarbons, phtalates, etc.) in the natural biochemical cycles. The biofixation technology fixes alive microorganisms in the microscopic capillaries of the porous supports that have been selected for their particular characteristics. The cells of the different micro-organisms are "attached to" the walls of the capillaries, partly occupied by water molecules, through an ion exchange process. After this phenomenon, the cell membrane undergoes a modification that facilitates the enzymatic excretion and guarantees an increased external protection to the cell; the biofixed micro-organisms can act in very polluted

sites. The availability of surfaces to attach to allows them to compete and remain longer in the environment. The blending of a significant number of bacteria and fungi inside the same support allows to obtain products with a wide degradation spectrum. Furthermore, the porous material fixes heavy metals and attracts the pollutants that are dispersed in the contaminated waters.

FIGURE 1. Application of the biotechnological product by means of a properly equipped watercraft

Before carrying out the treatment itself, it is necessary to characterise the site and to collect some information and data that are necessary to plan accurately the product application. The real and complete treatment therefore includes all the following steps :

- Site localisation
- Characterisation (sampling and analysis)
- Choice of the micro-organisms
- Seeding of the selected micro-organisms
- Water and sediment monitoring
- Data processing
- Gradual improvement
- Operational optimisation

RESULTS AND DISCUSSION

Here below we briefly describe the results of some successful and promising applications :

Pisa sewer collector

In the sewer collector flowing in Pisa historical center a significant accumulation of solid organic matter was observed. This resulted in both reduction and putrefaction phenomena with consequent production of foul odour.

Of the ~ 50.000 inhabitants that can be evaluated as organic charge of the Pisa North basin, only about 50% are connected to the sewer system that flows into the two municipal waste water treatment plants. The remaining part of the habitations discharge in the canals network of the rain waters, once the sewage have been treated in septic tanks which do not always meet the required standards or are not adequately managed.

The performed treatment aimed at reducing both odours and sediment layer, while improving at the same time the effluent quality and the conditions of the receiving water bodies. The degradation carried out by properly selected micro-organisms (dosed monthly in two successive additions) produced a reduction of the sediment layer from a minimum of 7 cm to a maximum of 80 cm, in comparison with the initial thickness, demonstrating the biodredging efficiency (Figure 2).

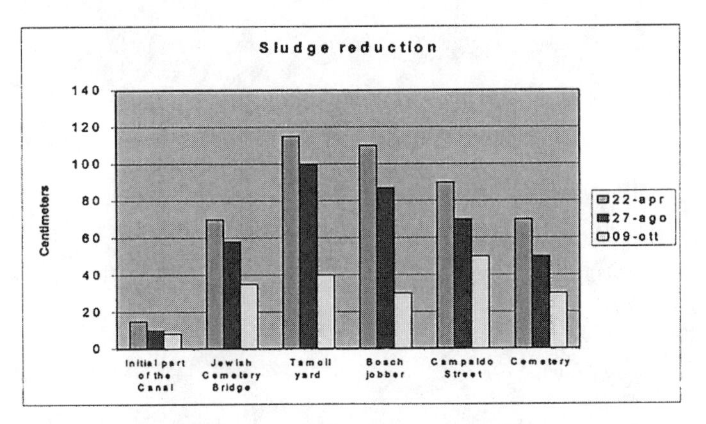

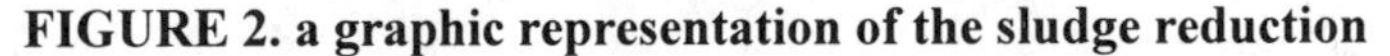

FIGURE 2. a graphic representation of the sludge reduction

Also the analysis performed on the collected samples during the treatment have shown a significant reduction of the main pollutants. Furthermore, the biological treatment resulted in a significant reduction in the environmental conditions responsible for the foul odours development, producing positive benefits for the local environment. Details of this can be accessed online at www.comune.pisa.it/doc/ambiente/progetti/deod1.htm.

Genoa Ancient Harbour

In the Ancient Harbour of Genoa, the old activities, the presence of the Historical Center and the existing infrastructures, that have been added to the natural hollow morphology of the coast, have lead to the formation of very low bottoms (max. 5 m), so that the surface transport and the consequent exchange from the bottom are absent in practice. This created an accumulation basin isolated from the open sea, where the effects of the discharges didn't undergo a dilution but a temporal increase in the sediments. In addition to the presence of stratified organic sediments, other related problems were encountered due to:

- Unpleasant odours and high turbidity of the water column in a tourist area;
- Pollution of the water and of the sediments by light, heavy and aromatic hydrocarbons;
- Disappearance of the typical fauna and flora;
- Need of avoiding or minimising more drastic mechanical interventions, due the image of the area itself.

The treatment has been carried out in co-operation with the Harbour Authority and with the University of Genoa, in the area shown in the following figure (Figure 3):

FIGURE 3. A bird's-eye view of the area selected to carry out the treatment

The first analytical data allowed us to select the ideal biotechnological product to be employed (enriched with micro-organisms efficient to degrade the aromatic polyciclic hydrocarbons) and to plan the dosage rates of the two following applications. Within six months the environmental remediation produced the following results : a reduction of the settled sludge (Figure 4), with high degradation of the hydrocarbon fraction, corresponding to 30 cm (i.e. 2 hectares of water body or biodredging of 6.000 m^3 of sediment). This is a significant improvement in the water quality; a reduction of the unpleasant odours and the reappearance of animal and plant species local to the treated area.

FIGURE 4. The sludge layer reduction

CONCLUSIONS

Several successful treatments have shown the enormous potential that bioremediation technology can offer when utilised appropriately (Carozzi, 1997; Cetara, 2000; Minella, 2000; Rancich and Pesce, 1996). Successful uses of bioremediation resulted in a increase of the biodiversity, the reinstatement of complex sensitive environmental balances (biogeochemical cycles), the digestion of the organic sediments (biodredging) responsible for the reduction of the water body and for the reduction of the putrefying phenomena that produced unpleasant odours. On the basis of the positive results and aspects of the reported application applications and its environmental friendly nature, the Udine province majority has recently decided to provide the Marano municipality with a financing for the realization of an experimental research pilot project directed to the in situ sludge biodegradation, to the sediment decontamination and to the Marano lagoon bioremediation (Branca, 2001; Cordioli, 2001). The use of this environment-friendly natural technologies can also have economical and ecological advantages to environmental problem (ecosystems, water quality, sludge thickness, deodorisation), they have lower cost less use of mechanical equipment and avoid expensive disposal of waste by-products with all the related problems of authorisation. Furthermore, the use of biological compounds can be considered promising since it has been successful in many other fields (www.idrabel.it), such as : municipal sewer collectors (Rancich and Casarino, 2000a; Rancich and Casarino, 2000b), septic tanks, Imhoff tanks, grease traps, waste water treatment plants, hydrocarbon storage tank and hydrocarbon polluted soils (Basciani and Rancich, 2000a; Basciani and Rancich, 2000b; Rancich and Casarino, 2000c; Rancich and Casarino, 2001a; Rancich and Casarino, 2001b; Rancich and Casarino, 2001c; Rancich and Casarino, 2001d).

REFERENCES

Basciani A. and I. Rancich. 2000a. "Il Trattamento Sistematico dei Fanghi Provenienti dall'Industria Petrolifera" *Inquinamento* 19:71-72.

Basciani A. and I. Rancich. 2000b. "Il Trattamento dei Fanghi Provenienti dall'Industria Petrolifera attraverso l'utilizzo di Prodotti Biotecnologici" *L'Ambiente* 4:32-33.

Branca M.R. 2001. "La Provincia finanzia un Progetto per Ripulire il Fondale della Laguna di Marano. Saranno usati Microrganismi per Rendere Innocui Metalli e Idrocarburi" *Gazzettino del Friuli*

Carozzi G. 1997. "Batteri Dragaporto. Seminati in Acqua per Pulire i Fondali" *Il Secolo XIX* 29.

Cetara G. 2000. "Arenzano, con le Alghe Fossili l'Acqua sarà più Limpida" *Il Secolo XIX* 33.

Cordioli G. 2001. "I Batteri per Ripulire la Laguna di Marano" *Messaggero Veneto* XIV

Lantero A.M. and I. Rancich. 2001. "Biotecnologie nel Recupero Ambientale" In UTET Libreria (Eds). *Recupero ambientale - Tecnologie, Bioremediation e Biotecnologie*, pp. 431-442.

Minella M. 1999. "Canale di calma di Prà: un Microbo Ghiotto di tutti gli Scarichi che si Depositano sul fondo ti Salverà"

Munizzi A. and I. Rancich. 2000. "Processi Biotecnologici Avanzati per Bonifiche Ambientali in Aree Portuali, Litorali e Canali" *L'Ambiente* 6: 30-32.

Rancich I. and A. Munizzi. 2000. "Processi Biotecnologici Avanzati per Bonifiche Ambientali in Aree Portuali, Litorali e Canali" *Inquinamento* 20: 82-85.

Rancich I. and L. Pesce. 1996. "Interventi Graduali e Ipotesi di Soluzioni Progettauli Finalizzate al Miglioramento dei Corpi Idrici della Provincia di Savona" In *Interventi per la Salvaguardia dell'Ambiente Marino e Costiero*, pp. 17-43.

Rancich I. and P. Casarino. 2000a. "Pre-Trattamenti in Fognatura. Ottimizzazione della Funzionalità di una Rete Fognaria e degli Impianti di Depurazione mediante il Biotrattamento delle Acque" *Hi-Tech Ambiente* 9:28-30.

Rancich I. and P. Casarino. 2000b. "Trasformazione dei Collettori Fognari da Sistema Passivo in Sistema per il Pre-Trattamento o Pre-Depurazione dei Liquami" *L'Ambiente* 2:42-43.

Rancich I. and P. Casarino. 2000c. "Batteri per Fanghi Petroliferi" *Hi-Tech Ambiente* 12:14.

Rancich I. and P. Casarino. 2001a. "I Biosurfattanti per il Miglioramento Ambientale nell'Industria Petrolifera, Petrolchimica e nell'Ecosistema" *Siti Contaminati* 2:58-62.

Rancich I. and P. Casarino. 2001b. "Biotecnologie per Affrontare gli Sversamenti di Prodotti Petroliferi" *ICP* 3:78-81.

Rancich I. and P. Casarino. 2001c. "Idrocarburi Recuperati. Esempi e Risultati di Applicazioni Pratiche di Prodotti Ramnolipidici nei Fanghi e Fondami di Derivazione Petrolifera" *Hi-Tech Ambiente* 1:51-52.

Rancich I. and P. Casarino. 2001d. "Trattamenti Biotecnologici Innovativi per il recupero degli Idrocarburi" *ICP* 1:115-118.

Rocchiccioli E. 2000. "L'Applicazione delle Biotecnologie nel Trattamento Depurativo delle Acque Reflue, Bonifica di Fiumi, Laghi e Canali, Biodragaggio, Compostaggio". Degree Thesis, IDRABEL Italia S.r.l., Centro di Biotecnologie Avanzate, Università degli Studi di Genova, Genova, Italia.

www.comune.pisa.it/doc/ambiente/progetti/deod1.htm

www.idrabel.it

DEGRADATION AND REMOVAL OF SEDIMENT PCBs USING MICROBIALLY GENERATED IRON SULFIDE

J.H.P. Watson and D.C. Ellwood (Department of Physics and Astronomy, University of Southampton, Southampton UK)
Bruno Pavoni, L. Lazzari, and L. Sperni (Dipartimeno di Scienze Ambientali, Universita degli studi di Venezia, Venice, Italy)

ABSTRACT: This paper describes a method of decontaminating anaerobic sediments from organic micropollutants (polychlorobiphenyls, PCBs) using the controlled growth of the sulfate-reducing bacteria consortia present in the sediment. If iron sulfate is introduced into the sediment, under controlled anaerobic conditions, the sulfate-reducing bacterial consortia will build up over a period of time and produce large quanitities of FeS, which has been shown to be an excellent adsorbent for halogenated hydrocarbons and heavy metals. Evidence is presented that PCBs are first adsorbed, and then electron transfer leads to their breakdown in situ. As the FeS adsorbent is strongly magnetic, it an be removed from the sediment using high-gradient magnetic separation (HGMS) together with the remaining halogenated hydrocarbons and heavy metals which have been adsorbed. HGMS can be done cheaply and on a large scale. The goal of this research is to reduce the polluted sediment disposal and to investigate treatment methods suitable for in situ and in-plant remediation.

INTRODUCTION

Early studies of immobilisation of heavy metals and halogenated hydrocarbons

Immobilisation of metal ions from solution by the sulfate-reducing bacteria(SRB), for example *Desulfovibrio*, has been studied by Watson & Ellwood [1, 2]. In this work, it was found that heavy metal ions with insoluble sulfides are removed from solution to residual levels less than 1 ppb. Further, each bacterium can in many cases take on 3 to 4 times its own wet weight of metal sulfides such as iron or uranyl.

On the basis of the initial results, a study was undertaken of several effluent streams from the precious metal industry. The experiments were conducted on a laboratory scale. The samples were incubated with an actively growing culture of the laboratory strain of *Desulfovibrio* at 37°C under nitrogen and aliquots taken after 5 d for examination by ICP-MS. In general the metal ions that formed sulfides at pH 7.5 were removed from solution, e.g. Ag, Hg, Pb, Cu, Zn, Sb, Mn, Fe, As, Ni, Sn, Al. However other metals such as Rh, Au, Ru, Pd, Os, Pt, Cr were also removed. Silicon was reduced by about two thirds. Elements such as Mg and Sr were not usually affected. The remarkable fact is that a large number of the ions removed do not have insoluble sulfides, so the conclusion was that another powerful immobilisation mechanism or adsorption process was present in addition to the precipitation of insoluble sulfides. In all cases, the precious metals together with many others metals were collected when FeS was produced by SRB. This suggested that the precious metals, apart possibly from iridium, could be collected by adsorption using FeS. This insight led to the study of FeS as an adsorbent for heavy metals [3-9] and a number of different effluents containing precious metals, rare earth metals, actinide metals, transition metals and toxic metals have been examined in this way. In the early work, it was discovered that the FeS material also adsorbed halogenated hydrocarbons as shown in Table 1.

TABLE 1. A solution containing standard materials (concentrations in ng per liter) used in the agricultural industry incubated with adsorbent at a concentration of 0.7 g/litre for a contact time of 24 hours.

Substance	Initial Concentration	Concentration After 24h	% Removal
Hexachlorobenzene	1000	60.0	94
Heptachlor	1000	<10.0	>99
Heptachlorepoxide (cis)	1000	20.0	98
Aldrin	1000	<10.0	>99
Endosulphan A	1000	<10.0	>99
Endosulphan B	1000	<10.0	>99
O,P'-DDE	1000	<10.0	>99
P,P'-DDE	1000	80.0	92
O,P'-TDE	1000	<20.0	>99
P,P'-TDE	1000	260	74
O,P'-DDT	1000	<20.0	>98
P,P'-DDT	1000	<20.0	>98
Endosulphan sulfate	1000	<20.0	>98
Carbetamide	1000	340	66
Chlorotoluron	1000	370	63
Fluoranthene	100	<5.0	>95
Benzo(ghi)porylene	20	<2.0	>90
Benzo(u)fluoranthene	20	<2.0	>90
Ideno(123 cd)pyrene	20	<2.0	>90
Benzo(b)fluoranthene	20	<2.0	>90
Benzo(a)pyrene	20	<2.0	>90

The structure of iron sulfide materials

The reason for this is that iron sulfide FeS is known to be a compound of variable composition behaving rather as a phase than as a stoichiometric compound. The structure of the FeS system is complex. On the S-rich side, a wide range of compositions of the form $Fe_{1-x}S$ can exist and the cationic defects can provide active adsorption sites. These structural factors in $Fe_{1-x}S$ are clearly reflected in the magnetic properties of the sulfides of iron which are quite variable depending on the exact nature of the Fe/S ratio [8, 10].

On the Fe-rich side of 50 at.% is mackinowite $Fe_{1+x}S$ which is paramagnetic and in which the Fe is tetrahedrially co-ordinated [11]. At 50 at. % Fe is FeS, troilite, in which all the Fe sites are individually magnetic. All the available sites for Fe are occupied. Troilite is antiferromagnetic below the Néel temperature T_N=325°C. In each unit cell there is zero magnetic moment, as the moments of the two Fe ions are well compensated. On the S-rich side of 50 at.% lies the region of "FeS", more accurately written $Fe_{1-x}S$. The generic pyrrhotite formula signifies iron cation vacancies and x varies from 0 to 0.13. The region is quite complex. In addition to the ferrimagnetic monoclinic Fe_7S_8 (x = 0.125, 46.67 atomic % Fe), there is a variety of closely related hexagonal and monoclinic crystal structures which are linked to the number and spatial order or disorder of the Fe cation vacancies (which become mobile in the 200-300°C range). Randomly frozen vacancies produce hexagonal pyrrhotite which will be approximately antiferromagnetic. Ordered vacancies such as at Fe_7S_8 produce a monoclinic structure which is ferrimagnetic. The material has an effective magnetic susceptibility of approximately $1.3(10^{-2})$(SI).

At the chemical composition Fe_3S_4 a strongly ferrimagnetic material occurs, named griegite [12]. This material has the spinel structure. Ideally, for easy and cheap recovery of the adsorbent material from suspension, it is necessary for the microorganisms to produce either Fe_7S_8 or Fe_3S_4. Pyrite FeS_2 occurs at 33 at. % Fe. The Fe ions do not possess a localised moment.

The occurrence of sulfate-reducing bacteria in nature and the consequences for remediation

Sulfate-reducing bacteria (SRB) are found in anoxic regions throughout the world and here our main interest is the occurrence of SRB in sediments and the implications this has for heavy metal immobilisation [13]. From the work presented previously [14-18], it is clear that if heavy metals, with insoluble sulfides, are present, in a suitably liberated form, they will be precipitated, at values of pH ~7, onto the surface of the bacteria [17]. As the build-up proceeds, much of the metal sulfide can become detached from the bacterial surface and form mineral particles within the sediment.

Many surveys of the heavy-metal content of sediments have been conducted in rivers, salt marshes bays, lagoons and harbours throughout the world and many of the heavy-metal deposits are associated with sulfur and have been attributed to the action of sulfate-reducing bacteria [14, 16, 19-33].

The occurrence of *Desulfovibrio* in nature has been studied by Leclerc et al. [34]. A total of 609 samples of water, sludge and human stools were examined for sulfate-reducing bacteria. Of these, 45 samples were of river water in various states of pollution. In the case of river water two maxima of the distribution curve were observed, one at $10^2 - 10^4/cm^3$ and the other at $10^5 - 10^6/cm^3$. The first level is that of river water downstream of a waste water treatment plant. The second is that of polluted water with a very low value of oxidation potential which, under these conditions sulfate-reducing bacteria (SRB) can have a higher concentration than in sewage. The highest levels were found to be in sludge which concentrated all bacteria of aquatic origin. The 45 samples of sludge showed two maxima in the distribution for SRB between $10^4 - 10^5/cm^3$ and between $10^6 - 10^7/cm^3$. From this work therefore we can conclude that there are enough SRB present in normal sediments in order to accomplish significant immobilisation.

Extensive surveys of heavy metal pollution in the Venice Lagoon by Pavoni et al. [24] and in the northern Adriatic by Donazzolo et al. [35] found that many metals are present which can directly precipitated by SRB and others which can be adsorbed by microbially-produced material such as iron sulfide. If sediments are left undisturbed these immobilisation processes are extremely beneficial by preventing the heavy metals from entering the food chain. Problems occur when the sediments must be moved, for example, to improve harbour facilities or navigation. In the case of Venice, in some inner canals of the city, at low water, it is difficult to provide access for accident and medical services.

EXPERIMENTAL

Production of FeS adsorbent using sulfate-reducing bacteria from the Venice Lagoon

The procedure used to obtain the adsorbent was very similar to that described previously [4, 36], with two basic differences:

1. The adsorbent was produced in flasks rather than in a chemostat. The adsorbent oxidised very easily, so much greater care was taken to exclude oxygen.
2. No feed-back system was necessary because all the absorbent produced was magnetic enough to be separated by a low magnetic field using permanent magnets.

Phase 1: The flask was first filled with 2 L nutrient solution (the basic recipe was: KH_2PO_4 at 0.01 g/L, NH_4SO_4 at 0.078 g/L, $FeSO_4$ at 3.23 g/L, $Fe_2(SO_4)3$ at 0.536 g/L, sodium lactate (70% w/w) at 5 g/L) and then incubated with a 30-g aliquot of anaerobic sediment (collected from an inner canal of the Lido island, April 1999). This sediment was analysed using HCl extraction to determine the total heavy metal concentrations (ppm), which were as follows:
Pb: 2.3, Cd: 5.8, Cu: 4.0, Ni: 15, Co: 11, Cr: 3, Zn: 120, Fe: 3700.

Phase 2: The adsorbent was produced using a 500-ml sterile, conical flask. The flask was maintained at 28ºC by placing the flask in a temperature-controlled water bath. The flask was filled with 50 mL nutrient solution, and oxygen-free nitrogen gas was bubbled through. At 2- to 3-hour intervals, small aliquots of nutrient solution (10 – 20 ml) were added. No previous pH or redox potential adjustments were necessary. The system was continuously stirred and no oxidization was observed.

Phase 3: 2-g aliquots of adsorbent were separated from the growing culture by a rough magnetic separation method using permanent magnets.

Adsorption of polychlorobiphenyls by microbiologically produced iron sulfide

Aliquots of an equilibrated standard solution of PCBs (Aroclors 1242) were added, under agitation and in aerobic conditions, to a well-characterised suspension of the FeS adsorbent material. The adsorption of PCBs was measured after the contact times of 12 and 24 hours. After the specified contact period, the adsorbent material was removed by filtration and then freeze-dried prior to analysis. The PCBs were extracted from the freeze-dried adsorbent using a mixture of organic solvents (n-hexane and dichloromethane, 4:1 v/v) in an ultrasonic bath. A multilayered clean-up column was used to remove unwanted organic compounds and sulfur. An additional purification step with metallic mercury was necessary for a complete elimination of sulfur as the presence of sulfur can interfere with the analysis for chlorine. The final determination was made using a gas chromatograph equipped with an electron capture detector (ECD) and an internal standard (CB-155). Each experiment was conducted in triplicate. A more detailed list of analytic phases follows.

1. **Determination of adsorbent solution concentration:** 10 ml of slurry were filtered with a GF/F 0.7 mm filter which was freeze-dried prior to weighing.
2. **Contact phase:** 100 ml of PCB solution (Aroclor 1242, 10 ng/ml in Milli-Q water) were added to different aliquots of adsorbent suspension, namely 5, 10 and 15 ml. The Aroclor solution was prepared by mixing and equilibrating the constituents for 12 hours: 10ml of concentrated standard (Aroclor 10 ng/ml in cyclohexane) and 90 ml of Milli-Q water. Contact was carried out under agitation and in aerobic conditions with 100ml of solution containing 100 ng of PCBs. Different contact times of 12 and 24 hours were used.
3. **Filtration:** The solution of adsorbent and Aroclor was filtered under vacuum, and the residual water-based solution was collected in a vial prior to analysis. The adsorbent solution concentration was verified by weighing the freeze-dried filter.
4. **Extraction and purification phase:** Freeze-dried adsorbent material was extracted through a mixture of organic solvents (n-hexane and dichloromethane, 4:1 v/v) in an ultrasonic bath. The extraction was carried out with three aliquots of the solvent mixture for 3 separate periods of ultrasonic treatment lasting two hours each. The three aliquots were then combined and reduced to 1 to 1.5 ml by a rotary evaporator at a low temperature and a partial vacuum. Purification was performed with a glass column packed (from bottom to top) with:

    - 6g Silica gel RS (0.005-0.2 mm)
    - 1.5g Florisil RS (60-100 mesh)
    - sodium sulfate
    - activated copper powder
    - 8g neutral aluminium oxide (0.063-0.2 mm) deactivated with 1.5% Milli-Q water

The PCBs were eluted with 90 ml of n-hexane. Further purification was necessary to remove all traces of sulfur. This purification was carried out by treating the fraction with metallic mercury. The fraction, after being concentrated to about 0.5 ml by means of a rotary evaporator, was pipetted into a glass vial and (after room temperature evaporation to dryness) added to 200μl of isooctane and spiked with the internal standard (CB-155).

5. **Gas-chromatographic determination:** Determinations were performed with a gas chromatograph (Hewlett-Packard 5890 Series II) equipped with a ^{63}Ni electron capture detector (CC-ECD), an HP Ultra 2 (crosslinked 5% Ph Me Siloxane) column (0.2 mm i.d. and 50m long) and a splitless injector. The injection temperature was 250°C; the column temperature was 110°C for 1 minute, raised to 141°C at 9° C/min, to 271°C at 5°C/min, to 280 at 2°C/min; this temperature was maintained for 15 min. The total run time was 49.9 minutes. From 1 to 2 μl of sample were manually injected into the column with the detector temperature at 300°C.

RESULTS

The first series of experiments were conducted during August-September 1998. In the first series, the adsorption of PCBs was tested over a 24-hour period only, and the PCB concentration in the residual water was not measured. The adsorption of PCBs was tested using 5, 10 and 15 ml of adsorbent solution (Table 2). The tests at each volume were carried out in triplicate.

TABLE 2. First series of analysis, August-September 1998 (100 ng PCBs initially). Contact time 24 hours.

Sample	Adsorbent Suspension Volume *ml*	Adsorbent Weight *g*	Adsorbent Concentration *g/ml*	Adsorbed PCBs *ng*	Adsorption Rate *ng/g*
DSV1	5	0.1553	0.0311	34	219
DSV2	5	0.1636	0.0327	36	222
DSV3	5	0.1577	0.0315	36	227
DSV4	10	0.3104	0.0310	63	205
DSV5	10	0.3030	0.0303	65	215
DSV6	10	0.2939	0.0294	69	234
DSV7	15	0.4584	0.0306	93	203
DSV8	15	0.4569	0.0305	89	194
DSV9	15	0.4552	0.0303	86	188
		Mean	**0.0308**		**212**
		St. Dev.	**0.0009**		**16**

The mean adsorption rate was 212 ±16 ng/g (adsorbed PCBs/gram of adsorbent).

A second series of analyses were conducted in February 1999. The same volume of adsorbent suspension (5 ml) was tested for contact times of 12 and 24 hours and, in this case, the PCB content of the residual water collected during the filtration phase was included. Table 3 summarises the results concerning the 24-hour contact time. There is a good agreement with the first series of results (as shown in Table 2) for both the adsorbent solution concentration (0.0304 against 0.0307) and the adsorption rate (198 ± 29 against 212 ± 16).

TABLE 3. Second series of Analysis, February 1999 (100 ng PCBs initially). Contact time 24 hours.

Sample	Adsorbent Suspension Volume *ml*	Adsorbent Weight *g*	Adsorbent Concentration *g/ml*	Adsorbed PCBs *ng*	Adsorption Rate *ng/g*
DSV17	5	0.1515	0.0303	27	181
DSV18	5	0.1511	0.0302	35	232
DSV16	5	0.1537	0.0307	28	181
		Mean	**0.0304**		**198**
		St. Dev.	**0.0002**		**29**

The results for the 12-hour contact time are given in Table 4. From these results the mean adsorption rate was found to be 367 ± 44 ng/g, approximately 50% greater than for the 24-hour contact time.

TABLE 4. Second series of Analysis, February 1999 (100 ng PCBs initially). Contact time 12 hours.

Sample	Adsorbent Suspension Volume *ml*	Adsorbent Weight *g*	Adsorbent Concentration *g/ml*	Adsorbed PCBs *ng*	Adsorption Rate *ng/g*
DSV10	5	0.1867	0.0311	62	330
DSV11	5	0.1509	0.0302	44	316
DSV13	5	0.1501	0.0300	63	422
DSV14	5	0.1520	0.0304	59	391
DSV15	5	0.1523	0.0307	57	375
		Mean	**0.0304**		**367**
		St. Dev.	**0.0008**		**44**

The results obtained from determination of the concentration of PCBs in the residual aqueous phase collected during the filtration phase for both the 12- and the 24-hour contact time experiments are included in Table 5.

Previous experiments on the recovery percentages of the analytical methods suggested a recovery percentage of 89% for the ultrasonic extraction, purification and determination of PCBs; a recovery percentage of 70% was indicated for the liquid-liquid extraction, purification and determination of PCB.

In the context of these results, it is possible to consider the following, as given in Table 6.

TABLE 5. Residual water determinations, 24-and 12-hour contact times, February 1999.

Sample	Residual water volume *ml*	Contact Time *hr*	PCB content *ng*
DSV13	110	12	8
DSV14	110	12	7
DSV16	110	24	5
DSV17	110	24	6

TABLE 6. Balance for the total PCB content (adsorbed, not adsorbed, not accounted for). DSV 13 & 14, 12 hours contact; DSV 16 & 17, 24 hours contact.

Sample	Total PCBs in solution *(ng)*	Total PCBs extracted from adsorbent *(ng)*	Total PCBs extracted from adsorbent (corrected value) *(ng)*	Total PCBs extracted from residual water *(ng)*	Total PCBs extracted from residual water (corrected value) *(ng)*	Total PCBs not accounted for *(ng)*
DSV13	100	63	71	8	11	18
DSV14	100	59	66	7	10	24
DSV16	100	28	31	5	7	62
DSV17	100	27	30	6	9	61

DISCUSSION AND CONCLUSIONS

The experimental results suggest that, with a contact time of 12 hours, the adsorbent adsorbs about 1.5 times more than with a 24-hour contact time, as shown in Table 6. As the uptake in both cases must be identical after 12 hours, one suggestion could be that there is a certain residence time for the adsorbed species and after that time the PCBs are released to some extent. However, the level of PCBs remaining in the residual water appears to be slightly larger for the 12 hour contact time than for the 24 hours, rendering this explanation very unlikely.

The results on the uptake of the adsorbent for PCBs based on the extraction method could be interpreted by the supposition that, although the uptake after the 12 hour contact time is the same as the 24-hour contact after 12 hours, the subsequent 12 hours allows the PCBs to become more strongly bound. We therefore conclude that there are molecular processes occuring which render the adsorbed PCBs tightly bound or transformed and perhaps escaping into the residual water as a transformed molecular species and, in consequence, not detected. From previous work it is known that the iron sulfide is an excellent reducing agent and will readily transfer electrons to reduce an adsorbed species together with a change of the Fe^{2+}/Fe^{3+} ratio within the adsorbent. At this point it is not clear how the neutralisation of the adsorbent occurs; much more work is required to elucidate the interactions that occur between the PCBs and adsorbent.

Further supporting evidence for this conclusion has been presented by Adriaens et al. (37) using well-attenuated suspensions of *Ceobacter metallireducens* and biogenic magnetite produced by this same strain, tetrachloromethane (CT) transformation by the magnetite surfaces (abiotic) was approximately two orders of magnitude faster than the cell mediated reaction (biotic). Moreover, the abiotic reaction could be differentiated from the biotic reaction on the basis of methane and carbon monoxide production. Similarly, under sulfidogenic conditions, formation of "tell-tale" products allowed discrimination between biotic and abiotic dechlorination reactions. Ferrous sulfide mediated dechlorination of tetrachloroethene (PCE) resulted in the production of acetylene, whereas biotic dechlorination by the sulfate-reducing bacteria *Desulfovibrio vulgaris* ssp *vulgaris* proceeded via hydrogenolysis forming tri- and di-chloroethenes.

Although the method of desorption of the adsorbed PCBs described above was performed for different environmental matrices, (sewage sludges, sediments, vegetables, marine organisms) and not with the adsorbent used here, it is still consistent with the conclusions drawn above. Moreover, the total PCB content extracted from the residual water arising from the filtration of the adsorbent solutions confirms that the adsorbent is able to adsorb almost the same rate of total PCBs for both 24-hour and 12-hour contact times, in both cases about 90%.

Based on this work, a method of decontaminating anaerobic muds and sludge from heavy metals and halogenated carbon compounds and, in particular, PCBs can be suggested.

By innoculating the sediment with iron sulfate and any limiting nutrient over a period of time, the population of SRB colonies will increase and the iron and other heavy metals precipitated and immobilised in the colonies of the substantial quantities of the FeS adsorbent compound produced in the sediment. This adsorbent will adsorb heavy metals and, as shown in Table 1, the halogenated hydrocarbons and PCBs. After the sediment is dredged the application of high gradient magnetic separation will concentrate a substantial fraction of the heavy metals and the halogenated carbon compounds into a small magnetic fraction by weight at the very high throughputs of 250-500 tonnes/hour which can easily be achieved with high gradient magnetic separators presently in use.

REFERENCES

1. D. C. Ellwood, M. J. Hill, and J. H. P. Watson, "Pollution Control using Microorganisms and Magnetic Separation," in *Microbial Control of Pollution,*

Symposia of the Society of General Microbiology, J. C. Fry, G. M. Gadd, R. A. Herbert, C. W. Jones, and I. A. Watson-Craik, Eds. Cambridge, UK: Cambridge University Press, 1992, pp. 89-112.
2. J. H. P. Watson and D. C. Ellwood, "Biomagnetic separation and extraction process for heavy metals from solution," *Minerals Engineering*, vol. 7, pp. 1017-1028, 1994.
3. J. H. P. Watson, D. C. Ellwood, Q. Deng, S. Mikhalovsky, C. E. Hayter, and J. Evans, "Heavy metal adsorption on bacterially produced FeS," *Minerals Engineering*, vol. 8, pp. 1097-1108, 1995.
4. J. H. P. Watson, D. C. Ellwood, and C. J. Duggleby, "A chemostat with magnetic feedback for the growth of sulfate reducing bacteria and its application to the removal and recovery of heavy metals from solution," *Minerals Engineering*, vol. 9, pp. 973-983, 1996.
5. J. H. P. Watson, D. C. Ellwood, A. K. Soper, and J. Charnock, "Nanosized strongly-magnetic bacterially-produced iron sulfide," presented at International Conference on Magnetic Nanostructured Phases, San Sebastian, Spain, 1998.
6. J. H. P. Watson and D. C. Ellwood, "The removal of heavy metals from solution using bacterially-produced adsorbents," presented at Environmental strategies for the 21st century-An Asia Pacific conference., Pan Pacific Hotel, Singapore, 1998.
7. J. H. P. Watson, D. C. Ellwood, A. K. Soper, and J. Charnock, "Nanosized strongly-magnetic bacterially-produced iron sulfide," *Journal of Magnetism and Magnetic Materials*, vol. 203, pp. 69-72, 1999.
8. J. H. P. Watson, B. A. Cressey, A. P. Roberts, D. C. Ellwood, J. M. Charnock, and A. K. Soper, "Structural and magnetic studies on heavy-metal-adsorbing iron sulfide nanoparticles produced by sulfate-reducing bacteria," *Journal of Magnetism and Magnetic Materials*, vol. 214, pp. 13 -30, 2000.
9. J. H. P. Watson, I. W. Croudace, P. E. Warwick, P. A. B. James, J. M. Charnock, and D. C. Ellwood, "Adsorption of radioactive metals by bacterially-produced strongly magnetic iron sulphide nanoparticles," *Separation Science and Technology*, vol. 36, pp. 2571-2607, 2001.
10. L. F. Power and H. A. Fine, "The iron-sulphur system, Part 1, The structures and physical properties of the compounds of the low-temperaturephase fields," *Minerals Sci. Engng.*, vol. 8, pp. 106-128, 1976.
11. G. A. Waychunas, M. J. Apted, and G. E. Brown Jr., "X-ray K-edge Absorption Spectra of Fe Minerals and Model Compounds: Near-Edge Structure," *Physics and Chemistry of Minerals*, vol. 10, pp. 1-9, 1983.
12. B. J. Skinner, R. C. Erd, and F. S. Grimaldi, "Greigite, the thio-spinel of iron; a new mineral," *The American Mineralogist*, vol. 49, pp. 543-555, 1964.
13. J. M. Odam, "Industrial and Environmental Activities of Sulfate-Reducing Bacteria," in *The Sulfate-Reducing Bacteria: Contemporary Perspectives*, Brock/Springer Series in Contemory Bioscience, J. M. Odam and R. Singleton Jr., Eds. New York, N.Y., USA: Springer-Verlag, 1993, pp. 189 - 210.
14. B. B. Jørgensen, "Bacterial Sulfate Reduction within Reduced Microniches of oxidized marine sediments," *Marine Biology*, vol. 41, pp. 7-17, 1977.

15. J. Henrot, "Bioaccumulation and Chemical Modification of Technetium by soil bacteria," *Health Physics*, vol. 57, pp. 239-245, 1989.
16. N. W. Revis, J. Elmore, H. Edenborn, T. Osborne, G. Holdsworth, C. Hodden, and A. King, "Immobilization of Mercury and other heavy metals in soil, sediment, sludge and water by sulfate-reducing bacteria," in *Biological Processes*, vol. 3, Innovative Hazardous Waste Treatment Series: W.H. Freeman & Co., San Francisco, CA, 1991, pp. 97 - 105.
17. J. H. P. Watson, D. C. Ellwood, E. O. Hamilton, and J. Mills, "The removal of heavy metals and organic compounds from anaerobic sludges," presented at Congress on Characterisation and Treatment of Sludge, 26 February-1 March, Gent, Belgium, 1991.
18. J. R. Lloyd, G. H. Thomas, J. A. Finlay, J. A. Cole, and L. E. Macaskie, "Microbial reduction of technetium by Escherichia coli and *Desulfovibrio* desulfuricans: Enhancement via the use of high-activity strains and effect of process parameters," *Biotechnology and Bioengineering*, vol. 66, pp. 122-130, 1999.
19. J. Hilton, J. P. Lishman, and J. S. Chapman, "Magnetic and Chemical Characterisation of a diagenetic magnetic mineral formed in the sediments of productive lakes," *Chemical Geology*, vol. 56, pp. 325-333, 1986.
20. J. Hilton, G. J. Long, J. S. Chapman, and J. P. Lishman, "Iron mineralogy in sediments: A Mossbauer study," *Geochimica et Cosmochimica Acta*, vol. 50, pp. 2147-2151, 1986.
21. J. Hilton, "A simple model for the interpretation of magnetic records in Lacustrine and Ocean sediments," *Quaternary Research*, vol. 27, pp. 160-166, 1987.
22. J. Hilton, "Greigite and the magnetic properties of sediments," *Limnology and Oceanography*, vol. 35, pp. 509-520, 1990.
23. T. Arakaki and J. W. Morse, "Coprecipitation and adsorption of Mn(II) with mackinawite (FeS) under conditions found in anoxic sediments," *Geochimica et Cosmochimica Acta*, vol. 57, pp. 9-14, 1993.
24. B. Pavoni, Marcomini, A. Sfriso, R. Donazzolo, and A. A. Orio, "Changes in Estuarine Ecosystems : The Lagoon of Venice as a case study," in *The Science of Global Change*, American Chemical Society Symposium Series No. 483, D. A. Dunnette and R. J. O'Brien, Eds. Washington D.C., USA: American Chemical Society, 1992, pp. 287- 305.
25. V. V. Panchanadikar and R. N. Kar, "Precipitation of copper using *Desulfovibrio* sp.," *World Journal of Microbiology and Biotechnology*, vol. 8, pp. 280-281, 1993.
26. B. M. Tebo, "Metal Precipitation by marine bacteria:Potential for Biotechnological Applications," in *Genetic Engineering*, J. K. Setlow, Ed. New York, N.Y., USA: Plenum Press, 1995, pp. 231-263.
27. K. L. Verosub and A. P. Roberts, "Environmental Magnetism: Past, present and future," *Journal of Geophysical Research*, vol. 100, pp. 2175-2192, 1995.
28. A. P. Roberts and G. M. Turner, "Diagenetic formation of ferrimagnetic iron sulphide minerals in rapidly deposited marine sediments, South Island, New Zealand," *Earth and Planetary Science Letters*, vol. 115, pp. 257-273, 1993.

29. A. P. Roberts, "Magnetic properties of sedimentary greigite (Fe3S4)," *Earth and Planetary Science Letters*, vol. 134, pp. 227-236, 1995.
30. A. P. Roberts, R. L. Reynolds, K. L. Verosub, and D. P. Adam, "Environmental magnetic implications of Greigite(Fe3S4) formation in a 3 m.y. lake sediment record from Butte Valley, Northern California," *Geophysical Research Letters*, vol. 23, pp. 2859-2862, 1996.
31. D. C. Cooper and J. W. Morse, "Biogeochemical Controls on Trace Metal Cycling in Anoxic Marine Sediments," *Environmental Science & Technology*, vol. 32, pp. 327-330, 1998.
32. D. C. Cooper and J. W. Morse, "Extractability of Metal Sulfide Minerals in Acidic Solutions: Application to Environmental Studies of Trace Metal Contamination within Anoxic Sediments," *Environmental Science & Technology*, vol. 32, pp. 1076-1078, 1998.
33. D. C. Cooper and J. W. Morse, "Selective Extraction Chemistry of Toxic Metal Sulfides from Sediments," *Aquatic Geochemistry*, vol. 5, pp. 87-97, 1999.
34. H. Leclerc, C. Oger, H. Beerens, and D. A. A. Mossel, "Occurence of sulphate reducing bacteria in human intestinal flora and in the aquatic environment," *Water Research*, vol. 14, pp. 253-256, 1980.
35. R. Donazzolo, O. H. Merlin, V. Menegazzo, and B. Pavoni, "Heavy Metal Content and Lithological Properties of Recent Sediments in Northern Adriatic," *Marine Pollution Bulletin*, vol. 15, pp. 93-101, 1984.
36. J. H. P. Watson and D. C. Ellwood, "GB 9516753.2 Feedback Chemostat," British Patent Application, vol. GB 9516753.2, 1995.
37. P. Adriaens, M. J. Barcelona, Kim F. Hayes, M. L. McCormick, and K. L. Skubal, "Biotic and Abiotic Dechlorination in iron-reducing and Sulfidogenic Environments," presented at In Situ and On-Site Bioremediation, San Diego, California, USA, 2001.

BIOREMEDIATION OF *p*-NITROPHENOL USING SOIL MICROCOSMS

Om Vir Singh, Gunjan Pandey and ***Rakesh K. Jain***
Institute of Microbial Technology, Sector 39-A, Chandigarh-160036, India

ABSTRACT: The microorganisms were isolated from a variety of Indian soils, i.e. agriculture field soil, garden soil, and pond soil and from unfertilized soil. Higher percentages of *p*-nitrophenol (PNP) degraders were obtained by direct soil isolation method from vegetable field soil while low percentages were obtained from garden soil and pond soil whereas no PNP degrader could be recovered from unfertile soil. Further characterization of soils resulted in isolation of three different PNP degraders designated as ORJ-1, ORJ-2 and ORJ-3. In addition, a PNP-degrading strain *Arthrobacter protophormiae* RKJ100 available in the laboratory was employed for PNP degradation studies in broth and in soil microcosms. In these studies, approximately 95-99% PNP depletion was achieved with concomitant release of nitrite molecules in broth and thereafter in static soil microcosms. Natural degradation capacity of *Arthrobacter protophormiae* RKJ100 were performed in soil microcosms on various PNP concentration (0.01 mM, 0.1 mM, 0.5 mM, 1.0 mM and 1.5mM). The results indicated that 0.5 mM PNP could easily be degraded by the organism. Further, the residual level of PNP was also degraded from the natural vegetable soil sample in static soil microcosm conditions by *Arthrobacter protophormia* RKJ100.

INTRODUCTION

Nitroaromatic compounds are widely used as insecticides, pesticides, fungicides and explosives. The hydrolysis of several organophosphorous insecticides such as parathion or herbicide nitrophenol builds up nitrophenols in soil (Ramanathana and Lalithakumari, 1999). Various nitrophenols are toxic to plants, fishes and many other organisms and are on the list of United States Environmental Protection Agency (USEPA) as priority toxic pollutants. By virtue of their well-known toxicity for man and animals their occurrence in the environment causes serious problems. Soil microorganisms collectively decompose various xenobiotic compounds and return elements to the mineral state for the utilization of plants. The hydrolytic product of parathion, PNP, has been described for the microbial degradation by several genera including *Flavobacterium, Pseudomonas, Moraxella, Arthrobacter* and *Bacillus* (Zaidi and Mehta, 1995; Zaidi et al., 1996; Bucheli and Egli, 2001). There are numerous investigations focused on the isolation of microorganisms which are able to partly degrade or to mineralize these toxic xenobiotics (Zaidi et al., 1996; Chauhan et al., 2000; Nunez et al., 2001)

Objective. The objectives of this study is (i) to isolate microorganisms capable of degrading PNP in pure culture from the sites contaminated with parathion (ii) to bioremediate PNP from contaminated soil by *Arthrobacter protophormiae*

RKJ100 in soil microcosm and determination of PNP degrading capacity of this strain and (iii) to treat the residual level of PNP from parathion sprayed soil in static microcosms.

MATERIAL AND METHODS

Soil Samples. The soil samples for bacterial isolation were procured from Uttar Pradesh and Punjab, India of vegetable field soil and garden field, pond and unfertile soil.

Bacterial Cells. Various microorganisms were isolated from vegetable field soil, garden soil, pond and unfertile soil by direct isolation method. The strain *Arthrobacter protophormiae* RKJ100 (Chauhan et al. 2000), previously isolated and identified in this laboratory, was used for all soil microcosm studies. Cells for microcosm were prepared by growth of organism in pre defined minimal medium (MM) (Prakash et al., 1996) containing 1% glucose at 250 rpm for 24 h of incubation at 30^0C followed by induction with 0.2 mM PNP at log phase (14-16h) for inducing the PNP degradative enzymes. Thereafter cells were centrifuged at 10,000 rpm for 15 min. at room temperature, cells were washed, centrifuged and resuspended in minimal medium to the cell density of 2 X 10^8 colony forming unit (CFU/ml). PNP depletion in broth was carried out as carried out earlier (Chauhan et al., 2000).

Direct Isolation of Bacteria from Soils. One gram soil samples were suspended in 10 ml of sterile double distilled water. The undissolved matter was allowed to settle down and the supernatant was diluted in MM. Thereafter, 100 μl sample from each dilutions were plated out on (i) Nutrient Agar and (ii) MM + 0.2 mM PNP + 2% (W/V) agar plates using PNP as a sole sources of carbon and energy. The growth of colonies was analyzed upto 120 h in each case. The level of PNP degraders from various soil samples were calculated as the total numbers of PNP degraders with respect of total vital count obtained on nutrient agar plate.

Preparation of PNP Spiked Soil. Collected soil samples from garden soil were air dried and sieved (2 mm). 400 ml of double distilled water containing varying concentration of PNP (0.01 mM, 0.1 mM, 0.5 mM, 1.0 mM and 1.5 mM) were added to 1000 g of each soil and mixed thoroughly. After air drying for 24-48 h, the soils were pulverized and used for degradation studies. For the degradation of residual PNP, vegetable field soil was directly applied in microcosm after it was air-dried and sieved (2 mm).

Degradation of Soils Containing Residual PNP and Spiked PNP in Microcosms. Degradation of residual PNP and spiked PNP in uninoculated and inoculated conditions was studied in parallel sets of two containers each for variable time point. Twenty grams of PNP-spiked soil was added to each container. To one set (uninoculated) 6.0 ml of mineral medium, and to another (inoculated) 6.0 ml of mineral medium containing 2 X 10^8 CFU of bacterial cells were added and mixed. All containers were covered with perforated aluminum foil and incubated at 30^{0}C for 20-25 days. Distilled water was added weekly to compensate for evaporational loss of water. Soil samples were taken after every 24 h of time interval upto 7 days and thereafter samples were collected on increased time period (10, 12, 15, 20 and 25 days).

Extraction and Analysis of Residual PNP from Soil. One gram of wet soil was removed from the incubating soil samples at the sampling times mentioned above for different treatments. The soil sample was transferred to a test tube and vortexed with 5% aqueous NaOH solution for 5 min, thereafter, the soil suspension was centrifuged at 5000 rpm for 20 min, at room temperature. The supernatant was collected and extracted by ethyl acetate followed by acidification with concentrated HCL. Residual PNP was quantitated by high performance liquid chromatography (HPLC) using a Waters liquid chromatograph (Milford, Mass, USA) equipped with two model 600A solvent pumps, a model 996 variable-photodiode array (PDA) multiple wavelength detector set at 315 nm, with advance data interpreter millenium software. The mobile phase was Trifluoroacetic acid : Acetonitrile (80:20 V/V). Aliquots of 20 ml were injected onto a Waters C-8 Spherisorb (5 mm, 4.6 X 250 mm) analytical column at room temperature. The flow rate was 1.0 ml/min.

RESULTS AND DISCUSSION

Isolation and Analysis of PNP Degraders. In search of novel PNP degraders, various kinds of soil i.e. vegetable field, garden field, pond and unfertile soil were selected and tested for total PNP degraders. All soil samples were divided into 5 sets (duplicate) of experiments with five different kinds of microorganisms on the basis of colour and morphology etc. and their behavior on PNP plates were screened after every 24 h of interval upto 120 h. Based on the above consolidated results the higher percentage (1-2%) of PNP degraders were recovered from vegetable field soil (Figure 1) which had been earlier sprayed with the pesticide methylparathion.

Garden and pond soil contain very less number of PNP degraders 0.5-1.0% (Figure 1) whereas no PNP degraders were obtained from unfertile soil (no pesticide sprayed and the land not under farming). The selection of soil samples was made on the basis of pesticide use. Thus the pond soil was collected due to the run off water in the particular pond from agriculture fields. General microbial ecology in soils has been extensively studied and reviewed in past (Morgan and Watkinson, 1989). However, it is pertinent to consider those aspects that are fundamentally applicable to considerations of biodegradation in soil. Various other soil bacteria i.e. *Arthrobacter* sp., *Ralstonia* sp., *Burkholderia* sp., *Mycobacterium* sp. etc. have also been isolated for the degradation of certain other nitroaromatic compounds (Zhao and Ward, 1999; Bhushan et al., 2000; Chauhan et al., 2000). The strain of *Pseudomonas* sp. has also been isolated in past by soil enrichment technique that was able to degrade methylparathion, malathion and diazinon (Ramanathan and Lalithakumari, 1999). A total three bacteria were selected on the basis of removal of yellow colour from PNP plates. Thus, three isolates were found the efficient PNP degraders. These organisms were designated as ORJ 1, ORJ 2 and ORJ 3. Further studies on identification and the degradation capacity of these isolates are underway.

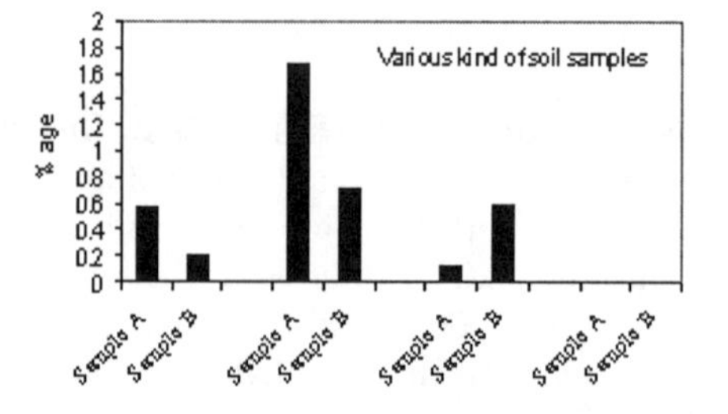

FIGURE 1. PNP degraders from various kind of soil sample i.e. Garden Soil, Vegetable Soil, Pond Soil and Unfertile Soil, respectively by Direct Soil Isolation Method.

Degradation of PNP in Broth and Soil Microcosms. The environmental fate of organic pollutants in soils is influenced by various biotic and abiotic factors. The degradation of PNP by *Arthrobacter protophormiae* RKJ100 in liquid culture shows yellow colour of PNP (0.5 mM, final concentration) gradually changed into colourless indicating its utilization by the microorganism. The oxidative degradative pathways are indicated by the release of the nitro group of nitroaromatic compounds and nitrite reductive pathways by the release of nitro group in the reduced form as ammonia. In case of *A. protophormiae* RKJ100 the release of nitrite molecule in medium suggesting the degradation of PNP and no traces of ammonia could be detected at any stage of growth (Figure 2). The stoichiometric amount of nitrite molecule was released when RKJ100 was grown in the presence of another nitrogen source such as ammonium sulphate (data not shown). In the absence of any nitrogen sources, the release of nitrite during PNP degradation utilized as the sources of nitrogen (Chauhan and Jain, 2000).

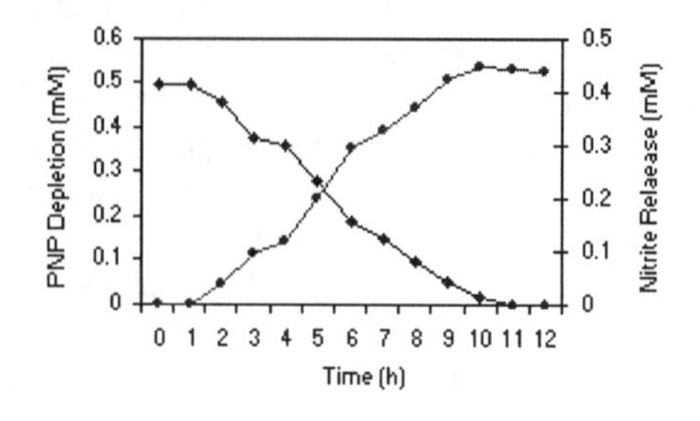

FIGURE 2. Degradation of PNP in broth by *A. protopharmiae* RKJ100 and concomitant release of nitrite molecules (■, PNP depletion; •, Nitrite released) (Adapted from Chauhan and Jain, 2000)

Washed cells of *Arthrobacter protophormiae* RKJ100 were employed in various concentrations of PNP spiked soils i.e. 0.01 mM, 0.1 mM, 0.5 mM, 1.0 mM and 1.5 mM in microcosm studies to determine its degradation. Two equal sets of microcosms, with induced and inoculated cells of *A. protophormiae* RKJ100 whereas the other without the addition of any PNP degrader. The cells of *Arthrobacter protophormiae* were tremendously degraded lower concentration (0.01-0.5 mM) of PNP in soil microcosm and 99.8% PNP degradation was achieved
with in 3-4 days of incubation (Figure 3a-c). Whereas, only 30-40% PNP depletion was measured when cells were applied at 1.0 mM of PNP soil microcosm (Figure 3d), almost no depletion occurred from 1.5 mM of PNP microcosm study (Figure 3e).

As shown in Figures 3a-c, same degradation of PNP occurred even in uninoculated condition which may be because of some physicochemical properties of soil and already present microbial community (Awasthi et.al., 2000).

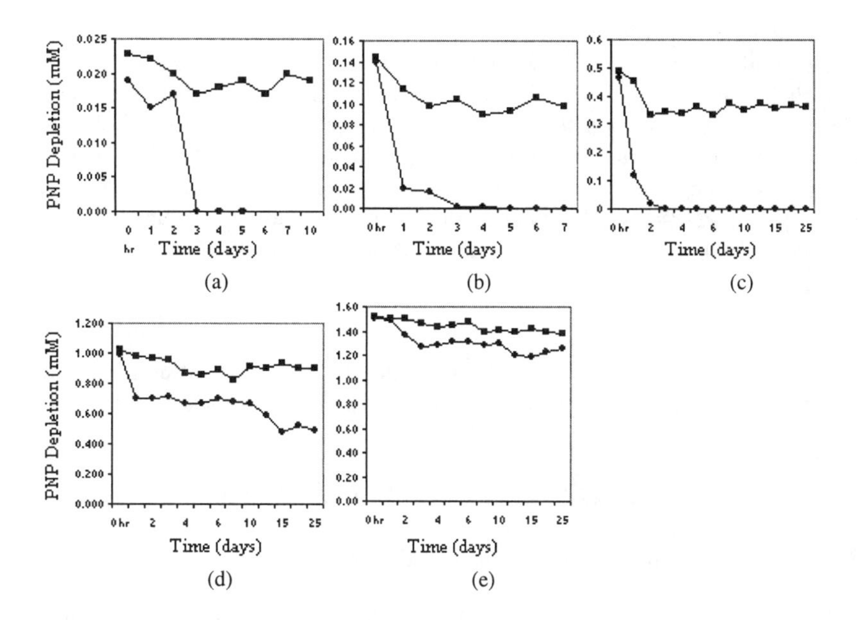

FIGURE 3a-e. PNP in soil microcosm by *Arthrobacter protopharmiae* RKJ100 at different concentration (•, PNP depletion in inoculated sample; ■, PNP depletion in uninoculated samples. Concentration of PNP used: a, 0.01 mM; b, 0.1 mM; c, 0.5 mM; d, 1.0 mM; e, 1.5 mM)

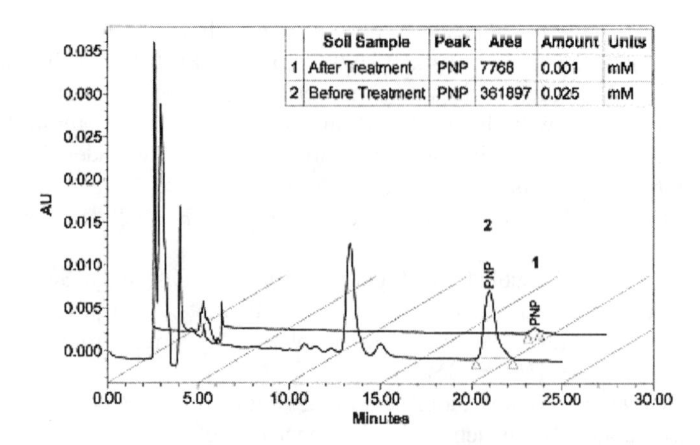

FIGURE 4. HPLC profile of vegetable field soil before and after treatment of residual PNP in soil microcosm by *A. protophormiae* RKJ100.

In another experiment, RKJ100 was also used in microcosm showed to test the degradation of PNP in vegetable soil, which contain 0.025mM residual level of PNP (Figure 4). The results obtained after 4 days of incubation showed the degradation of PNP to a level of 99.5% (Figure 4). It has been shown in present study that bacterium *Arthrobacter protophormiae* RKJ100 was not only very efficient to degrade the lower concentration of PNP but also can degrade upto 1.0 mM of PNP in soil. Further studies of present case are in underway to potentially apply this organism for PNP bioremediation in field.

REFERENCES

Awasthi, N., R. Ahuja, and A. Kumar. 2000. "Factors Influencing the Degradation of Soil-applied Endosulfan Isomers." *Soil Biol. Biochem. 32*: 1697-1705.

Bhushan, B., S. K. Samanta, A. Chauhan, A. K. Chakraborthi, and R. K. Jain. 2000. "Chemotaxis and biodegradation of 3-Methyl-4-Nitrophenol by *Ralstonia* sp. SJ98." *Biochem. Biophys. Res. Commun. 275*: 129-133.

Bucheli W. M., and T. Egli T. 2001. "Environmental fate and microbial degradation of aminopolycarboxylic acids." *FEMS Microbiol. Rev. 25*(1): 69-106.

Chauhan, A., A. K. Chakraborthi, and R. K. Jain. 2000. "Plamid Encoded Degradation of *p*-Nitrophenol and 4-Nitrocatechol by *Arthrobacter protophormiae*." *Biochem. Biophys. Res. Commun. 270*: 733-740.

Chauhan, A., and R. K. Jain. 2000. "Degradation of o-Nitrobenzoate via Anthranilic acid (o-Aminobenzoate) by *Arthrobacter protophormiae*: A plasmid-Encoded New Pathway". *Biochem. Biophys. Res. Commun. 267*: 236-244.

Morgan, P., and R. J. Watkinson. 1989. "Hydrocarbon Degradation in soils and Methods for Soil Biotreatment." *Crit. Rev. Biotechnol. 8*(4): 305-333.

Nunez, A. E., A. Caballero, and J. L. Ramos. 2001. "Biological Degradation of 2,4.6-Trinitrotoluene." *Microbiol. Mol. Biol. Rev. 65*(3): 335-352.

Prakash, D., A. Chauhan, and R. K. Jain. 1996. "Plasmid encoded Degradation of *p*-nitrophenol by *Pseudomonas cepacia.*" *Biochem. Biophys. Res. Commun. 224*: 375-381.

Ramanathan, M. P., and D. Lalithakumari. 1999. "Complete Mineralization of Methylparathion by *Pseudomonas* sp. A3." *Appl. Biochem. Biotechnol. 80*(1) 1-12.

Zaidi, B. R., and N. K. Mehta. 1995. "Effect of Organic Compounds on the Degradation on *p*-Nitrophenol Lake and Industrial Wastewater by Inoculated Bacteria." *Biodegradation 6*(40): 275-281.

Zaidi, B. R., S. H. Imam, and R. V. Greene. 1996. "Accelerated Biodegradation of High and Low Concentration of *p*-Nitrophenol (PNP) by Bacterial Inoculation in Industrial Wastewater: The Role on Inoculum Size on Acclimation Period." *Curr. Microbiol. 33*: 292-296.

Zhao, J. S., and O. P. Ward. 1999. "Microbial Degradation of Nitrobenzene and Mono-Nitrophenol by Bacteria Enriched from Municipal Activated Sludge." *Can. J. Microbiol. 45*: 427-432.

DEVELOPMENT OF A COMBINED IN SITU TECHNIQUE FOR THE TREATMENT OF CONTAMINATED SUBAQUATIC SEDIMENTS

Juergen Thomas, Evangelos Gidarakos
(Battelle Ingenieurtechnik GmbH, Eschborn, Germany)

ABSTRACT: World wide, harbor and/or river sediments are often contaminated by a dangerous mixture of organic and inorganic pollutants creating severe eco-toxicological problems. Conventionally, less expensive methods of disposal of dredged material like off shore dumping and upland placement without pretreatment/treatment are in use. Available pre-treatment and treatment technologies are focused on reducing the volume of dredged material by dewatering, separation, decomposition, and isolation (fixation). Based on the experience in the field of in situ bioremediation techniques for contaminated soil and groundwater the advantages of these technologies in combination with other injection techniques will be used to create an in situ treatment technology for the stabilization and biodegradation of contaminated subaquatic sediments.

INTRODUCTION

World wide, harbor and/or river sediments are often contaminated by a dangerous mixture of organic and inorganic pollutants creating severe eco-toxicological problems. Although only a minor volume of the total sediment volume stored in harbors, belts, and rivers is contaminated these materials cause great handling problems. The problems become evident both in the origin area of accumulation (e.g. harbor, river, channel, belt etc.) and during dredging and on site treatment (Koethe, 1995).

In the first case, the ecosphere will be influenced by the contamination directly. Mainly, soluble substances can pollute the water and will be found in the biosphere inclusive humans (e.g. drinking water pathway, marine foodstuff). Additionally, a direct impact will be given on citizens by volatile substances in the atmosphere resulting from degradation and transformation processes in the contaminated subaquatic sediment. All these inputs reduce the quality of life and may cause extended medical costs.

Furthermore, a direct contact with the contaminated material will take place during dredging of contaminated sediment material. Dredging means in principle the removal of sediments from the river or harbor bottom above the navigable depth which is defined by the standard profile. This maintenance dredging of inland and coastal waterways or harbors produces a huge amount of material consisting usually of fine-grained suspendible solids which carry most of the contaminants and/or the coarser bed-load material. For example, the annual Hamburg harbor dredging produces about 2 Million cubic meters of harbor sludge, which corresponds to 2/3 of the amount dredged from the entire network of German Federal inland waterways. Additionally, a high amount of dredged material is generated during capital dredging for river development and new installations.

The problem in capital and maintenance dredging is that the contaminated portion of the material may not be dumped again in the waters. Depending on the degree of contamination, and volume and physical/chemical properties of the dredged material several options of beneficial use or forms of disposal are defined, but practically are not available and effective at each time and for every place.

Conventionally, less expensive methods of disposal like off shore dumping and upland placement without pre-treatment/treatment are in use as shown in Figure 1. In fact, for contaminated sediments these methods of disposal cause many geo-ecological problems in the dumping/disposal areas. As soon as sediment material is taken out of the water it has to be considered as a waste an the relevant laws have to be obeyed.

In some countries, like Germany and other EU-states, legal regulations require to avoid wastes in advance, to utilize them if they occur, and to dispose the remaining material in a stabilized form.

Available pre-treatment and treatment technologies are focused on reducing the volume of dredged material by dewatering, separation, decomposition, and isolation (fixation). These technologies do not reduce the contamination directly but concentrate single substances in one portion of the treated material. This heavily contaminated material creates high costs for treatment (e.g. soil washing, incineration, vitrification) or final disposal. According to the enormous volume of dredged material world wide the development of a cost effective and applicable treatment technology for contaminated subaquatic sediments is necessary.

OBJECTIVE

Based on the experience in the field of in situ bioremediation techniques for contaminated soil and groundwater (Gidarakos et al., 2001) the advantages of these technologies in combination with other injection techniques will be used. The main goal of the study is to create an in situ treatment technology for the stabilization and biodegradation of contaminated subaquatic sediments applicable world wide, finally. This treatment option can be added as the new treatment pathway 0 visualized in Figure 2. The basic advantages of this new treatment pathway can be summarized as follows:

- Contaminated material will remain in its origin position
- Additional spreading of contamination will be avoided
- The potential of intrinsic bioremediation will be used
- Direct improvement of the eco-toxicological situation and living conditions
- Minimizing of up land treatment procedure of dredged material and related costs

After appropriate treatment and possible degradation of contaminants the material can remain at its former position or if necessary, conventional dredging can be conducted without spreading the contamination. According to the in situ pre-treatment up land treatment, use or disposal of the dredged material is much easier than before.

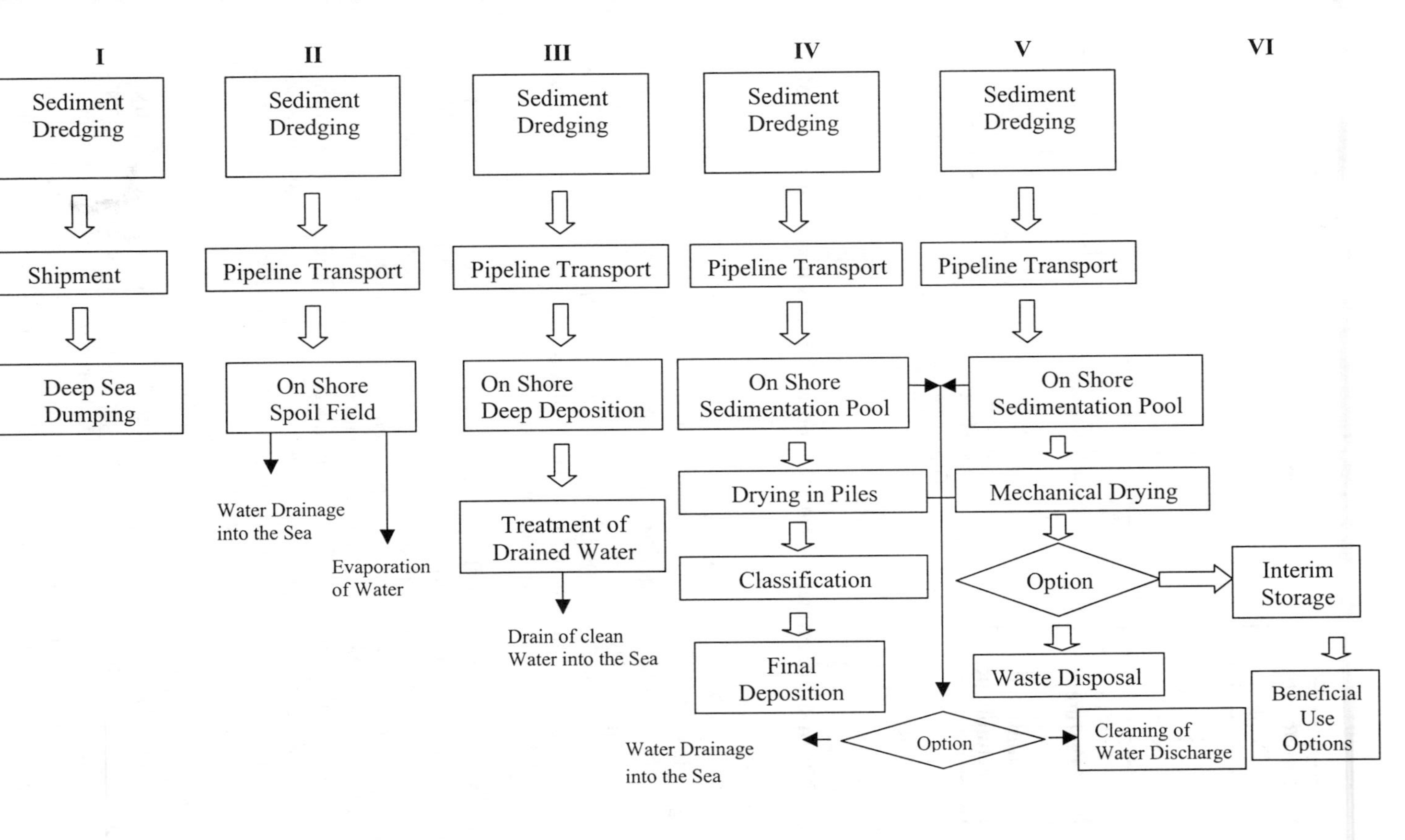

FIGURE 1 Conventional treatment and disposal pathways for dredged material

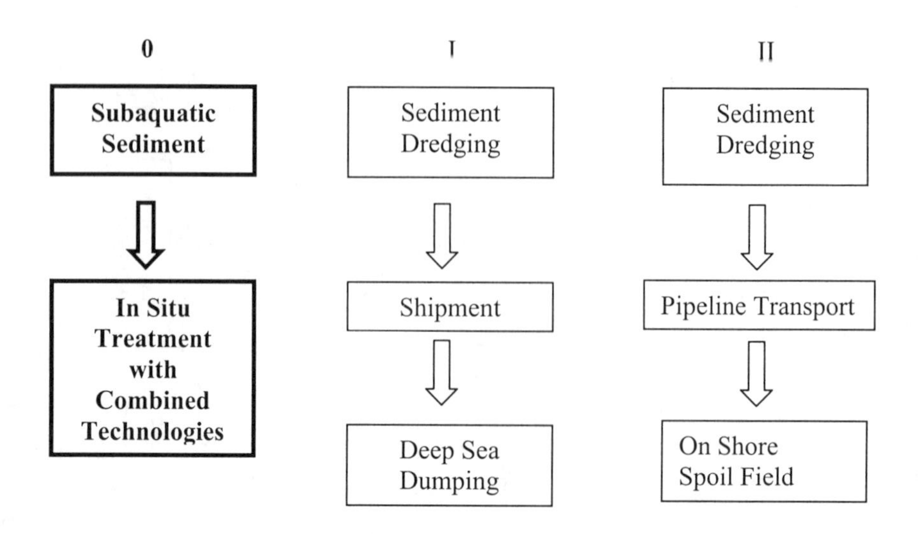

FIGURE 2: In situ treatment option for contaminated subaquatic sediments

INVESTIGATION

To meet this main goal in the R&D project already started a stepped proceeding is planned:

- In a first phase laboratory investigations and technical-scale tests will be conducted.
- In a second phase the results and findings will be utilized for the development, design and installation of a pilot system. This system will be operated at a field demonstration site.

Figure 3 schematically shows the design and configuration of a pilot-scale field test unit. The test platform primarily is designed for first tests in a harbor site working under aquatic and marine conditions. Based on access and availability a working platform will be constructed positioned on the water table of an identified contaminated harbor area. From the platform several wells for monitoring, injecting and aerating will be drilled and installed in the aquatic harbor sediments. During a detailed monitoring program data will be sampled to evaluate mass balance and efficiency. Experiences during installation will be transferred to an installation and operation instruction manual.

PROJECT STATUS

In May, 2001 the R&D project was granted by the German Ministry of Education and Research (BMBF) through the administrative project supporter "Water Technology" of the Research Center Karlsruhe, Germany. Within this project, laboratory and field studies will be conducted for the basic understanding of the reaction of contaminated sub-aquatic sediments during in situ treatment.

The investigations will be performed as a joined project between Battelle Ingenieurtechnik GmbH, Eschborn, Germany and Institute for Technical Chemistry (ITC) of the Research Center Karlsruhe, Germany. The ITC will be responsible for chemical analysis and laboratory-scale tests in a first part of the project. According to the basic data and findings a pilot-scale treatment unit will be constructed. The treatment unit will be installed, operated and monitored at a field demonstration site. Battelle, Germany will be responsible for this second part of the project.

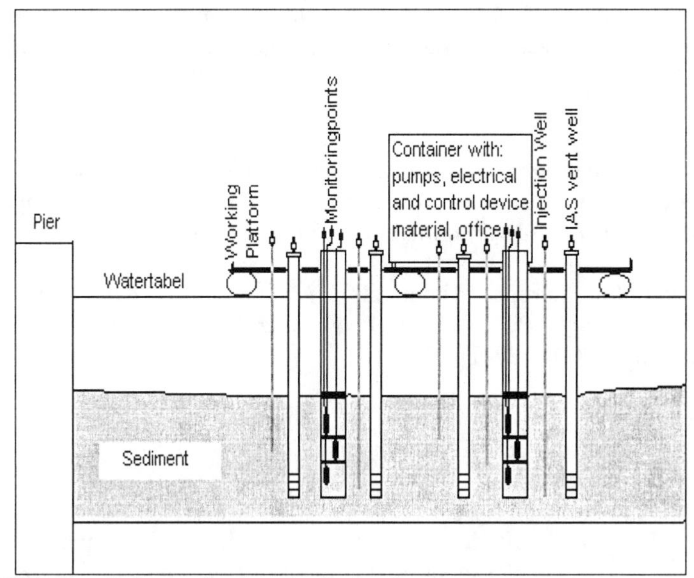

FIGURE 3 Schematic sketch of a harbor treatment area cross section

First Tasks. During identification of an appropriate demonstration site first samples of harbor sludge and dredged material were taken and chemically analyzed according to the LAGA list (1997) and recommendations of the EU (Commission of the EU, 2000, 2001). Then tests were conducted to check the potential of biological degradation and the behavior of heavy metals. After injection of different volumes of H_2O_2 and oxygen, concentration of heavy metals were measured in the suspension. At the same time the reaction of the main part of the organic compounds was recorded. First results showed, that heavy metal concentrations were not increased significantly in the liquid phase. 90 % of the organic compounds were degraded after inoculation of microbes commonly used in public sewage treatment plants. These first findings have to be confirmed by more tests with samples taken from different geological background and with different chemopysical behavior.

In the next step more sediment material will be used in long term Lysimeter-tests. Here the effects of a combined treatment by aeration with ambient air and injection of several oxidants will be studied. Experiences will be gained for the construction of the pilot-scale treatment unit.

Demonstration Site. Possible sites have to meet some pre-conditions before they can be used as a demonstration site, like access, thickness and position of the sediment layer, pedo-chemical parameter of the sediments, measurable concentration of organic and inorganic contaminants, permits of the owner and user of the site and last but not least of the local government. To show the effectivity of the in situ treatment options the identification and selection of the location for the operation of the pilot-scale unit have to be conducted carefully. Since now, three river harbors at the river Main nearby Frankfurt, Germany and several hot spots at the Piraeus Ports, Greece were selected for a more detailed investigation.

REFERENCES

Commission of the European Union. 2000. *Vorschlag fuer eine Entscheidung des Europaeischen Parlaments und des Rates zur Festlegung der Liste prioritaerer Stoffe im Bereich der Wasserpolitik.* KOM(2000) 47 final.

Commission of the European Union. 2001. *Mitteilung der Kommision an den Rat und das Europaeische Parlament zur Umsetzung der Gemeinschaftsstrategie fuer Umwelthormone.* KOM(2001) 262 final.

Gidarakos, E., Thomas, J., Gaglias, I., Lambrinoudis, D. 2001. "Bioslurping Treatment for Subsurface Contamination." *Petroleum Technology Quarterly*, Summer 2001:43-49.

Koethe, H.F., 1995. "Criteria for the beneficial Use of Contaminated Dredged Material in the Federal Republic of Germany". *Proceedings of the Fourteenth World Dredging Congress*, Amsterdam, Volume I:459-474.

LAGA-Laenderarbeitsgemeinschaft Abfall. 1997. "Anforderung an die stoffliche Verwertung von mineralischen Reststoffen/Abfaellen – Technische Regeln". *Mitteilung 20 der Laenderarbeitsgemeinschaft Abfall.*

PHYTO-ENGINEERING APPROACHES TO CONTAMINATED DREDGED MATERIAL

Charles R. Lee (USACE-ERDC-WES)
Thomas C. Sturgis and Richard A. Price (USACE-ERDC-WES)
Michael J. Blaylock (Edenspace Systems Corporation, Inc.)

ABSTRACT: The U. S. Army Corps of Engineers dredges 280 million m^3 of dredged material annually from waterways to maintain navigation. Some of the dredged material is contaminated from various discharges into waterways. Dredged material not suitable for open water disposal is normally placed in confined disposal facilities (CDFs). These facilities fill up. New CDFs are getting extremely difficult to establish. Storage space for future dredging is becoming a real problem. Consequently, the U.S. Army Corps of Engineers' Engineer Research and Development Center, Waterways Experiment Station (USACE-ERDC-WES) has been evaluating innovative technologies for reclaiming contaminated dredged material from CDFs and producing beneficial reuse products. One reclamation technology is Phyto-Engineering or the use of plants to achieve engineering objectives. Incorporated in Phyto-Engineering is other technologies such as recycled soil manufacturing technology (RSMT) that allows a fertile topsoil to be manufactured from contaminated dredged material, residual cellulose, residual biosolids and other residual waste materials. Once topsoil is manufactured from contaminated dredged material, plants can be grown to phyto-remediate contamination and phyto-reclaim the dredged material from contamination. Phyto-Stabilization uses plants to stabilize the dredged material and controls soil erosion. Phyto-Extraction of metals uses specific metal hyper-accumulator plants to extract metals from the reconditioned dredged material to clean up the dredged material for beneficial use. Phyto-Degradation of organic contaminants uses specific plants and root associated microorganisms to bio-degrade organic chemicals such as PAHs and PCBs in dredged material. These Phyto-Engineering technologies can be applied to contaminated soils as well as dredged material. Examples include application to Superfund sites, Brownfield Sites, Minelands and Landfills.

INTRODUCTION

The U.S. Army Corps of Engineers (USACE) has the responsibility to maintain navigation in all Harbors and waterways across the U.S.A. This task requires the dredging of more than 280 million cubic meters of sediment from the waterways annually. Places to dispose of these vast amounts of sediment are required. Dredged material that is not suitable for open water disposal usually is placed in confined disposal facilities (CDFs). Current CDFs are filling up and new CDFs are difficult to locate. In an effort to provide storage capacity for future dredging, dredged material in CDFs is being evaluated for beneficial uses. Reclamation of contaminated dredged material and the manufacture of soil products are a major thrust of a Phyto Engineering Research and Development Effort at the U.S. Army Corps of Engineers' Engineer Research and Development Center located at the Waterways Experiment Station (USACE-ERDC-WES) in Vicksburg, MS. Cooperative research and development agreements (CRDAs) have been established between USACE-ERDC-WES and specific innovative technologies to develop and demonstrate the application of these technologies to the reclamation and reuse of dredged material from existing CDFs. CRDAs have been established with Recycled Soil Manufacturing Technology (RSMT), N-Viro International (reconditioned sewage sludge biosolids), Bion Technologies, Inc. (reconditioned animal manures) and Advanced Remedial Mixing, Inc. (blending equipment for reclamation of dredged material and other residuals). Examples of the reclamation and reuse of dredged material include manufactured topsoil for landscaping, acid mine drainage remediation, mineland restoration, Brownfield redevelopment, Superfund restoration, landfill cover and constructed wetlands for water quality improvement.

Phyto-Engineering is an approach that uses plants to achieve engineering objectives. The USACE-ERDC-WES has been applying this approach to solve environmental pollution problems for the past 25 years. Initial applications included the use of plants to filter dredged material slurry placed in confined disposal facilities and overland flow systems to treat wastewater and produce purified surface water. Recently, specific plants have been used to either phyto-stabilize contaminated dredged material, phyto-extract contaminants such as metals and arsenic, and phyto-degrade organic contaminants in dredged material. The USACE Dredging Operations Environmental Research Program (DOER) is evaluating phyto-reclamation and beneficial use of contaminated dredged material (Lee et al. 1999, Price et al 2000) as well as USACE Districts who have a need to remove dredged material from filled CDFs. Bench-scale greenhouse tests have been conducted as well as pilot-scale field tests. In some cases, recycled soil manufacturing technology has been used to improve the fertility of contaminated dredged material to grow plants more readily to phyto-remediate contaminants in the dredged material. Examples of the application of phyto-engineering include New York Harbor, NY sediment, Pearl Harbor, HI and Wilmington, NC.

PROCEDURES AND RESULTS

Three examples of reclamation and beneficial use of dredged material will be discussed: New York Harbor, NY, Pearl Harbor, HI and Wilmington, NC.

New York Harbor, NY. New York Harbor requires dredging of approximately 3-4 million cubic meters of contaminated sediment annually. No CDFs are available at present. Reclamation of dredged material was evaluated using a combination of RSMT and Phyto-reclamation technologies. RSMT had been used successfully on uncontaminated freshwater dredged material (Lee et al. 1998) and the USACE New York District wanted to apply the technology to a contaminated estuarine dredged material from New York Harbor. Fresh dredged material from New York Harbor with a salinity of 35 parts per thousand was blended with yardwaste and Bionsoil™ (from reconditioned dairy cow manure) according to RSMT procedures (Sturgis and Lee, 1999). The manufactured topsoil was planted with three types of plant species to demonstrate three approaches to Phyto-Engineering as described by Cunningham and Lee, 1995. One planting consisted of plants (colonial bentgrass, red fescue, birdsfoot trefoil) that tend to stabilize the topsoil and control soil erosion without purposely effecting contaminants (Contaminant containment/Phyto-Stabilization). Another planting consisted of plant species (tall fescue) known to take up metals for Contaminant Uptake/Phyto-Extraction. The third planting consisted of plants (bermudagrass, wheatgrass, hard fine fescue) known to phyto-degrade organic contaminants such as PAHs. PAHs were phyto-remediated within 14 months from 24.3 mg/kg Pyrene to less than 1 mg/kg and Fluoranthrene from 28.6 mg/kg to less than 1 mg/kg as examples (Lee et al. 2001). Phyto-degradation of PCBs and Dioxin were not observed for the 14 month period of the demonstration. The topsoil produced from New York Harbor still contained contaminants and would have restricted uses such as landfill cover, mineland restoration, and/or Superfund restoration. This topsoil should not be used in areas of exposure to humans until the remaining contaminants are remediated.

Pearl Harbor, HI. Pearl Harbor required dredging for military activities. The dredged material was to be placed on land. An evaluation of manufacturing a topsoil according to RSMT and then Phyto-remediating PAHs contained in the dredged material was conducted in cooperation with the US Navy and the U.S. Army Corps of Engineers, Hawaii District, and the University of Hawaii, Monoa. The dredged material was estuarine with a salinity of 28 parts per thousand and a pH of 7.0. Most contaminants were low, except PAHs. Bench-scale screening tests were conducted according to RSMT and blends of topsoil using different sources of cellulose such as Navy green yardwaste, Navy shredded paper, Navy compost, and Air Force green land clearing tree waste (Hue et al., 2001). Biosolids included swine manure, Class B sewage sludge biosolids and Class A N-Viro processed sewage sludge biosolids. Blends that produced plant growth equal to or better than fertile potting soil controls or local agricultural soil were recommended for a large scale demonstration. PAHs were phyto-degraded by approximately 30-50% from the original PAH contamination in the dredged material. The topsoils produced could be used for appropriate uses depending on the quality of material produced. For example, topsoils manufactured from Class B biosolids would be restricted to landfill cover and other low human exposure uses. Topsoils manufactured from Class A biosolids could be used unrestricted for landscaping and other uses with higher human exposure.

Wilmington, NC. Wilmington Harbor and the Cape Fear River require annual dredging of approximately 1 million cubic meters of brackish dredged material with low levels of contamination. The Eagle Island CDF is almost completely full and there are limited alternatives for future disposal of dredged material. The USACE Wilmington District asked ERDC-WES to evaluate removal and beneficial use of dredged material in order to provide storage space in the existing CDF for future dredging. Bench-scale tests were conducted applying RSMT using the dredged material, storm debris waste resulting from hurricanes and swine manure Bionsoil™. Blends produced plant growth equal to a commercial fertile potting soil, but contained 10 mg/kg soil arsenic, which was of concern to State regulators. Before the blended topsoil could be used for landscaping homes in Wilmington, the level of soil arsenic was required to be reduced. In order to reduce soil arsenic, phyto-extraction using specialized ferns were evaluated. Additional bench-scale tests were conducted using the "Brake Fern", *Pteris vitatta*. After indications that the fern could extract arsenic from the dredged material, a pilot field test was established at the CDF on the dredged material. The field demonstration evaluated three "Brake Ferns". Three plant densities were evaluated, 15 cm, 30 cm and 45 cm spacings. Two lime levels were also evaluated in an attempt to enhance plant uptake of arsenic, pH 6.0 and 7.0. Preliminary indications show that the ferns are phyto-extracting soil arsenic. The field demonstration will continue through the winter months to evaluate survival and the potential for continued soil arsenic extraction during winter conditions under appropriate insulated conditions.

CONCLUSION

Phyto-Engineering approaches can be applied to a large number of contaminated dredged materials/soil situations. These sites can include either metal or organic contaminants or both. Once the dredged material is phyto-reclaimed from contamination, RSMT topsoil applications include mineland restoration, Brownfield redevelopment, landfill restoration, Superfund remediation, and acid mine drainage remediation. A combination of RSMT and phyto-reclamation can be a less costly, effective approach to recycling dredged material and providing additional storage space for future dredging requirements.

Acknowledgements. This research has been supported by USACE New York, Hawaii and Wilmington, NC Districts, USACE DOER Program, USEPA Region 2, CRDA partners RSMT, Bion Technologies, N-Viro International, Inc. and Advanced Remedial Mixing, Inc. Technical assistance was provided by H. Banks and J. Cole, Dyntel Corporation, Vicksburg, MS; and Dr. N. Hue and J. Fong, University of Hawaii-Monoa, Honolulu, HI.

REFERENCES

Cunningham, S. D. and C. R. Lee. 1995. *Phytoremediation:Plant-Based Remediation of Contaminated Soils and Sediments*. In Bioremediation, Ed. by H. A. Skipper, Society of Soil Science Special Publication Number 43, American Society of Agronomy, p 145-156.

Hue, N. V., C. R. Lee, S. Campbell, Q. X. Li, J. Fong, R. A. Price, and A. J. Palazzo. 2001. *Demonstration of Agriculture-Based Bioremediation Technologies Relevant to Pacific Island Ecosystems*. Miscellaneous Paper EL-01- __, U.S. Army Engineer Waterways Experiment Station, Vicksburg, MS. (in press).

Lee, C. R. , T. C. Sturgis, W. Cadet, P. Adams, and C. Drill. 1998. *A Manufactured Soil from Dredged Material, Yardwaste, and EQ Biosolids.* Proceedings of the Water Environment Federation 12th Annual Residuals and Biosolids Management Conference, 12-15 July. Bellevue, WA. Pages 674-704.

Lee, C. R., T. C. Sturgis, K. Johnson, N. Murray, and A. J. Palazzo. 2001. *Demonstration of Manufactured Topsoil from NY/NJ Harbor sediment, cellulose and biosolids.* Miscellaneous Paper EL-01- __, U.S. Army Engineer Waterways Experiment Station, Vicksburg, MS (in press)

Lee, C. R. 1999. *Implementation Guidance for Selected Options for the Reclamation and Beneficial Use of Contaminated Dredged Material.* USACE DOER Technical Notes Collection (TN DOER C7) U.S. Army Engineer Research and Development Center, Vicksburg, MS.

Price, R. A., C. R. Lee. 2000. *Findings of a Working Group on Phyto-Reclamation of Contaminated Dredged Material.* USACE DOER Technical Notes Collection (TN DOER C3) U.S. Army Engineer Research and Development Center, Vicksburg, MS.

Sturgis, T. C., and Lee, C. R., 1999. *Manufactured Soil Screening Test.* USACE DOER Technical Notes Collection (TN DOER C6) U.S. Army Engineer Research and Development Center, Vicksburg, MS.

WILLOWS GROWN ON CONTAMINATED SEDIMENT: POSSIBILITIES FOR PHYTOREMEDIATION

P. Vervaeke (Laboratory of Forestry, Ghent University, Ghent, Belgium)
F.M.G. Tack (Laboratory of Analytical Chemistry and Applied Ecochemistry, Ghent University, Ghent, Belgium)
N. Lust (Laboratory of Forestry, Ghent University, Ghent, Belgium)

ABSTRACT: A field trial was designed to assess the impact of planting a fast growing willow stand on the dissipation of organic contaminants (mineral oil, EOX and PAHs) in dredged sediment. In addition, the accumulation of heavy metals (Cd, Cu, Pb and Zn) in the biomass tissue was determined. A significant decrease of 57 % in the mineral oil concentration of the planted depot was observed while mineral oil degradation in the fallow depot was only 15 %. The mineral oil degradation under willow was most pronounced (80 %) in the root zone of the stand. In the fallow depot there was a significant reduction of the total PAH content by 32 % after 1.5 years compared to a 23 % reduction in the planted depot. The moderate and selective metal uptake measured in this study, limits the prospects for phytoextraction of metals from dredged sediment.

INTRODUCTION

As a result of a range of pollutant sources, sediments dredged from our waterways are enriched with various organic and inorganic contaminants. As a result, the possibilities for re-use of dredged sediment in for instance public works and agriculture are limited. Currently, the disposal on land is one of the few options and the most common approach these days in Flanders is landfilling and complete containment. The disposal of the dredged sediment results in the establishment of fertile but contaminated sites with little beneficial uses. In addition the shortage of land for disposal is becoming an increasingly important problem.

If however the dredged sediment is planted with fast growing willow clones for energy production in Short Rotation Forestry systems (SRF), this product can be valorized as a useful and fertile substrate with minimal risks for the environment. Previous research indicated that dredged sediment can be appreciated as a suitable substrate for willow wood production (Vervaeke et al., 2001). Over the last years, research has focussed on planting fast growing willow tree species for biomass production on derelict land often contaminated with heavy metals and other pollutants, on agricultural land affected by sewage sludge and waste water disposal, on mining wastes and on industrial wasteland (Perttu et al., 1997; Hasselgren, 1998). Willow stands are hereby appreciated as effective vegetation filters, which can stabilize polluted substrates, recycle nutrients and accumulate pollutants in their tissues. As a result of a high evapotranspiration, water use efficiency and interception, metal and nutrient percolation through the soil profile of these sites can be reduced significantly (Schnoor, 2000). In addition, it is recognised that certain of these fast growing tree species have a strong potential for heavy metal uptake and that SRF systems may be used to remove heavy metal contamination from the soil through a repeated harvest of the willow wood for energy purposes (Duncan et al., 1995; Punshon et al., 1997). Very little information is available on willow trees planted on substrates contaminated with organic pollutants. In general, with the introduction of a dense root system, aeration and microbial biomass can be enhanced in the root zone, which can result in an increased organic contaminant degradation (Shimp et al., 1993). Remediation of contaminated substrates may thus be enhanced by plants because they function like solar-driven pumps which have degrading activity in their rhizosphere (Cunningham et al., 1996).

However, the planting of a willow vegetation cover on wet substrates such as hydraulically confined dredged sediment is difficult, as the surface is impassable for both man and machinery. Through applied research at the Laboratory of Forestry of the Ghent University, the SALIMAT technique was developed as a practical, low-cost and mechanized planting method for willow species on wet and inaccessible substrates (De Vos, 1994). The SALIMAT technique is based on the vegetative reproduction of willow from horizontally inlayed willow rods. Willow rods with a length of 2 meter are tied together with biodegradable string and subsequently rolled around a central disposable tube. The SALIMATS are unrolled by dragging them across the hydraulically filled lagoon with a crane and dragline.

Thus, because land disposed dredged sediment is rich in nutrients, occupies extended areas and often is unsuitable for use in agriculture and public works due to the presence of contaminants, planting fast growing willow species can be a possibility for the revalorisation of this material. Especially when combined with the uptake and/or degradation of contaminants from dredged sediment, the production of biomass could offer multiple advantages over the current strategies for dredged sediment disposal.

METHODS AND MATERIALS

In July 1999, two small dredged sediment disposal sites (20 x 20 m) were established at the experimental site in Menen (Belgium) and filled with sediment from the adjacent river Leie. After hydraulic filling, one depot was planted with the commercial biomass clone Salix viminalis 'ORM' using the SALIMAT technique, the other depot was kept fallow. At the same time, both depots were sampled at 8 grid points, each at 3 depths (0-20 cm, 40-60 cm and 80–100 cm). Each of the samples was analysed for pH, particle size distribution, carbon content, nutrient (N, P, K, S, Ca, Mg) and heavy metal (Cd, Pb, Zn, Cu) contents. The investigated organic parameters were: mineral oil (on FT-IR), extractable organic halogens (EOX) and PAHs (on GC-MS). In August 2000, the different biomass compartments of the willow stand were sampled. Several trees were cut at each of the 8 grid points, after which roots, wood and leaves were separated and sub sampled. Litter was collected in October 2000 using litter traps. Biomass samples were analysed for nutrients (N, P, K, S, Ca, Mg) and heavy metals (Cd, Pb, Zn, Cu). After one full growing season, in November 2000, the sampling of the sediment and organic contaminant analysis were repeated. Biomass production and stand density were assessed in November 1999 and November 2000 using allometric relations between tree diameter and weight.

RESULTS AND DISCUSSION

Sediment characteristics. Although the sediment in both depots originated from the same river there were some differences in the physical and chemical characteristics (Table 1). The planted depot featured a higher sand content and a lower clay content. This was reflected in the element and organic compound concentrations. Contents were all slightly elevated in the fallow depot. No significant differences could be found in organic contaminant concentrations as measurement in each depot showed a very high variability horizontally over the area of the depot as well as vertically through the soil profile. All metals (except for Cu) and organic contaminant concentrations exceed the Flemish VLAREA threshold values for the use of the dredged sediment as soil.

TABLE 1: Characteristics of the dredged sediment in the planted and fallow depot (* indicates significant differences with $\alpha = 0.05$)

Depot		Planted	Fallow			Planted	Fallow
# samples		24	20			24	20
> 50 µm	%	43	28.3 *	min.oil (IR)	mg/kg	245.2	364
20-50 µm	%	38.5	51.3 *	EOX	mg/kg	3.8	6.2
10-20 µm	%	3.1	3.1				
2-10 µm	%	4.5	3.9	Naphthalene	mg/kg	0.47	0.78 *
< 2 µm	%	10.8	13.5	Benzo(a)pyrene	mg/kg	0.89	1.08
				Phenanthrene	mg/kg	0.98	2.03
C	%	2.9	3	Fluoranthene	mg/kg	2.35	3.56
N (tot)	mg/kg	1236.7	1391	Benzo(a)anthracene	mg/kg	0.65	0.85
P	g/kg	2.4	3.3 *	Chrysene	mg/kg	0.73	0.87
K	g/kg	18.5	20.7 *	Benzo(b)fluoranthene	mg/kg	0.88	1.14
S	g/kg	3.2	5.1 *	Benzo(k)fluoranthene	mg/kg	0.38	0.38
Mg	g/kg	1.9	1.2 *	Benzo(g,h,i)perylene	mg/kg	0.44	0.53
Ca	g/kg	25.7	22.6	Indeno(1,2,3-c,d)pyrene	mg/kg	0.52	0.67
Cd	mg/kg	3	4.6 *				
Pb	mg/kg	142.9	230.1 *	Total PAH	mg/kg	8.29	11.89
Zn	mg/kg	437.3	662.2 *				
Cu	mg/kg	72.5	97.7				

Stand establishment. One week after planting, the rods in the SALIMATS started to sprout and a high density wood grass willow stand developed. After a very short growing season of 3 months in 1999, the stand produced 3.7 ton DM/ha and a shoot density of 755,000 shoots/ha was counted. Shoot density after the second growing season was 417,000 shoots/ha, indicating intense natural thinning due to competition in the wood grass stand. Biomass production during this second growing season was 11.5 ton DM/ha. Leaf biomass was 2.4 ton DM/ha in 2000.

Organic contaminant degradation. When the initial organic contaminants concentrations (1999) are compared with the concentrations measured after the second growing season, a significant decrease of 57 % in the mineral oil concentration of the planted depot was observed while mineral oil degradation in the fallow depot was only 15 % (Table 2). The mineral oil degradation is most pronounced at the surface of the planted depot. About 80 % less mineral oil was measured after the second growing season in the 0-20 cm samples. With increasing depth in the profile the dissipation of mineral oil decreased (Table 3). In the fallow depot there was a significant reduction of the total PAH content by 32 % after 1.5 years compared to 23 % in the planted depot. Total PAH concentrations in the 0-20 cm zone were reduced by 29 and 47 % in respectively the planted and fallow depot (Table 3). Only in the 0-20 cm samples of the fallow depot significant reductions of individual PAHs were noticed. Phenanthrene exhibited the most significant degradation in the fallow depot. With increasing depth in the profile the degradation of PAHs in both depots decreased.

TABLE 2: Comparison of organic contaminant concentrations in the planted and fallow depot between July 1999 and November 2000 and the percentage degradation. (n = # samples,* = significant difference with α = 0.05)

	Planted (n = 24)			Fallow (n = 20)		
	1999 (mg/kg)	2000 (mg/kg)	Degradation (%)	1999 (mg/kg)	2000 (mg/kg)	Degradation (%)
Mineral oil (IR)	245.2	104.8 *	-57	364	307.8	-15
EOX	3.8	4.2	+10	6.2	4	-36
Naphthalene	0.47	0.38	-19	0.78	0.53 *	-32
Benzo(a)pyrene	0.89	0.49	-44	1.08	0.7 *	-35
Phenanthrene	0.98	0.74	-25	2.03	0.9	-56
Fluoranthene	2.35	1.73	-26	3.56	2.05 *	-42
Benzo(a)anthracene	0.65	0.54	-17	0.85	0.69	-19
Chrysene	0.73	0.64	-12	0.87	1.07	+23
Benzo(b)fluoranthene	0.88	0.81	-8	1.14	1	-12
Benzo(k)fluoranthene	0.38	0.36	-5	0.38	0.56	+47
Benzo(g,h,i)perylene	0.44	0.27	-39	0.53	0.39	-26
Indeno(1,2,3-c,d)pyrene	0.52	0.38	-27	0.67	0.45 *	-33
Total PAH	8.28	6.37	-23	11.88	8.11 *	-32

TABLE 3: Mineral oil and PAH degradation percentages at different depths in the planted and fallow profile. *: indicates significant differences between July 1999 and November 2000 values ($\alpha = 0.05$)

	0-20 cm		40-60 cm		80-100 cm	
	Planted	Fallow	Planted	Fallow	Planted	Fallow
Mineral oil (IR)	-79 *	-21	-44 *	-4	-31	-15
Naphthalene	-17	-57 *	-21	-40	-25	+23
Benzo(a)pyrene	-56	-48 *	-43	-30	-32	-26
Phenanthrene	-17	-63 *	-37	-71 *	-12	-6
Fluoranthene	-14	-60 *	-34	-50	-26	-11
Benzo(a)anthracene	-25	-20	-20	-38	+4	+6
Chrysene	-38	-18	+9	0	-2	+104
Benzo(b)fluoranthene	-22	-31	+3	-7	+1	+6
Benzo(k)fluoranthene	-26	-7	+2	+2	+20	+89
Benzo(g,h,i)perylene	-49	-42 *	-33	-25	-29	-9
Indeno(1,2,3-c,d)pyrene	-39	-46 *	-22	-29	-21	-16
Total PAH	-29	-47 *	-25	-39	-14	+5

The growth of willow on dredged sediment thus had a positive effect on the degradation of mineral oil but did not enhance the degradation of PAH. On the contrary, the reduction in PAH concentrations in the upper parts of the profile was highest in the fallow depot. As microbial degradation is the most important mechanism of PAH degradation it was expected that the introduction of a dense root system would stimulate microbial activity and increase PAH degradation. Aprill et al. (1990) described a faster PAH degradation under the influences of grass. A possible explanation for the reduced degradation could be that the introduction of a very dense willow cover resulted in a slower photodegradation of PAHs by shading of sunlight and by introducing a cooler microclimate on the surface of the depot. Compared to the planted depot, the fallow substrate featured a much higher density of cracks, which would result in a better aeration of the profile and hence an increased microbial activity.

Heavy metal accumulation. The metal concentrations of the biomass compartments of the sampled Salix viminalis 'Orm' stand are presented in table 4. The normal range of metals in plants are: 0.1–2.4 mg Cd/kg, 1-400 mg Zn/kg, 0.2-2.0 mg Pb/kg and 5-20 mg Cu/kg (Kabata-Pendias et al., 1992). Concentrations of Pb and Cd in this study exceeded these values for all compartments. Cd can be appreciated as the only metal that is truly accumulated as it features a BCF > 1 in all compartments. Several studies already indicated that there is a pronounced uptake of cadmium into the shoots of Salix plants. Contents of 0,4 - 3,9 mg/kg stem-wood have been measured (Ledin, 1998). Punchon et al. (1997) reported very high Cd concentrations of 44 and 76 mg/kg in leaves and stems respectively, of Salix cinerea growing on mining spoil.

TABLE 4: Mean metal concentrations and bioconcentration factors (BCF) in the biomass compartments of the Salix viminalis 'Orm' stand (n = 8; BCF = plant concentration/soil concentration).

	Roots		Wood		Leaves		Litter	
	Conc. (mg/kg)	BCF	Conc. (mg/kg)	BCF	Conc. (mg/kg)	BCF	Conc. (mg/kg)	BCF
Cd	3.2	1.1	3.6	1.2	4.3	1.4	4.9	1.6
Cu	15.1	0.2	5.3	0.1	7.6	0.1	15.9	0.2
Pb	17.7	0.1	12.7	0.1	2.9	0.02	23.9	0.2
Zn	243.0	0.5	146.1	0.3	362.5	0.8	331.3	0.7

If the stand would be harvested after three years with an annual biomass production of 11.5 ton DM/ha, a mineral mass of 5 kg Zn/ha and 0.12 kg Cd/ha could be exported from the site. Assuming a soil density of 1.3 ton/m^3, one hectare of substrate contains 11.7 kg Cd in the top 30 cm. A target concentration can be identified from the VLAREA legislation which postulates threshold levels for the use of dredged sediment as soil as 2 mg Cd/kg or 7.5 kg Cd/ha. The period to achieve this target would be 33 years, or 11 successive 3-year rotations. Metal concentrations of the other metals are too low to extract them in substantial quantities in this time frame. It has to be mentioned that the investigated Salix viminalis 'Orm' was not selected for metal uptake, and therefore higher amounts of metals could be extracted when metal accumulating species or clones are applied.

Rather high concentrations of metals were found in the litter layer of the stand compared to the concentrations in the leaves. However, the amounts of metals that reach the stand surface with leaf fall are of the same order of magnitude as the annual atmospheric metal deposition in rural reference areas in Flanders (Table 5). The risk of contamination of the wider environment and the food chain through leaf fall can thus be considered as minimal.

TABLE 5: Amounts of metals that annually reach the stand surface with leaf fall (2.4 ton DM/ha) compared with the annual atmospheric deposition in rural reference areas in Flanders (Van Grieken, 1996).

	Conc. (mg/kg)	Amount in litter (g/ha/y)	Atmospheric deposition (g/ha/y)
Cd	4.9	11.8	7.3
Cu	15.9	38.2	73
Pb	23.9	57.4	110
Zn	331.3	795.1	440

CONCLUSIONS

The planting of dredged sediment with willows using SALIMATS resulted in an effective 'green capping' of the dredged sediment depot with a high density fast growing willow stand with minimal risks for the environment. Contaminated dredged sediment, which previously had to be discarded, can thus be valorized for the production of biomass for energy purposes. While the introduction of a willow stand resulted in an increased degradation of mineral oil, a slower PAH dissipation was observed in the planted depot. This study was not able to target the factors responsible for the differences in PAH degradation under fallow and vegetation. As it was expected that the introduction of a dense root system would stimulate microbial activity and PAH degradation, further investigation on laboratory and glasshouse scale is needed to provide more insight.

The moderate and selective metal uptake limits the prospects for phytoextraction of metals from dredged sediment. However, metal removal by willow would be of value if it depletes the fractions of bio-available or environmentally active metals in the substrate. In addition, the establishment of a dense willow cover could result in the immobilisation of metals as a result of rhizosphere processes and increased evapotranspiration. Further research at our research group will focus on the impact of willow stands on the speciation and mobility of metals in contaminated substrates. This will be assessed in field and pot trials by comparing the metal fractionation in planted and unplanted substrates using sequential as well as single extraction procedures.

ACKNOWLEDGEMENTS

Research was performed by the Laboratory of Forestry, University Ghent, in the scope of the IWT project nr. 960201 in cooperation with the dredging company Jan De Nul N.V. and the analytical laboratory Environmental Research Center N.V..

REFERENCES

Aprill, W., and R. C. Sims. 1990. "Evaluation of the Use of Prairie Grasses for Stimulating Polycyclic Aromatic Hydrocarbon Treatment in Soil." *Chemosphere. 20*: 253-265.

Cunningham, S. D., T. A. Anderson, A. P. Schwab, and F. C. Hsu. 1996. "Phytoremediation of Soils Contaminated with Organic Pollutants." *Adv. Agron. 56*: 55-114.

De Vos, B. 1994. "Using the SALIMAT Technique to Establish a Willow Vegetation Cover on Wet Substrates." In P. Aronsson, and K. Perttu (Eds.), *Willow Vegetation Filters for Municipal Wastewaters and Sludges: a Biological Purification System.* pp. 175-182. Sveriges Lantbruksuniversiteit, Uppsala.

Duncan, H. J., S. D. McGregor, I. D. Pulford, and C. T. Wheeler. 1995. "The Phytoremediation Of Heavy Metal Contamination Using Coppice Woodland." In

W. J. van den Brink, R. Bosmans, and F. Arendt (Eds.), *Contaminated soil '95*, pp. 1187-1188. Kluwer Academic Publishers.

Hasselgren, K. 1998. "Use of Municipal Waste Products in Energy Forestry: Highlights from 15 Years of Experience." *Biomass and Bioenergy. 15*: 71-74.

Kabata-Pendias, A., and H. Pendias. 1992. *Trace Elements in Soils and Plants.* CRC Press., Boca Raton.

Ledin, S. 1998. "Environmental Consequences when Growing Short Rotation Forests in Sweden." *Biomass and Bioenergy. 15*: 49-55.

Perttu, K. L., and P. J. Kowalik. 1997. "Salix Vegetation Filters for Purification of Waters and Soils." *Biomass and Bioenergy. 12*: 9-19.

Punshon, T., and N. M. Dickinson. 1997. "Mobilisation of Heavy Metals Using Short-rotation Coppice." *Aspects of Applied Biology. 49*: 285-292.

Schnoor, J. L. 2000. "Phytostabilisation of Metals Using Hybrid Poplar Trees." In I. Raskin and B. D. Ensley (Eds.), *Phytoremediation of Toxic Metals: Using Plants to Clean Up the Environment*, pp 133-150. John Wiley & Sons.

Shimp, J. F., J. C. Tracy, L. C. Davis, E. Lee, W. Huang, L. E. Erickson, and J. L. Schnoor. 1993. "Beneficial Effects of Plants in the Remediation of Soil and Groundwater Contaminated with Organic Materials." *Environ. Sci. Technol. 23*: 41-77.

Van Grieken, R. 1996. "Distribution of Heavy Metals." (in Flemish). *Environmental Report Flanders 1996.* Flemish Environmental Agency.

Vervaeke, P., S. Luyssaert, J. Mertens, B. De Vos, L. Speleers, and N. Lust. 2001. "Dredged Sediment as a Substrate for Biomass Production of Willow Trees Established Using the SALIMAT Technique." *Biomass and Bioenergy, 21*(2): 81-90.

REMEDIATION OF MARINE MUCK SEDIMENTS, ST. LUCIE ESTUARY, FLORIDA

Patricia L. Sime, South Florida Water Management District, Stuart, FL, USA
Patricia Goodman, South Florida Water Management District, Stuart, FL, USA
Gary N. Roderick, FL Dept. of Environmental Protection, Port. St. Lucie, FL, USA
Peter J. Stoffella, University of Florida, IRREC, Ft. Pierce, Florida, USA

ABSTRACT: The Indian River Lagoon (IRL) Restoration Feasibility Study was initiated in July 1996 as part of the Comprehensive Everglades Restoration Project. Two of the goals of this larger restoration project are to remediate for existing muck sediments in the estuary and to control further deposition. A pilot study was initiated to explore possible beneficial use of marine muck sediments. A literature review component of the study was originally intended to support and refine a study of agricultural uses, such as a compost additive. New data from muck sediment cores collected early in the pilot project contradicted past studies with respect to the likelihood of beneficial agricultural uses. In particular organic matter and nutrient contents present in the muck were found to be much lower than in some earlier studies; and the sodium content in the muck was relatively high when compared with other soil amendments and plant tolerance levels. Alternative beneficial uses being explored within the IRL Plan context include using the muck as a sealant to retard seepage within storage reservoirs; as a substrate in stormwater treatment areas; and as a pasture grass herbicide and soil amendment for conversion of fallow pasture to native habitat.

INTRODUCTION

Background. The St. Lucie Estuary (SLE), located in Martin and St. Lucie Counties, FL. (Fig. 1) has undergone significant changes over the last 100 years. Though there is evidence that there were temporary episodes of an ephemeral inlet, until 1892 the St. Lucie was predominantly a freshwater system. With the opening of an artificial inlet in 1892, the system became estuarine (Woodward-Clyde Consultants, Inc. 1994). Since that time, changes have occurred in the watershed that feeds the SLE. A series of drainage canals that discharge into the SLE were constructed between 1920 and 1960 to facilitate agricultural and urban development (Figure 1). These canals increased the size of the watershed and are believed to have contributed to the accumulation of fine-grained, sediment, commonly referred to as "muck" in the estuary (SFWMD and USACOE 1999).

In 1991, the Indian River Lagoon was listed as an estuary of national significance and included in the National Estuary Program (NEP) by the United States Environmental Protection Agency (USEPA). The 775-sq. mi. (2,010 sq. km) SLE watershed is a major tributary at the southern end of the IRL. Land use intensification and drainage modifications in the watershed have dramatically increased wet-season flows, and significantly reduced dry-season flows to the estuary (SFWMD and USACOE 1999). Today, rapid decreases in salinity and enormous sediment loads result from high volume, nutrient laden, freshwater flows to the estuary through canals constructed as part of the Central and Southern Florida Flood Control Project.

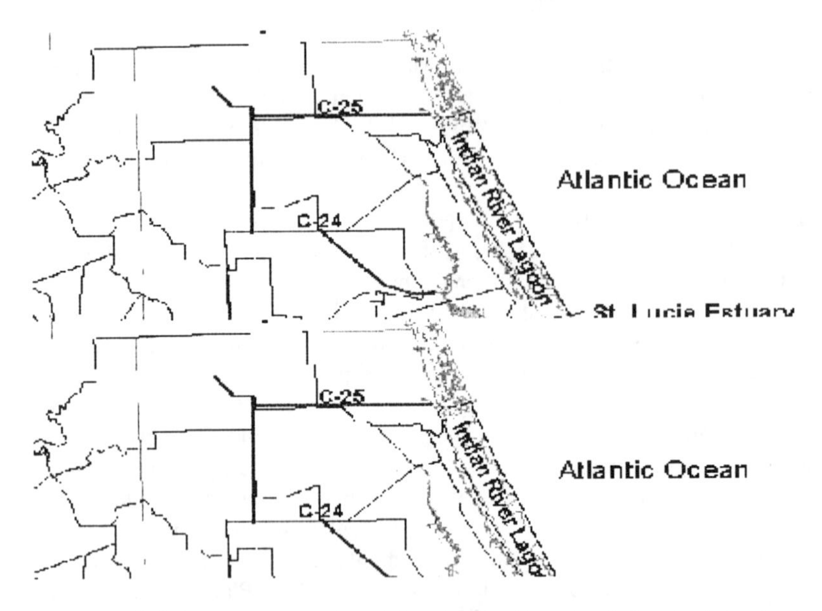

FIGURE 1. Central and South Florida Flood Control Project.

The increase in nutrient and sediment loading has contributed to the build-up of fine-grained, nutrient-rich muck in the estuary (Haunert 1988). The high nutrient levels support algal blooms which in turn, further contribute to the sediment problem. The resultant changes in aquatic communities within the estuary include an increase in pollutant tolerant benthic organisms and declines in the seagrass and oyster communities. Habitat and species diversity is affected by the degradation in water and sediment quality (SFWMD and ACOE 1999).

Accumulation of Sediments in the St. Lucie Estuary. Muck sediments naturally accumulate in estuaries like the St. Lucie. Past studies in the North Fork, and comparisons between 1884 and 1984 bathymetry (Morris 1986), indicate that the deeper muck sediments pre-date any impacts by man on the river basin. (Trefry 1996a) concludes that these deposits offer protection to the Lagoon by sequestering excess fines and nutrients, as only the upper layer of sediment is available for significant interaction with the water column. Several estimates of muck sedimentation rates and effects of upland sediment sources have been reported (Pitt 1972, Davis and Schrader 1984, Schrader 1984). Pitt (1972) concluded that canals C-23 and C-24 deposited 4500 and 9000 tons (4080 and 8160 metric tons) respectively of sediment annually to the North Fork. Schrader (1984) estimated sediment deposition rates of 0.2 to 1.0 inches (0.508 to 2.54 cm) per year in the North Fork since man's influences began on the basin, with average annual accumulations of 2 to 3 inches (5.08 to 7.62 cm) of fine sediments.

The general sedimentation pattern is one where coarse sand accumulates in areas of higher current velocity, with increasingly fine sediments deposited as currents decline. The areas of greatest muck sediment deposition are where the narrow areas of the North and South Forks widen, and where the Middle Estuary widens east of the Roosevelt Bridge (Figure 1). Haunert (1988) illustrated excellent maps of muck sediments. Cross-sections depicting type and accumulation depths can be viewed in Schrader (1984).

Muck sediment accumulations during this century average 2 to 3 feet (0.6 to 0.9 m) in most of the Estuary that has a depth greater than 6 feet (1.9 m). However, some local areas, such as the mouth of Poppleton Creek, now average a depth of only 3 feet (0.9 m) deep, and are choked with deep muck deposits. A flocculent surface layer is associated with many of these muck sediments but is not reported in bathymetric studies, as it cannot be discerned from the water itself except by visual observation underwater. Schropp et al. (1994) estimated this layer to be 1 to 2 feet (0.305 to 0.61 meters) deep in the wide North Fork.

Environmental Degradation. The accelerated buildup of these sediments has resulted in well-documented negative impacts to water quality and estuarine habitat (SFWMD and USACE 1999). The surface of muck sediments is typically hypoxic due to the high biochemical oxygen demand of the organic fractions and they tend to be a very soft substrate. These two properties generally preclude colonization by most desirable benthic organisms, resulting in large areas of the bottom of the estuary being essentially devoid of biological activity.

Early results of muck removal from Crane Creek in Brevard County, FL conducted in 1998 indicate that improved water quality resulting from reduced turbidity has aided seagrass bed recovery and support this strategy for estuarine environment improvement (Trefry et al. 1989). The natural shoreline vegetation once stabilized the substrate, filtered storm water runoff, and provided quality habitat. Perhaps if the extent of shoreline mangrove and seagrass could also be restored, additional water quality and habitat improvements would follow including an increase in the population of the American oyster (*Crassostrea virginica)* whose filtration capacity could also greatly aid in water quality improvements.

MATERIALS AND METHODS

A pilot study was initiated in preparation for the large-scale muck remediation project. The initial purpose of the pilot study was to explore possible beneficial use of marine muck sediments to offset the enormous cost of muck remediation as planned under the Southern Indian River Lagoon (SIRL) Restoration Feasibility Study. Approximately $59 million of the $920 million budget has been allocated for muck remediation. Assumptions based on prior research led to the beneficial uses study, and supported the hypothesis that agricultural uses were the most plausible.

Volume calculations. United States Army Corps of Engineers (USACE) performed a muck survey in the SLE. Four areas in the estuary were identified where muck remediation was believed to be most advantageous: two in the North Fork, one in the South Fork, and one in the Mid-Estuary. These sites correlate closely with potential muck accumulation "hot spots" identified by Haunert (1988).

The total quantity of muck in these four areas was estimated using data from a dual-frequency hydrographic survey. The survey was conducted 13-16 September 2000. A total of 79 transects were performed. The transects for each area of interest were 500 feet apart, and data points were taken every 25 feet along each transect. Depth of muck was determined from the difference of two simultaneous operated depth transponders, one at a frequency that measured the top of the muck, the other at a penetrating frequency which measured the distance to the bottom of the muck. The resulting data allowed estimates of the total muck volume (Table 1).

TABLE 1: Volume calculations in cubic yards (cy)(cubic meters)

Areas	**Volume (cy)**	**(cubic meters)**
North Fork northern area	1,159,486	(886,491)
North Fork southern area	807,509	(617,385)
South Fork	430,581	(329,203)
Mid-Estuary	3,116,987	(2,383,108)
Total	**5,514,563**	**(4,216,187)**

Muck Sample Collection and Processing. The objective of this portion of the project was to extract muck samples from the SLE, assess the components of, and determine the potential beneficial uses for the muck. Core borings of muck samples were collected by the (USACOE) from the SLE using 5-foot sections of coring tube that was pounded into the soft sediment. Additional five-foot sections were added on in deep deposit areas until the entire column was sampled at each location. The samples were then delivered to the University of Florida (UF), Indian River Research and Education Center (IRREC). The samples are indicated in Figure 2. These core samples were stored in a freezer at IRREC after their arrival and prior to use. Before analysis, each sample was transferred from a tube to a plastic bag, mixed thoroughly in the bag, air-dried, and ground to pass a 2-mm mesh sieve for measurement of pH and electrical conductivity (EC). Sub-samples from different depths (0-1.5, 1.5 to 3.0, 3.0 to 4.5 m) if available at each location (Table 1) were analyzed. A portion of the air-dried sample was further ground to pass a 0.15 mm mesh sieve for analysis of total organic carbon (C), nitrogen (N), potassium (K), phosphorus (P), magnesium (Mg), calcium (Ca), zinc (Zn), copper (Cu), manganese (Mn), aluminum (Al), iron (Fe), cadmium (Cd), lead (Pb), nickel (Ni), arsenic (As), and silver (Ag).

The pH was measured in water using a muck:water (wt:wt) ratio of 1:1 utilizing a pH/ion/conductivity meter. The (EC) was measured in water using a sediment:water ratio of 1:2 also utilizing a pH/ion/conductivity meter. The total concentrations of organic C and N in the muck samples were determined using a CNS-Analyzer. For determination of total concentration of macro- and microelements in muck, the finely ground samples were digested in a solution mixture of 9 ml HNO_3, 3 ml HF and 0.5 ml HCl using a Microwave Digestion System.

The concentrations of K, P, Mg, Ca, Zn, Cu, Mn, Al, Fe, Cd, Ni, and Ag in the acid digests were determined using an Inductively Coupled Plasma Argon Emission Spectrometer (ICPAES, JY 46 P). The EPA method 200.7 (EPA, 1998) was followed for these analyses. The As, Pb and Mo were determined using an Atomic Absorption Spectrometer (Perkin Elmer, Simaa 6000).

Data analyses were conducted using SAS program procedures (Version 6.12) (SAS, 1985)(He et al., 2001).

TABLE 2. Vertical variation in pH, electrical conductivity (EC), and element concentrations in sediments (He et al., 2001).

Properties		Depth (m) 0-1.5	1.5-3	3-4.5	Mean
PH	1:1 H2O	7.48A*	6.90B	6.72B	7.15
EC	us/cm	1996A	2071A	2043A	2030
Organic C	%	4.181A	4.760A	4.365A	4.41
OM	%	7.208A	8.206A	7.525A	7.60
N	%	0.242A	0.278A	0.264A	0.26
Ca	g/kg	19.597A	16.697A	19.529A	18.60
Mg	g/kg	3.553A	2.245A	2.110A	2.87
Al	g/kg	26.769A	30.820A	32.212A	29.07
Fe	g/kg	12.909C	18.606B	27.438A	17.30
P	g/kg	1.009B	1.122AB	1.351A	1.11
K	g/kg	3.423A	3.997A	4.038A	3.72
Na	g/kg	37.205A	39.350A	34.020A	37.39
Cd	mg/kg	12.183B	14.181B	17.030A	13.68
Co	mg/kg	18.217A	19.423A	21.226A	19.14
Cr	mg/kg	30.561C	43.209B	63.085A	40.36
Ag	mg/kg	3.278A	3.252A	3.599A	3.32
Cu	mg/kg	28.514A	24.785A	24.321A	26.54
Mn	mg/kg	111.60A	114.84A	138.02A	117.18
Ni	mg/kg	34.194A	36.514A	40.018A	35.97
Zn	mg/kg	36.976A	42.924A	44.395A	40.25
As	mg/kg	6.123B	8.078B	13.869A	8.10
Pb	mg/kg	10.660A	9.691A	10.219A	10.26
Sample (no.)		29	20	10	59

*** Mean followed by different letter within the same row indicates difference at 5% significance level.**

RESULTS AND DISCUSSION

Literature Review. The literature on muck sediments specific to the St. Lucie Estuary is excellent in quality but limited in extent. This literature was recently summarized thoroughly with respect to physical, chemical and biological characteristics by Woodward-Clyde (1998). Their primary purpose in reviewing the literature was to establish potential for recovery of healthy populations of oysters and submerged aquatic vegetation (SAV) within the Estuary. Schropp et al. (1994) defined muck sediments as "sediment with greater than 60% silts and clays, greater than 6% total organic C and greater than 75% water." This definition is as useful as any in the literature.

A study of properties and beneficial uses of SLE muck completed by Lee in 1998 was uncovered during the literature review. This study represented a thorough examination of the potential uses of these sediments as agricultural soil amendments. The study concluded that unless diluted to 10% or less of the total soil mass the muck had detrimental effects on subsequent plant growth. High Na content, which proved very difficult to modify, was considered the main problem (Henderson for SFWMD in prep 2001).

Muck Sampling and Analysis. New data from muck sediment cores collected early in the pilot project by the USACE contradicted earlier studies with respect to the likelihood of beneficial agricultural uses. In particular, the percentages of organic matter and nutrients present in the muck were reported to be much lower than in some earlier studies; and the percentage of Na in the muck was relatively higher than other soil amendments and plant tolerance levels.

CONCLUSIONS

Muck sediments in the St. Lucie Estuary have accumulated rapidly within the past 70 years. These fine sediments readily suspend due to wind, current, and boat shear forces, increasing turbidity and precluding re-establishment of historic seagrass beds. The organic content of these sediments, once thought to average around 30-40%, is actually closer to 10%. Low organic and nutrient contents and high salinity in interstitial water limit potential beneficial agricultural uses. These data, and prior studies of fine sediments in the St. Lucie, indicate that while more than 80% the sediments pass the 200 sieve, and a variety of metals are slightly higher than would be expected in a pristine environment, none present toxic or hazardous constituent (heavy metal) concentrations. The same properties that support easy re-suspension of the sediments, and associated turbidity within the estuary, also make removal and consolidation within spoil sites very difficult. High concentrations of hydrated cations within the mineral portions of these fine grained sediments result in unusually low sediment density and very slow consolidation. Consolidation in situ may be further reduced by slow but consistent anaerobic decay and out-gassing of the organic sediment fraction, further reducing density at the sediment/water interface and contributing to re-suspension. Physical removal and disposal methods are severely limited by these physical properties. The most cost effective methods of managing these sediments remains elusive.

Approximately 3058 cubic meters of muck will be dredged from the North Fork of the SLE using either a shielded clamshell or a precision cutterhead hydraulic dredge. Beneficial uses being explored within the IRL Plan context include using the muck as a sealant to retard seepage within storage reservoirs; as a substrate in stormwater treatment areas; and as a pasture grass herbicide and soil amendment for conversion of fallow pasture to native habitat.

The test will include areas that are flooded to examine wetlands recovery with and without muck; testing sealant properties; and simple disposal within a diked berm measuring consolidation rates and native revegetation characteristics on the spoil pile. Data from the demonstration plots will be analyzed to evaluate effectiveness of the muck as a soil amendment to improve subsequent plant growth of native species, and soil fertility and any negative impacts. The applicability to the larger scale IRL Feasibility Study Restoration work will be analyzed.

Reuse on Pastureland. This portion of the project will: a) investigate beneficial use of the muck as soil amendment to suppress the growth of Bahia grass (*Paspalum notatum* Flugge) and to enhance subsequent growth of native species and b) minimize the environmental impact and disposal costs of the muck.
The objectives will be accomplished by a combination of field measurements at selected sampling sites and laboratory tests on samples of the muck, soil, and plants from the field.

The muck will be dredged from the SLE and transported to a site on a publicly owned parcel of land, which had previously been used as pastureland for cattle. The intent of the project is to demonstrate beneficial use of St. Lucie Estuary muck sediments as a biological suppression of Bahiagrass in native wetland restoration areas.

Application of the Pilot Study. The SIRL Integrated Feasibility Report and Environmental Impact Statement (EIS), a component of the larger Comprehensive Everglades Restoration Plan (CERP), is an investigation of options to improve the quantity, timing, quality and distribution of freshwater flows to the SIRL. Another very important part of the restoration plan is the improvement of current conditions in the estuary itself. This includes floodplain restoration, and estuarine habitat enhancement through the removal of large volumes of muck sediment and the addition of hard bottom substrate such as artificial reefs and oyster shell. Improving substrate conditions in the estuary is expected to jumpstart fisheries recruitment and speed up water quality improvements by increasing dissolved oxygen levels as well as decreasing turbidity and nutrient flux from the bottom sediments. The study analyzed a series of components, which when combined into alternatives, produced solutions to the environmental degradation in the SIRL while maintaining the existing functionality of the C&SF project system.

REFERENCES

Davis, R.A. and D. L. Schrader 1984 Sedimentation in the St. Lucie Estuary, Martin and St. Lucie Counties, FL, report to South Florida Water Management District, West Palm Beach, Fl.

Haunert, D.E. 1988. Sedimentation Characteristics and Toxic Substances in the St. Lucie Estuary, Florida. Tech pub 88-10, SFWMD, West Palm Beach, FL.

He, Z. L., M.K. Zhang, P. J. Stoffella, D. V. Calvert, C. Wilson, 2001. Beneficial Reuse of Marine Muck Sediments St. Lucie River, Final Report. Final report to SFWMD, West Palm Beach, FL.

Henderson, K., 2001 in prep. Beneficial Reuse of Marine Muck Sediment St. Lucie Estuary. Report to SFWMD, West Palm Beach, FL.

Lee, C. L., T. C. Sturgis, J. Miller, and B. W. Bunch. 1998. Evaluation of environmental effects of muck removal from the St. Lucie Estuary and the potential beneficial uses of the muck. Miscellaneous paper EL-97-, US Army Corps of Engineers Waterways Experiment Station, Vicksburg, MS.

Pitt, W.A. Jr. 1972. Sediment loads in canals 18, 23, 24 in Southeastern Florida, USGS Open file 72013.

SAS Institute. 1985.SAS user's guide: Statistics. 5th ed. SAS Inst., Cary, NC.

Schropp, S.J., W.F., McFetridge. R.B. Taylor, 1994. St. Lucie Estuary muck removal demonstration project. Unpublished final report to SFWMD.

Shrader, D.C. 1984. Holocene sedimentation in a low energy microtidal estuary, St. Lucie River, Florida. Masters thesis, Department of Geology, University of South Florida, Tampa, Florida.

SFWMD and USACOE. 1999. Central and Southern Florida Project: Comprehensive Review Study. West Palm Beach, Florida.

Trefry, J. H. 1996a&b Indian River Lagoon muck: characteristics, origins, distribution, and impacts. Presentation at the Indian River Lagoon Muck Workshop, SJRWMD, Palatka, FL.

Woodward-Clyde Consultants, Inc. 1994. Physical features of the Indian River Lagoon. Project Number 92F274C. Prepared for IRLNEP, Melbourne, Florida.

ELECTROCHEMICAL REMEDIATION TECHNOLOGIES FOR GROUNDWATER, SOIL AND SEDIMENTS

Falk Doering, Electrochemical Processes L.L.C., Valley Forge, PA
Niels Doering, P2 Soil Remediation, Inc. Stuttgart, Germany
Don Hill, Weiss Associates, Emeryville, CA

ABSTRACT: Direct current technologies (DCTs) use direct-current electricity passed between at least two electrodes in either an ex situ or in situ mode to effect the remediation of soil, sediment and/or the groundwater. DCTs in line with U.S. terminology comprise electrochemical remediation technologies (ECRTs) and electrokinetics. The primary distinction between ECRTs and electrokinetics is the level of power input and the mode of operation. For over 11 years, ECRTs have been applied successfully in more than 50 remediation projects, primarily in Europe.

ECRTs involve the integration of geophysical, hydrogeological, and electrochemical processes with electrical engineering and colloidal chemistry. Employing low-energy, proprietary AC/DC current, ECRTs generate reduction/oxidation (redox) reactions that remediate at the pore scale. The ECRTs relate to two primary types of processes. Electrochemical geooxidation (ECGO) addresses organics, which are mineralized to basically carbon dioxide and water, whereas induced complexation (IC) treats inorganic substances, enhancing the mobilization and precipitation of heavy metals on electrodes of both polarity.

THE SIMPLIFIED MODEL OF ECRTS

ECRTs primarily are in situ processes that apply electrical current to electrodes driven into the ground to rapidly address a wide range of both organic and inorganic compounds. The voltages applied between electrodes, installed at distances of 6 to 20 m, are relatively low, about 35 to 45 DC with maximum voltages below 100 V DC.

The electrical field generated by two electrodes (Figure 1) comprises an anodic peak and a cathodic depression with steep flanks indicating a high transition resistance, whereas the area between the electrodes develops an electrical field characterized by a small resistance providing for a voltage gradient slightly inclined towards the cathode. This field depends on the specific resistivity of the system and the amperage and voltage applied.

Electrochemical applications basically use an electrochemical cell with two electrodes submerged in an electrolyte. In ECRTs, the principles of electrochemistry are applied to soils, with two major distinctions: the pore water assumes the functions of an electrolyte and the soil particles take on the properties of the electrodes. The electrochemical properties of the soil particles are discussed in detail below.

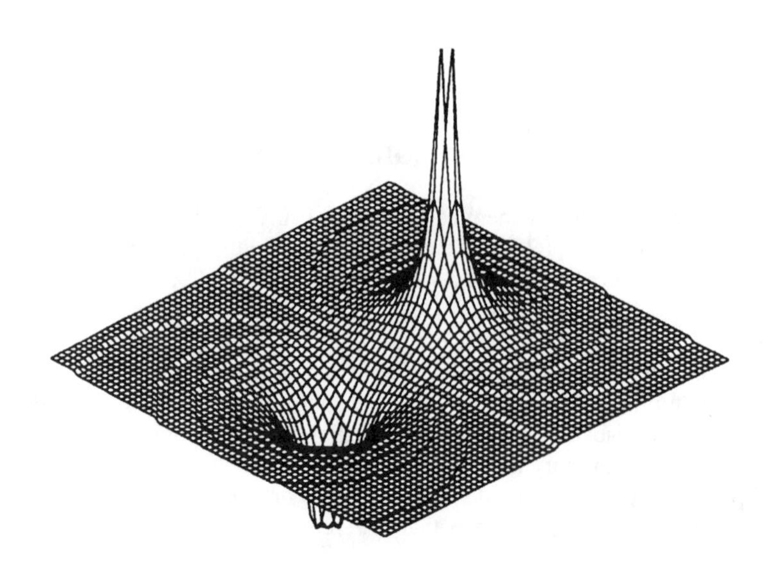

FIGURE 1. Example of electrical field generated by ECRTs using 10-m long by 20-cm nominal diameter tube electrodes (anode and cathode) at an 8.2 m separation.

Hydrogeology has shown that all soil particles, once in contact with water, obtain a water shell formed by electrostatic forces (e.g., Coulomb, Van-der Waals and/or London forces). The water shell is structured as a double layer. The inner layer is called hygroscopic water, about 1 to 10 water molecules thick; the middle layer, called hydration water, is about 200 water molecules thick; and the outer layer, the so-called captive water, is the transition area to the bulk solution of the capillary water. The outer limit of the captive water is defined by a pressure where the water becomes available to plants; this point sometimes is called the withering point and has a pressure of about 1.5 Mpa (Figure 2).

Comparing this double-layer model with the those related to colloid (Stern/Nernst model) and to the electrode/electrolyte system, we find similarities in the structure and in the description of the forces attracting the fluid (in this case water) to the solid (Figure 2). In colloidal chemistry and surface chemistry, we distinguish three layers. The inner layer is called Inner Helmholtz (or Stern) Layer (IHL), the middle one is called Outer Helmholtz Layer (OHL), and the outside one is called Diffuse Layer. Making the terms compatible, the hygroscopic water corresponds to the Inner Helmholtz Layer, the hydration water to the Outer Helmholtz Layer, and the captive water to the term Diffuse Layer.

With caution, we can combine hydrogeological definitions (and properties) with those of colloid chemistry. The conclusion is that all soil particles have a water shell of the colloidal type and between each layer we find electrochemically active interfaces, the so-called Inner Helmholtz Plane and the so-called Outer Helmholtz Plane. Additionally, because electrochemical reactions take place in the Outer Helmholtz Plane, electrochemical reactions must take

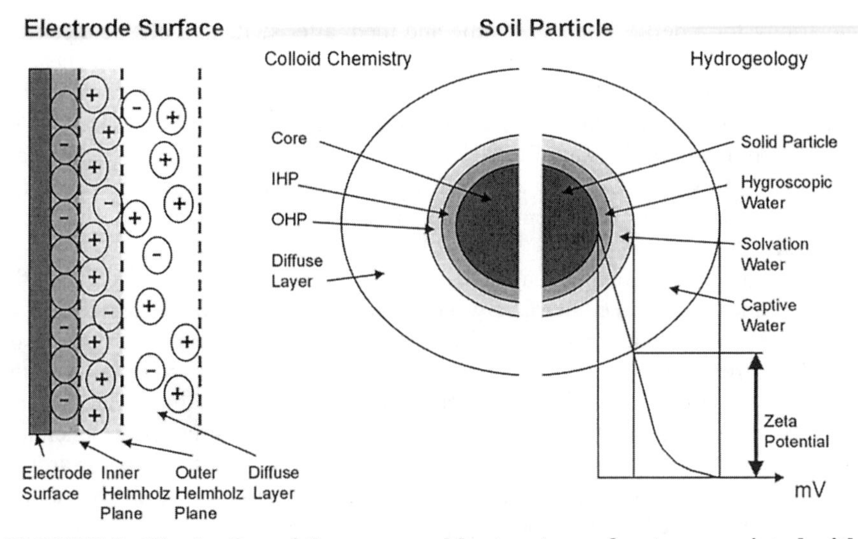

FIGURE 2. Illustration of the comparable structure of water associated with electrode surfaces, colloids, and soil particles.

place at the soil particles when the system is sufficiently energized. In standard laboratory and industrial electrochemical processes, only a few electrodes are used in the electrochemical cell. In soils, however, a plentitude of electrodes exist, namely all soil particles. This has an impact on electrode surfaces: in a standard 2-electrode array made up of two tubular electrodes 10 m long, 172-mm outer diameter, installed at 8.2 m distance (center to center) and groundwater as electrolyte, the surface of the traditional electrodes is 5.4 square meters per electrode, whereas in an average clay (illite) soil, the surface of the clay minerals in the electric field generated by the electrodes totals 168,000 square kilometers, which is greater than the size of the U.S. state of Pennsylvania. Because the rate of electrochemical reactions depends partly on the electrode surface area, we find that electrochemical decomposition of pollutants by ECRTs is faster in clays than in sands, and faster in sands than in gravel.

To complete this simplified model of ECRTs, we need to consider existing natural electric fields in soils, which have voltages between 15 mV dc and about 1,000 mV dc. In line with the terminology created by Conrad Schlumberger, the natural polarization of the soil particles is called "spontaneous polarization" (SP). That is, the polarization is generated instantaneously, without any obvious external source, such as the ECRT power supply.

The question now is, what happens when a direct current is introduced into the soil? In this report we refer to a direct electric current rectified from a three-phase alternating current by a 6-pulse rectification having a ripple (alternating current) of ±5 We believe that the geophysical exploration method, called induced polarization, provides the best explanation for the energization of the soil particles during ECRTs. IP is based on transport through and exchange of ions and electrons at (double layer) interfaces on pore and micro-pore scale in soils. Classical IP measurements are effected by first charging the soil with electricity for a defined period of time and then, after switching off the electrical current, by measuring the pattern of how the electrical charge in the soils decays. The characteristic signals detected reflect the effects of polarization of metallic conductors and polarization of electrochemical interfaces in pore systems. The distinction between IP and ECRTs is that ECRTs maintain the steady-state condition for several months after the initial electrical charging, whereas the energy source of the IP system is switched off once the steady state has been reached.

Whenever a voltage is applied to the soil, energy is required to cause the current flow across the interface. This energy barrier can be considered to constitute an electrical impedance. Grahame (1951, 1952) describes two paths by which current may pass the interface between the solid particle and the electrolyte: the faradiac and nonfaradiac F paths. Current passage in the faradiac path requires a conductive soil particle; the passage causes electrochemical reactions (redox reaction) and also comprises the diffusion of the ions towards or off the interface. In the nonfaradiac case, namely when the soil particles are non-conductive, charged particles do not cross the interface; the current is carried by charging and discharging of the double layer. The double layer then behaves like a capacitor, varying its impedance with the frequency of the ripple. In the faradiac path, the ion diffusion is customarily referred to as Warburg impedance (*W*), and its magnitude varies inversely with the square root of the electrical frequency. In soils, the distinction between the two paths is academic. Empirical evidence indicates that both types occur simultaneously.

We can confirm the postulation presented above through the use of oscilloscope readings during the application of ECRTs. The oscillogram example provided in Figure 3 is derived from a standard ECGO system consisting of two electrodes (tube electrodes of 172-mm outer diameter, 10-m length, an 8.2-m distance between the electrodes), charging the soil with about 27 V dc and 27 A. The upper curve in Figure 3 represents the voltage, measured at the receptacles of the power generator. The AC sine of the ripple as measured is about 2 V AC at a frequency of about 329 Hz. The lower curve depicts the amperage, measured again at the positive terminal. The scale differs from that of the voltage: 1 mV = 100 ma, i.e., the measured value of the signals related to the ripple totals about 16 A. These data were obtained in silty to clayey soils in the southwestern part of Texas, USA.

In the oscillogram, the impedance is expressed as phase displacement, i.e., the amperage lags by an angle of $\theta \sim 5°$ behind the voltage. The storage of electrons at the double layer around non-conductive substances must yield discharge signals, which can be seen in the voltage curve as occurring as spikes at about 500 mV. The spikes are directly related to a drop of amperage at the receptacles, i.e., the AC/DC converter supplies less current to the soil. On the other hand, the voltage curve goes up and continues to follow the sine-wave pattern, which indicates that there must be another supplier of electrons. The oscilloscope power supply is filtered to reduce (power) line noise. Soil energy (capacitive) energy storage remains as sole alternative explanation for the voltage noise spikes and current reversals observed in the oscillogram. The reading of the

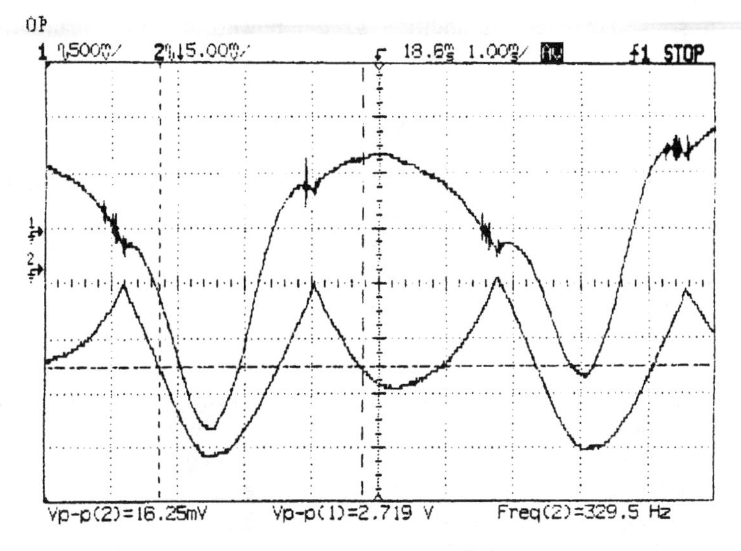

FIGURE 3. Voltage and amperage of an ECGO System in silty to clayey soils.

parabolic part of the amperage curve is 9 mV = 9 A over 2.8 ms which indicates, under consideration of the frequency, an energy discharge of ~ 22.7 A-sec (Coulombs), collected at the power electrodes.

Empirical ECGO field remediation data of rapid mineralization of organic contaminants strongly suggests that the secondary current released via soil electrical discharges provides at least the activation energy for the ensuing redox reactions. Additionally, it is suspected that trace metals in the soil may act as catalysts, reducing the activation energy required for the redox reactions. The quantification of these effects remains to be completed. The dissociation energy of the most common pollutants is between 45 $kcal \cdot mol^{-1}$ and 120 $kcal \cdot mol^{-1}$.

The principal electrode reactions in electrochemical cells are half reactions comprising of the anodic oxidation or the cathodic reduction. The oxidizing or reducing agents, namely O_2 and H_2, are generated at the electrodes by water electrolysis. Typical half reactions are:

Reduction:
- reduction of carbon-carbon double bonds
- reduction of carbonyl-groups
- reduction of nitrile or nitro groups
- reduction of halogenated bonds
- reductive coupling
- reductive cleavage

Oxidation:
- oxidation of hydrocarbons
- oxidation of functional groups
- oxidative coupling and
- oxidative substitution.

In soils, however, in addition to the power electrode reactions, the complete system of redox-reactions also takes place simultaneously at every soil particle. Any and all reactions related to an oxidation are combined with a reduction: the electrons of the oxidized substance (provided by the reducing agent as electron donator) must be received by the reduced substance (by the oxidizing agent as electron acceptor). The standard redox potential (E_o) determines the direction whether electrons are accepted or donated. Having analyzed more than 2000 GC-MS chromatograms arranged in more than 400 sequences over a period of up to 180 days, the reactions and products correspond to those in the electrochemical literature.

For the treatment of inorganic substances, by induced complexation, the small range of voltages where ECRT still works, but phases out, and Geokinetics start to work, must be used. We have stated that at this transitional point, which we suggest naming the electrokinetic point (EP), the electric discharge signals fade out and simultaneously the resistance substantially increases, i.e., the electrical conductance is determined now by the ionic conductivity of the groundwater and no longer by the colloidal conductivity of the double-layer structure. IC effects the mobilization of immobile heavy metal, which are converted into compounds, which can be mobilized. We suggest to name the processes taking place in that range “induced complexation (IC)”, because the chemical transformation mainly covers the conversion of immobile heavy metal compounds into complex ions. The precipitation of heavy metals occurs on both anode and cathode.

Basis of IC is the precipitation of metal ions, complex ions, and colloids onto the cathode in compliance with Faraday’s laws. Monovalent mercury, for example, will precipitate at a rate of 7.484 g per A-h onto the cathode. The precipitated metals form a rigid layer on the electrodes. Heavy metals at the end of the site remediation can be recovered from the electrodes and recycled when the electrodes are removed from the soil.

This approach is now complemented by the effects of IC, converting the inorganic compounds by electrochemical reactions as follows:

- Insoluble compounds are split into their elemental constituents, such as HgS into elemental mercury (Hg°) and colloidal sulfur. This phenomenon has also been observed for Pb_3O_4 and other compounds
- There is a high probability that heavy metal ions or colloids have been converted into complex ions. In the field, we have observed the process of complexation such as the formation of chloro-complex ions (in saline water), hydrated complex ions, hydroxo-complex ions and ammine salts (in areas of intensive agriculture). These complex compounds commonly precipitate onto anodes. A structural X-ray analysis also indicates that the most common complex compounds with two and more central atoms are generated in analogy to natural geochemical processes caused by weathering. We have also observed elements (such as arsenic and tin) which normally do not precipitate onto electrodes, form these complexes precipitate onto the anodes.

- Certain metals and half-metals such as Sn, As, Sb, Se and Te react with the hydrogen ions or radicals to hydrides. These hydrides are volatile and can be extracted from the soil by vacuum-technologies.

REFERENCES FOR SEDIMENT REMEDIATION

1. Elemental and Organic Mercury in the Sediments of a Scottish Canal

A mercury remediation demonstration project was conducted in the sediments of a Scottish canal. The canal contains brackish water (total dissolved solids = 800 mg/l) and is 10 m wide x 1.1 m deep. The canal is almost completely filled with silt. The silt contains both elemental and organic mercury (about 10% of total) originating from an upstream detonator factory.

The area remediated was 220 m^3 (20 m x 10 m x 1.1 m working depth). Two electrode pairs were placed in the silt along the banks of the canal. There were six sampling locations within the remediation cell and one outside the cell. Initial average total mercury concentration within the remediation cell was 243 mg/kg (Table 1), with a total mercury concentration range of 1,570 mg/kg to 33 mg/kg.

Table 1 presents the average total mercury concentration soil field sampling results at remediation days 1, 12 and 26.

TABLE 1. Mercury sediment concentrations, mg/kg.

Days	1	12	26
T1 anode	33	204	11
T6 anode	218	417	9
T3 middle	102	36	11
T5 middle	282	48	6
T2 cathode	98	45	4
T4 cathode	156	9	0,7
Outside	809	73	4

A total of 76 kg (168 lbs) of mercury was plated on the electrodes after 26 days. Total average mercury concentrations decreased from the average of 243 mg/kg to 6 mg/kg. The cleanup objective was 20 mg/kg. A mass balance was completed based on the field data, and a reasonably good agreement exists between what was plated on the electrodes and the initial and post-remediation concentrations in the sediments. The mercury precipitated on both anode and cathode.

2. PAH in the Sediments of a German Creek

Until 1991, a manufactured gas plant in eastern Germany polluted a small creek with PAH, TPH, and other by-products originating from the manufacture of gas. The sediments of the creek, which feeds a small pond, were blackish and had a strong PAH odor.

The test section was about 15 m long and about 5 m wide. The creek was cut into the alluvial plane by about 1 m. The average velocity of the creek was about 0.8 m/s; the maximum pollution of the sediments, having an average thickness of .5 m has been measured in the (prospective) test area at 4,111 mg/kg of PAH.

Taking into account that a rebound was likely to occur, the parties to the test agreed that the test should be accepted when "a significant" reduction of the PAH could be evidenced. Significant was internally defined as a reduction of PAH to 50% of the initial value. As method, DIN ISO 1387, Process B has been selected. The duration of the test was fixed at 120 days and took place between August 1998 and December 1998.

Two steel electrodes have been installed horizontally in a trench of about 75 cm depth (bottom of the trench) excavated by an excavator on both banks of the creek. The distance between both electrodes was about 10 m. The steel tubes had a nominal diameter of 15 cm. The energy delivered to the electrodes totaled 0.73 kW. The results are shown in Table 2.

TABLE 2. PAH Test Results for a German Creek, Expressed as mg PAH/kg d.m.

Parameter	Baseline (21 Aug 1998)	Mid-Term	End of Project (2 Dec 1998)
naphthalene	130	34.8	1.1
acenaphthylene	220	21.8	1.7
acenaphthene	110	51.7	1.5
fluorene	290	53.2	2.8
phenanthrene	950	206.9	21
anthracene	390	101.3	14
fluoranthene*	740	132.3	54
pyrene	310	100.0	23
benzo(a)anthracene	230	63.0	26
chrysene	210	52.5	18
benzo(b)fluoranthene*	120	50.6	16
benzo(k)fluoranthene*	75	27.7	12
benzo(a)pyrene*	160	38.7	25
dibenzo(a,h)anthracene	24	7.3	3.8
benzo(g,h.i)perylene*	64	10.0	6.8
indeno(1,2,3-c,d)pyrene*	85	5.5	1.2
Σ of PAH (EPA 1-16)	4111	957.3	227.9

**carcinogenic*

3. Field Test on the Elimination of Heavy Metals in the Waste Water Lagoon of a Rolling Mill

For about 100 years, a rolling mill has produced sheets, profiles, and tubes from steel, aluminum, copper, and brass. The waste water has been and is discharged into flat lagoons coming from abandoned clay pits. The concentration of pollutants is extremely inhomogeneous and varies up to 5,000%, over short distances. The lagoons have sizes of about 1 acre to 2 acres at about 1.1 m depth filled with a dust comprising of blasting sands and finest metal particles, mainly iron and copper. The sediments were dust dry. Attempts to irrigate the sediments caused a cloud of dust. IC was invited to perform a test in that area for 30 days. The electrodes were 2-m^2 sheet electrodes; the anode was made of copper, the cathode of steel; the plate electrodes were incorporated into the soil at a distance of 8 m. Most time of the test was spent on the development of a technology to wet the sediments, having an initial resistance of greater than 320 Ω. Step by step, the resistance had been reduced to 19,6 Ω at the end of the project. The technology comprised of a special surfactant/water mixture injected at the anode and driven to the cathode by electroosmosis. By control of the resistance, the progress of the wetting technology has been examined. At the end of the test, currents of about 10 A were being driven through the sediments.

Measurement of the metals precipitated onto the electrodes has been hampered by the corrosion of the copper anode. Nevertheless, approximately 8.5 kg of heavy metals (mainly lead, copper and cadmium) have precipitated on both electrodes, in the rate of 38% (anode) to 62% (cathode). The rates are in line with the calculated rates of precipitation at an electric current efficiency of 92.2%. The most surprising effect was that the iron injection lances placed at the anode were plated with zinc.

Because of the hydrophobia of the sediments, no realistic data on the removal of heavy metals in the sediments within the 30 days of test could reasonably be expected. The evaluation of the process therefore focused on the effects on the groundwater, where a substantial cleanup of the groundwater has been reached, as shown below.

TABLE 3. Heavy Metal Elimination Field Test in a Rolling Mill Waste Water Lagoon. (All figures are in µg/l.)

	Anode		**Cathode**		**Center**	
	Baseline	after 30 days	Baseline	after 30 days	Baseline	after 30 days
Pb	37	<15	31	24	65	< 15
Cd	< 1.8	< 1.8	< 1.8	< 1.8	< 1.8	< 1.8
Cr	<6.0	<6.0	< 6.0	< 6.0	6.6	< 6.0
Cu	48,000	3,500	2,400	670	49,000	4,200
Ni	530	170	290	120	340	110
Hg	< 0.1	< 0.1	< 0.1	< 0.1	< 0.1	< 0.1
Zn	4,600	700	4,900	1,200	3,100	840
As	< 10	< 10	< 10	< 10	< 10	< 10

STATUS

Four validation projects have been scheduled to start in the U.S. this year: One project evaluates ECRTs in marine sediments polluted by PAH, phenols, and mercury, examining ECRTs under the conditions of a highly chlorinated environment. Special attention is given to the impact of electricity onto special species of fish, one hunting by electroreceptors (such as sharks and rays) and the other emitting electrical signals (certain flat fish). A second project deals with dredged port sediments polluted by a mix of inorganic pollutants, PCBs and PAH. A third project (in progress) tests off-site the decontamination of high concentrations of PCBs in river sediments. A fourth project is devoted to an in situ test in a river treating mercury, silver and copper with PCBs, DDD and low level radionuclides largely co-located with the main heavy metals.

CONCLUSION

ECRTs have been shown effective for organic and metal remediation through the ECGO and IC processes. Two million metric tons of soil have already been remediated.

REFERENCES:

V. Vacquier, Ch.R. Holmes et al: Prospecting for Groundwater by Induced Electrical Polarization; *Geophysics*, vol. XXII, No. 3, July 1957, p.666

D.J. Marshall, Th.R. Madden: Induced Polarization, a Study of Its Causes; *Geophysics*, vol. XXIV, No. 4, October 1959,p. 790

J.R. Wait (ed.): Overvoltage Research and Geophysical Applications, Pergamon Press, London, New York et al., 1959

E. Warburg: Ueber die Polarisationscapacitaet des Platins; *Ann. Physik*, 4. F., Vol. 6, H. 9, 1901, p. 125

E. Warburg: Lehrbuch der Experimentalphysik, Verlag Theodor Steinkopff, Dresden/Leipzig 1929 (21 and 22 ed.)

D.C. Grahame:The Role of the Cation in the Electrical Double Layer; *Journal Electrochem. Society*, vol. 98, 1951, p. 343

D.C. Grahame: Mathematical Theory of the Faradiac Admittance (Pseudocapacity and Polarization Resistance), *Journal Electrochem. Society*, vol. 99, 1952, p. 370C

W.J. Wedenejew, L.W. Gurwitsch, W.H. Kondratjew, W.A. Medwedew, E.L. Frankewitsch: Energien chemischer Bindungen, Ionisationspotentiale und Elektronenaffinitäten; VEB Deutscher Verlag für Grundstoffindustrie, Leipzig 1971

K. Scott: Electrochemical Processes for Clean Technology; the Royal Society of Chemistry, Cambridge (UK), 1995

HP Jordan, H.-J. Weder: Hydrogeologie, VEB Deutscher Verlag für Grundstoffindustrie Leipzig, 1988

P.A. Domenico, F.W. Schwartz: Physical and Chemical Hydrogeology; John Wiley & Sons, Inc., New York, Chichester, Weinheim et al., 1998

K. Scott: Electrochemical Processes for Clean Technology; The Royal Society of Chemistry, Cambridge (UK), 1995

HP. Jordan, H.-J. Weder: Hydrogeologie; VEB Deutscher Verlag fur Grundstoffindustrie Leipzig [GDR], 1988 H.J. Roesler, H. Lange: Geochemische Tabellen [Geochemical Tables], Ferdinand Enke Verlag Stuttgart, 1976

ELECTROKINETIC REMEDIATION OF CONTAMINATED SEDIMENTS

Steve Granade, U.S. Navy, Point Mugu, California, USA
David Gent, U.S. Army Corps of Engineers, Vicksburg, Mississippi, USA

ABSTRACT: A field scale treatability study has been in progress since March 1998 to evaluate the use of electrokinetics (EK) to perform in-situ remediation of metals and chlorinated solvent contaminated sediments. The primary goal of the project is to demonstrate that EK can be used to remediate sediments to below naturally occurring levels. The secondary goal is to determine if EK can be used to deliver electron donor to stimulate reductive dechlorination of chlorinated solvents. Sediment samples taken in June 1999 showed that chromium levels in 77.5% of the sediment mass between one anode/cathode pair had been reduced to below natural background levels. Evidence indicating that reductive dechlorination is occurring includes the detection of electron donor metabolites, depression of naturally occurring sulfate levels, and reduction of chlorinated solvent levels by up to 99.0%.

INTRODUCTION

Industrial operations such as electroplating and metal finishing have contributed to large amounts of metals and chlorinated solvent contamination in sediment, soil, and groundwater. Efforts to identify remediation methods that are effective and affordable continue throughout the world. Electrokinetic remediation has been identified as a possible in-situ technology to extract metals from sediment and soil.

Electrokinetic remediation is a process in which an electrical field is applied within a sediment or soil matrix by applying a low-voltage direct current (DC) between electrodes placed in the soil. When DC current is applied to the electrodes, an electrical field develops between the anodes and cathodes. The application of the electric field has several effects on the sediment, water, and contaminants. These effects include electromigration, electroosmosis, changes in pH, and electrophoresis. Of the electrokinetic phenomena discussed, electromigration is expected to be the main process that mobilizes the metals in the sediment and transports electron donor to support anaerobic dechlorination of chlorinated solvents in the groundwater. An organic acid, citric acid, is added at the cathodes to neutralize the cathodic electrolysis reaction and act as a chelation agent to enhance the transport and extraction of cationic species at the cathodes. The citric acid can also act as an electron donor supporting anaerobic dechlorination of chlorinated solvents. It is added at the cathodes to maintain a pH of 4 (~6 to 8 percent by weight).

A field scale pilot-test using electrokinetics to remediate metals and tetrachloroethylene (PCE) has been ongoing since 1998. The results of the first 3 years of operation are presented here.

Objective. The objective of this study is to evaluate the effectiveness of electrokinetics to remediate metals and PCE contaminated sediment. The goals are to achieve removal of metals from the sediment to below naturally occurring levels and reduce levels of PCE and daughter products to below drinking water standards.

Site Description. The project is being performed at the Site 5 area at Point Mugu, California. The Site 5 area encompasses an area of approximately 2000 m^2 in and around two former waste lagoons. These unlined lagoons were used between 1947 and 1978 to receive wastewater containing chromium, cadmium, copper, lead, silver and PCE. The two waste lagoons are located within the boundary of the 10 km^2 Mugu Lagoon. Mugu Lagoon is the largest remaining tidal marsh located between San Francisco, California and the California border with Mexico.

The earthen dikes surrounding the waste lagoons are approximately 1.8 m above mean sea level. Intertidal wetlands to the east, west, and north and road to the south border the waste lagoons. Site near surface sediments consist of fine to medium grained sand with lenticular and laterally continuous gravel, silt and clay strata from 0 to 2 m below ground surface (LB&M, 1997). The water in the waste lagoons is tidally influenced.

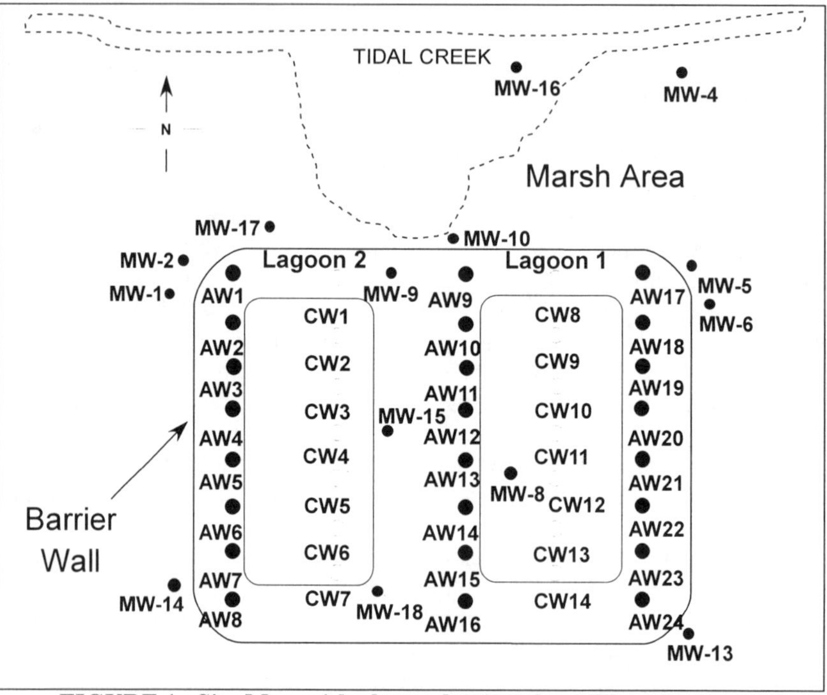

FIGURE 1. Site Map with electrodes (anodes AW, cathodes CW), monitoring wells (MW), and barrier wall.

MATERIALS AND METHODS

The pilot test system contains 3 rows of anode wells (24 total wells) and 2 rows of cathode wells (14 total wells) laid out in an array pattern over the two waste lagoons. The anode and cathode wells are spaced 2 m apart along each row. The anode and cathode rows are spaced approximately 4.5 m apart. The entire electrode array is surrounded by 80-mil high-density polyethylene sheet pile wall extending from ground surface to a depth of 6.1 m (Figure 1). The sheet pile wall was designed to isolate the pilot system from the surrounding marsh. It also reduced the hydraulic gradient within the electrode array to nearly zero.

The anode electrode wells are 4-inch (101.6-mm) diameter slotted polyvinyl chloride (PVC) well casings. The anode well casings are wrapped in a tightly woven linen fabric. The well packing is a kaolinite and sand mixture. The cathode wells are 3-inch (76.2-mm) diameter porous ceramic well casings with a kaolinite and sand packing mixture. All electrode wells were installed to a depth of approximately 1.8 m.

The anode electrodes are rod shaped and constructed of titanium with an iridium oxide coating. The cathode electrodes are constructed from approximately 2-inch (50.8-mm) wide strips of stainless steel mesh. All electrodes were initially installed to a depth of 3.05 m. Later the cathodes were shortened to a depth of 1.8 m in an effort to focus the electric field closer to ground surface.

Twenty, 2-inch (50.8 mm) monitoring wells were installed in and around the pilot test site (Figure 1). The monitoring wells were installed to varying depths with several different screen intervals (Table 1). The monitoring wells were sampled to look for breakout of target metal compounds and for evidence of anaerobic dechlorination.

TABLE 1. Breakout wells sampled with their drilled and screened depths.

Piezometer	Depth (m)	Screen Interval (m)
MW-1	6.7	6.1-6.7
MW-2	1.5	0.9-1.5
MW-3	1.5	0.9-1.5
MW-4	1.5	0.9-1.5
MW-5	6.7	6.1-6.7
MW-6	1.5	0.9-1.5
MW-7	3.7	3.1-3.7
MW-8	3.7	3.1-3.7
MW-9	6.7	6.1-6.7
MW-10	6.7	6.1-6.7
MW-11	3.0	2.4-3.0
MW-12	6.7	6.1-6.7
MW-13	6.7	6.1-6.7
MW-14	6.7	6.1-6.7
MW-15	1.5	0.9-1.5
MW-16	1.5	0.9-1.5
MW-17	5.5	4.9-5.5
MW-18	6.7	6.1-6.7

System Operation. The electrokinetic pilot test system has been operational during three separate periods since March 1998. Period 1 began in March 1998 and continued through October 1998. Total operation time during this period was 22 weeks. The initial current density applied to the array during this period was 0.1 mA/cm^2. This was increased to 0.33 mA/cm^2 in May 1998. The second period began in January 1999 and continued through June 1999. Total operation time during the second period was 22 weeks. The initial current density applied to the array during this period was 1.0 mA/cm^2 and reduced to 0.9 mA/cm^2 in April 1999 for the remainder of the period. The third period began in November 2000 and is scheduled to continue until November 2001. The initial current density applied during this period was 0.14 mA/cm^2. The current was slowly ramped up until a current density of 0.3-mA/ cm^2 was achieved by February 2001. The target current density for this phase in 1.0 mA/cm^2.

Sediment Sampling and Analysis. A set of 240 sediment samples was collected within the pilot test electrode array in August 1997. These samples were collected from 20 different continuous core holes to a depth of 3.66 m. Twelve samples were analyzed from each core at 0.3 m intervals. Sediments were analyzed for metals using U.S. EPA SW-846 Method 6010/7000. Determinations of the distribution of chromium by species using U.S. EPA SW-846 Method 3500-Crd. A second set of 80 sediment samples was collected in June 1999. This sample set was collected from 16 core holes located in between Anode Well 10 (AW-10) and Cathode Well 9 (CW-9) to a depth of 2.5 m. Five samples were analyzed for metals from each core at 0.5 m intervals.

Water Sampling and Analysis. Water samples are taken with a weekly and monthly frequency from the monitoring wells at the test site. Water samples are analyzed by the following methods: for metals using U.S. EPA SW-846 Method 6010/7000; for volatile organic compounds (VOC) using U.S. EPA SW-846 Method 8260; for sulfate using U.S. EPA SW-846 Method 300.0; dissolved gases (methane, ethane, and ethene) using U.S. EPA modified Method RSK 175; and for organic acids with high pressure liquid chromatography.

RESULTS AND DISCUSSION

Sediment Metals Analysis. The results of the pre-treatment metals analyses performed in August 1997 indicated that cadmium and chromium were the 2 main metallic contaminants of concern. The results for cadmium and chromium from the cross section from AW-10 to CW-9 are presented in Figures 2 and 3. The majority of the chromium contaminant is located in the top sections of sediment nearest the cathode. The minimum chromium concentration is this 180 mg/kg nearest the anode and the maximum was 1100 mg/kg near the surface at the cathode. 100 % of chromium results are above the 109 mg/kg local chromium background value (PRC, 1997). The majority of the cadmium contaminant is located in the top sections of sediment evenly distributed between the anode and the cathode. The minimum cadmium concentration was 5 mg/kg located at a depth of 10 ft. and the maximum was 20 mg/kg near the surface. 100% of cadmium results are above the 3.1 mg/kg local cadmium background value.

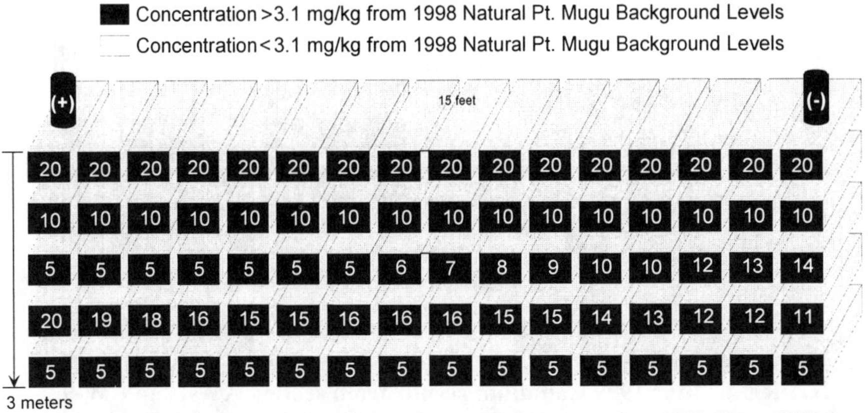

Concentration > 3.1 mg/kg from 1998 Natural Pt. Mugu Background Levels
Concentration < 3.1 mg/kg from 1998 Natural Pt. Mugu Background Levels

(+) 15 feet (-)

20	20	20	20	20	20	20	20	20	20	20	20	20	20	20	20
10	10	10	10	10	10	10	10	10	10	10	10	10	10	10	10
5	5	5	5	5	5	5	6	7	8	9	10	10	12	13	14
20	19	18	16	15	15	16	16	16	15	15	14	13	12	12	11
5	5	5	5	5	5	5	5	5	5	5	5	5	5	5	5

3 meters

FIGURE 2. August 1997 Cadmium results from section AW-10 to CW-9.

Concentration > 109 mg/kg from 1998 Natural Pt. Mugu Background Levels
Concentration < 109 mg/kg from 1998 Natural Pt. Mugu Background Levels

(+) 15 feet (-)

200	240	260	300	300	340	370	400	650	800	850	900	900	950	1000	1100
185	190	190	190	200	215	230	245	260	285	300	320	340	360	380	400
180	185	190	190	195	200	215	230	250	275	300	325	350	375	400	410
550	560	580	600	600	600	600	600	600	575	550	525	500	490	485	475
200	210	210	200	200	200	200	205	205	205	210	210	215	215	220	220

3 meters

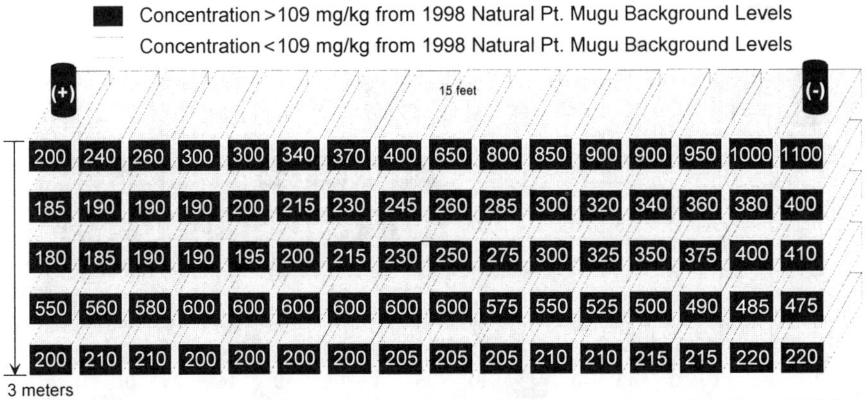

FIGURE 3. August 1997 Chromium results from section AW-10 to CW-9.

The results from the June 1999 sediment sampling indicate that most of the chromium and cadmium contamination has moved upward and toward the cathode (Figures 4 and 5). Chromium was not detected in the sections closest to the anode to over 2.4 m toward the cathode. The maximum result for chromium increased to8226 mg/kg near the surface by the cathode. Compared to the August 1997 samples, 77.5% of the sediment section had chromium levels reduced below local background levels. Cadmium was not detected in most of the sections closest to the anode to over 2.4 m toward the cathode. The maximum result for cadmium increased to 78 mg/kg near the surface by the cathode. Compared to the August 1997 samples, 78.8% of the sediment section had cadmium levels reduced below local background levels. Both cadmium and chromium exhibited similar trends with the contaminant mass moving to within 2.1 m from the cathode and upwards to within 1.2 m from the ground surface.

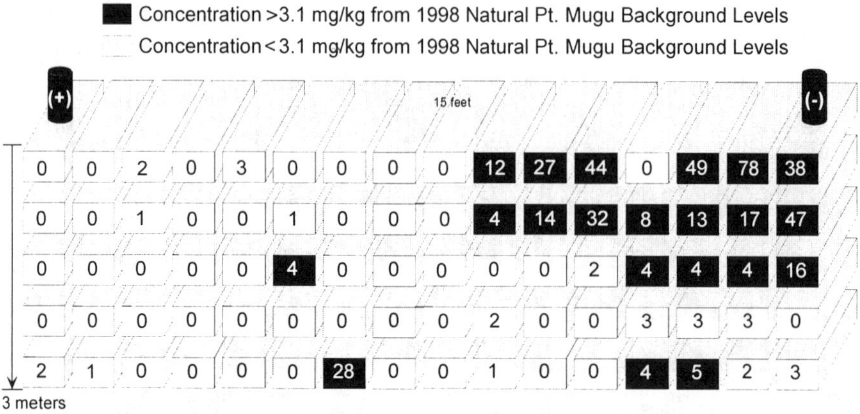

Concentration >3.1 mg/kg from 1998 Natural Pt. Mugu Background Levels
Concentration < 3.1 mg/kg from 1998 Natural Pt. Mugu Background Levels

(+) 15 feet (-)

0	0	2	0	3	0	0	0	0	12	27	44	0	49	78	38
0	0	1	0	0	1	0	0	0	4	14	32	8	13	17	47
0	0	0	0	0	4	0	0	0	0	0	2	4	4	4	16
0	0	0	0	0	0	0	0	0	2	0	0	3	3	3	0
2	1	0	0	0	0	28	0	0	1	0	0	4	5	2	3

3 meters

FIGURE 4. June 1999 Cadmium results from section AW-10 to CW-9. Zeros have been substituted for non-detect values for ease of viewing.

Concentration >109 mg/kg from 1998 Natural Pt. Mugu Background Levels
Concentration <109 mg/kg from 1998 Natural Pt. Mugu Background Levels

(+) 15 feet (+)

1	0	0	0	147	5	0	0	318	1351	1935	3222	5163	5052	8226	2072
0	0	90	0	25	0	0	0	236	122	340	510	216	278	344	2419
0	0	0	0	0	0	0	0	37	0	0	47	30	79	0	58
76	0	0	0	0	0	0	0	0	0	0	0	40	33	0	0
43	0	13	0	13	0	542	81	35	1	8	0	14	2	0	0

3 meters

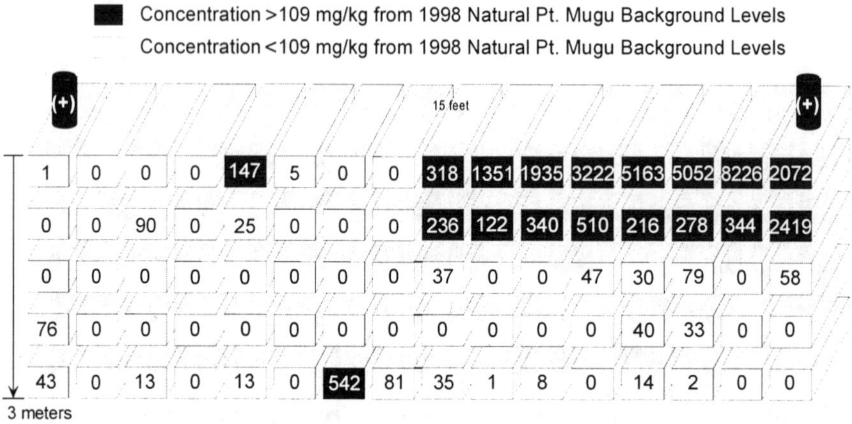

FIGURE 5. June 1999 Chromium results from section AW-10 to CW-9. Zeros have been substituted for non-detect values for ease of viewing.

These results show that electrokinetics can be used to dramatically reduce the volume of in-situ metal contaminated waste that requires cleanup. Phase 3 of the pilot test is attempting to improve the volume reduction efficiency to 95% wherein the removal electrodes wells with their well packing will remove all the metallic contamination.

Water Analysis. Samples were collected from anodes wells during the first operation period were analyzed for citric acid to estimate the transport rate through the test cell. Initial breakthrough of citric acid at anode wells was measured at an average of 10 mg/L on April 1, 1998. This was just 15 days after the beginning of the pilot test and represents a nominal transport rate of 0.3 m/day. The average concentration of citric acid in anode wells increased to 220 mg/L by October 1998. The average concentration of acetic acid in anode wells was 625 mg/L in October 1998. The presence of acetic acid, a biodegradation product of citric acid, is indicative that the pilot test site was biologically active.

Select wells were used during the second operation period to evaluate sulfate reduction and dehalogenation of PCE with the test site. Monitoring well MW-16 is representative of the select wells which included MW-4, MW-6, and MW-8. VOC, sulfate, and organic acid data for MW-16 are presented in Table 2. PCE concentrations declined from 1100 ug/L to <1 ug/L. Trichloroethene (TCE) concentrations declined from 86 ug/L to <1 ug/L. Cis-1,2 dichloroethene (DCE) increased from 120 ug/L to 460 ug/L and then decreased to 11 ug/L. Vinyl

TABLE 2. Monitoring well MW-16: VOC, sulfate, and organic acid data.

Sample date	PCE ug/L	TCE ug/L	DCE ug/L	Vinyl Chloride ug/L	Sulfate mg/L	Acetic acid mg/L	Propionic acid mg/L	Citric acid mg/L
15-Mar-98	1100	86	120	<1	2300	no data	no data	no data
15-Feb-99	51	32	460	770	no data	no data	no data	no data
05-May-99	5	9	86	220	710	no data	no data	no data
02-Jun-99	9	11	70	337	510	36.4	1.1	<1
28-Jun-99	9	12	127	88	1040	4.3	<1	<1
27-Jul-99	13	9	83	34	1400	no data	no data	no data
15-Sep-99	9	6	16	<1	2200	3.6	<1	<1
01-Mar-00	4	9	22	<1	2350	<1	<1	<1
08-Jun-01	<1	<1	11	<1	1900	no data	no data	no data

Chloride increased from <1 ug/L to 770 ug/L and then decreased to <1 ug/L. DCE and vinyl chloride concentrations increase simultaneously with the decrease of PCE and TCE concentrations. These trends are all indications of sequential dechlorination.

Sulfate levels decreased from 2300 mg/L to 510 mg/L and then increased back to 2350 mg/L. The lowest sulfate concentration occurred simultaneously with the highest detected concentrations of organic acid. This is indicative of sulfate reducing conditions. Sulfate reducing microorganisms have been previously shown to sequentially dechlorinate TCE using lactic acid as an electron donor at Point Mugu (OHM, 1998).

The correlation of sequential dechlorination under sulfate reducing conditions achieved during this pilot test indicates that citric acid is an effective electron donor. The initial and final concentrations of chloroethenes were converted to moles of chlorine. On a mole basis, 99% of the available chlorine had been transformed from chloroethene into a non-hazardous byproduct.

Two items were identified during review of water analysis data that merit some additional discussion. Dechlorination is occurring at the site in bulk sulfate reducing conditions. However, the levels of sulfate that remained in the groundwater during active dechlorination were much higher than most reported literature values. Chloroethenes were degrading even though sulfate was not reduced below 510 mg/L and even when sulfate levels increased from 1400 mg/L to 2200 mg/L. Secondly, the groundwater was analyzed for the presence of ethene, the expected byproduct of vinyl chloride. However, ethene has not been detected at this site. Vinyl chloride may be degraded by some mechanism other than microbiologically based anaerobic dechlorination at this site.

Pilot test problems. Groundwater analysis did indicate one complicating issue with using electrokinetics to remediate sediment and groundwater at the site. Naturally occurring chloride in the groundwater caused chlorine gas to be generated at the anodes at levels up to 1,700 mg/L. The chlorine gas reacted with naturally occurring dissolved organic carbon and the added electron donor to form trihalomethanes (THMs). Chloroform was the most commonly formed trihalomethane, reaching as high as 28,600 ug/L. Various control strategies were tried to control the formation of THMs with only limited success. These included air sparging of the anodes, periodic replacement of fluid in the anodes with fresh water, and reduction of the power applied to the system. The addition of sodium thiosulfate as a chlorine scavenger is currently being evaluated. Preliminary results are encouraging, with residual chlorine levels being reduced to 0.2 mg/L.

REFERENCES

LB&M Associates, Inc. 1997. *Technical Data Report for the Initial Characterization Event covering the In Situ Electrokinetic Remediation Demonstration at Site 5 (Old Shops 6 Area) Naval Air Weapons Point Mugu, California.* U.S. Army Environmental Center Technical Report, SFIM-AEC-ET-CR-97039, U.S. Army Environmental Center, Aberdeen Proving Grounds, MD

OHM Remediation Services Corp. 1998. *Technical Memorandum, IRP Site 24, Naval Air Weapons Station, Point Mugu, California.* U.S. Navy Technical Report, Document Control No. SW4628, Naval Facilities Engineering Command, San Diego, CA.

PRC Environmental Management, Inc. 1997. *Technical Memorandum, Ambient Metals Concentration in soils Naval Air Weapons Station (NAWS) Point Mugu.* U.S. Navy Technical Report, Naval Facilities Engineering Command, San Bruno, CA.

ELECTROKINETIC REMEDIATION OF A CONTAMINATED RIVER SEDIMENT

F. García Herruzo, J.M. Rodríguez Maroto, R.A. García Delgado and C. Gómez Lahoz (Universidad de Málaga, Málaga, Spain)
C. Vereda Alonso (Technical University of Denmark, Lyngby, Denmark)

ABSTRACT: Electrokinetic remediation was selected to clean-up river sediment samples containing metallic contamination. A number of experiments were carried out at a lab scale with the purpose of testing the performance of this technique in a 2 dimensional arrangement and establishing the base for future studies. The sediments were obtained from Palmones River and were characterized before use in the experiments. Six metals (Cd, Cu, Fe, Ni, Pb and Zn) were analyzed in the sediments before and after the treatment. The evolution of the electrical potential difference between the electrodes was monitored along the duration of each experiment. The electrical potential drop increased with time, reaching values from 5 to 10 times the initial values. Sometimes, the formation of precipitates in the sediment areas close to the cathode and anode could be observed a few hours after the beginning of an assay. After a few days of operation, pH values decreased below 2 in the anode region and increased above 12 in the cathode region. A near symmetrical distribution of the metals concentration relative to the imaginary line connecting the electrodes was found. Iron presented a different distribution pattern than metals like Zn, Cu and Ni .

INTRODUCTION

Sediments have been used to detect sources of contamination and to evaluate the effects of different land-use practices on the fluvial environment. (Birch et al., 2000)

Anthropogenic activities related with the industry have significantly enriched the bed sediment of the rivers, estuaries and bays in heavy metals that may represent a threat to biota.

The electrokinetic remediation is a promising technology that can be used to remove metals from saturated or unsaturated porous media with low permeability (in unsaturated soils, control of the medium's water content is essential since this technique requires the presence of liquid in the soil pores).

The main processes by which contaminant transport takes place in sediment under an applied electric field are: electromigration, (the ions present in the liquid, including H^+y OH^- ions from water electrolysis, move towards the electrodes at a speed proportional to the electrical field intensity times the charge of the ion and its mobility), electroosmosis (which refers to the drag of liquid in the sediment pores caused by capilarity due to the migration towards the cathode of the cations present) and electrophoresis (movement of electrically charged particles towards the electrodes) (García-Herruzo et al., 2000; Maini et al., 2000).

Figure 1 summarizes the mechanisms of contaminant transport in sediments treated by electroremediation.

Therefore, the intrinsic nature of the sediment, together with its characteristics of low permeability and saturation may feasible its treatment by electroremediation after being dredged and land-disposed. If no treatment were applied this sediment would pose, as a major environmental concern, the contamination of surface and ground water contamination by trace metals leached from it.

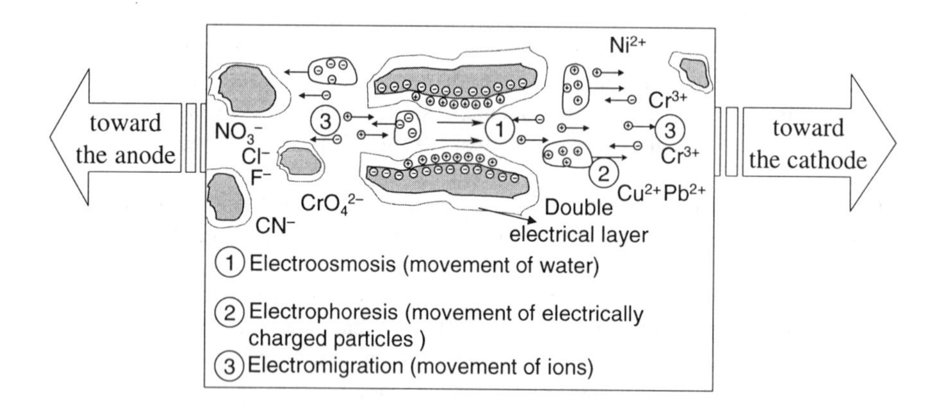

FIGURE 1. Mecahnisms of contaminant transport in soil and sediment during electroremediation.

MATERIALS AND METHODS

Soil Sampling and Analysis. The contaminated sediments were obtained from Palmones River close to Bay of Algeciras (Southern Spain). The place is an industrial zone with petrochemical, energy and steel activities and some of them use the river for effluent discharge to the Mediterranean sea. The first 15 cm of sediment bed were sampled. The sediment was dried in oven at 65 °C for over 24 hours. Then it was crushed and sieved rejecting the fraction with size over 2 mm. The sediments were characterized before being used in the experiments. Several properties were measured including density, porosity, field capacity, hydraulic conductivity, organic matter content, etc. Texture was determined by the bouyoucos densimeter method. Samples of this sediment were analyzed in a X-ray difractometer and a differential thermal analysis was conducted too.

Six metals (Cd, Cu, Fe, Ni, Pb and Zn) were analyzed in the sediments. The determination of the total content and the leacheable aliquot by aqua regia dissolution was carried out by atomic absorption spectrometry (AAS) after a microwave digestion of the sediment samples.

Electrokinetic experiments. The experiments were performed in a small plastic container of inner dimensions 12 x 12 cm and 1 cm of depth using platinized electrodes. These electrodes were arranged in the line that divides the compartment into two symmetrical parts, placed perpendicularly to the soil surface and separated by a distance of 10 cm. The experiments were performed in 180 g of sediment mixed with distillated water up to saturated conditions, yielding a water content around 55% (v/v).

The assays were conducted at a constant current intensity (12 mA and 24 mA) that was provided by a DC power supply Iso-Tech model IPS601A. These current intensities correspond to current densities of 1 and 2 mA/cm^2 referred to the transversal surface of the lab cell. The evolution of the electrical potential difference between the electrodes was monitored along the duration of each experiment.

At the end of each experiment the soil surface was divided into a regular 2 dimensional Cartesian grid of 5 rows and 6 columns. The weight, pH values, water content and concentrations of the metals of interest were measured in each one of these 30 cells.

RESULTS AND DISCUSSION

Sediment characterization. Table 1 summarizes the results obtained from the soil characterization process. The Bouyoucos densimeter measurements suggest that the sediment texture is corresponded to a sandy loam type with a low content in clay. The X-ray difractometer analysis indicates silica as the main constituent of the sediment, so the cationic exchange capacity of the sediment is probably due to the clay fraction.

The differential thermal analysis conducted on 0.0673 g of air dried sediment indicates a water content of 2% (at 100°C). The mass loss around 2.5% up to 500 °C is attributed to the organic matter, which is in agreement with the organic matter analysis determined with H_2O_2 where the mass loss was around 2.9 %. The mass loss between 500 and 1000 °C was around 8 % corresponding presumably to carbonates.

TABLE 1. Sediment characterization.

Bouyoucos analysis	Sand (%)	62.5
	Lime (%)	27.5
	Clay (%)	10.0
	Total	100
Sediment density (g/cm^3)		2.54
Porosity (v/v)		0.56
Water holding capacity (v/v)		0.46
Hydraulic conductivity (m/s)		$2.1\ 10^{-6}$
Organic matter content (%)		2.9
Cationic exchange capacity (meq/100g)		14

Table 2 shows the initial concentration expressed on dry base in the Palmones river sediment of the six metals that were selected for this study. Although iron is not a hazardous metal, his relatively high concentration value makes advisable his study in this work, since it can affect the efficiency of the electroremediation technology on the other metals in lower concentrations.

TABLE 2. Averaged initial concentration of the metals of interest in the sediment.

Metal	Concentration (mg/Kg)
Fe	16,300 ± 600
Zn	97 ± 2
Ni	58.5 ± 1.7
Cu	35.6 ± 1.7
Pb	20.2 ± 2.0
Cd	2.0 ± 0.2

Electrokinetic experiments. Figure 2 shows the evolution of the electrical potential difference between anode and cathode along two lab assays of electroremediation conducted on the Palmones river sediment at different constant electrical intensity values. The net amount of charge that has passed through the soil up to each moment is represented in the x-axis of this graphic. In this way, results from experiments of different duration can be compared, as occur in this case, where the experiment conducted at 12 mA lasted almost twice than the one at 24 mA.

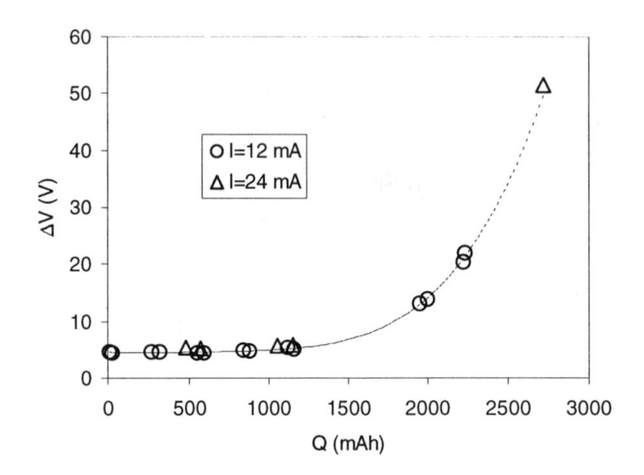

FIGURE 2. Evolution of the electrical potential difference along two lab assays of electroremediation.

The graph shows that the variation of the electrical potential drop in both experiments is very similar, almost all experimental point from both assays are located in the same tendency line. This indicates that the changes of the electrical resistance of the system occur in a similar way in both assays and mainly depend on the net amount of charge that has passed through the system.

A graphic showing the pH distribution in the sediment at the end of one electrokinetic assay is presented in the figure 3. The initial pH of the sediment at the beginning of the electrokinetic experiment and under saturated conditions is 7.4. When the experiment is finished, it can be seen that the pH values decrease below 2 in the region of the anode and increase above 12 at the zones close to the cathode. This is due to the electrolysis reactions of water at the electrodes, which generate protons at the anode and hydroxyl ions at the cathode acording to the equations:.

At the anode: $$2\,H_2O \rightarrow 4\,H^+ + O_2\uparrow + 4e^- \quad (1)$$

At the cathode: $$4\,H_2O + 4e^- \rightarrow 4\,OH^- + H_2\uparrow \quad (2)$$

An acid front moving from the anode towards the cathode meets with the basic front that is moving in the opposite direction originating zones of low conductivity. This causes the increase of the electrical potential drop between the electrodes showed in figure 2.

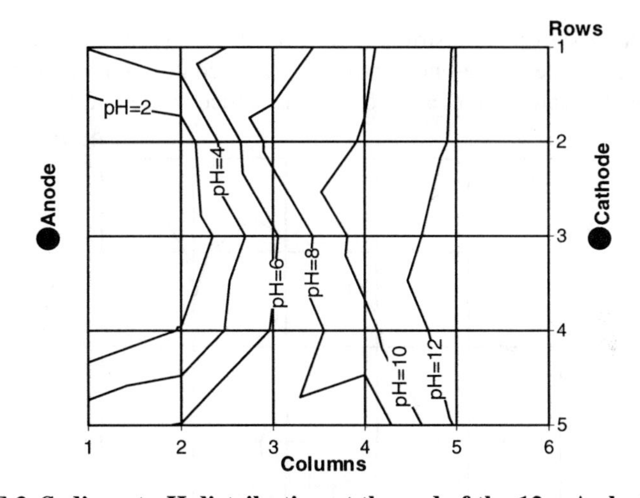

FIGURE 3. Sediment pH distribution at the end of the 12 mA electrokinetic assay

As already commented in the introduction section, the electroosmosis is one of the transport processes that can take place during an electrokinetic remediation. Figure 4 clearly demonstrates that this phenomenon occurs in the assays presented in this work. This figure shows the water content (w/w) distribution in the sediment at the end of one assay, in this case, the initial water content was around 28 %. It can be seen that the water moved from the anode towards the cathode where the water content increased above 30 % whereas decreased up to 24 % in the anode region.

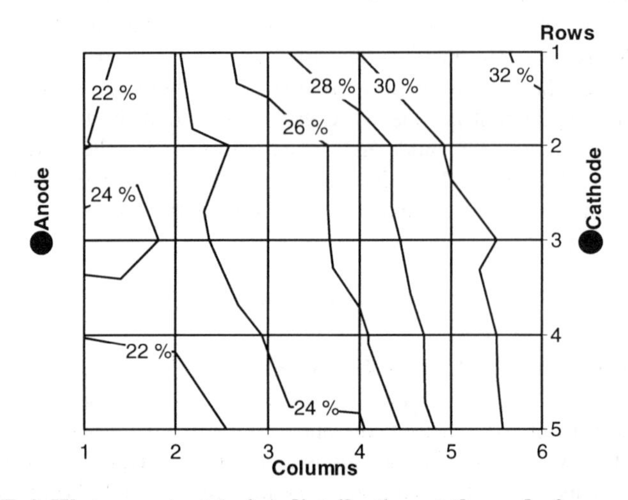

FIGURE 4. Water content (w/w) distribution at the end of an experiment conducted at 24 mA

Figure 5 shows the distribution of iron at the end of one assay carried out at 24 mA. The values represented in this graph as contour lines correspond to the iron relative concentration referred to its initial concentration in the sediment.

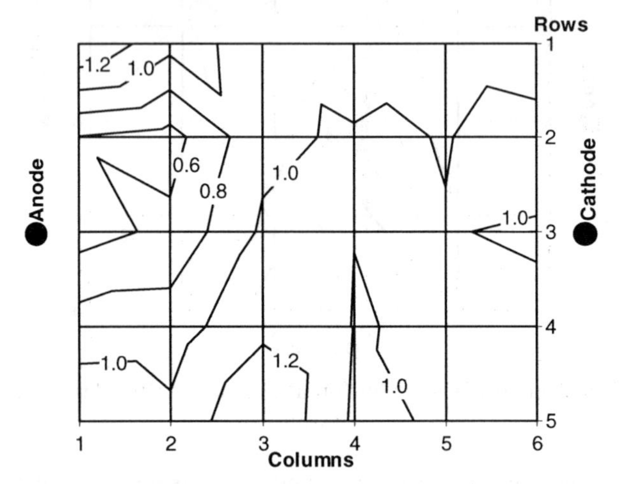

FIGURE 5. Iron relative concentration distribution at the end of an experiment conducted at 24 mA

It can be seen that the iron concentration decreases in the region of the anode. The low pH in this zone probably allows the electromigration of iron ions in solution from the anode toward the cathode, nevertheless when these ions reach the basic zones the precipitation of the hydroxide doubtless occurs. In fact, the formation of a brown front surrounding the anode was clearly noticeable during the assays. However, the concentration values in the basic zones of the sediment remain practically unchanged or slightly higher than the initial iron concentration.

Nevertheless, the cases of zinc, copper and lead are different from the iron because these metals are capable of forming negatively charged soluble complexes with the hydroxyl ion in basic solutions, $Me(OH)^{-3}$. The instability constants of these complexes are 1.3 10^{-14}, 2.6 10^{-15} and 10^{-17} for zinc, lead and copper. For example, assuming chemical equilibrium in a solution where the pH is 10 the concentration of the zinc complex is 80 times higher than the concentration of Zn^{2+}. So, in spite of the precipitation of zinc hydroxide takes place in basic solution, the redissolution of the hydroxide is possible if the pH is very high. All this could explain the accumulation of zinc in the middle regions of the lab cell that can be seen in the figure 6, where zinc relative concentration distribution at the end of one electrokinetic assay is presented.

As it can be seen in this graph, the zinc, probably as Zn^{2+}, leaves the region of the anode, with pH below 6, toward the cathode. At pH 7 the solubility product of zinc is overcome for the initial concentration, so in this zone zinc hydroxide precipitates. Nevertheless, in the region of the cathode where the pH is high (above 10) the zinc is mobilized as the negatively charged complex that migrates toward the anode and precipitates in the middle zone as the hydroxide. Probably this is the reason of the accumulation of zinc around the third column.

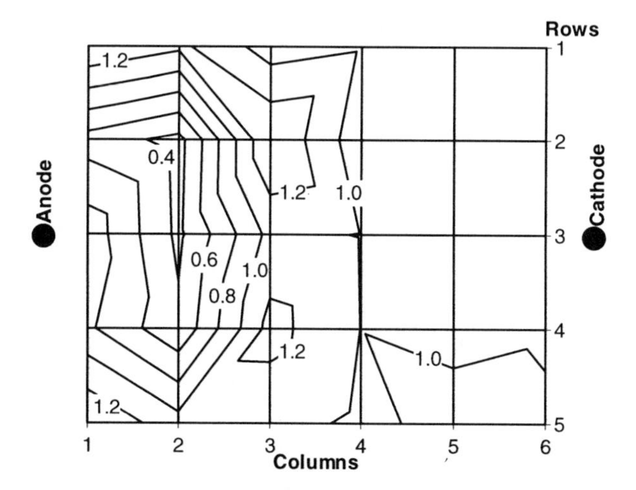

Figure 6. Zinc relative concentration distribution at the end of one experiment conducted at 24 mA

ACNOWLEDGMENTS

Vereda Alonso gratefully acknowledges the postdoctoral grant obtained from the Spanish Dirección General de Enseñanza Superior e Investigación

Científica and the invitation of the Dpt. of Geology and Geotechnical Engineering to stay in this department.

REFERENCES

Birch, G.F., E. Robertson, S. E. Taylor and D.M. Mcconchie. 2000. "The Use of Sediments to Detect Human Impact on the Fluvial System." *Environmental Geology 39*(9): 1015-1028.

Maini, G.; A. K. Sharman, C. J. Knowles, G. Sunderland and S.A. Jackman. 2000. "Electrokinetic Remediation of Metals and Organic from Historically Contaminated Soil." *J. of Chem. Technol. and Biotechnol. 75*(8): 657-664.

García-Herruzo, F. , J.M. Rodríguez Maroto, R.A. García-Delgado, C. Gómez Lahoz and C. Vereda Alonso. 2000. "Limpieza de suelos contaminados por electrodescontaminación. I. Fundamentos y aspectos básicos". *Ingeniería Química* XXXII(369): 215-220.

FRACTIONING OF CONTAMINATED MARINE SEDIMENTS

Lars J. Hem (Aquateam AS, Oslo, Norway)
Eilen A. Vik (Aquateam AS, Oslo, Norway)
Hans-Jacob Svensen (Alfa Laval As, Oslo, Norway)
Tore Lundh, NET (Sandefjord, Norway)

ABSTRACT: Physical separation of marine sediments into fractions has been studied in lab-scale and pilot-scale. The results from the laboratory study showed that the distribution of the pollutant in the various particle size fractions of the sediments was not a function of the particles surface area. The total mass of the various fractions, the organic content, and the origin of the particles before accumulation in the sediments, seem to be of importance to the distribution of the pollutants. The content of total PAH and high chlorinated PCB-congeners was highest in the fraction >60 µm, while the content of other contaminants analysed was evenly distributed in the particle size fractions.

The sediments can be fractioned by centrifugation. The distribution of dry solids between concentrate and centrate can be adjusted by changes in the operational parameters, in particular the G value.

INTRODUCTION

Whenever contaminated sediments shall be treated to reduce the environmental impact of the pollutants, the costs will be highly dependent on the amount of sediments concerned. Any action that can reduce this amount will therefore be economic beneficial. Such actions could be investigations for identifying more or less contaminated parts of the sediments, or physical or chemical treatment that can separate the more contaminated part of the sediments from the less contaminated. This requires, however, thorough knowledge about the sediments and the site-specific contamination. An example of the physical treatment is separation of the coarser particles from the smaller ones, if the pollution is mainly associated with one particle size fraction. Sand can be removed by wet sieving, while other separation technologies like settling or hydrocyclones can be used for separation of clay from silt (Francingues, 2000). This has been used in Hamburg, where they used hydrocyclones to separate less contaminated sand from more contaminated silt (Detzner, 1995).

Norwegian sediments are typically contaminated by heavy metals, PAH and/or PCB. The sediments are typically mixtures of clay, silt and sand, with variable content of organic carbon. These particles can only be separated if technologies like sedimentation, centrifugation or hydrocyclones are used.

This paper includes results from a study looking into the possibility of separating highly contaminated sediments from the harbour of Sandefjord from less contaminated sediments. The idea behind the project was that a high concentration of pollutants in one size fraction would make it possible to separate the sediments into more and less contaminated parts, and hopefully reduce the amount of sediments that need further treatment. Most of the contaminated marine sediments consist of clay, silt and fine sand, and the separation of the sediments into fractions with different particle size cannot be done with screens. Separation of such sediments must be based on techniques like sedimentation, centrifugation or hydrocyclones. The project had two separate parts; a laboratory study of the distribution of selected contaminants in fractions with different particle size and organic content, and a pilot study where a centrifuge was used for separation of the sediments.

MATERIALS AND METHODS

The sediments chosen for the laboratory tests and for the pilot study was from Kilen in the Sandefjord harbour. In this harbour area there are restrictions on intake of seafood due to PCB content in the livers of the fish. Several samples were analysed for particle size distribution, and one of the more average was chosen for lab-scale fractioning. The sample for the pilot scale tests contained more fines than the one used for lab-scale tests.

Initially, in the lab-scale tests, the particle size fractioning was performed with either a batch centrifuge or by successive filtration through membrane filters. When using a batch centrifuge a considerable part of the fines was found among the coarser particles, possibly due to the batch-operating mode. The longer the centrifugation time, the more fines was found together with the coarser particles. Due to this lab-scale effect, filtration was used for the fractioning. The sediments were separated into the following fractions for further analysis; <5 μm, 5-11 μm, 11-30 μm, 30-60 μm, 60- 120 μm and $>$ 120 μm. Fractioning based on particle density was performed with sedimentation in a diluted sodium tungstate solution with a density of 2 g/cm^3, using a method developed by US Army Corps of Engineers (Olin-Estes, 2000).

The pilot study was performed with a pilot scale centrifuge from Alfa Laval with a load of 400-800 l/h. Dredged sediments from the harbour with dry solids (DS) content of approximately 35 % were diluted 1:2 prior to the test to achieve optimal conditions for the operation of the centrifuge. The solids content of the supplied liquid-sediment mixture was approximately 10 % DS. No chemical additives were used.

The various fractions of the sediments were analysed for particle size distribution, DS, volatile dry solids (VDS), polycyclic aromatic hydrocarbons (PAH_{16}), polychlorinated biphenyls (PCB_7), copper and lead. The particle size distribution was measured with a Malvern Mastersizer particle counter. Determination of easily extractable organic contaminants was performed by extraction in tetrahydrofuran.

RESULTS

Lab-scale tests. In the various bulk samples collected from the Sandefjord harbour, the volume of the particles with a diameter less than 1 μm was 0.2-20 % of the total volume of particles, and the volume of the particles smaller than 10 μm was 25-70 % of the total volume. The volume of the particles larger than 100 μm was 0-20 % of the total volume.

For the sediment tested, the major part of the particles was smaller than 120 μm, and approximately 2/3 of the DS was related to particles passing a 60 μm filter. The amount of particles passing an 11μm membrane filter, and the contaminants related to these particles, was marginal. As a result this discussion focuses on the fraction passing a 30 μm filter, the fraction passing a 60 μm filter but not a 30 μm filter, and the fraction captured on a 60 μm filter.

Figure 1 shows the particle size distribution for the test sediment from the Sandefjord harbour, and the particle size distribution is shown for fractions from successive membrane filtration. Of the particles not passing the 60 μm filter, 35 % were smaller than 60 μm, showing an acceptable separation with this filter. When separating with a 30 μm filter, the amount of particles smaller than 30 μm not passing the filter was 70 %, showing a less distinct separation.

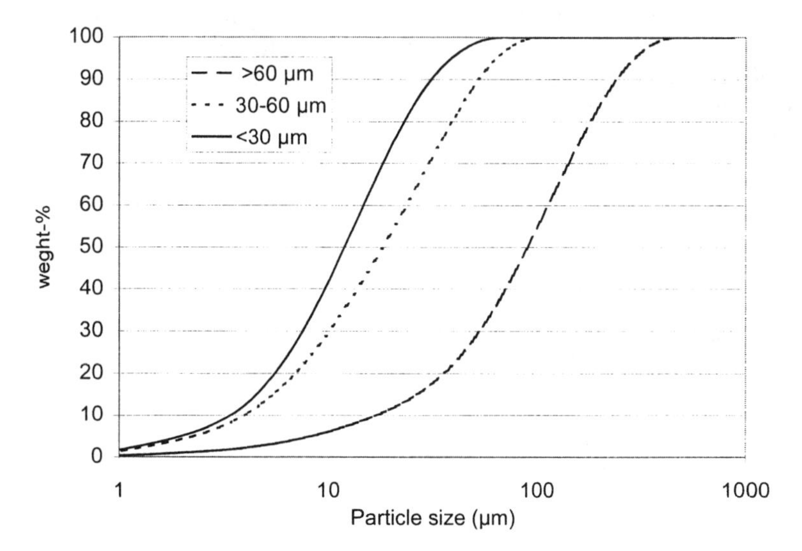

FIGURE 1. Particle size distribution for the test sediment and for fractions from successive membrane filtration

The DS content in the test sediment was approximately 35 wt %. 10-12 wt % of the DS was volatile. The concentrations of copper and lead were 200-330 mg/kg DS and 220-340 mg/kg DS, respectively.

Table 1 shows the PAH content in the sediments. The PAH_{16} was dominated by soot related PAHs, while only 1,3 % of the PAH_{16} was naphthalene. The PAH content was mainly associated to the particles not passing a 60 μm filter, while only approximately 30 % of PAH_{16} was associated to the smaller particles.

TABLE 1. PAHs in fractions obtained from successive membrane filtration

Contaminant	Concentration (mg/kg DS)		
	Passing 30 µm	Passing 60 µm, but not 30 µm	Not passing 60 µm
PAH $_{16}$	2.0	3.7	14.4

Figure 2 shows the concentrations of PCB congeners in fractions obtained from successive membrane filtration. The concentrations in the two fractions with the smallest particles were equal, while the concentrations of high chlorinated PCBs was higher for the particles not passing a 60 µm filter. The difference between the concentrations associated with the largest and the smaller particles is similar to the PCB profile for Aroclor 1262. In 1998, the PCB content was analysed in purple snails sampled in Kilen in the Sandefjord Harbour (Kelley, 1998). Approximately 70 % of the PCB_7 in the snails were the high-chlorinated congeners 138, 153 and 180, which are the same congeners that were found associated to the largest particles. This may indicate that the largest particles, between 60 and 120 µm, play an important role for the uptake of PCB in snails, but how and why this occurs, and even whether it may be a coincidence, is yet to be answered.

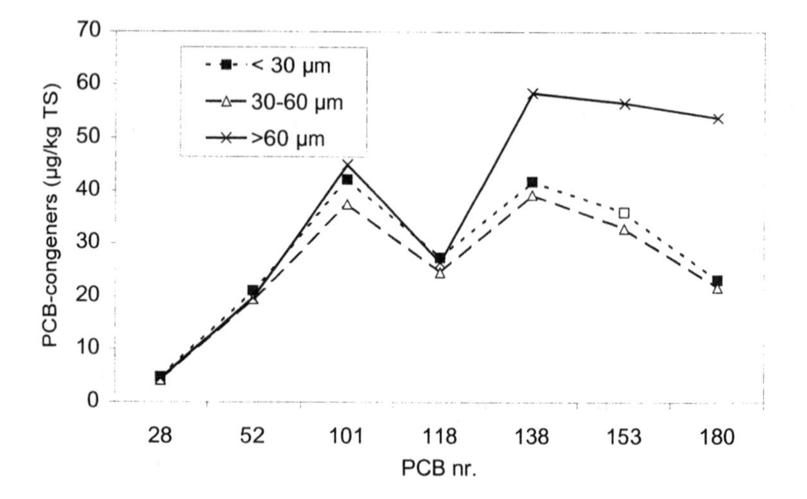

FIGURE 2. Concentration of PCB congeners in fractions obtained from successive membrane filtration

The distribution of the contaminants and DS between the fractions obtained from successive membrane filtration is shown in Table 2. The DS was almost equally distributed between the three fractions. The volatile part of the DS was slightly higher for the largest particles than for the smaller ones. The copper, lead and PCB_7 were distributed similar to the DS, while the concentrations of high-chlorinated PCBs were considerably higher for the largest particles than for the smaller ones. 70 % of the benzo(a)pyrene is associated to the particles not passing a 60 μm filter, but this part of the benzo(a)pyrene is not easily extractable.

TABLE 2. Distribution of DS, VDS and contaminants in fractions obtained from successive membrane filtration

Compound	Passing 30 μm (%)	Passing 60 μm, but not 30 μm (%)	Not passing 60 μm (%)
DS	29	31	40
VDS	28	25	48
Lead	32	27	42
Copper	30	26	44
PAH_{16}	10	18	72
Benzo(a)pyrene	12	17	71
Easily extractable benzo(a)pyrene	29	58	12
PCB_7	33	23	44
PCB no 180	19	19	62

Table 3 shows the result from separation with sedimentation in a sodium tungstate solution. Approximately 30 % of benzo(a)pyrene were associated to the heavier, assumed inorganic particles, while the remaining was associated with lighter, assumed organic or coal, particles. A damaging fire in the town approximately 100 years ago may be the origin for a great part of the soot related PAH, and this may explain the association with organic particles. The DS, however, was mainly (82 %) associated with the heavier particles.

TABLE 3. Benzo(a)pyrene and DS associated with particles with density above or below 2 g/cm^3

Particle density	Benzo(a)pyrene	DS
% of total sample associated with particles with density < 2 g/cm^3	70	18
% of total sample associated with particles with density > 2 g/cm^3	30	82

The distribution of contaminants was not associated with particle surface. While copper and lead were associated to the particles mass, the PAH and PCB distributed cannot be explained with association with neither mass nor particle surface. The PAH was mainly related to larger particles, possibly organic matter or coal particles. The distribution of PCB can be explained if assuming that at least the high chlorinated PCB is associated to particles when transported to the sediments from onshore sources, and still remains associated to the same particles. The distribution of the contaminants in the sediments of the Sandefjord harbour then shows the importance of knowing and analysing the origin of the contaminants, and previous and present transportation mechanisms.

Physical separation of the sediments in the Sandefjord harbour, with further treatment of only one of the fractions, will give a substantial reduction in the treatment costs. Surprisingly, in the Sandefjord harbour there will only be an environmental benefit from such an action if the larger particles are sent for further treatment. The benefit would be due to a situation where high-chlorinated PCB, and possibly some PAHs, are the target contaminants. The benefit would only be valid if the contaminants released when discharging the lighter particles to the sea do not increase in bioavailability, a potential problem that is presently being studied.

Pilot scale tests. Three parameters were varied in order to manage to vary the amount of particles recover in the concentrate/sludge, and the size of these particles. The differential speed of the centrifuge when below 35 rpm did not influence the recovery. The hydraulic load had a minor impact on the recovery for the highest loads tested (800 l/h). The variation of the G value had a significant influence on the recovery.

Figure 3 shows the amount of particles recovered in the concentrate/sludge from pilot scale separation with a centrifuge for different particle size fractions and with two different G values. The particle content in the inlet, measured as DS minus salinity, was approximately 10 weight %, and 40-50 weight % in the concentrate/sludge phase after pilot scale centrifugation. All the particles larger than 60 μm were recovered in the sludge, while the recovery of the smaller particles varied due to the G value.

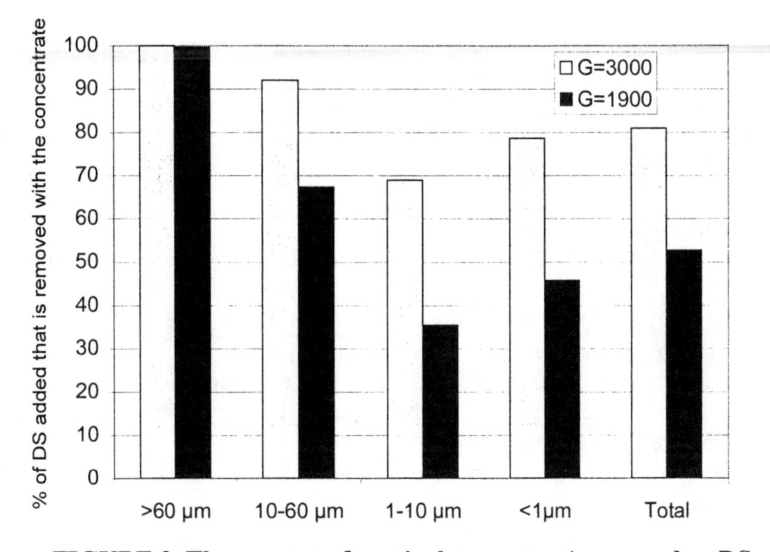

FIGURE 3. The amount of particulate matter (measured as DS minus salinity) recovered in the concentrate/sludge from pilot scale centrifugation. Average data for hydraulic loads between 400 and 800 l/h, and differential speeds between 18 and 35 rpm.

The pilot study results show that it is possible to use a centrifuge to separate fine sediments into fractions with different particle sizes. The results also verify that varying the operational parameters for the centrifuge can change the amount and size of the particles in each fraction. This is important for the possibility of using such an approach in remediation of sediments.

The content of organic contaminants is shown in Table 4. 53 % of the particulate DS was recovered in the concentrate. The distribution of PCB_7 between concentrate and centrate was close to the distribution of particulate DS. However, 55-60 % of the high-chlorinated PCBs and 30-40 % of the lower chlorinated PCBs were recovered in the concentrate, confirming that the different congeners were not associated to particles of different size in quite similar ways. 75 % of the PAH_{16} was recovered in the concentrate, confirming that PAH to a high degree was associated to the larger particles.

TABLE 4. Organic contaminants in fractions obtained from pilot scale separation with a centrifuge. G value=1900. Load=600 l/h.

Contaminant	Concentration (mg/kg DS)		Distribution (%)	
	Concentrate	Centrate	Concentrate	Centrate
PAH_{16}	29	11	75	25
PCB_7	0.27	0.34	48	52

The results from the pilot study show that for this specific sediment, and with PCB as the target contaminant, fractioning of the sediments does not give any particular benefit. This is a contradiction to the lab-scale results on sediments from another site in Kilen in the Sandefjord harbour. The bioavailability of PCB and PAH in inlet, concentrate and centrate, is presently being studied.

CONCLUSIONS

- The contaminants in the sediment are neither associated to particle surface nor mass only
- The PAH and PCB are mainly associated to the larger particles
- It is possible to use a centrifuge to separate fine sediments into fractions with different particle sizes, and the amount and size of the particles in each fraction can be changed by varying the operational parameters for the centrifuge.
- There will be limited benefit in fractioning the sediment used in the pilot study

ACKNOWLEDGEMENTS

The project was financed by the Nordic Industrial Fund, NET AS, Alfa Laval AB and Aquateam AS.

REFERENCES

Detzner, H. D. 1995. "The Hamburg project METHA; large scale separation, dewatering and reuse of polluted sediments." *European Water Pollution Control,5*(5), 38-42.

Francingues, N. 2000. Physical separation. Workshop on remediation of marine sediments, June 8-9, Sandefjord, Norway.

Kelley, A. 1998. *PCB analyses in purple snails.* Analytical report 255/98, KM Lab AS, Oslo, Norway.

Olin-Estes, T. 2000. Personal communication. US Army corps of engineers.

BENTONITE EFFECTIVENESS ON THE CESIUM RETENTION IN CEMENTED WASTE PRODUCTS

Clédola Cássia Oliveira de Tello (CNEN/CDTN, Belo Horizonte, MG, Brazil)
Elias BasileTambourghi (UNICAMP - FEQ – DESQ, Campinas, SP, Brazil)

ABSTRACT: The cementation is a process to immobilise wastes, reducing their potential for migration or dispersion of contaminants. To assure that this treatment is satisfactory, some tests should be done, mainly in the waste form. Leaching test indicates the capacity of the waste form to retain the contaminants. Since 1984 a research has been conducted at Centro de Desenvolvimento da Tecnologia Nuclear (CDTN/CNEN, Brazil) to evaluate the performance of the bentonites in fixing cesium in waste forms. Mixtures are prepared with different bentonite proportions, and submitted to some tests. The results of 10-years leaching test and the modelling of them are presented. The data demonstrated that the bentonite is effective to retain this radionuclide, and the modelling suggests a partition mechanism for the cesium release.

INTRODUCTION

All human activities produce wastes. Some of them are hazardous due to the concentration of toxic elements, so that they should not be released without previous treatment. Therefore the main objective on the management of these wastes is to dispose them in such way that the risks remain as low as reasonably possible. Different waste management strategies can be adopted concerning to technological processes, storage and disposal options.

The immobilisation process consists on the conversion of wastes to solid forms in order to reduce the potential for migration or dispersion of contaminants from the wastes by natural means. The cementation is one of the useful techniques to immobilise radioactive and hazardous wastes using cement as matrix. The cemented waste product should have some properties to assure its handling, storage and disposal. Low leachability is generally considered one of the most important properties in the evaluation of this product since it represents the first barrier to the release of radionuclides to the environment over long periods. However some contaminants are very soluble and there isn't any chemical or physical interaction with the cement matrix and therefore it is necessary to add some materials to increase their retention in the final product.

Centro de Desenvolvimento da Tecnologia Nuclear (CDTN) is one research centre on nuclear technology of Brazilian National Commission of Nuclear Energy (CNEN/Brazil). Since 1979 a research has been carried out at CDTN to study the effectiveness of Brazilian natural materials, including bentonites, on the contaminant retention without jeopardising the process and other characteristics of the waste product. This paper describes the research performed to evaluate the effectiveness of bentonite in reducing the cesium leachability from cemented waste forms, and the modelling of the data from these experiments.

CEMENTATION

Cementation is one process to treat radioactive waste and it consists of mixing the waste with cement and other materials, if it is necessary, to modify the waste properties so that the cemented waste product can be handled, transported, and stored safely (International Atomic Energy Agency, 1983). Portland cement is the most common matrix used for solidification of wastes because it is easily available, inexpensive, and it is processed at low temperatures. The cemented waste forms are structurally stable, radiation resistant, and leach resistant if correctly formulated.

The waste form is a free-standing monolith solid resulting from the waste cementation process. In the multiple barrier concept it is considered the first barrier, and it has the function of retaining the radioactive components, allowing that significant radioactive decay occurs before they are released to the environment. Some properties are very important to guarantee that the waste form can resist the aggressiveness of the weather and the environment.

The waste forms should have high mechanical strength, chemical and leaching resistance. These properties can be measured through several tests, e.g., compressive strength and leaching tests (International Atomic Energy Agency, 1983). The main objective of the research in this domain is to improve the cementation process efficiency and the waste form quality. It means to increase the waste/product ratio while maintaining the quality of the waste form.

The use of additives in the cementation process is an option to improve the waste form properties. These additives can play different roles, for instance, the clays are helpful in the radionuclide retention (Tello, 2001). Bentonite is a mineral that is composed essentially by montmorillonite, and characterised by its high absorptive power and active colloidal properties. It swells, increasing its volume by as much as a factor of twelve, when wetted with water and has strong adsorption properties.

EXPERIMENTAL

This research was initiated in 1984, and its main objective was to verify the effectiveness of four types of Brazilian bentonites regarding to the retention of the cesium in the waste form.

Materials. Ordinary Portland Cement, CP 320, was used as matrix. Four Brazilian bentonites, named G, F, N and S were selected, and their amount in the mixture varied from 0 to 150 g/kg (0 to 15 % by weight). Stable cesium was used as tracer. To simulate the waste it was prepared a solution with nitrates (Fe, Ni, Cr, Ca, Mn, Mg, and Al), sodium salts, TBP, kerosene, and detergent. (Tello, 1989; Rudolph et al., 1982). The cement and bentonite were mixed with the solution and they were homogenised for 5 min.

Tests. The viscosity was measured in a Brookfield viscometer. The measurement was made in the just prepared paste (Tello, 1988). Setting time was measured with a Vicat needle automatic apparatus in the fresh paste (Associação Brasileira de Normas Técnicas, 1991). Compressive strength resistance was measured in cylindrical samples (5.0 cm-diameter and 7.0 cm- height) after 28 days under dry curing at room temperature using an hydraulic strength machine (Associação Brasileira de Normas Técnicas, 1991a).

Evaluation of the Bentonites Absorption Capacity. Pastes with 70 g/kg (7% bentonite), simulated wastes and cement were prepared, and after curing for 28 days they were crushed. The particles smaller than 1 mm were put in contact with cesium solutions, in different concentrations, doped with Cs-137. Aliquots from the solution were taken and the Cs-137 activity was measured and returned to the flasks until the Cs concentration stayed constant within the range of experimental error. The same procedure was used for pure bentonite (Tello, 1989).

Leaching. The leaching experiments were performed based in the Long-term Leaching Test ISO 6961 (International Standards Organisation, 1979) using cylindrical specimens (5.0 cm diameter and 7.0 cm height) with 28 days curing time. The whole surface of the specimens were put in contact with the leachant (deionized water at room temperature). The leachant was completely replaced, daily in the first five days, twice in the second week, weekly in the following four weeks, monthly in the following months till one year experiment time. After the first year, the leachant was replaced mostly yearly. The cesium concentration was determined through atomic absorption analysis.

Modelling. The Accelerated Leach Test Computer Program (ALT) (Fuhrman et al, 1990; ASTM, 1996) was developed to accompany the leach test, and to compare experimental data to theoretical curves generated by four models. This program attempts to link a physical test method with a set of mechanistic models that can be used to interpret the test results. The leaching mechanisms described by these four models are: (a) Diffusion through a semi-infinite medium (for low CFL); (b) Diffusion through a finite cylinder (for high CFL); (c) Diffusion plus partitioning of the source term, and (d) Solubility limited leaching.

RESULTS AND DISCUSSION

The viscosity of all mixtures permitted good workability, and initial and final setting times above 1.0 hour and below 24 hours, respectively. The compressive strength of the specimens varied from 10.9 till 25.2 MPa.

The results of the absorption tests are presented in the Figure 1. From this figure it can be observed that pure bentonite could absorb almost all the cesium added to the solution. At the other hand when they were mixed to the cement this capability was reduced. One reason can be the competition for sorption sites among some cations existing in the cement and the cesium in solution.

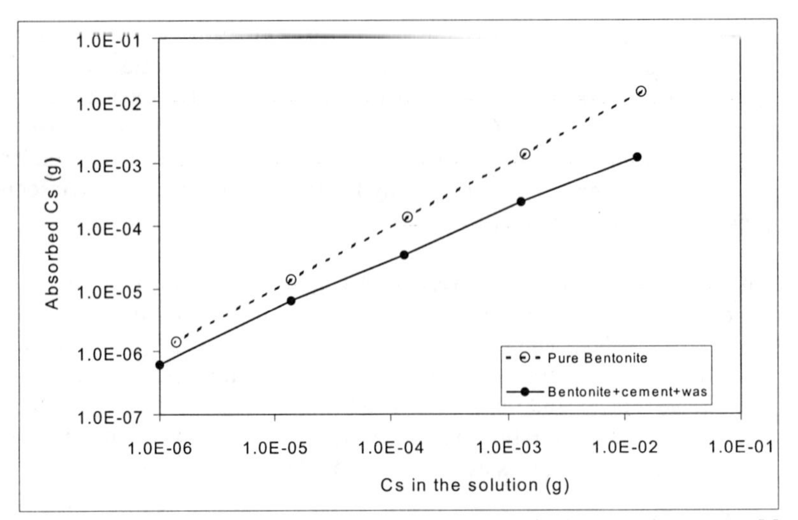

FIGURE 1. Mass of cesium absorbed by bentonite/cement/waste and by pure bentonite from the solution

To study the results of the leaching tests two parameters were calculated: the Incremental Fraction Leached (IFL), and the Cumulative Fraction Leached (CFL) (Fuhrman et al, 1990; ASTM, 1996).

The *IFL* is calculated by the following equation:

$$IFL = \frac{{}_i a_n}{{}_i A_0} \qquad (1)$$

Where:

${}_i a_n$ = the quantity of species *i* observed in the leachate at any given time interval;

${}_i A_0$ = the source term, the total original quantity contained in the leaching specimen at the beginning of the test.

The *CFL* is calculated as:

$$CFL = \frac{\Sigma_i a_n}{{}_i A_0} \qquad (2)$$

Where:

$\Sigma_i a_n$ = the sum of the quantity of species *i* leached during the experiment at any given time interval;

${}_i A_0$ = the source term, the total original quantity contained in the leaching specimen at the beginning of the test.

The Figures 2 and 3 show IFL and CFL of samples B-A, G7, G10, G12, and G15 (with 0, 7, 10, 12 and 15% Bentonite G, respectively) for 10-years leaching experiments. From the Figure 2 it is noticed that the product without bentonite loses Cs quicker than the others, mainly in the beginning of the test.

From the CFL curves is observed that the use of bentonite reduced the release of the cesium. The samples without bentonite released all the cesium, and the best results are obtained when 120 g/kg of bentonite (G-12) was used in the mixture, where the Cs release was around 40%.

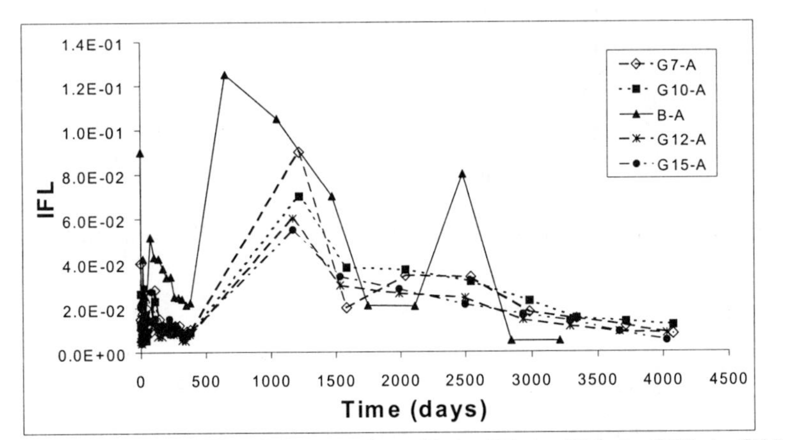

FIGURE 2. Cesium IFL of the samples: B-A; G7-A; G10-A; G12-A; G15-A

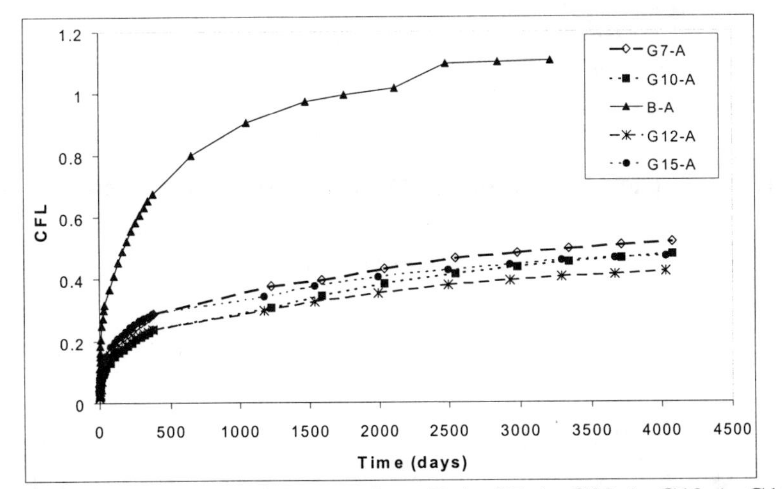

FIGURE 3. Cesium CFL of the samples: B-A; G7-A; G10-A; G12-A; G15-A

The data of the leaching experiments were processed using the ALT Program and the Effective Diffusion Coefficient (D_e) was determined in order to describe the Cs release mechanisms. The Figures 4, 5 and 6 exemplify the results of the program. The curve for the partition model (Figure 5) presented the best fit, in comparison with the other ones modelled for the diffusion (Figure 4) and dissolution (Figure 6) mechanisms. It means that for these experiments the Partition Model is more suitable to explain the release mechanism.

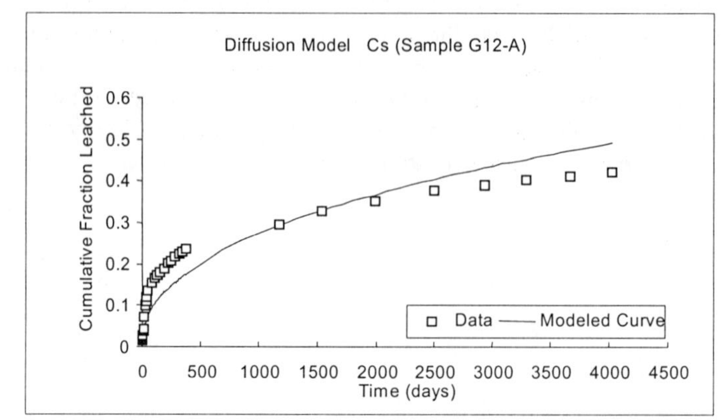

FIGURE 4. CFL vs. Time and the Modelled curve (Diffusion model; G12-A)

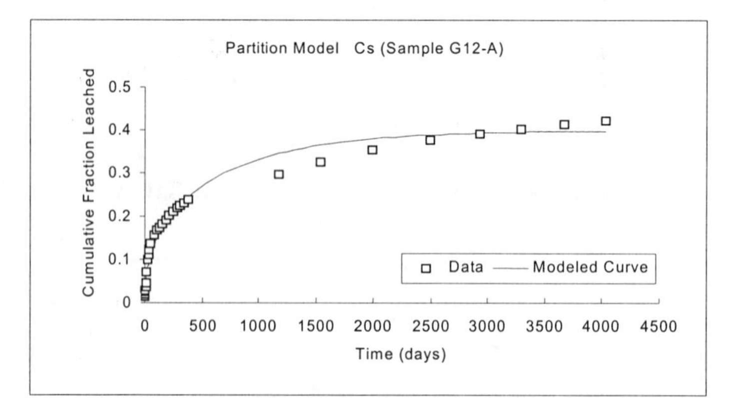

FIGURE 5. CFL vs. Time and the Modelled curve (Partition model; G12-A)

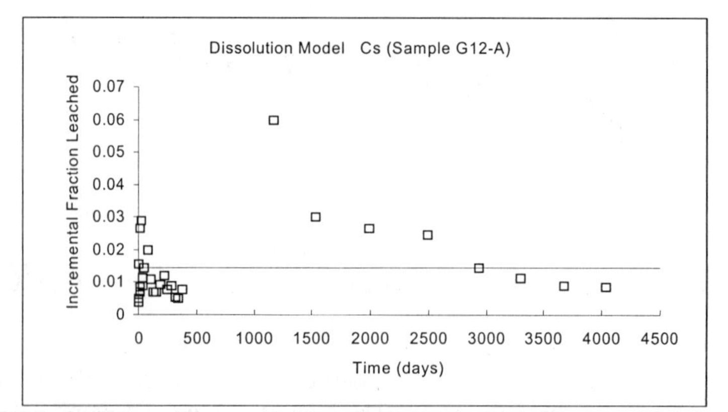

FIGURE 6. IFL vs. Time and Modelled curve (Dissolution model; G12-A)

The Figure 7 shows the relationship between D_e, calculated by the Partition Model, and the amount of bentonite in the mixture. It can be observed that the addition of bentonite in the range from 5 to 10 % presented the better results, and the bentonites G and F retained more cesium than the others

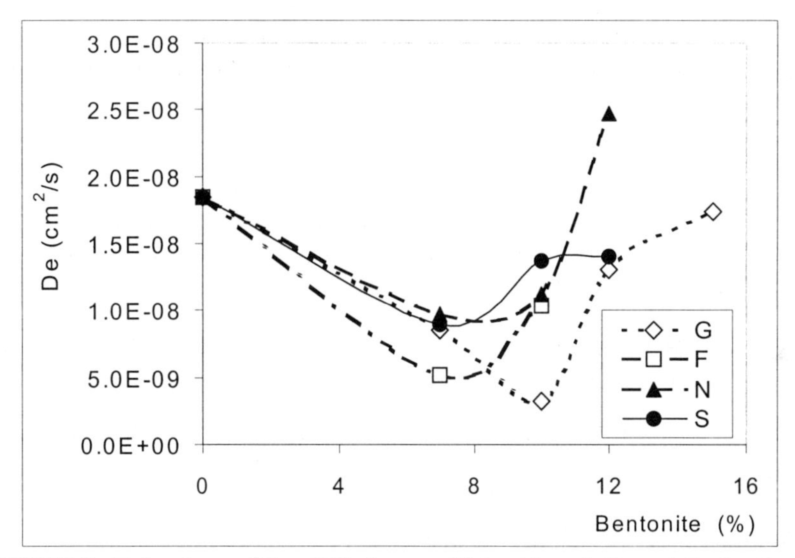

FIGURE 7. Behaviour of the diffusion coefficient in relation of the bentonite amount in the mixture

CONCLUSIONS

The nuclear energy has been used for the human development in different areas, as in the medicine, in the agriculture, in the industry and in the environmental protection, besides the electricity generation. The residues generated on the nuclear energy uses are considered radioactive wastes when the contaminant content can bring a potential negative impact in the human health and in the environment. In general the waste processing consists in a volume reduction followed by solidification and/or conditioning.

A study to verify the effectiveness of four types of bentonite is carried out at CDTN. Many mixtures were prepared with simulated waste, cement and bentonite in different proportions. The leaching resistance is the most important parameter in the product evaluation, because it indicates the capacity of the matrix to retain the radionuclides when the product is in contact with the water.

Since 1985 samples without and with bentonite have been leached. The results show that bentonite are efficient in the cesium retention. The leaching data were analysed using the ALT computer program. The model that describes the partition mechanisms presented the best fit for the experimental data. It indicates that, besides the diffusion, there is (are) other process(es) affecting the cesium leaching. One possibility is the absorption of it by the bentonites, as it was showed in the absorption tests. The D_e calculated for this model was between 10^{-8} and 10^{-10} cm^2/s. The lower D_es were presented by the products with 7 and 10 %

of bentonite G. It was observed that, independent of the bentonite type, there was a positive effect on the cesium retention in comparison of the mixture without this clay.

These results can be used to study the long-term behaviour of cemented wastes containing bentonite in the repository safety assessment. During this research some of the results were used to help in the solution of some problems, for example during the management of the Goiania's Radiological Accident with a cesium source.

REFERENCES

American Society Testing and Materials. 1996. *Standard Test Method for Accelerated Leach Test for Diffusive Releases from Solidified Waste and a Computer Program to Model Diffusive, Fractional Leaching from Cylindrical Waste Forms.* C 1308 – 95. ASTM, New York, NY.

Associação Brasileira de Normas Técnicas (ABNT). 1991. *Portland Cement: Determination of Setting Time (In Portuguese).* NBR 11582. Rio de Janeiro.

Associação Brasileira de Normas Técnicas. (ABNT). 1991a. *Portland Cement: Compressive Strength* Test *(In Portuguese).* NBR 7215. Rio de Janeiro.

Fuhrman, M., J. H. Heiser, R. Pietrzak, E. Franz and P. Colombo. 1990. *User's Guide for the Accelerated Leach Test Computer Program.* BNL 52237. Brookhaven National Laboratory, Upton, NY.

International Standards Organisation. 1979. *Long-term Leach Testing of Radioactive Waste Solidification Products.* ISO, Genève. (Draft ISO 6961).

International Atomic Energy Agency. 1983. *Conditioning of Low- and Intermediate-level Radioactive Wastes.* IAEA Technical Report Series No. 222. IAEA, Vienna.

Rudolph, G., S. Luo, P. Vejmelka, R. Köster. 1982. *Untersuchungen zum Abbindeverhalten von Zement-suspension.* KfK 3401 Kernforschungszentrum Karlsruhe, Karlsruhe.

Tello, C. C. O. 1988. *Determination of Viscosity of Cement Pastes with Bentonite and Simulated Waste.* (In Portuguese). CDTN, Belo Horizonte, MG.

Tello, C. C. O. 1989. "Evaluation of National Bentonites as Additive in the Radioactive Waste Cementation" (In Portuguese). M.Sc. Thesis, Universidade Federal de Minas Gerais, Belo Horizonte, MG.

Tello, C. C. O. 2001. "Bentonite Effectiveness on Cesium Retention in Cemented Waste Forms" (In Portuguese). Ph.D. Thesis, Universidade Estadual de Campinas, Campinas, SP.

DETOXIFICATION OF PCBs-POLLUTED SOIL BY HIGH-ENERGY MILLING

Michele Aresta (Centro Ricerche METEA, Bari, Italy)
Luca De Stefano (ENEL Produzione Ricerca, Brindisi, Italy)
Tiziano Pastore (INCA, Venezia, Italy)

ABSTRACT

In this paper we report some results from an ongoing study on the mechanochemical treatment of PCBs contaminated soil. A sample of soil from a controlled landfill, has been mixed with $NaBH_4$ powder, in the solid state. The reaction carried out by high energetic milling, changes the chemic-physical characteristics of the soil only by impact and compression between milling bodies. The efficiency of the dehalogenation reaction has been studied as a function of the milling time (3.5 up to 30 hours). During the mechanochemical process we have measured the total PCBs content and the production of chloride. Complete abatement has been obtained for a PCB load of about 2600 mg/kg.

INTRODUCTION

PCBs are a category of chemical compounds known to be among the most persistent and widely distributed pollutants in the global ecosystem. They are produced by chlorination ($1<n_{Cl}<10$) of biphenyl ($C_{12}H_{10}$). The two-ring structure allows the formation of 209 congeners (Hutzinger et al., 1974). Their well known physical and chemical stability, along with their excellent dielectric properties, led to the widespread commercial use of these compounds as insulating materials. The production of PCBs was banned from OCSE in 1973 (after 50 years), but their production really ended in 1979. From 1929 up to now about 950.000 t of PCBs have been dispersed in the environment so that PCBs are common contaminants of industrial soil and harbour sediments. Differently from air and water, soil represents a complex and heterogeneous environmental matrix, with a low velocity of contaminant diffusion. In many cases the characteristics of soil contamination is the evidence of past and present industrial activity on site. In Italy, from a normative point of view, until 1982 (D.P.R. 915), the PCB's contaminated soil were disposed to landfill as a special category of toxic-harmful waste. The recent enforcement of the Ronchi's law (art. 17 - D.Lgs. 22/97) and following regulation D.M. 471/99, introduce a limitation to landfill practice and fixes at 5 mg/kg the maximum PCB's concentration limit for decontaminated soil for reuse. Innovative treatment technologies are, then, very important. "High energy milling" (the technology used in this work) is used for obtaining fine particulate of gypsum, limestone, talcum, coal. The physical fundaments of this process are quite simple: impacts and compression between grinding bodies and particles permit to obtain fine powders with different physic-chemical properties with respect to the original matrix. The milling action carried out by impact of milling bodies and the powders can supply energy up to 20 g. Many reactions occur during the treatment: only a little part of the mechanical energy transferred from milling bodies to the solid system, changes into dissipated heat and most is used to induce breaks, stretches and compression at micro and macroscopic level. The modifications of the molecular structure of solids are plastic- reticular stretches and fractures. Beyond obvious industrial application in alimentary, plastic, chemicals, "high energy milling" is the base of process like "mechanical alloying", "mechanical-synthesis", "solid state vitrification" (Tyagin et al., 1998). Recently, this process has been applied with success to environmental problems (Hall et al.,1996; Loiselle et al., 1997; Monagheddu et al., 1999). In this work we report the results of an experimentation on mechanochemical treatment of PCBs contaminated soils, based on our previous experience (Caramuscio et al., 2000; Aresta et al., 2001) and on previous results obtained by mixing soil with sodium hydride, NaH.

MATERIALS AND METHODS

The ring mill used in the experimentation has the following characteristics:

- Trade-name: FRITSCH
- Model: Pulverisette 9
- Total mass of rings: 3637 g
- Volumetric capacity: 350 c.c.
- Rotational rings velocity: 750/1000 rpm
- Alimentation: 230 V.
- Energetic consumption: 0.6 kW

The granulometry and composition of the soil sample, coming from a controlled landfill of toxic-harmful waste, are reported in Table 1 and 2. Table 1 gives the high total PCB's concentration and other important parameters of the soil sample. The data in Table 2 show a typical sandy soil (more than 90% is classifiable as fine sand).

TABLE 1. Soil compositional characteristics

Parameter	Unit	Result
Total PCBs	mg/kg	1520
Total PCBs *after 3 h of milling*	**mg/kg**	**2600**
BTEX	mg/kg	< 0,1
Total IC (C >12)	g/kg	15,88
pH		6,84
Cr VI	mg/kg	< 1

TABLE 2. Soil granulometric characteristics

Granulated diameter (mm)	Partial detained (%)
4.75	0.5
2	1.4
0.85	2.4
0.425	3.5
0.18	71.3
0.075	18.1
< 0.075	2.8

The analytical methods used during the experimentation are:

- *Determination of chlorides:* Method IV-2 "Metodi Ufficiali di Analisi Chimica del Suolo" (Italian official methodology).
- *Extraction of PCBs*: EPA-3540.
- *Extract Clean up:* EPA- 3620 - 3630 – 3665.
- *Determination of PCBs:* EPA- 8082.

The soil sample has been homogenized and dried at 50 °C for 20 h. The amount used in each test (100 g) represents about 1/36 of the total weight of the mill's ring and occupies about 2/3 of total mill's volume.

To verify the efficiency of abatement of PCBs samples have been withdrawn at intervals of time (3.5, 11, 18, 23, 30 h) and both total PCBs and chlorides were determined. Before adding the hydride ($NaBH_4$) to the soil, the sample was milled for about 3 h. After this operation, the PCBs total content was determined in order to verify the efficiency of solvent extraction which depends on the size of the particles.

We have demonstrated that the amount of extracted PCBs strongly depends on the degree of milling and three hours of preliminary milling can assure a correct extraction, due to the great specific surface of the ground soil exposed to the solvent.

Finally $NaBH_4$ as pure powder is added to the soil. Two ratios were used: 5:100 (hydride : soil) and 2.5:100, 5% and 2.5% w/w respectively. The mixtures were homogenized by hand and introduced into the ring mill.

The general reaction of hydrogenation/dechlorination occurred during the mechanochemical treatment is:

$$C_{12}H_xCl_y + MH \xrightarrow{E\ um.} C_{12}H_{10} + MCl \qquad (1)$$

Where:

$C_{12}H_xCl_y$ = *polychlorobiphenyl with $0 < x < 10$ and $1 < y < 10$*
MH = *hydrogen donor compound (hydrides in our case)*
E_{um}. = *energy given by milling*
$C_{12}H_{10}$ = *biphenyl*
MCl = *chloro salt.*

RESULTS AND DISCUSSION

The progressive abatement of PCBs in soil mixed with $NaBH_4$ is clearly show in Figure 1 and 2, where the chromatograms obtained at different milling time are reported.

Figure 1-A and 2-A are relative to PCBs concentration in the soil as it is. Figures 1-B, 1-C and 1-D are relative to soil ground for 3.5, 11 and 18 hours after the addition of $NaBH_4$.

Figure 2 shows the chromatograms with the progressive evolution of PCBs during the treatment with $NaBH_4$ at 2.5% w/w.

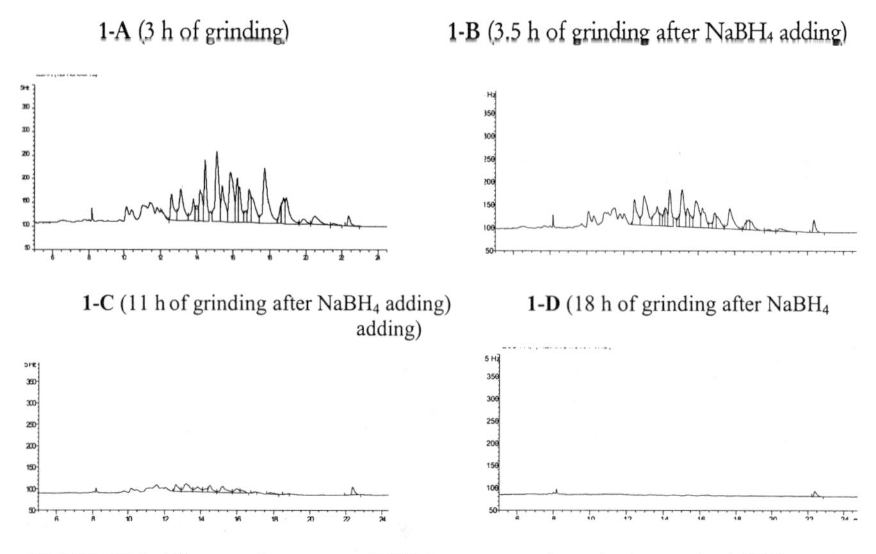

FIGURE 1. Chromatograms of PCBs progressive abatement at different milling time (after $NaBH_4$ adding at 5% w/w).

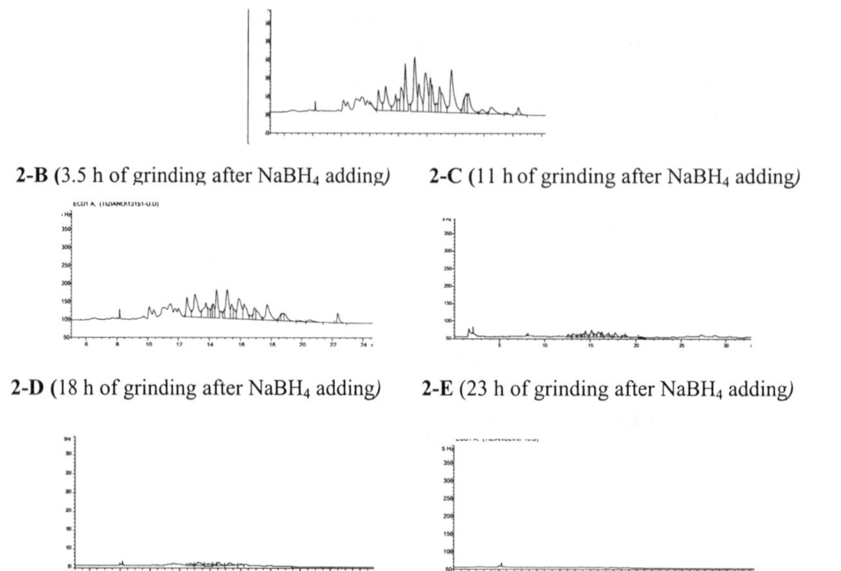

FIGURE 2. Chromatograms of PCBs progressive abatement at different milling time (after $NaBH_4$ adding at 2.5% w/w).

Every PCBs peak is reduced increasing the milling time even if a great abatement efficiency is obtained for high chlorinated congeners rather than the low chlorinated ones. Figure 3 represents the chloride and total PCBs trends as a function of milling time. The contaminants degradation increases with the chloride concentration. For milling time above 18 h the PCBs concentration decreases from 2600 mg/kg to zero and, correspondingly, the chloride concentration increases from 60 to 430 mg/kg.

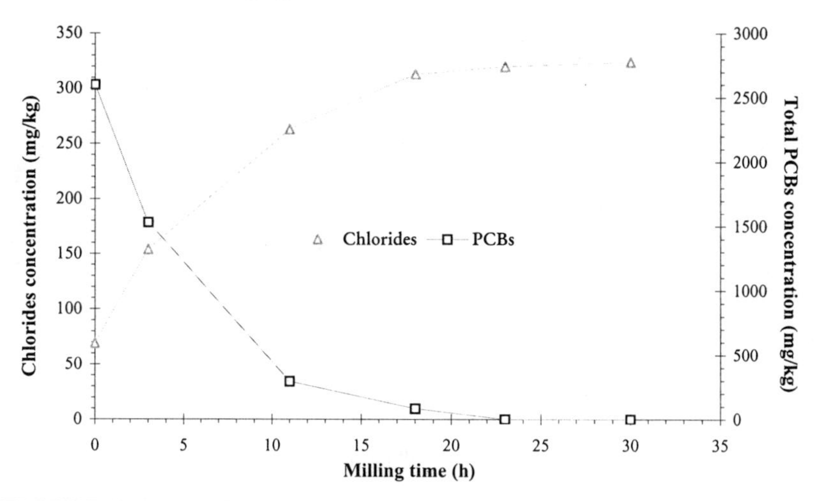

FIGURE 3. Total PCBs and chlorides concentration in the soil sample added with 5% of $NaBH_4$ (w/w) as function of milling time.

Figure 4 shows the results obtained adding a lower quantity of $NaBH_4$ to the contaminated soil.

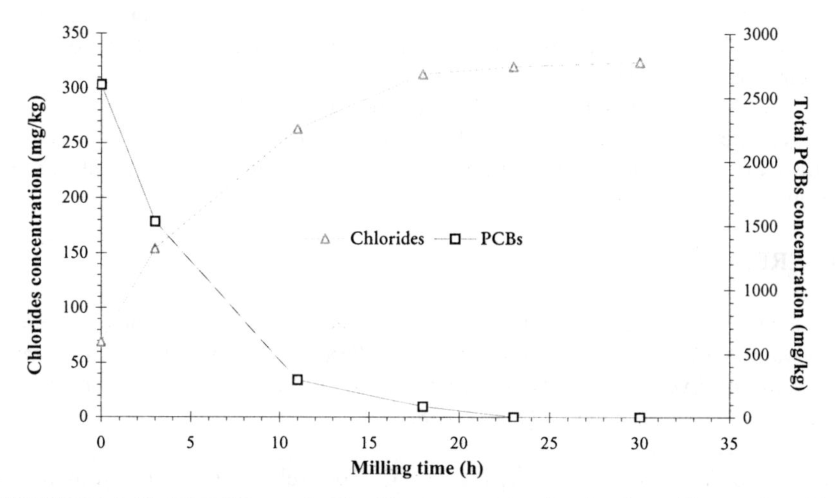

FIGURE 4. Total PCBs and chlorides concentration in the soil sample added with 2.5% of $NaBH_4$ (w/w) as function of milling time.

Even if the ratio between the soil and the hydride is reduced to half (from 5 to 2.5 :100 w/w), the milling time for total abatement of PCBs concentration is increased by only 5 hours.

In Figure 5 the PCBs abatement efficiency, for both cases, are compared.

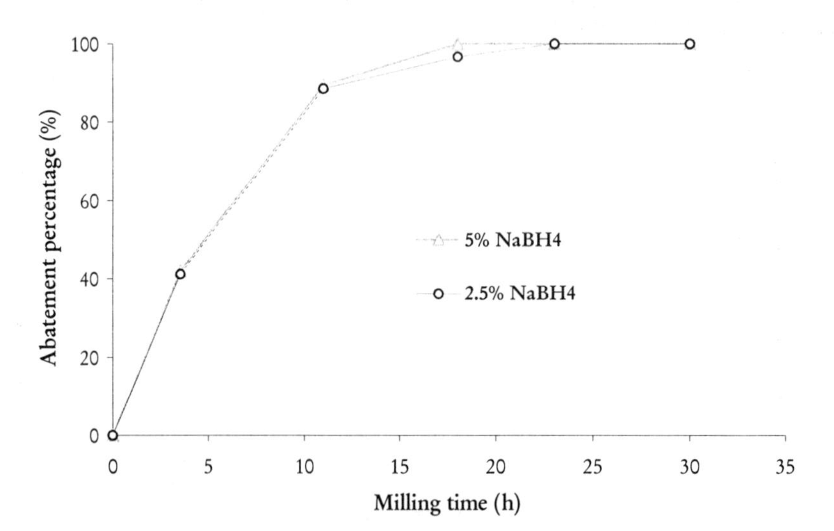

FIGURE 5. PCB's abatement efficiency in two studied cases.

CONCLUSIONS

Starting from a previous experience (Caramuscio et al., 2000), about high energy milling of contaminated soil with NaH, in these work we have experimentally demonstrated the detoxification of PCBs contaminated soil by a mechanochemical process.

We have used $NaBH_4$ in different concentration (5% and 2.5% w/w) and achieved the total abatement of PCBs in 18 h and 23 h respectively. For its intrinsic simplicity (low cost and zero emission), this kind of treatment is very attractive as remedial technology.

REFERENCES

Aresta M., P. Caramuscio, L. De Stefano, T. Pastore. 2001. "Mechanochemical Removal of PCBs in Contaminated Soil." *Proceedings of III Congress and International Exhibition "Added Value and Recycling of Industrial Waste" – VARIREI 2001 - L'Aquila - Italy*: 154-158.

Caramuscio P., L. De Stefano, T. Pastore, C. Tortorella. 2000. "Un Trattamento innovativo per i terreni contaminati da Policlorobifenili." *Siti Contaminati, Ranieri Ed. n° 1:* 26-29.

Hall K., M. Harrowfield, J. Hart, G. 1996. McCormick. "Mechanochemical Reaction of DDT with calcium Hydride." *Env. Sc. & Tech. Vol. 30, n° 12.*

Hutzinger O., S. Safe and V. Zitko. 1974. "The Chemistry of PCBs." *CRC Press. Cleveland, Ohio.*

Loiselle S., M. Branca, G. Mulas G. Cocco. 1997. "Selective Mechanochemical dehalogenation of chlorobenzenes over calcium hydride." *Env. Sc. & Tech. Vol. 31, n°1.*

Monagheddu M., G. Mulas, S. Doppiu, G. Cocco, S. Raccanelli. 1999. *Reduction of Polychlorinated Dibenzodioxins and Dibenzofurans in Contaminated Muds by Mechanically Induced Combustion Reactions.* Env. Sc. & Tech. Vol. 33, n°14.

Tyagin B., P. Yu. 1998. "Activate State in Mechanochemical Alloying of Brittle Materials." *Met. Trans. A-Phy. Met. Mat. Sc. 19 A, :* 2867.

SEDIMENT IMPROVEMENTS WITH IN-STREAM WATER MANAGEMENT

Henry O. Pate, Amy L. Swiecichowski, John A.D. McArthur, & Felix Fernandez,
Battelle Memorial Institute, Columbus, OH, USA

ABSTRACT: The large amount of organic sediments and nutrients deposited in waterbodies, when coupled with poor circulation, has contributed to the worldwide increase in hypoxia. From September 2000 to August 2001, three InStreem™ water units were installed in Wilson Bay, North Carolina, USA to enhance circulation and develop an aerobic sediment cap, as part of a multi-phase effort to restore Wilson Bay. Prior to InStreem™, the black, septic sediments only supported macroinvertebrates in the cooler months of the year. Physically, surficial sediments changed from black to a light brown and the septic smell was eliminated. Apparent redox potential discontinuity (RPD) layer depth increased from zero to several millimeters within one month of InStreem™ operation. In August 2001, the methodology changed to benthos evaluation when RPD layers at most sites were too complex for visual tabulation. Benthos in the areas of enhanced flow compared well to the positive control areas – areas with natural flow. Hypoxic events (12-hour averages of dissolved oxygen near the bottom) decreased nearly three-fold with InStreem™ use. The results of this study indicate that in-stream water management offers an effective addition to integrated waterbody restoration efforts.

INTRODUCTION

In-stream water management uses devices and techniques to directly treat and restore affected waterbodies. Wilson Bay, a shallow, estuarine 0.32-km^2 bay on the New River in Jacksonville, North Carolina, USA was the site of such an integrated approach. The Wilson Bay Water Quality Initiative was established in 1998 to remediate Wilson Bay. Its efforts include wetland recovery, bivalve planting, education, and monitoring (Levine, 2001). In 2000, flow enhancement and aeration were introduced into the bay with an InStreem™ system (15-hp). InStreem™ water units (floatable, 5-hp units; see Figure 1) drag water into motion by rotating 10 vertical discs (1.37-m diameter).

FIGURE 1. InStreem™ units in Wilson Bay.

Objective. The objective of this demonstration was to enhance circulation within Wilson Bay and develop an aerobic sediment cap capable of sustaining benthic life. This yearlong study focused on the effectiveness of using a three-unit

InStreem™ system to improve sediment and water quality of a large eutrophied waterbody. Evaluation of other Wilson Bay efforts (i.e., bivalve planting) is anticipated (Levine, 2001).

Site Description. Prior to World War II, Wilson Bay was used for recreational activities and served as a nursery area for a variety of aquatic species. From the late 1940s to 1998, the bay received discharge from a wastewater treatment plant. Runoff from the growing city of Jacksonville enters Wilson Bay mainly from four urban creeks that load the bay with sediments, nutrients, and organic matter. To amplify the problem of eutrophication, the river channel bypasses the bay, and the tidal influence is minimal with only 0.3 m amplitude. Over the years the bay has lost most benthic life forms and its anaerobic sediments have been releasing methane, ammonia, hydrogen sulfide, and phosphates.

An InStreem™ evaluation began with selection of sampling sites within Wilson Bay and adjoining waters. Sampling sites included existing Hydrolab stations and points along transects for benthic grabs and cores (See Figure 2). All sampling sites and drifter movements were located using a differential global positioning system (GPS) with accuracy to 1-5 m (Garmin Corp, Olathe, KS, USA). Four transects were selected with a total of 14 sites for core samples which were collected on Days 1, 19, and 48 from system start. Transect 1 and the Yacht Club, areas upstream and removed from the influence of the units, served as control sites; Transects 2, 3, and 4 were within Wilson Bay. Benthic grabs were collected near the Yacht Club and Old Dock Hydrolabs, downstream of each of the units and at sites 1-2, 2-4, 3-1, 3-2 and 4-3 (a total of 10 samples). Wilson Bay Water Quality Initiative provided Hydrolab readings from 1999 through August 2001 (Donovan-Potts, 2001). This data provided a way to compare bay water quality with conditions prior to system initiation on September 8, 2000.

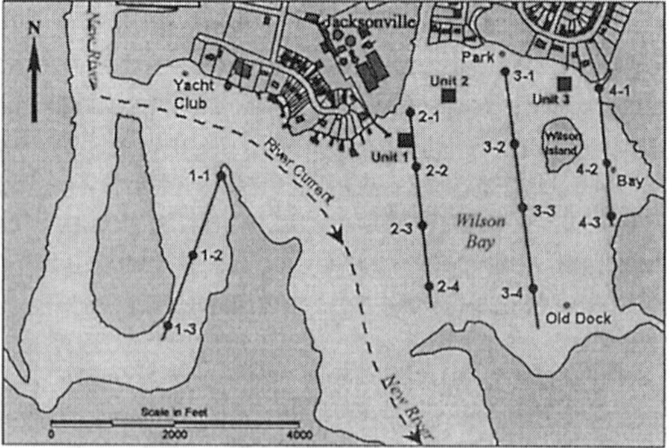

FIGURE 2. Wilson Bay sampling sites and Hydrolab locations.

MATERIALS AND METHODS

Wilson Bay's progress was evaluated with sediment sampling, water current measurements, and water quality data. For analysis, sites were grouped according to their distance from system influence (see Table 1).

Sediment RPD. Single core samples (20-cm depth) were collected with an acrylic hand corer (89-mm diameter) at each of the 14 stations, using equipment and methods similar to Grizzle and Penniman (1991). Sediment profile images were recorded photographically to document the apparent redox potential discontinuity (RPD) layer. The RPD layer is an important estimator of benthic habitat quality (Grizzle, 1991). This relation is in accordance with the classical concept of RPD depth, which associates sediment redox state with sediment color (Fenchel, 1969; Rhoads, 1986). In this study, RPD layer measurements were taken using 35mm photographs and video equipment. If the top sediment layer was black or dark brown, RPD depth was recorded as zero.

Benthos. A Petite Ponar sediment sampler was used to collect composite samples at each of 10 stations. The sampling plan included eight Wilson Bay sites and two sites upstream of Wilson Bay (see Table 1). Benthic samples were sieved through a 0.5 mm mesh screen, preserved with 10% formalin, stained with rose bengal, and finally stored in 70% ethanol. All benthic macroinvertebrates were carefully sorted and identified to major taxonomic groups (i.e., class or order). Benthic sample collection, handling, and faunal identification followed standard guidelines (Clesceri, 1998; Gosner, 1971; Stachowitsch, 1992). Cluster analysis (Boesch, 1977; Bloom, 1997) was performed on the faunal data to examine between-station differences within the study area.

TABLE 1. Suggested grouping of sampling sites. Bold indicates benthic grab sites.

Location	Grouping	Sampling Sites
Wilson Bay	InStreem™ units' outlets	Within stream flow of units: **unit 1**, **unit 2**, **unit 3**
Wilson Bay	Under InStreem™ influence	2-1, 2-2, **3-1**, 4-1, 4-2, **4-3**, Park, Bay
Wilson Bay	Reduced influence	2-3, **2-4**, **3-2**, 3-3, 3-4, **Old Dock**
Upstream	Areas outside influence	Oxbow: 1-1, **1-2**, 1-3; **Yacht Club**

Water Velocity Measurements. Due to the minimal velocity currents (often < 3 cm/s) and shallow depths (0.6 m - 1.8 m) within Wilson Bay, conventional current meters were not always able to detect and measure velocities within the bay. As an alternative, Lagrangian-style holey sock drifters were built to measure Wilson Bay currents. This type of device is able to move with the currents and measures its velocity and path, regardless of how minimal (Geyer, 1989). The drifters were 90-cm diameter cylinders, made with lightweight nylon fabric in two depths: 61-cm for deep sections and 30-cm for shallow areas. Three or four plastic floats counteracted the slight negative buoyancy of each drifter. At the time of deployment and subsequent sampling times (10-40 min), differential GPS coordinates, time, wind information, tidal phase, and surface water conditions were recorded.

Dissolved Oxygen. The Wilson Bay Water Quality Initiative operates and maintains eight Hydrolabs in and around the bay. The Hydrolabs are deployed within 15 cm of the bottom for two weeks or more to collect data every 15 minutes. The dissolved oxygen (DO), temperature, salinity, turbidity, and depth data is downloaded to individual files for each site (Wilson Bay Initiative, 2001). The raw data was averaged into 12-hour periods roughly correlated to periods of light and periods of darkness. Frequency charts were created to show hypoxic events between sites and to show the comparison between periods with and without InStreem™ influence (Analyse-it, 2000). The longer the duration and extent of hypoxia (DO < 2.0 mg/L), the more degraded the habitat is for most aerobic life.

RESULTS AND DISCUSSION

Sediment RPD. Sediment profile images on or before Day 1 (September 9, 2000) of the study uniformly lacked apparent RPD layers over the entire sampling area. The majority of those samples (11 of 15) taken on or before Day 1 revealed black or deep brown upper layers indicative of extended periods of low oxygen (See Figures 3 and 4). RPD depth in Wilson Bay went from zero at the beginning of the study, to an average of 5.5 $\pm$ 1.2 mm (mean $\pm$ sd) at InStreem™-influenced sites (see Table 1) after 19 days of operation. Reduced influence sites averaged 3.1 $\pm$ 1.2 mm during this same time period. Control sites out of the influence of InStreem™ (1-1, 1-2, 1-3) did not have an RPD layer. At 48 days, control sites averaged 1.2 $\pm$ 1.5 mm and InStreem™-influenced sites averaged 6.7 $\pm$ 2.9 mm. Reduced influence sites averaged 2.6 $\pm$ 3.7 mm. After nearly a year of operation, odors were reduced, but cores sediment profiles were too complex for visual tabulation. The evaluation of sediment conditions shifted from RPD measurements to benthos analysis.

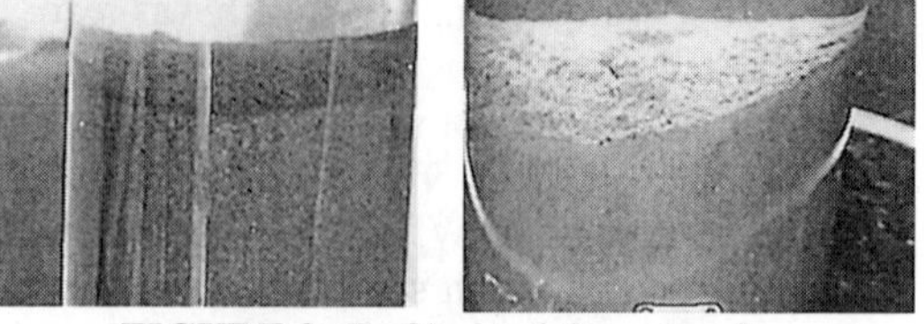

FIGURE 3. (Left) site 4-3 on day 0. (Right) site 4-3 on day 19.

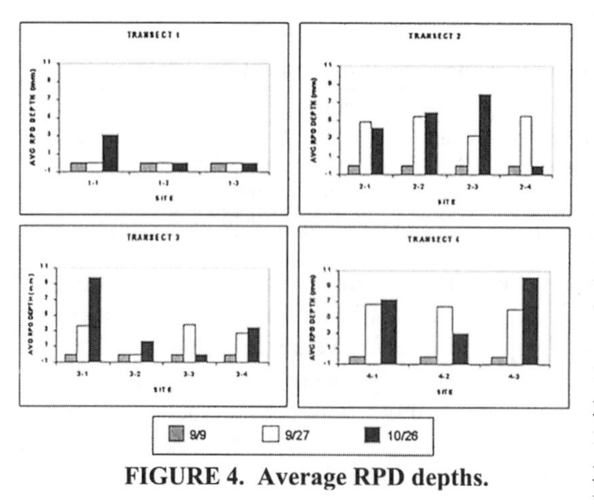

FIGURE 4. Average RPD depths.

Benthos. When cores were taken on Day 0 of this project, no benthic macro-invertebrate life was noted during visual inspection at any of the sampling stations in the bay. Nearly a year later benthos were found at sites in the bay that follow the currents created by InStreem™. A total of six major taxa were identified from the 10 stations. Faunal composition and abundance ranked from highest to lowest across the stations as follows: 1) controls, 2) InStreem™ outlets, 3) areas under InStreem™ influence, and 4) areas of reduced system influence (Figure 5). A dendrogram shows the results of cluster analysis on the benthic data from the 10 stations (Figure 6). Results can be interpreted at a two-group level. One group

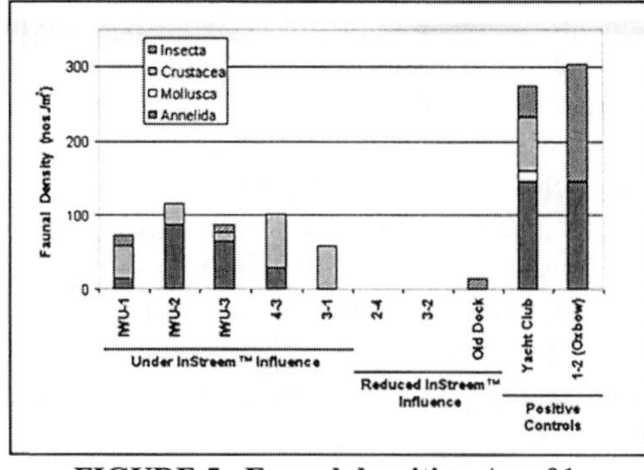

FIGURE 5. Faunal densities, Aug 01.

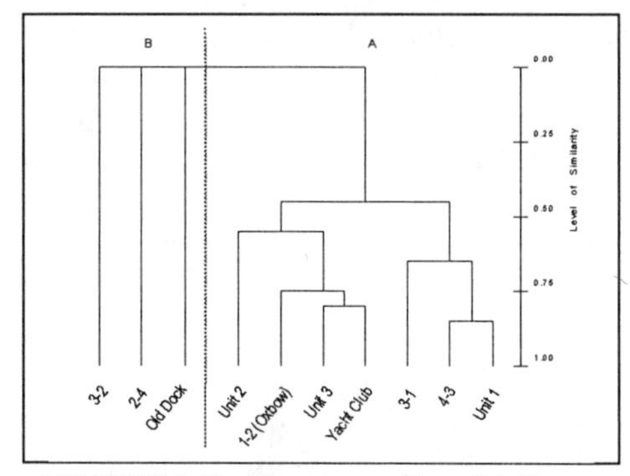

FIGURE 6. Dendrogram of benthic fauna similarities between sites, Aug 01.

(A) contained all controls, InStreem units' outlets, and sites under InStreem™ influence (> 0.45 similarity level). The second group (B) consisted of three bay sites considered to have reduced InStreem™ influence (2-4, 3-2, Old Dock). The clear and marked differences in faunal composition and abundance and grouping from cluster analysis among the 10 sampled sites suggest strong environmental and ecological conditions at work, ranging from fair to good at group A sites to poor at group B sites. Although results provide a cursory study of the benthos in the bay, the Wilson Bay Water Quality Initiative has been conducting separate benthic sampling since 1999. A comprehensive benthic community evaluation of Wilson Bay and surrounding waterbodies will be published in the future (Levine, 2001).

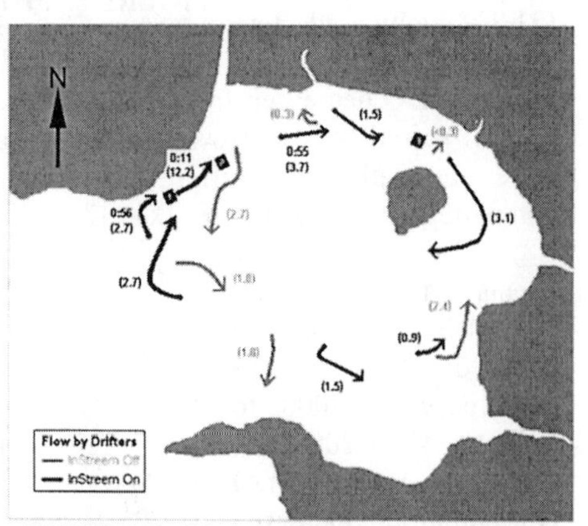

FIGURE 7. Circulation patterns with and without InStreem™, Feb 01.

Water Velocity Measurements. Circulation was measured using drifters throughout the bay with and without InStreem™ operating, gathered 2 days apart, at the same phase of a falling tide. Figure 7 shows drifter flow under these two conditions, with average velocities (cm/s) given in parenthesis, where average deployment duration was 1:45 hrs (± 10 min). Time (hh:mm) is shown by a drifter's path if its deployment duration was considerably shorter than 1:45 hrs, this being due to their proximity to system influence. When the system was not operating, minimal circulation was observed within the N and NE sectors of the bay (< 0.3 cm/s, without a particular direction). Southerly flow was observed within the W sector of the bay (bay mouth). This southerly current (~2 cm/s) may have been augmented by the falling tide and prevailing winds. Current in the SE sector was also significant (2 cm/s) with a sporadic counterclockwise direction. When all three units were operating significant changes were observed, with increased magnitudes and flow reversal. Now within the N and NE sectors, velocities ranged over 2-4 cm/s, with clockwise flow around Wilson Island. Drifters were observed to travel from unit 3 around the island nearly to sampling site 3-3. In the W sector of the bay, flow direction was reversed, drawing and projecting river water NE into the bay, with average velocities of 3 cm/s. Operation of units did not appear to affect flow in the SE sector of the bay. Averaging all drifter data throughout the bay, velocity increased over twofold with the units operating (On: ~3.9 cm/s vs. Off: ~1.5 cm/s).

Dissolved Oxygen. This study focused on the extended summer periods of May 1 through September 30. This time period historically showed temperatures above 25°C and lower DO concentrations. Due to the limited availability of 2001 data, the period was truncated to May 1 through July 24. See Figure 2 for the Hydrolab locations. The Yacht Club was used as a control because it is upstream of the bay, out of the influence of the InStreem™ system, and benthic and aquatic life thrive there. The occurrence of 12-hour hypoxia events at the Yacht Club (see Figure 8) increased in 2001 (31.2% compared to 22.4% in 2000) and the mean DO decreased from 3.34 mg/L to 2.74 mg/L (statistically different: $p < 0.0001$, paired t-test). However, the average DO concentrations during hypoxia were not statistically different (2000: 1.39 mg/L; 2001: 1.46 mg/L) when comparing the two years ($p = 0.1837$, paired t-test).

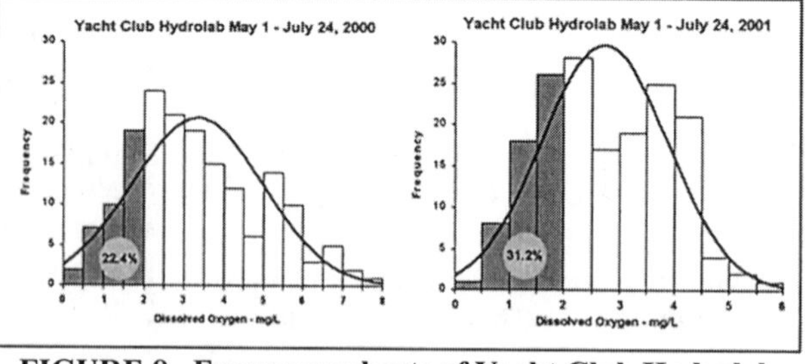

FIGURE 8. Frequency charts of Yacht Club Hydrolab dissolved oxygen. Dark bars indicate hypoxia.

However hypoxic events at the Park Hydrolab (Figure 9) dropped from 28.8% occurrence in 2000 to only 10.1% in 2001 and the overall DO concentration increased from 3.31 mg/L to 3.64 mg/L (significantly different, $p = 0.0391$, paired t-test of 12-hour periods, $n = 169$). The average DO concentration, during hypoxia, also improved from 0.80 mg/L to 1.25 mg/L (statistical significance $p <$

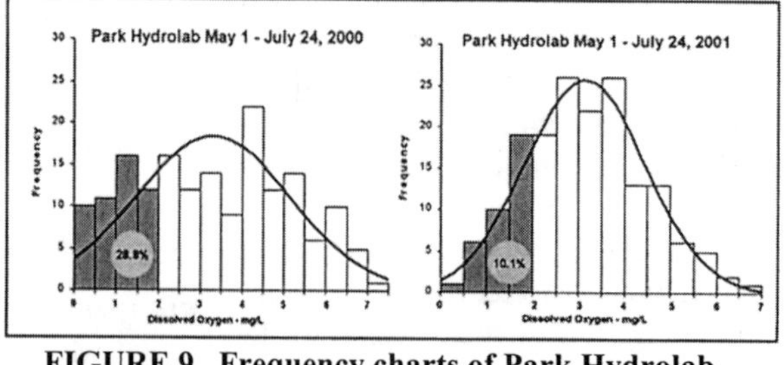

FIGURE 9. Frequency charts of Park Hydrolab dissolved oxygen. Dark bars indicate hypoxia.

0.0001, paired t-test of 15-min hypoxic events, n = 679). These results indicate that the duration and effect of hypoxia was reduced.

Discussion and Conclusions. Despite unprecedented spending on water management by the U.S., some 40% of U.S. waters are unfit for human or ecosystem use (EPA, 1998; Stamm, 1998). Regional failures in water quality have occurred in areas like the Chesapeake, Neuse Valley, Imperial Valley, Everglades, and Mississippi Delta. Worldwide the problems are much worse, with some 2/3 of the population facing "water stress" by the year 2025 (U.N., 2000). These examples highlight the shortcomings of conventional approaches to water management, whose capital and infrastructure needs require decades and many billions of dollars to implement on a regional basis. Vast reservoirs of nutrients and other pollutants are stored in waterbody sediments, of which Wilson Bay is an example. These conditions promote hypoxic events that are increasingly common in estuarine and coastal ecosystems around the world (Diaz, 1995). Currently, dredging and capping with sand seem to be the only treatment alternatives, but have not been widely applied due to the huge areas and volumes involved. InStreem™ offers a promising alternative since formation of a surficial oxic layer above anoxic sediments has been shown to increase decomposition rates of organic material by factors of 3 to 10, while reducing nutrient flux that promotes eutrophication (Diaz, 1995; Kristensen, 1995; Hulthe, 1998).

Operation of the InStreem™ units in Wilson Bay during September 2000 – August 2001 has had a positive impact on the bay ecosystem. Fewer and shorter periods of hypoxia were observed in 2001 than in 2000. The rapid formation of measurable RPD layers 19 days after the units were turned on indicate that sediment oxygen demand at the surface can be met and attributed to unit operation. During this same period, RPD depths remained at zero in areas outside the system's influence (sites 1-1, 1-2, 1-3). Nearly one year later, benthic grabs revealed a more diverse population within areas of enhanced circulation, while sites with reduced system influence were markedly less diverse. Perhaps the most compelling evidence is anecdotal: just like the RPD layers, fishing activity has gone from zero (Summer 2000) to daily catches of crabs and fish by commercial fishermen. These results show that with modest amounts of energy, large eutrophied water bodies similar to Wilson Bay can be managed in-situ.

REFERENCES

Analyse-it Software, Ltd. 2000. Analyse-it for Microsoft Excel, version 1.62. Leeds, England, UK.

Bloom, S.A. 1997. The Community Analysis System, software version 5.0. Ecological Data Consultants, Archer, FL.

Boesch, D.F. 1977. Application of Numerical Classification in Ecological Investigation of Water Pollution. USEPA Report 60/3-77-033, Corvallis, OR.

Clesceri, L.S., A.E. Greenberg, and A.D. Eaton. 1998. *Standard Methods for the Examination of Water and Wastewater*. 20th ed. APHA, Washington, D.C.

Diaz, R.L., and R. Rosenberg. 1995. "Marine Benthic Hypoxia: A Review of Its Ecological Effects and the Behavioral Responses of Benthic Macrofauna." *Oceanography and Marine Biology. 33*: 245-303.

Donovan-Potts, P. 2001. Hydrolab Water Quality Raw Data Files, 1999 – 2001. Wilson Bay Water Quality Initiative. Jacksonville, NC.

EPA. 1998. http://www.epa.gov/305b/98report/98summary.html.

Fenchel. 1969. "The Ecology of Marine Microbenthos." *Ophelia 6*: 1-182.

Geyer, R.W. 1989. "Drifter Observations of Coastal Surface Currents." *Atmospheric and Oceanic Technology. 6*(2): 333-342.

Gosner, K.L. 1971. *Guide to Identification of Marine and Estuarine Invertebrates.* John Wiley & Sons, New York, NY.

Grizzle, R.E., and C.A. Penniman 1991. "Effects of Organic Enrichment on Estuarine Macrofaunal Benthos: A Comparison of Sediment Profile Imaging and Traditional Methods." *Mar. Ecol. Prog. Ser. 74*: 249-262.

Hulthe, G., S. Hulth, and P.O.J. Hall. 1998. "Effect of Oxygen on Degradation Rate of Refractory and Labile Organic Matter in Continental Margin Sediments." *Geochimica Et Cosmochimica Acta. 62*(8):1319-1328.

Kristensen, K., S.I. Ahmed, and A.H. Devol. 1995. "Aerobic and Anaerobic Decomposition of Organic Matter in Marine Sediment: Which is Fastest?" *Limnology and Oceanography 40*(8): 1430-1437.

Levine, J.F. 2001. Personal communication. Wilson Bay Water Quality Initiative. Jacksonville, NC.

Rhoads, D.C., and J.D. Geranamo. 1986. "Interpreting Long-term Changes in Benthic Community Structure: A New Protocol." *Hydrobiologia 142*: 291-308.

Stachowitsch, M. 1992. *The Invertebrates: An Illustrated Glossary.* John Wiley & Sons, New York, NY.

Stamm, K. 1998. "Government Initiatives for the Control of Nonpoint Source Pollution from Animal Waste." American Institute of Chemical Engineers.

U.N., 2000. United Nations Environmental Program Geo-2000: http://www.unep.org

EX SITU TREATMENT TO DETOXIFY HIGHLY MIXED POLLUTED SEDIMENT

Luigi Volpe (Ambiente, S.Donato Milanese -MI, Italy)
Ciro Cozzolino (Snamprogetti, Fano -PS, Italy)
Roberta Miglio (Snamprogetti, S.Donato Milanese -MI, Italy)
Sergio Moratto (Aquater, S.Lorenzo -PU, Italy)

ABSTRACT: A polluted sediment was fed to a pilot-scale thermal treatment unit for removal of organic and volatile contaminants combined with a leaching treatment for heavy metals extraction. The sediment was highly contaminated by persistent toxic substances, with PCDD/Fs, PAHs, PCBs and chlorinated pesticides among the most abundant, followed by chlorinated volatile organics and toxic heavy metals such as Cr, Ni, Pb, Zn, Cu, Hg, Cd, and As. Desorption efficiency for Hg and PCDD/Fs (indicator compounds) from the experimented matrix was confirmed even at 450° C. The metals extraction that resulted was efficient at low pH value. The combination of thermal desorption with solubilization treatment of residual heavy metals produced a sediment completely detoxified from organic and inorganic pollutants with potentially beneficial use in restoration or for end-products such as topsoil and bricks. The two aforementioned main steps were completed with auxiliary units for effluents treatment in order to guarantee "zero" emission (level of the best technology available) in any compartments and to design the integrated approach for decontamination.

INTRODUCTION

The most widely used procedure for reducing the contamination effects of lagoon and sea sediment is dredging and disposal of the dehydrated and possibly stabilised materials in controlled dumps. This method is becoming impracticable because it is increasingly difficult to find adequate space. In addition, the problem is not definitively solved but is only transferred to areas where pollution causes less damage. Therefore, it is necessary to have technologies for the removal of these pollutants. The selection of the various operating alternatives is generally based on the choice of technologies that enable the quality criteria to be reached, have the best engineering and economic guaranties, and cause as little impact as possible on the environment. Inertization is a process applied on matrices of a prevalently inorganic nature, but solid end residues have a weight higher than the original and have proved to be not suitable and not very efficient in the case of the immobilisation of pollutants of an organic nature. When the mud is contemporaneously contaminated by various groups of substances, it is generally necessary to apply several decontamination technologies. The vitrification process, bringing the raw material to a temperature higher than the softening point, is very efficient and simultaneously produces both the incineration of the organic substances and the stabilization of the inorganic fraction to an inert mass of a vitreous consistency. However, it does not make it possible to return the treated materials to their original sites. Among thermal treatments, the desorption technique is extremely versatile. It offers the advantage of effecting the treatment at moderate temperature (450° to 700° C) and generating a minimum volume of fumes to be purified, with a consequent reduction in the dimensions of the whole equipment. The greatest advantage of the process is, however, that the natural characteristics of the solid remain unaltered.

Thermal desorption is a well-known and frequently applied solution to detoxify organically contaminated soils, but persistent toxic substances removal levels are strongly affected by the natural organic content of the original matrix. Setting up an effective treatment requires an experimental step, which also represents a suitable tool for focusing the critical aspects relevant to the contaminants balance and fluxes: solids, liquid phases, and off-gases. At the basis of the thermal treatment is a full knowledge of the raw material, implying an accurate determination of all the physical-chemical parameters that determine the project characteristics of the plant: dimensions, operative temperatures, residence times, and type of blasting of the gaseous effluents. Inorganic compounds, such as heavy metals, evidently are not destroyed and are found in the solid residue. The last at this point can be subjected to an extraction process for the solubilization of the toxic heavy metals by chemical attack to produce the fully detoxified sediment. The selection of the reagent is mainly linked to the type and levels of each pollutant species. The aqueous solution in which the metals have been dissolved can be separated by decanting from the solid and then be sent to a specific conventional reprecipitation/inertization of the metals or selective reprecipitation for recovery of some species.

Objective. The aim of the experimental research was to obtain the technical feasibility for completely detoxifying a mixed polluted sediments derived from the Venice Lagoon. It shows no sign of life, due to its high level of contamination and this has a negative impact on surrounding environment and human life. These sediments, classified as "dangerous" by Italian law (DL 22/1997), should be treated in order to meet the acceptance levels for some "specific land uses" (DM 471/1999). That means application of technologies characterised to have a minimal environmental impact (no liquid or gaseous emissions) and produce a final solid residue which, as quality criteria, can be reused for restoration of the Lagoon morphology in areas permanently in contact with water. (class A of Venice Sediment Management Criteria D.L. 360/91).

Site description. A representative quantity of natural sediment was dredged from Brentelle di Porto Marghera (Figure1). The industrial area of Porto Marghera faces the central part of the Venetian Lagoon where the "first" and the "second" industrial zones are located. The first was created during the post- First World War period when the Venetian economy was strongly affected by war damage. A plan to create a new industrial zone was envisaged to link the industrial zone and the harbour. The first base industries (production of glass, sulfuric acid, P-fertilisers, pesticides, mechanical engineering, a dockyard, a refinery and related oil deposits) were settled before the 1930s. During the 1930s and the 1940s, the nonferrous (zinc and aluminium) metallurgical industries were developed. Moreover, one of the biggest plants for the ammonia production from synthesis gas was built. After reconstruction of the plants destroyed during the Second World War, a second industrial area was planned for the petrochemical industries and vegetable oil refining, power production and fine framing. At that time, the area southeast of Canale Brentelle, not yet industrialised, was made of a flat lagoon environment with lacustrine and marsh areas that were progressively under reclamation and where the construction of petrochemical plants began in the early 1950s. Between 1951 and 1960, the production lines for PVC, acetic cyanicdes, acid, fertilisers, caprolactam, and inorganics were realised. A second stage of development began in 1961, and new production lines were constructed (methacrylates and terephtalates, upgrading of the inorganic and florurates production lines). In the late 1970s, ethylene and propylene cracking facilities were built. At the same time, production lines for toluenediisocyanate, HCl, CVM, PVC, and carbon tetrachloride and the electrolytic plants for chlorine production were built.

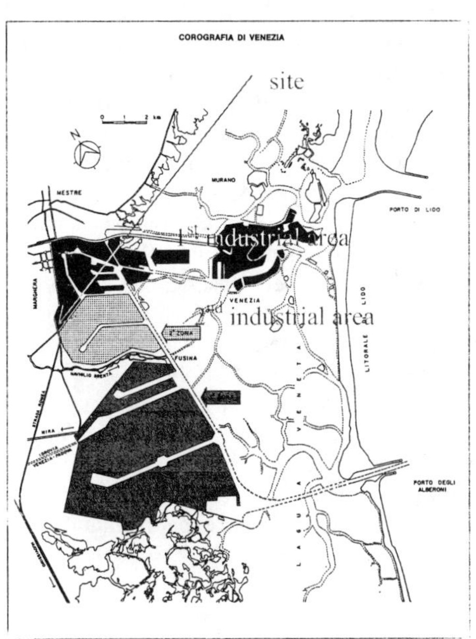

FIGURE 1. Sediment location

The present morphology of the industrial area is similar to the one observed in 1971. A third industrial area was planned but never realised.

Before 1970, there were virtually no legal regulations on industrial expansion. The industries discharged part of their wastes directly into the Lagoon or disposed them in uncontrolled dumps. Moreover, the state of the art technologies for effluent treatment was obviously rudimental. Only in the 1980s were waste regulation and quality control criteria for liquid effluent issued and purification plants constructed.

Since 1990 extensive studies have been carried out on the Venice Lagoon (Alcock R.E., 1999; Bellucci L.G., 1999: Di Domenico, 1998) in order to plan an integrated approach to restore and protect its unique ecosystem. Now, a detailed picture of the industrial canals pollution has been achieved, whether from the geographic or historical point of view. More than 1,500,000 m^3 of sediment needs to be detoxified for beneficial uses inside the Lagoon.

MATERIAL AND METHODS

Analysis of chlorinated compounds were carried out using established methods. Sediment samples were spiked with solutions of labelled PAHs, PCBs, PCDD/Fs internal standards to correct for analytical losses and ensure quality control. PAHs and PCBs analytical determinations were performed with HRGC-MS (TS250 Micromass) at a resolution of 1000 r.p. (10% valley). Recoveries of added quantification standards ranged between 45 and 90%. All the 2.3.7.8 substituted PCDD/Fs and the total homolog groups were quantified. The analytical determination was performed using isotope dilution HRGC–HRMS (AutospecX Micromass) at a resolution of 10000 r.p. Detection limits for the various 2.3.7.8 substituted PCDD/Fs varied between 0.5 – 5 pg injected. Chlorinated solvents were analized by GC-FID, pesticide by GC-ECD. Heavy metals quantitative determinations were carried out with ICP-AES, fitted with a CID (Charge Injection Device) detector, and confirmed by alternative analitycal tecniques (i.e., spectrophotometry of A.A.)

Preliminary characterisation and evaluations were performed with laboratory equipment (System for Thermal Diagnostic Studies [STDS]; infrared analysis of evolved gases and dioxin analysis of solid before and after thermogravimetric tests [TG] and suggested the operative conditions ranges leading to a complete removal from the solid phase and destruction of evolved pollutants (Bassetti et al., 1999)

The pilot plant with a capacity of 10 kg/h consisted of a rotary kiln indirectly heated, followed by a cyclone to separate most of the entrained particulate matter and a fired afterburner chamber to complete combustion of vaporised organic components in the purge gas. Figure 2 is the pilot plant flow diagram chosen for the experimentation.

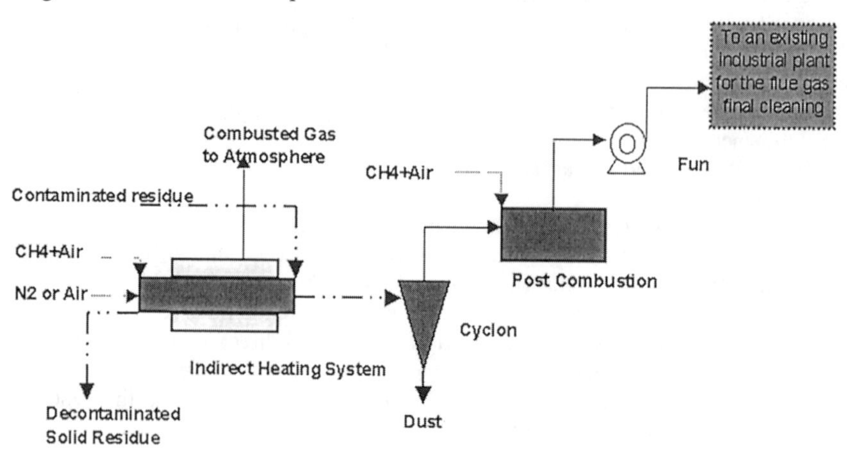

FIGURE 2. Pilot Plant Flow Diagram

The contaminated sediment in crude form, with a water content of 57%, is fed to the kiln via a cochlea in countercurrent flow direction in respect to the moving direction of a purge gas. The kiln was operated at nominal temperatures in the range of 450-650 °C, with inert or oxidative reacting atmosphere in contact with the solid. The kiln slope and the rotating velocity were regulated to assure

kiln solids residence times of about 25 minutes. Solid samples were collected from both the treated sediments and from the cyclone dusts. Off-gases also were sampled periodically upstream of the afterburner chamber.
Test run conditions are reported in Table 1.

TABLE 1. Test running conditions for Desorption step

RUN	*T1*	*T2*	*T3*	*T4*	*T5*	*T6*	*T7*	*T8*
T (°C)	650	500	650	450	550	450	460	480
Rotating Speed (g/min)	3	3	3	4	4	4	4-2.5	2.05
Reacting Atmosphere	N2	N2	Air	N2	Air	N2	N2	Air
Gas flow rate	5	5	5	6	4	6	7	7

Leaching tests were performed on raw and organically decontaminated material with acetic and hydrochloric acids, producing in the latter case a very aggressive aqueous environment for attempting to exalt the extraction of heavy metals which were present in the original matrix. In a screening phase other strong acids, (e.g., H_2SO_4 and HNO_3), and chelating compounds (e.g., EDTA) were tested. Figure3 is the Block Diagram of the process.

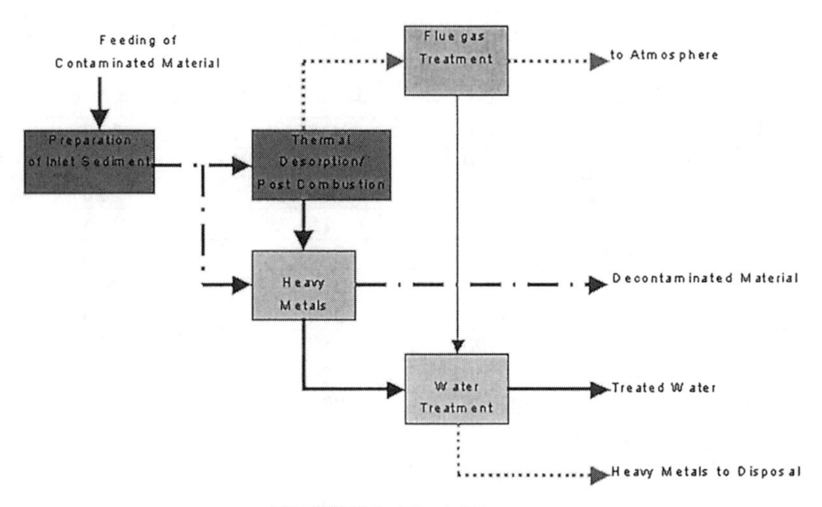

FIGURE 3. Block Diagram

RESULTS AND DISCUSSION

Test feed material characterisation is reported in Table 2. Chemical parameters of the sediment cause it to be classified as dangerous/toxic in reference to Italian law. The main chemical parameters for sediment classification inside the Lagoon (class A, B and C) and for the Italian law are reported in Table 3.

TABLE 2. Pilot plant feed – main characteristics of dried solid

Water content (if not dried)	57%		
Density	1,32 t/m^3		
Particle Size (median)	25 μm		
C 7.4 %		Hg	160 mg/kg
H 0.6 %		Pb	211 mg/kg
Cl 3.5 % (75% inorganic)		Cd	2 mg/kg
S 0.9 %		Cu	123 mg/kg
		Cr	86 mg/kg
PAH	170 mg/kg	Zn	556 mg/kg
PCB	25000 μg/kg	Ni	80 mg/kg
PCDD/F (TEQ)	600 ng/kg	As	20 mg/kg

TABLE 3. Sediment classification

		Toxic Waste (DPR 915/82)	Dump Confinement 2B (DPR 915/82)	Industrial area Tab. B (DM 471/99)	Residential area Tab. A (DM 471/99)	No contact with Lagoon C (L360/91)	Limited contact with Lagoon B (L360/91)	No restrictions A (L360/91)
Hg	mg/kg	**100**	test	**5**	**1**	**10**	**2**	**0.5**
Pb	mg/kg	5000	test	1000	**100**	500	**100**	**45**
Cu	mg/kg	5000	test	600	**120**	400	**50**	**40**
Cd	mg/kg			15	2	20	5	**1**
Cr	mg/kg	100	test	800	150	500	100	**20**
Zn	mg/kg			1500	**150**	3000	**400**	**200**
Ni	mg/kg			500	120	150	**50**	**45**
As	mg/kg	100	test	50	20	50	25	**15**
PAH	mg/kg	500		**100**	**10**	**20**	**10**	**1**
PCB	ug/kg	500000		**5000**	**1**	**2000**	**200**	**10**
PCDD/F	ngTE/kg			**100**	**10**			
2378-TCDF	ng/Kg	**1000**	10					

All thermal desorption tests performed resulted in effective removal of organic contaminants; test results are summarised in Table. PAHs and PCBs are almost completely removed (> 98%) at 450°C. Meanwhile it is important to note that the removal of PCDD/Fs in the temperature range of 450-650°C is always above 95%, both in oxidative and inert atmospheres. After thermal treatment, PAHs and PCBs remaining in the treated sediments were less than 10 μg/kg and <150 μg/kg, respectively, and the residual amount of PCDD/Fs as Toxic Equivalent Quantity ranged between 1 and 13 ngTEQ/kg of dried matter. This equates to the environmental levels measured in low antropic impact areas and for land suitable for general use.

Desorption efficiency is confirmed by elemental analysis data. Chlorine concentrations in the feed and outlet streams were equal to 3.5% and below 2%, respectively, with the latter being all inorganic. Carbonium concentrations in the feed and outlet streams were equal to 7,4% and 1% respectively, the latter being all inorganic carbonium; therefore, the carbonate fraction of the original matrix does not decompose under the temperature interval identified for micropollutants desorption. Greater understanding of the reactions involved is the basis for development of more effective equipment design. There are no major differences between the patterns of isomers produced on decontaminated sediment and those patterns observed in the original sediments. The decontaminated solid has residual quantities of hepta- and octa-chlorinated isomers, which had higher

TABLE 4. Removal of PCDD/PCDF, PAH and PCB by the thermal desorption treatment

	T3	Rem(%)	*T4*	Rem (%)	*T7*	Rem(%)	*T8*	Rem(%)
PAH ug/kg	<10	100	<10	100	<10	100	<10	100
PCB ug/kg	<0,1	100	37	100	152	98	111	99
PCDD/F ngTE/kg	1.3	99	6,4	97	8	97	12,2	95

concentrations in the original matrix. Similarly off-gas and cyclone dust isomer patterns both show a significant presence of low chlorinated furans. The higher temperature of the explored range (650°C) and the oxidative reacting atmosphere give the advantage of producing less toxic gaseous and solid effluents — at this thermal level there is a superposition of a formation and of a greater destruction rate of PCDD/Fs. For complete treatment and final elimination of the micropollutants and all the vaporised organic components from the desorbed phase, a step in the afterburner chamber was provided.

As for inorganic contaminants, Hg is successfully desorbed (removal more than 98%) at 450° C with a residual level of less than 2 mg/kg. However, under these conditions Cd and As still remain in the solid stream. Leaching tests carried out with acetic acid on thermal treated sediments produced results nearly free from toxic heavy metals. It is important to underline the absence of Cr(VI) in the solid and in the leachate of those tests performed in oxidative atmosphere at the highest temperature of the explored range (650°C). This behaviour is ascribed to the high alkalinity of the original sediment for which the metallic species remain immobilised into insoluble forms (hydroxides/oxides). The residual content of toxic heavy metals classified the thermal treated sediment in class B of the aforementioned Venice Management Criteria.

Tests conducted utilising solutions of HCl and treating the materials under examination with acidic solutions for at least 20 hours, showed that metal extraction is really significant at very low pH value. Metals solubilization by acid attack (with H_2SO_4 and pH=2-4) is partial, with 80% efficiency for Mn; about 50% for Cu, Cd, and Zn; and 20% for Cr. An oxidation pretreatment with H_2O_2 before acid extraction seems to improve solubilization efficiency for all the controlled metals, with a value more than 90% for Cu and 30% for Cr. The treatment efficiency is strictly related to pH value; an appreciable solubilization of metals was observed with pH<2. The effect of the different concentrations of HCl (different pH values) is indicated in Figure 4 for some metals.

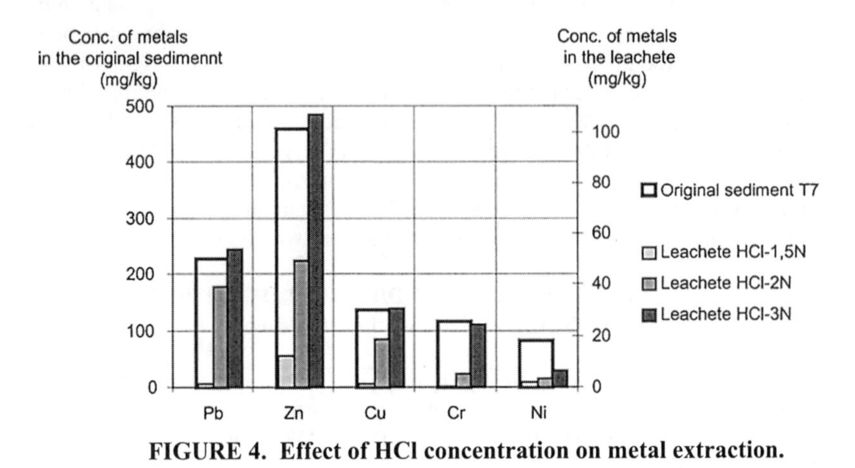

FIGURE 4. Effect of HCl concentration on metal extraction.

During treatment, the sludge/acid suspension grew in relation to the contact time up to a pH value of equilibrium, higher than the starting value, caused neutralization reactions, as carbonates degradation or basic salts dissolution. Acid consumption for undesired reactions is registered. Ca is already released from sludge in pure water washing, but a sharp increase is registered with a pH value of 2, confirming the necessity of acid pH in Ca and Mg dissolution. Na release is 50% using water only; the release is insensitive to pH variation, confirming the presence in sludge of Na soluble salt (NaCl, Na_2SO_4) as it is shown in Figure 5.

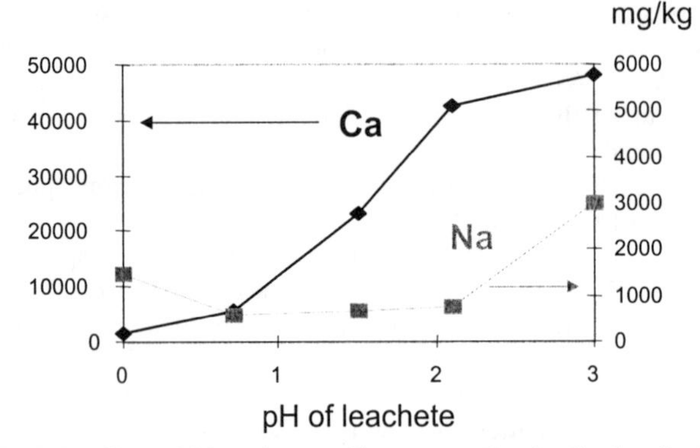

FIGURE 5. Ca and Na release: Concentration in the leachete (mg/kg) as a function of the treatment's pH

Toxic metals solubilization seems to perform contemporaneously with the dissolution for chemical attack of the calcareous component of sediments. The reduction of HCl-specific consumption produces less-efficient removal. Heavy metals solubilization efficiency seems to perform with pH<2, contemporary with optimal condition for carbonates degradation. To reach the Class A limits for sediments, it is necessary to operate at pH<2.

CONCLUSION

The experimental results confirm that low-temperature thermal treatment is an attractive option that could effectively remove micropollutants (PCDD/Fs, PAHs, PCBs) from the examined matrix sediment at a temperature as low as 450° C. This temperature level allows the carbonate fraction of the solid to not decompose. PCDD/Fs, which represent the strongest pollutants to be desorbed, ranged between 1-13 ngTEQ/kg of dried matter as residual level. This equates to the environmental levels measured in land suitable for general usage. Hence energy consumption in greatly reduced in comparison to other alternative high temperature processes such as incineration or vetrification. Therefore this *less destructive* thermal treatment makes it possible to produce a sediment with the same characteristics as the original, without heavy modifications of the natural matrix. The advantage of conducting the desorption in a indirectly heating kiln with only a purge gas stream is that a minimum quantity of off-gases remain which require treatment in the afterburner chamber and in the gas cleaning train, before they can be safely released to the atmosphere.

Solubilization and reprecipitation of heavy metals seems practicable and efficient, even with high chemical consumption caused by the polluted matrix. A more acid-resistant matrix should permit lower treatment costs. In any case, the same heavy metals solubilization tests, performed both in untreated and treated materials, showed a significant lower metals mobility in treated material; that means it can be considered a residue in which inorganic toxic pollutants are immobilized and leaching tests performed on this materials could give a leachete without toxic pollutants.

Combination of thermal desorption with solubilization treatment of residual heavy metals can produce a fully decontaminated sediment with potentially beneficial use in restoration without any restrictions, or for end-products like topsoil and bricks.

REFERENCES

Alcock, R.E., A. Marcomini, K. Jones. 1999. "PCDD/Fs and PCBs in Sediment Samples from the Venice Lagoon" *Organohalogen Compounds 43*:339-342.

Bassetti A., G. Mascolo, and R. Rausa. 1999. "Thermal Treatment of Sediments as a Function of Temperature and Reacting Atmosphere". *J. Anal. Appl. Pyrolysis 52*: 115-135

Bassetti A., E. Berneschi, C. Cozzolino, M.C. Cristofori, A. Gorni, R. Miglio, A. Rollo. 1999. "Fate of PCDD/Fs in Thermal Treatment of Polluted Sediments." *Organohalogen Compounds 43*: 387-391.

Bellucci, L.G., C. Carraro, M. Frignani, S. Raccanelli. 1999. "PCB in sediments of the Venice Lagoon and their toxicity with respect to other organic microcontaminants. " *Organohalogen Compounds 43*:295-298.

Bellucci, L.G., C. Carraro, M. Frignani, and S. Raccanelli. 2000. "PCDD/Fs in Surficial Sediments of the Venice Lagoon" *Marine Pollution Bulletin 40(1)*:65-76.

Di Domenico, A, L. Turrio Baldassari. et al. 1998 *Organohalogen Compounds 39*: 205-209.

MODELING OF STEAM INJECTION PROCESSES IN HOMOGENEOUS POROUS MEDIA WITH HIGH TEMPERATURE GRADIENTS

Sven Crone, Karl Strauss (University of Dortmund, Germany)

ABSTRACT: A new modeling approach for the description of steam injection in porous media is presented and discussed in detail. These processes are used for an effective heating of a porous medium or the removal of organic phases from a porous structure as in the tertiary oil recovery or the decontamination of soils with the Soil Vapor Extraction process. Due to the intensive condensation of the vapor a transient multicomponent, multiphase flow occurs which is characterized by the appearance of steep pressure and temperature gradients. Therefore, besides a detailed description of the heat transfer between the fluid and the solid phase, it is necessary to carefully model the momentum balance of the separate fluid phases. A thorough presentation of the new modeling approach and the numerical implementation is given. Finally some numerical results are presented.

INTRODUCTION

Over the last years multiphase transport phenomena in porous media have gained an increasing importance, since a variety of technical applications connected to these processes came up. The injection of saturated steam in a porous medium for an effective heating represents the basic case for this type of flows. The porous structure is initially filled with humid air at ambient conditions and steam is then applied to the porous medium through a pressure gradient from the top of the column. Due to the temperature difference between the solid and fluid phase an intensive condensation occurs. This condensation front propagates through the column heating the solid. According to the phase distribution of liquid and vapor 5 different regions can be detected, compare Figure 1. For a distinct moment the profiles of the temperature T, pressure p and the saturation s are plotted over the length of the column. The saturation gives the volume fraction of the pore space which is filled with the liquid phase:

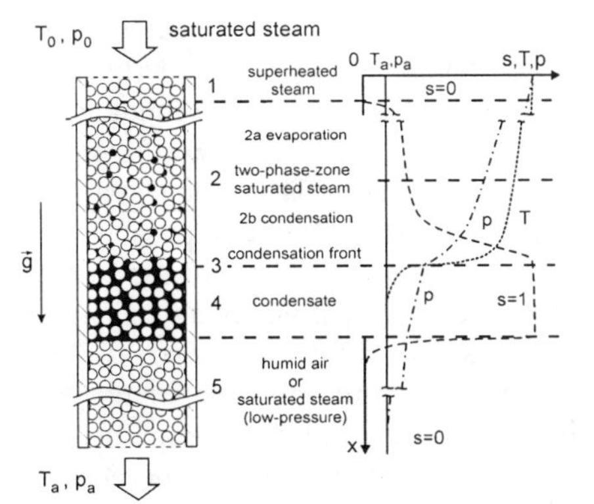

FIGURE 1: Model of steam injection

1) *Superheated steam (s=0):* After the condensation front (3) has passed, immobile liquid near the entrance is evaporated due to the expansion of the entering steam

2) *Two-phase-region:* In the two-phase region the fluid is in a saturated state. Saturation increases monotonously up to s=1 at the condensation front.

2a) *Evaporation region:* Due to the expansion of the entering steam immobile water can be evaporated. As a result the saturation decreases.

2b) Condensation region: Already upstream the actual front, considerable condensation takes place. A high pressure drop combined with a strong increase in saturation is observed immediately upstream of the condensation front.

3) Condensation front: At the condensation front, the condensation of the remaining steam occurs by transfer of its latent heat to the solid material. Energy is also transported downstream into the liquid phase region (4) by heat conduction.

4) Fluid region (s=1): Whereas the pressure drop is nearly linear because of the almost incompressible character of the liquid, the temperature downstream the condensation front decreases fast caused by heat transfer and heat conduction.

5) *Downstream region:* The downstream region is filled with inert gas at ambient temperature. The gas flow rate in this region is similar to that of the condensate in region (4). Therefore the pressure decreases slowly.

The dewatering of lignite by the mechanical/thermal dewatering (MTE) (Berger et al, 1999) should be mentioned as a technical application for this process. After heating a bulk of brown coal from ambient level to temperatures of about 200°C with steam the water can be removed in a liquid state by the application of mechanical forces. The application of steam injection for the extraction of organic fluids from a porous structure represents an extension of the process to a more complex multicomponent system. Besides the aqueous and the gas phase a separate organic phase, the non-aqueous phase liquid (NAPL), exists. These systems can be found in petroleum reservoir engineering as the tertiary oil recovery or in environmental engineering as the thermally enhanced Soil Vapor Extraction for the remediation of contaminated soils.

MODEL

For a deeper understanding and optimization of the processes a detailed model is necessary. Motivated by technical processes like the mechanical/thermal dewatering the present study is especially aiming at multicomponent multiphase flows with phase change with a high spatial and transient temperature gradient. A major requirement for the modeling of these processes is a detailed description of the heat exchange between solid and liquid. The assumption of local thermal equilibrium (LTE) for example limits the application of the model to small temperature differences, slowly propagating fronts or very small particles. A detailed description of the heat transfer should also incorporate the limitation of heat transfer to the solid by heat conduction inside the particles. Furthermore a main characteristic of the mechanical/thermal dewatering is the appearance of steep spatial and temporal temperature gradients. They lead to high phase velocities especially close to the condensation front. Therefore the application of the Darcy law is not possible, but the use of an extended momentum balance is necessary. On the one hand due to the high pressure gradients the flow is dominated by the convection of the multiphase mixture and any kind of capillary pressure effects can be neglected. In addition the change of fluid properties like viscosity or heat conductivity needs to be taken into account. For this purpose the fluid properties need to be implemented as exact functions of the intensive variables of state. Another difficulty in modeling these processes concerns the existence of different zones, compare Figure 1,- and as a consequence the appearance or disappearance of phases. Considering these special characteristics it seems that existing models, as they are summarized in (Miller et al. 1998) for example, are not suited to properly reproduce the physics of multicomponent multiphase processes with phase change with high temperature gradients. Therefore, the new model has been developed especially for two-phase flows as they occur during the steam injection for the purpose of heating. Now this basic model, compare (Crone and Bergins 2000), has been extended to multicomponent flows including a separate NAPL phase.

Multiphase/Multicomponent System. The model is suited for the description of a non-isothermal multiphase multicomponent system, which consists of a vapor(v), a liquid or aqueous(l) and an organic (o) phase with the three components water (w), air(a) and an organic chemical (c). The composition of the different phases is presented in Figure 2.For the further mathematical formulation of the model the phases will be denoted by the subscripts v,l and o whereas the components will be characterized by superscripts w, a and c. The organic phase Consists of a pure chemical component (c). Properties of some typical petroleum or halogenated hydrocarbons, like o-Xylole, Dodecane, Benzene and TCE have been implemented in component libraries which can be used to simulate different multicomponent systems. Whereas the solubility of water(w) in the organic phase is neglected a small amount of the organic component can be found in the aqueous phase. Both liquid phases contain no dissolved air. The vapor phase consists of air(a), water(w) and organic vapor(c).

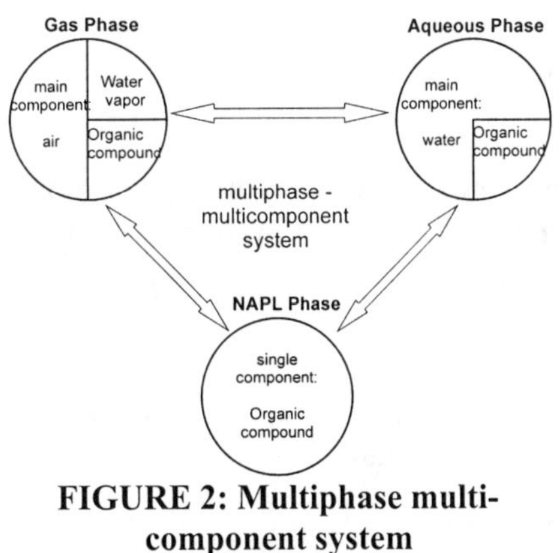

FIGURE 2: Multiphase multicomponent system

Multiphase Mixture. By the use of a mixture approach the number of governing differential equations decreases and phase transition terms are avoided. In this case all fluid phases are regarded as a multiphase mixture which can be characterized by a mixture density ρ and a mixture enthalpy h. The mixture density ρ is given by the single phase densities ρ_i and the individual saturations s_i:

$$\rho = s_L \, \rho_L + s_V \, \rho_V + s_o \, \rho_o \tag{1}$$

The saturation s_i gives the volume fraction of the pore space which is occupied by the phase i. Accordingly the specific mixture enthalpy h can be calculated from the densities ρ_i and enthalpies h_i of the single phases i, the saturations s_i and the mixture density ρ:

$$h = \frac{1}{\rho}\left(s_L \, \rho_L \, h_L + s_V \, \rho_V \, h_V + s_O \, \rho_O \, h_O\right) \tag{2}$$

Besides the mixture density ρ, which represents the total fluid mass, furthermore the quantification of the amount of chemical substance is necessary. For this purpose the species mixture density ρ^C for the chemical component is introduced. It can be calculated from the densities ρ_i of the single phases i, the phase saturations s_i and the mass-content of the chemical substance in the different phases w_i^C:

$$\rho^C = w_L^C s_L \rho_L + w_V^C s_V \rho_V + w_o^C s_o \rho_o \tag{3}$$

An effective mixture velocity u can be used to describe the convection of the fluid mixture. By introducing this variable the formulation of the conservation equation for mass can be simplified. The mixture velocity u results from the superficial velocities $u_{i,0}$ of the single phases i, which are calculated from separate momentum balances.

$$u = \frac{1}{\varepsilon}\frac{1}{\rho}\left(\rho_L u_{L,0} + \rho_V u_{V,0} + \rho_o u_{o,0}\right) \tag{4}$$

Analogous the species mixture velocity u^C for the chemical component and the velocity for the convective transport of energy w can be defined as:

$$u^C = \frac{1}{\varepsilon}\frac{1}{\rho}\left(w_L^C \rho_L u_{L,0} + w_V^C \rho_V u_{V,0} + w_o^C \rho_o u_{o,0}\right) \tag{5}$$

$$w = \frac{1}{\varepsilon}\frac{1}{\rho}\frac{1}{h}\left(h_L \rho_L u_{L,0} + h_V \rho_V u_{V,0} + h_o \rho_o u_{o,0}\right) \tag{6}$$

Conservation of mass. Using the effective mixture velocity u and the mixture density ρ the following differential equation of total fluid mass conservation is formulated:

$$\frac{\partial \rho}{\partial t} + \frac{\partial}{\partial x}(\rho u) = 0 \tag{7}$$

Similarly the mass conservation for the chemical component c can be expressed with the species mass density ρ^C and the species mixture velocity u^C as:

$$\frac{\partial \rho^C}{\partial t} + \frac{\partial}{\partial x}\left(\rho^C u^C\right) = 0 \tag{8}$$

The mass balance of the air as the inert vapor component is not expressed with mixture quantities, but with vapor phase properties and the void fraction ε:

$$\frac{\partial\left(\varepsilon s_V \rho^a\right)}{\partial t} + \frac{\partial}{\partial x}\left(\rho^a u_{V,0}\right) = 0 \tag{9}$$

Due to the high inertia of the modeled processes the propagation of the front is dominated by convection and diffusion can be neglected. On account of the mixture approach besides the equations 7, 8 and 9 no further balance equations are necessary for describing the mass conservation.

Momentum Balance. The description of the pressure loss of a phase i in a multiphase flow bases on the Ergun-equation. The porous medium is characterized by the permeability *K*, the passability η and the porosity ε, the fluid by the viscosity μ and the density ρ.

$$\frac{\partial p_i}{\partial x} = \underbrace{-\frac{\mu_j}{K \cdot k_{Ri}} \cdot u_{i,0}}_{\text{viscous drag}} \underbrace{-\frac{\rho_i}{\eta \cdot \eta_{Ri}} \cdot u_{i,0}^{\ 2}}_{\text{inertial drag}} + \underbrace{\rho_i \cdot g}_{\text{gravity}} \pm \underbrace{K_{int} \cdot (u_i - u_k)^2}_{\text{interfacial drag}} \qquad (10)$$

The viscous term describes the shear stress dominated flow (Darcy flow), the inertial term takes the effects at higher Reynolds numbers into account. In the present study the use of this extended momentum balance (Schulenberg, Mueller 1987) is necessary, since the vapor phase is accelerated towards the condensation front. The reduction of the free flow area of a phase due to the presence of another phase is considered by applying the concept of relative permeability and in the same way a relative passability η_{Rj} is introduced. By the use of an additional term called interfacial drag force in Eq.10 it is taken into account that interactions between the different phases lead to an additional pressure loss. The relative permeabilities and passabilities are mainly dependent on the saturation and have to be determined experimentally for each individual system. The simulations with the present model are currently validated with experiments using a bed of glass spheres and therefore the following correlations are chosen based on first experimental results:

$$k_{RL} = \eta_{RL} = \left(\frac{s - s_{im}}{1 - s_{im}} \right)^3 \qquad\qquad k_{RV} = \eta_{RV} = (1 - s)^3 \qquad (11)$$

Whereas the relative permeabilities and passabilities of the liquid and vapor phase are depend only on their own saturation the value of the relative permeability of the organic phase depends on all phase saturations and needs to be calculated from two-phase data for organic-water and organic-vapor systems. For this purpose the empirical Stone I method (Stone, 1970) which was normalized by (Aziz, Settari 1979) is used.

Energy Balance: Due to the appearance of high temperature gradients and the use of big particles the assumption of local thermal equilibrium is not appropriate and therefore separate energy balances for fluid and solid phase are required. Using the mixture enthalpy and the corresponding velocity w the *Fluid Phase Energy Balance* can be written as (Equation 12):

$$\frac{\partial (\rho h)}{\partial t} + \frac{\partial}{\partial x}(\rho h u) = \frac{1}{\varepsilon}\left[\frac{\partial}{\partial x}\left(\lambda_{eff} \frac{\partial T_F}{\partial x} \right) + (\alpha_{LS} a_{LS} + \alpha_{VS} a_{VS} + \alpha_{OS} a_{OS})(T_S - T_F) \right]$$

According to the "volume averaging concept" the effective thermal conductivity λ_{eff} indicates the effective thermal conductivity over the whole cross section of the bed, including the fluid multiphase mixture and the solid phase. The heat transfer between fluid and solid phase is calculated by the temperatures of both phases T_F and T_S, the heat transfer coefficients solid/liquid (α_{LS}) and solid/vapor (α_{VS}) as well as by the corresponding specific surface areas which depend on the different phase saturations. The heat transfer coefficients are calculated separately for each phase using the correlation from (Schluender, Tsotsas, 1988). The limited thermal conductivity of the solid material decreases the transferred amount of energy and leads to a radial temperature profile within the particles. Therefore the

Solid Phase Energy Balance leads to an additional spatial dimension and is formulated as follows:

$$\frac{\partial \rho_S c_{pS} T_S}{\partial t} = \frac{\partial^2 \lambda_S T_S}{\partial r^2} \tag{13}$$

Thermodynamic Relations: Due to the strong gradients of the intensive variables of state a drastic alteration of the fluid properties is taking place. Therefore in every iteration step it is necessary to update the properties of the fluid phases. The thermodynamic properties of pure water are calculated with the help of a new formulation of the steam table (Wagner et al.,2000). A numerical implementation of the VDI guideline (VDI4670) is used to represent the air and for the calculation of the pure chemical substance properties formulas from different authors are used. Furthermore the property functions are an essential part of the backwards calculation which represents the junction between the mixture modeling and the treatment of separate phases. The backwards calculation is aiming at the determination of the intensive variables of state from the quantities of the multiphase mixture. For this purpose the definitions of the mixture quantities and additional thermodynamic relations which describe the phase equilibria are used. Mathematically the backwards calculation represents a set of nonlinear equations and is solved by applying a multidimensional Newton procedure.

NUMERICS

Numerical Implementation: The described model was numerically implemented in the one-dimensional case. The time discretization of the differential equations is done in accordance with Crank-Nicholsen on two time steps. On a spatially staggered grid a finite differencing scheme is used, convective transport terms are represented by a first order UPWIND scheme, diffusive terms with central differences. The Thomas Algorithm is applied as the solver for the partial differential equations.

Boundary Conditions: The initial state is specified by the given intensive variables of state pressure, temperature and saturation and by setting the start concentration in the case of multicomponent systems. The thermodynamic state of the fluid is set by a Dirichlet-boundary condition for the intensive variables of state at the entrance to the porous medium. By this choice of boundary conditions, the entering mass flow rate is not required. Therefore the velocity of the condensation front is not set to a constant value. Consequently the flow is only driven by the impact of the initialized pressure difference.

RESULTS

The injection of steam in a homogenous bed of glass spheres was simulated. The porous medium is filled with air, a small amount of water and NAPL (Xylole, s_o=0.08) and is characterized by a porosity of ε=0.4 and a mean particle diameter of d_p=114µm. The entering vapor is saturated at a temperature of 413,15 K, whereas the porous medium initially exists at ambient level 303,15 K. Figure 1 shows the distribution of the field variables over the length of the column with the time as a parameter. Solving the partial differential equations first provides the profiles for the mixture values. The position of the condensation front can be detected as the minimum of the mixture enthalpy (a) and the maximum of the density (b), which is caused by the specific values of the water which is generated during the condensation. The profile of the species density ρ_C (c) visualize the removal of the NAPL phase, which is pushed towards the exit by the front.

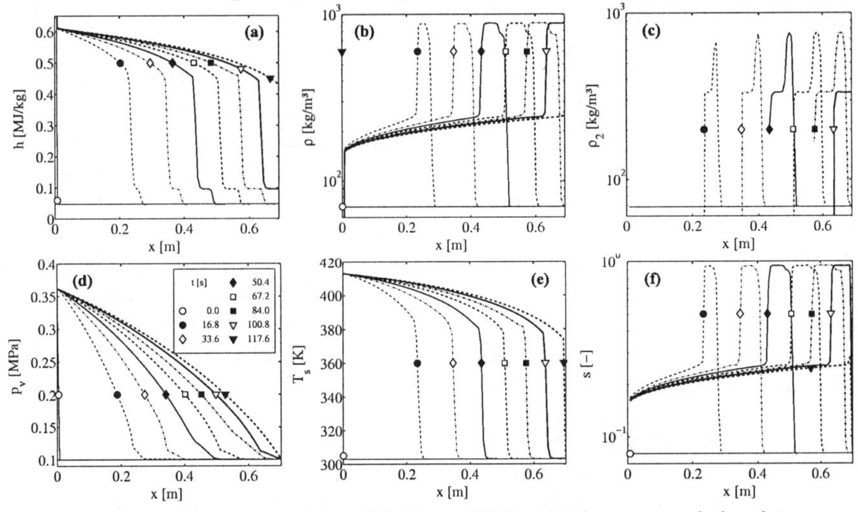

FIGURE 3: Distribution of field variables during steam injection

During the backwards calculation the intensive variables of state are calculated. The pressure profiles can be divided into two regions, one representing the flow of the compressible vapor, compare zone 2 in Figure 1. The remaining water in this section is immobile and only removed by evaporation whereas the organic substance in this part of the porous medium has been completely removed. Further down-stream a linear profile as an indication of the incompressible flow of a NAPL-water mixture, compare zone 4 in Figure 1, can be detected. The condensation front with a steep temperature gradient (e), which clearly indicates the excellent heat transfer conditions in this section, separates these two main sections. As a consequence the profile of the total saturation (f) results which shows a strong increase towards the condensation front. The NAPL and most of the condensed water is moved downstream the condensation front into the condensate region which as a consequence grows in size over time. The phase distribution in this region is analyzed in Figure 4 which shows the profiles of the normalized temperature and the liquid, organic and total saturation over the position for a distinct moment. Constant values for the organic and aqueous saturations can be observed in the condensate region. This effect is caused by the equality of the phase velocities for these two phases due to the dependency of the relative permeabilities from the saturation. Experiments confirm this phenomenon and for the same reason the total saturation does not exceed a maximum value of s=0.92. At this saturation level the organic and liquid phase and the gas phase have the same velocities.

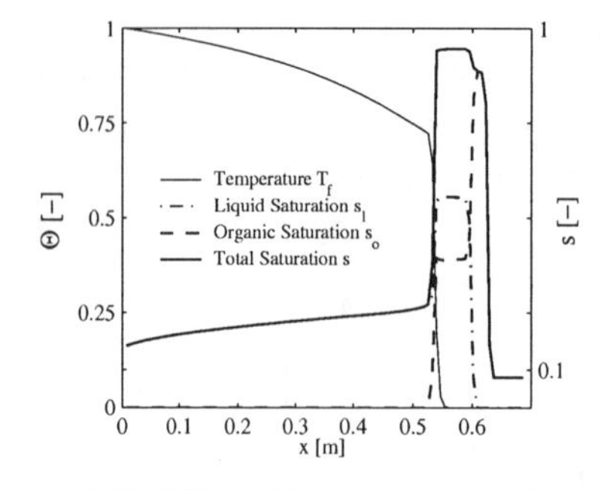

FIGURE 4: Phase Distribution

CONCLUSIONS

A new approach for the modeling of multicomponent multiphase flows with phase change which is especially suited for the description of processes with high temperature and pressure gradients is presented. These processes are dominated by convective transport mechanisms and require a detailed description of the heat and momentum transfer. The simulation results are in good agreement with experimental results during the development of the mechanical/thermal dewatering and show good numerical efficiency. Currently an extended experimental validation of the model is done and furthermore an extension of the model for three dimensional processes is executed, since this is necessary for the investigation of the uniformity of these flows in heterogeneous porous media.

ACKNOWLEDGEMENTS

The present research work was financially supported by the Deutsche Forschungsgemeinschaft (DFG) and by a research grant of the RWE Energie AG.

REFERENCES

Berger, S., C. Bergins, K. Strauß, F. B. Bielfeldt, R. O. Elsen and M. Erken. 1999. "Mechanical/Thermal Dewatering of Brown Coal", *VGB Power Tech.*79(2):44-49

Miller, C. T. G., G. Christakos, P. T. Imhoff, J. F. McBride, J. A. Petit and J. A. Tangenstein. 1998 "Multiphase Flow and transport modeling in heterogenous porous media: challenges and approaches", *Advances in Water Res.* 21(2): 77-120

Crone,S. and C. Bergins. 2000. "Transient Multiphase Flow of a multicomponent system with phase change in porous media" In Hahne, E.W.P., W. Heidemann and K. Spindler (Eds.). *3rd European Thermal Sciences Conference.*, pp. 497-502. Edizioni ETS, Pisa

Schulenberg, T. and U. Mueller. 1987. "An improved model for two-phase flow through beds of coarse particles". *Int. Journal of Multiphase Flow.* 13(1): 87-97

Stone, H.L. 1970 "Probability Model for Estimating Three-Phase Relative Permeability". *J. Pet. Tech.* 22(2): 214-218

Aziz, K. and T. Settari. 1979 *Petroleum Reservoir Simulation.* Applied Science Publishers, London

Wagner, W., J.R. Cooper, A. Dittmann, J. Kijima, H.-J. Kretschmar and A. Kruse. 2000 "The IAPWS Industrial Formulation 1997 for the Thermodynamic Properties of Water and Steam", *Transactions of the ASME* 122(2000): 150-182

Schluender, E.U and E. Tsotsas. 1988. *Waermeuebertragung in Festbetten, durchmischten Schuettguetern und Wirbelschichten.* George Thieme, Stuttgart

VDI. 2000. *Thermodynamische Stoffwerte von feuchter Luft.* VDI 4670

SEDIMENT DECONTAMINATION: EVOLUTION AND PRACTICAL APPLICATION OF BIOGENESIS℠ WASHING TECHNOLOGY

Mohsen C. Amiran (BioGenesis Enterprises, Milwaukee, Wisconsin)
Charles L. Wilde (BioGenesis Enterprises, Springfield, Virginia)

ABSTRACT: Research into the removal of organic and inorganic contaminants from soil and sediment, and the practical application of this research in field operations, has been in progress since 1991. The results described in this paper are the foundation of the BioGenesis℠ technical approach which uses both chemical and physical means to separate contaminated biomass from inorganic solids and then to clean the solids. Based on the results and subsequent bench and pilot testing, specialized equipment was developed to implement the washing process. At present, the decontamination washing process is being implemented in New York/New Jersey harbor with a capacity of 250,000 cubic yards per year (191,000 cubic meters per year).

INTRODUCTION

Soil washing is an ancient technology, and has been practiced in the mineral extraction industry for centuries. More recently, attrition scrubbing techniques have been used in Europe for treatment of soils consisting of gravel and sand which contain both organic and inorganic contaminants. However widespread application of washing technology has been limited by certain factors illustrated by the following excerpts from government sponsored technology assessments:

- *Fine soil particles (silts, clays) are difficult to remove from washing fluid.*
- *Complex waste mixtures (e.g. metals with organics) make formulating washing fluid difficult.*
- *High humic content in soil inhibits desorption* (U.S. EPA, 1993, *Remediation Technologies Screening Matrix and Reference Guide*).

and:

> *Unit processes have proven excellent for media such as sand and gravel, but need to be improved for fine-grained soils where more than 30-35% of particles are less than 0.063 mm* (NATO/CCMS, 1998).

BioGenesis constructed its first soil washing gondola in 1991, entered the U.S. EPA's Superfund Innovative Technology Evaluation (SITE) program, and demonstrated soil washing technology during 1992 on 3,800 tons of soil at an oil refinery site in Minnesota (U.S. EPA, 1993, *BioGenesis℠ Soil Washing Technology*). During plant operation, both the smaller grain sizes of silt and clay, about 11% of the total, and the more heavily weathered oil contaminated particles, were not cleaned as efficiently as the larger particles of sandy soil greater than 0.063 mm in size. Subsequently, BioGenesis sought ways to improve treatment of small particles, and working with Environment Canada, bench tested a new washing technique. The new approach is believed to be the first use of washing to effectively decontaminate fine grained silt and clay sediments contaminated with high levels (4,000 mg/kg) of poly-aromatic hydrocarbon (PAH) (Wastewater Technology Centre, 1993).

The research, reported for the first time in this paper, is the basis for the subsequent evolution of sediment washing equipment that has been successfully tested at large pilot scale (U.S. EPA Region 2, 1999), and is now being implemented at a full scale of 250,000 cubic yards per year in New York/New Jersey harbor.

Research Approach. In conjunction with the bench testing for Environment Canada, review of the washing literature showed that conventional washing approaches such as attrition scrubbing were less effective as the grain size of material decreased from sand to silt, and from silt to clay. Additionally, published writings indicated that washing efficiency also decreased as organic content of the soil/sediment increased. Missing from the literature, however, were answers to the questions that would allow improved contaminant extraction efficiency with full scale equipment:

1. What are the limitations of solvent extraction on organic contaminants in fine grained soil and sediment?
2. What is the role of biomass in absorbing organic contamination and preventing the contamination from being desorbed through washing?
3. How do inorganic contaminants partition in organic and inorganic solid particles?
4. How does the use of chemicals, impact pressure, and type of impact surface affect removal efficiencies in washing?
5. How fast does inorganic contamination transfer from liquid to solid, and how much contaminant is distributed between the inorganic solid and the organic solid?
6. What can be achieved with chemical extraction, what can be achieved with physical forces, and what is the optimum balance between the two?

Answers to these questions could lead to more effective chemicals and equipment for decontamination. Therefore the experiments in table 1 were performed, with results that will be discussed in turn.

Table 1. Experiment Summary

Experiment	Name	Purpose of Experiment
1	Solvent extraction of biomass	Evaluate chemical effectiveness in separating biomass from solids of differing soil grain sizes
2-1	Solvent extraction of #6 fuel oil tank bottoms plus biomass	Determine the correlation between TPH reduction and biomass reduction for differing soil grain sizes
2-2	Biomass absorption of TPH from contaminated water	Determine how biomass absorbs TPH from contaminated water for different soil grain sizes

Table 1. Experiment Summary		
Experiment	**Name**	**Purpose of Experiment**
3-1	Lead partitioning from water to the solid phase	Determine how lead in water partitions to differing soil grain sizes
3-2	Lead partitioning from water to the solid phase, biomass added	Determine the effect of biomass addition on partitioning of lead in water to differing soil grain sizes
4	Washing variables	Determine the effect of different surfaces, chemicals, and pressures on washing efficiency
5	Partitioning of lead among solid, organic, & water phases	Evaluate the partitioning of lead from the solid phase to the organic and water phases

MATERIALS AND METHODS

Composition of Test Soils. To enhance the reproducibility and repeatability of results, standard soil mixtures were used for all experiments. The soils were made by sieving a commercially available sand mixture into three size fractions. Grains between 2.0 to 0.15 mm were designated as sand; grains between 0.15 to 0.038mm were designated as silt. For the clay mixture, grains less than 0.038 mm were mixed with bentonite clay in a 40% fines - 60% clay ratio. Biomass was standardized through the use of a commercially available humus liquid.

Experiment 1-1, Solvent Extraction of Biomass. To evaluate chemical effectiveness in separating biomass from solids of differing grain sizes, 4% humus

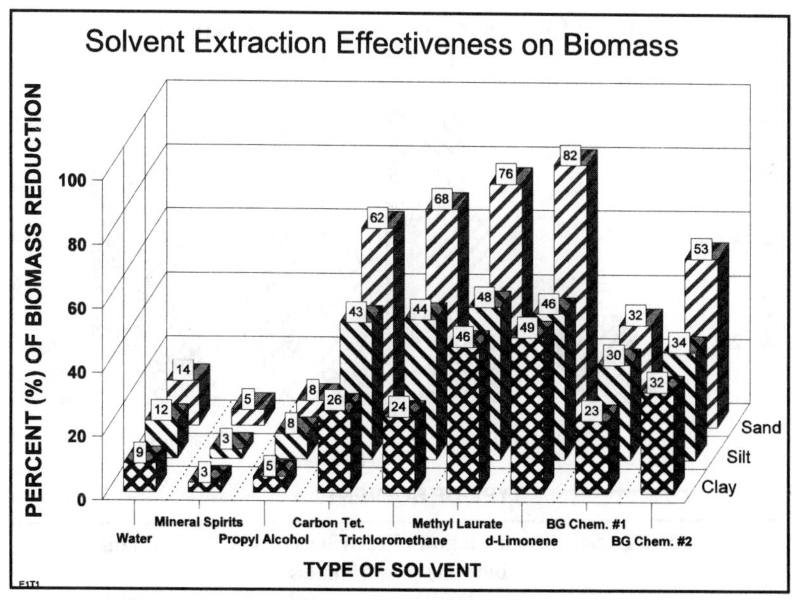

Figure 1. Solvent Extraction Effectiveness on Biomass.

was mixed with 100g each of sand, silt, and clay; mixed well with water; dried; and extracted for 30 minutes with a solvent. The solid phase was dried in a rotoevaporator, burned to remove the organic phase, and weighed before and after burning to determine the amount of organic that remained in the solid phase after the biomass extraction.

Figure 1 shows the results. Water, mineral spirits, and propyl alcohol were ineffective extracting chemicals. Carbon tetrachloride, trichloromethane, methyl laurate, and d-limonene had medium extraction rates with the notable exception that the chlorinated solvents had lower effectiveness on clay. Finally, BioGenesis water-based proprietary surfactant mixtures were competitive with the solvents, but with substantially lower handling hazards. This experiment demonstrated that biomass binds very tightly to silt and clay particles.

Experiment 1-2, Solvent Extraction of #6 Fuel Oil Tank Bottoms Plus Biomass. To determine the correlation between TPH reduction and biomass reduction for differing soil grain sizes, experiment 1-1 was repeated with the exception that 10% of #6 fuel oil tank bottoms was added to the mixtures. Tank bottoms were selected due to their high concentration of poly-aromatic hydrocarbon (PAH).

Figure 2 shows the results. Within experimental error, the results of the extractions were the same as for experiment 1-1. The result was surprising due to the large amount of TPH. This led to the conclusion that hydrocarbon mixed with biomass becomes an integral part of the biomass phase that binds very tightly to silt and clay particles.

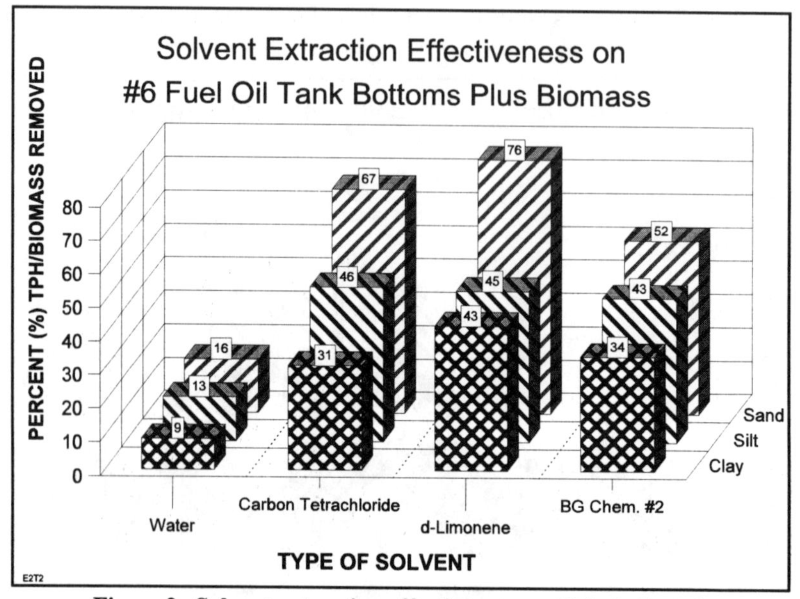

Figure 2. Solvent extraction effectiveness on #6 fuel oil tank bottoms plus biomass

Experiment 2, Biomass Absorption of TPH from Contaminated Water. To determine how biomass absorbs TPH from contaminated water for different soil grain sizes, 4% humus was mixed with 100g each of sand, silt, and clay. To the soil samples was added a mixture of 10% hydrocarbon in water. At 30, 60, 90, and 120 days, samples of the solids were dried in a rotoevaporator, burned at high temperature, and weighed before and after to determine the amount of hydrocarbon that had been absorbed by the sample.

Figure 3 shows the results. The sand, silt, and clay samples each absorbed TPH at a steady rate. However the clay/biomass sample absorbed 2.3 times the amount of TPH absorbed by the sand/biomass sample, and 1.5 times the amount of TPH absorbed by the silt/biomass sample. This illustrates the greater affinity of hydrocarbon for solids/biomass than for water.

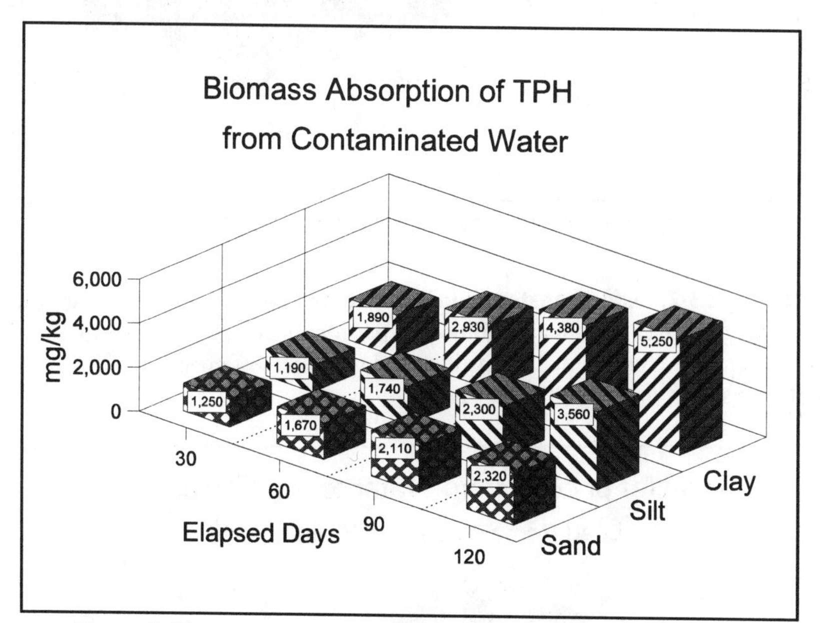

Figure 3. Biomass Absorption of TPH from Contaminated Water.

Experiment 3-1, Lead Partitioning from Water to the Solid Phase. To determine how lead in water partitions to differing soil grain sizes, a mixture of water containing 5,000 mg/kg lead nitrate was mixed with 100g each of sand, silt, and clay. At 15, 30, 60, 90, 120, and 180 days, the samples were mixed, and the lead content of the water phase was determined.

Figure 4 shows the results. The lead remaining in the water phase decreased gradually, and the clay and silt fractions exhibited the same general behavior in absorbing lead as was shown in figure 3 for TPH, though the absorbency differences between clay, silt, and sand were not as large.

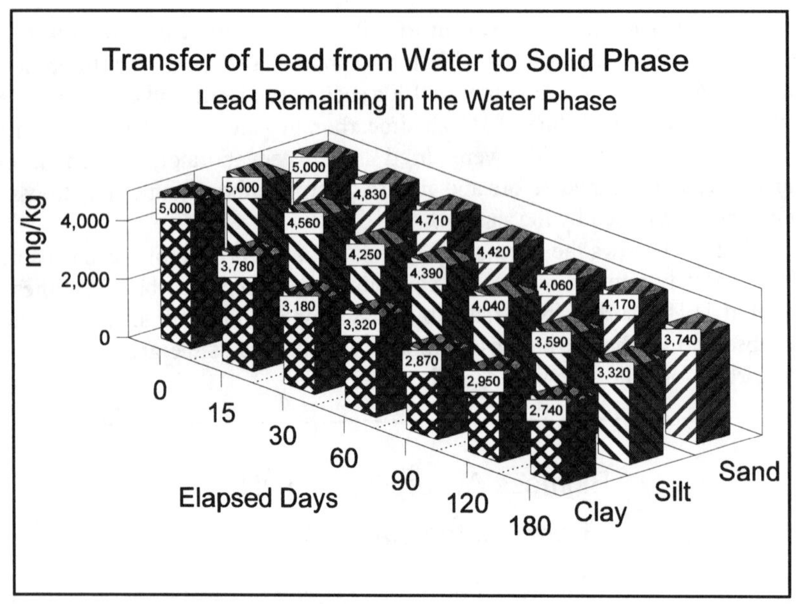

Figure 4. Transfer of Lead from Water to Solid Phase.

Experiment 3-2, Lead Transfer from Water to the Solid Phase, Biomass Added
To isolate the effect of biomass addition on partitioning of lead in water to differing soil grain sizes, experiment 3-1 was repeated with one difference. On the 15th day, 2% biomass was mixed with each sample.

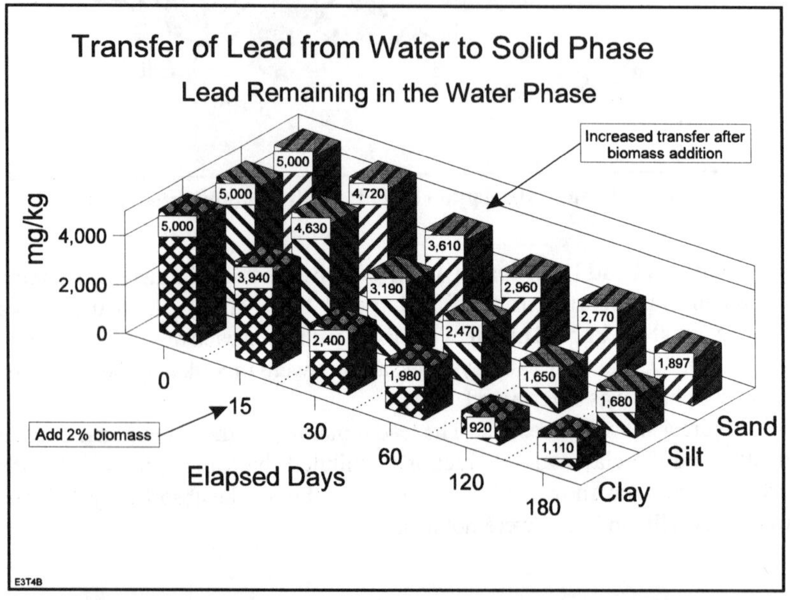

Figure 5. Transfer of lead from water to biomass/solid phase.

The results are shown in figure 5. Absorption of lead at day 15 was about the same as in experiment 3-1. However after biomass was added at day 15, the absorption rate increased to day 30, and then continued at a higher rate than in experiment 3-1. The conclusion is that biomass not only shows strong affinity for organic contaminants (experiment 2), but also for inorganic contaminants.

Experiment 4, Washing Variables. To explore the effect of different surfaces, chemicals, and pressures on washing efficiency, 2.5 mm (1 inch) tubes of aluminum, copper, carbon steel, and stainless steel, each 1.8 m (6 feet) long were filled with irregularly shaped pieces of the same material as the tube. The reason for filling the tubes was to provide multiple scrubbing surfaces in each tube for the sediment slurry to be pumped over. Two slurry mixtures were then prepared consisting of 5% silt, 5% clay, and 90% water. Both mixtures were contaminated with 5,000 mg/kg lead. One of the slurry mixtures also had 2% BioGenesis™ washing chemical added. Then both slurry mixtures were pumped through each of the four types of tubes at pressures of 3.4, 6.8, 34, and 68 bar (50, 100, 500, and 1,000 psi respectively).

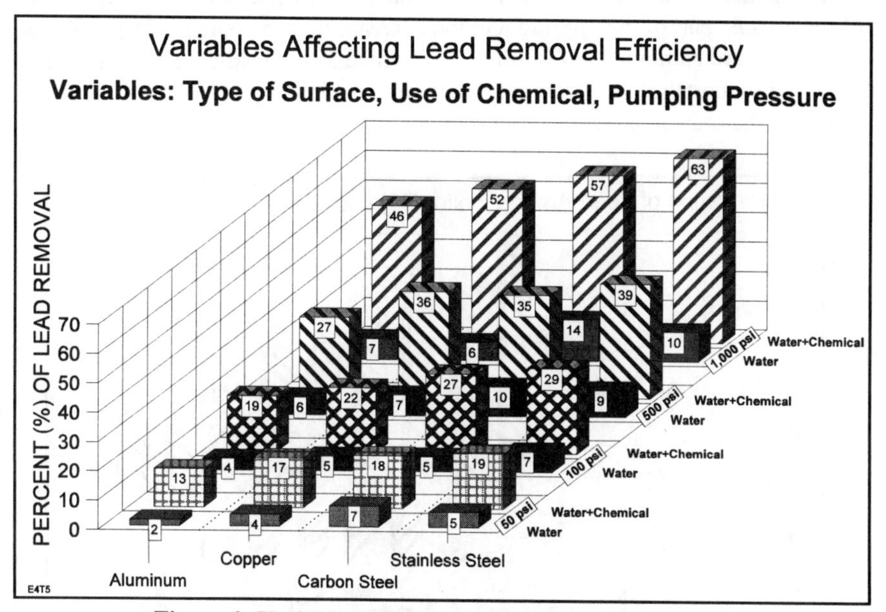

Figure 6. Variables Affecting Lead Removal Efficiency

Figure 6 shows the results. For all cases, use of chemical was clearly superior to water alone by factors of 200% to as much as 700%. In all cases, higher pressures were better at desorbing lead than lower pressures. In the washing chemical experiments, the largest gain in efficiency occurred when pressure was increased from 34 bar (500 psi) to 68 bar (1,000 psi). Assessing the impact of different metals was not conclusive because of experimental conditions where the size, distribution, and position of metal shapes in the tubes could not be controlled. Nonetheless, washing effectiveness apparently increased somewhat as the hardness of the metal tube increased. Additional research was not undertaken on this factor because stainless steel was already the clear choice for manufacturing equipment for durability reasons.

Experiment 5, Partitioning of Lead among Solid, Organic, & Water Phases. To evaluate the partitioning of lead from the solid phase to the organic and water phases, a mixture consisting of 30% sand, 30% silt, 40% clay, and 0.1% humus was prepared. To the mixture was added 5,000 mg/kg lead nitrate in water. Mixing was done very slowly in order minimize transfer due too mixing as a factor in the experiment. At 10, 20, 30, 60, 90, 120, 150, and 180 days, samples were taken, solvent extracted, and the extract tested for lead content. Aliquots of each sample were also analyzed to determine the lead content remaining in the soil mixture. The soil samples were then dried in a rotoevaporator, burned to remove the organics, and weighed before and after burning.

Figure 7 shows the results. Assuming the partitioning of lead throughout the organic phase of the solid is uniform (this is not proven), the increase in lead content of the organic part of the mixture and the decrease in lead content of the solid part of the mixture combined with little or no change in the lead content of the water phase, all support the hypothesis of inorganic contaminants partitioning to, and being stable, in the organic phase of sediments.

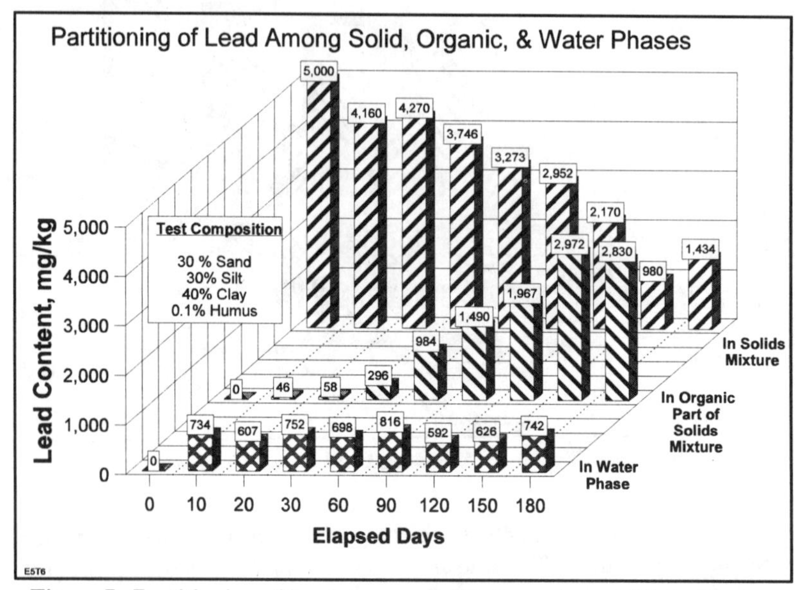

Figure 7. Partitioning of Lead Among Solid, Organic, and Water Phases.

IMPLICATIONS FOR FULL-SCALE SEDIMENT WASHING OPERATIONS

Table 2 summarizes the results of the experiments, their conclusions, and the practical implications for washing technology development.

Table 2. Experimental background for development of BioGenesis^{SM} Washing Technology

Experiment	Name	Conclusion	Conclusion: Practical Implication for Washing
1-1	Solvent extraction of Biomass	Biomass is strongly bound to silt/clay particles	If contaminants are in the biomass, chemical alone cannot clean completely
1-2	Solvent extraction of TPH & Biomass	Hydrocarbon mixed with biomass becomes an integral part of the biomass phase	Chemicals alone cannot clean completely
2	Biomass absorption of TPH from contaminated water	Hydrocarbon has more affinity for biomass and solids than for water	Should focus on separating the organic phase from solids
3-1	Partitioning of lead in the water phase	Lead partitions gradually to the solid over time; somewhat faster to smaller particles	Expect more contamination in smaller particles
3-2	Partitioning of lead to the organic/solid phase	Lead partitions much faster and more completely in the presence of biomass	Should focus on separating the organic phase from solids
4	Variables affecting removal efficiency	Chemicals and higher pressures enhance scrubbing effectiveness	Use chemicals at as high a pressure as possible
5	Solvent extraction of biomass	Lead partitions strongly to the organic phase	Should focus on separating the organic phase from solids

From the basic experiments described in this paper, BioGenesis evolved the concept of combining tailored chemicals with specialized sediment washing equipment that would result in desorbing contaminants and biomass containing contaminants from inorganic solids into the water phase. This concept is summarized in table 3, with the specialized equipment implementing the physical means and their purpose illustrated in figures 8, 9, and 10.

Table 3. BioGenesis^{OM} Washing Concept		
Means Employed	**Extract Contaminants Using**	**Destroy Contaminants Using**
Chemical Means	• Tailored Washing Chemicals	• Chemical Oxidation • Chemical Degeneration
Physical Means	• Preprocessor Equipment • Collision Chamber	• Cavitation

The BioGenesis^{SM} Sediment Preprocessor uses water at 680 bar (10,000 psi) pressure to fractionate sediments and separate biomass from inorganic solids. It processes 800 liters per minute (200 gpm) of a 35% sediment slurry.

Figure 8. BioGenesis^{SM} sediment preprocessor.

Figure 9. BioGenesis^{SM} Collision Chamber.

The BioGenesis^{SM} Collision Chamber uses water at 680 bar (10,000 psi) pressure to desorb contaminants from inorganic solid sediment particles. It processes 800 liters per minute (200 gpm) of a 35% sediment slurry, using one processing head. A second processing head is provided for use while the other is in maintenance.

The BioGenesis℠ Cavitation Unit uses induced cavitation and hydrogen peroxide to destroy organic contaminants. It processes slurry or water at rates of from 200 to 2,000 liters per minute (50 to 500 gpm) depending on the contaminant and its concentration in the slurry or water.

Figure 10. BioGenesis℠ Cavitation Unit.

CONCLUSION

This paper has described the experimental basis for BioGenesis℠ Washing Technology, and has shown how experimental results led to the development of specialized sediment processing equipment. Decontamination of sediments using washing technology is still an infant industry. However, BioGenesis has demonstrated that the decontamination of sediment is possible. All the results are promising and show that the washing process can be commercially viable. As commercial plants become more numerous and more practical operational data is obtained, the whole process can be optimized for different regions and contaminants.

REFERENCES

NATO Committee on the Challenges of Modern Society. 1998. *Evaluation of Demonstrated and Emerging Technologies for the Treatment and Clean Up of Contaminated Land and Groundwater, NATO/CCMS Pilot Study, Phase II Final Report*, p. 5-53. CCMS Report No. 219.

U.S. Environmental Protection Agency. 1993. *BioGenesis℠ Soil Washing Technology, Innovative Technology Evaluation Report*, EPA/540/R-93/510.

U.S. Environmental Protection Agency. 1993. *Remediation Technologies Screening Matrix and Reference Guide*, p. 43. EPA/542/B-93/005.

U.S. Environmental Protection Agency Region 2, U.S. Army Corps of Engineers, Brookhaven National Laboratory. 1999. *BioGenesis℠ Sediment Washing Technology, Full Scale, 40 cy/hr, Sediment Decontamination Facility for the NY/NJ Harbor Region, Final Report on the Pilot Demonstration Project.*

Wastewater Technology Centre. 1993. *Final Report, Bench Scale Studies of Thunder Bay Harbour Sediment, BioGenesis Washing Process.*

NEW SOLVENT EXTRACTION PROCESS FOR ORGANIC AND HEAVY METAL POLLUTANTS

Harry Bruning and Wim H. Rulkens
(Wageningen University, Wageningen, The Netherlands)

ABSTRACT: Research on laboratory scale has been executed into the feasibility of a new solvent extraction process to treat fine granular clayey sediments polluted with polyaromatic hydrocarbons (PAH), mineral oil and heavy metals. It was shown that with a water/acetone (40%) mixture as extracting agent, to which a small amount of EDTA was added as chelating agent, besides PAH and mineral oil, also substantial amounts of heavy metals could be removed. This process combines the advantages of a water/acetone mixture as extracting agent for almost complete removal of PAH and mineral oil and for efficient separation of the sediment particles from the extraction phase with the advantage of chelating agents such as EDTA to remove heavy metals. It may be expected that this process can be further optimised to make it feasible for practical application.

INTRODUCTION

Sediments contaminated with both organic pollutants, such as PAH and mineral oil, cause a serious environmental problem. This is especially the case if the sediment mainly consists of small clayey particles and contains a high percentage of organic matter. Clean-up of these sediments by bioremediation is difficult because the PAH are not or not easily biodegradable. Thermal treatment is possible. However, this process is not very costs effective due to the large amount of water in the sediment. A feasible alternative is solvent extraction (EPA, 1990; EPA, 1992[a]; EPA, 1992[b]; Rulkens et al, 1998) with a water-soluble organic solvent such as acetone. Acetone has several advantages:

- It has a high solubility for organic pollutants such as PAH and mineral oil, even when it is diluted with water.
- In the slurry of acetone/water/sediment the clayey sediment particles flocculate spontaneously to large porous particles, making the separation of the extraction medium and the sediment very easily compared with the separation of small clay particles from a water phase.
- Sediment organic matter hardly dissolves in an acetone/water mixture.
- Acetone can easily be completely recovered from the sediment and water phase and recycled in the clean-up process.

A general diagram of a solvent extraction process for wet dredged sediments using an acetone/water mixture as extracting agent, is given in Figure 1 (Rulkens et al, 1998). The process consists of four basic steps: extraction of pollutants, separation of sediment and extracting agent, removal of residual amounts of solvent from the treated sediment and recovery of the solvent for reuse. There exist several modifications of this process.

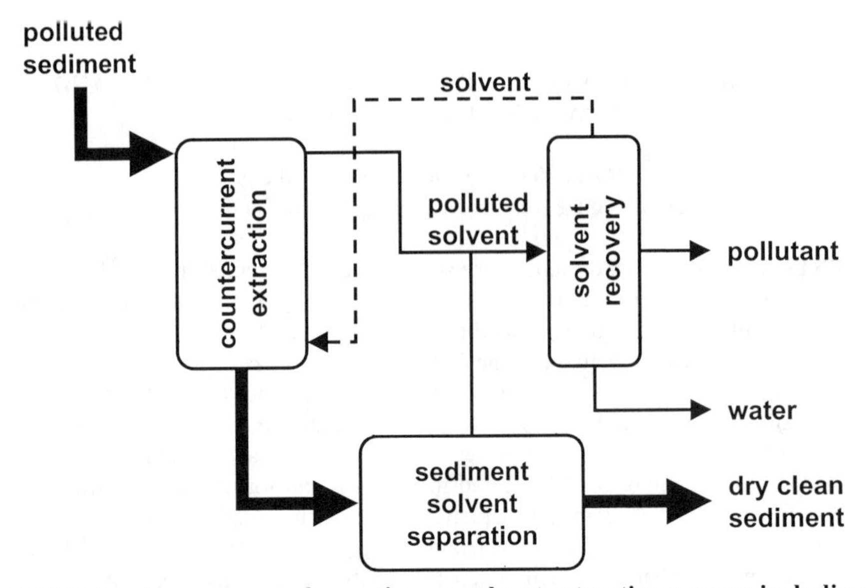

FIGURE 1. Flow scheme of a continuous solvent extraction process including extraction of sediment, sediment- solvent separation and solvent recovery.

A real disadvantage of conventional solvent extraction, but also of the two other clean-up processes, is that these processes can not remove heavy metals from sediments. Very often sediments contain, in addition to high levels of organic pollutants, substantial amounts of heavy metals. Acetone/water mixtures do not have much affinity for heavy metals. However, from a theoretical point of view it can be expected that due to the large percentage of water in such an extracting phase, addition of small amounts of chelating agents may be a solution for this problem. It is known from literature that chelating agents, present in a water phase, can be very effective for the removal of heavy metals from wastes and sludges (Peters and Young, 1988). If this property could be maintained in an acetone/water phase as extracting agent, than an interesting process can be developed in which all the advantages of the use of acetone/water as solvent could be combined with the advantage of the use of a chelating agent.

The objective of this study was to determine the feasibility of this approach. Therefore a lot of exploratory experiments on lab-scale were executed focused on the selection of suitable chelating agents, required amounts of chelating agents and measuring the achievable removal percentages of heavy metals.

EXPERIMENTAL

For the extraction experiments the fine granular fraction obtained as residue (sludge) from a washing plant for polluted soils was used as a model polluted sediment. This fine granular fraction comes free in the hydrocylone separation step of such a soil washing plant where the sand and silt fractions are separated from the fine clayey fraction. This clayey fraction in general contains the majority of the pollutants. It is more or less comparable with granular fine sediment. It contains often also a relatively large fraction of organic material.

The fraction obtained as sludge from the soil washing plant was subsequently pre-treated in a hydrocyclone in order to remove the particles with size > 50 µm. The composition of the remaining fine granular fraction used in the experiments is given in Table 1. Besides heavy metals this fraction contained also a substantially amount of PAH (16 EPA PAH: 310 mg/kg d.m.). The sludge was also polluted with a large amount of oil, estimated at approximately 1000 mg/kg d.m. From previous research (Hasselt, van et al., 1998; Rulkens et al., 1998) it was known that PAH and oil can easily be removed from this type of sludge by means of solvent extraction with a water/acetone mixture.

Two types of extraction experiments were executed. In order to determine the strength of the heavy metal binding in the sludge a sequential chemical extraction was applied to the sludge. Besides this sequential chemical extraction experiments extracting experiments with chelating agents dissolved in water or a water/acetone mixture were executed. The amount of soil and extracting agent in each of these extraction experiments was approximately 1 g and 10 ml respectively.

All experiments were carried out using a so-called end over end mixer at a temperature of 30 °C. The characteristic aspect of such a mixer is that a very intensive mixing is obtained by rotation and shaking.

Concentrations of heavy metals in the various samples obtained during the extraction experiments were analysed according to standard procedures.

Sequential chemical extraction. The sequential chemical extraction procedure comprises the use of a series of chemical extractions in a sequential order. In each extraction step, a specific chemical form of the metal is expected to dissolve. Several sequential chemical extraction schemes are discussed in literature. We have used the scheme according to Howard (Howard, 1995). With this scheme the following chemical forms in order of increasing binding strength of the metals can be analysed: free exchangeable metal ions, metal ions bound to carbonates, metal ions bound to manganate oxide, metal ions adsorbed to organic matter and metal ions bound to iron oxide.

The total amount of heavy metals was determined according to standard analytical procedures including a destruction step. From the total amount of heavy metals and the amounts measured in the sequential extraction steps the residual amount of heavy metals in the sludge, which can be considered as strongly bound to the sludge matrix, can be calculated.

Selection of chelating agents. In literature several extraction processes for heavy metals in wastes and sludges are described, in which an aqueous solution of an extracting agent in a water phase was used (Peters and Young, 1988). Most important agents are EDTA (ethylene-di-amine-tetra-acetic acid), PDA (pyridine-2,6-di-acetic acid) ADA (N-2-acetamido-imino di-acetic acid), NTA (nitrilo-tri-acetic acid) and citric acid. Best removal results were in general observed with EDTA. Regarding the behaviour of these chelating agents in acetone or a mixture of acetone and water, no literature information is available. Mainly on basis of results with aqueous solutions of chelating agents, mentioned in literature, three extracting agents were selected: EDTA, PDA and citric acid. This selection was supported by solubility experiments with these chelating agents in an acetone/water mixture (40% acetone).

Extraction experiments. With the selected chelating agents three types of extraction experiments were executed, using the end over end mixer as extraction equipment. The extraction liquid was always an acetone/water mixture (40% acetone) or water. The concentration of the chelating agent in the extracting liquid was 0 M, 0.005 M or 0.01 M. Extraction time amounted 2 hours or 20 hours.

RESULTS

Sequential chemical extraction. Some results of the sequential chemical extraction of the sludge are presented in Table 1. In this table the absolute values of the total amounts of each heavy metal initial present in the sludge (including As), the residual amount after completion of the sequential extraction procedure, as well as the percentage of heavy metals present in the sludge are given. It is clear from this table that the residual amount of heavy metal varies strongly with the type of heavy metal. From an environmental point of view the high concentrations of Pb and Zn are the most problematic.

TABLE 1 Results of sequential extraction.

Heavy metal	Total amount (mg/kg d.m.)	Residual amount (mg/kg d.m.)	Residual amount (percentage)
As	44	17	39
Cd	4.3	1.2	27
Cr	101	81	80
Cu	259	191	74
Ni	51	40	79
Pb	745	270	36
Zn	1197	327	27

Extraction with chelating agents. The results of the extraction experiments with EDTA are presented in Table 2. The table shows the removal percentages found for the heavy metals. Parameters are the composition of the extracting agent (water or water/acetone (40%) mixture), the concentration of chelating agent and the treatment time.

From the table it is clear that the removal percentage strongly depends on the type of heavy metal and the applied process conditions. From the results obtained after a treatment time of two hours can be concluded that Cd, Pb and Zn are removed readily. A maximum removal percentage of 48% is found for Pb. It is also evident that the removal efficiency increases with increasing amount of chelating agent. The observed removal percentages are slightly lower for water/acetone as extracting agent, than for water as extracting agent. The removal efficiencies for the other heavy metals (including arsenic) are relatively low. However, we can see that for these metals acetone has a positive effect on the removal percentage, especially for Cr and As. For water as extracting agent an increase in treatment time from two hours to twenty hours results in a substantially higher removal percentage of heavy metals. In this case the removal efficiency is less dependent on the concentration of the chelating agent.

TABLE 2. Removal percentages of heavy metals in water and 40% acetone/water at different EDTA-concentrations and extraction times.

	Water 0.005 M 2 hour	Water 0.01 M 2 hour	Water/Acetone 0.005 M 2 hour	Water/Acetone 0.01 M 2 hour	Water 0.005 M 20 hour	Water 0.01 M 20 hour
As	2	4	4	5	6	8
Cd	24	22	12	17	42	42
Cr	1	1	10	7	14	17
Cu	8	12	9	12	36	41
Ni	4	7	4	8	15	18
Pd	35	48	31	41	71	66
Zn	23	24	18	21	52	51

Similar results as observed for EDTA are found for PDA although the general tendency is that the removal efficiencies as an average are somewhat lower. There are small differences if the individual heavy metals are considered. Sometimes the removal percentage is slightly higher, sometimes slightly lower. With citric acid as complexing agent in general lower removal percentages are found compared with the percentages found for EDTA and PDA. It is remarkable that if PDA or citric acid is used as chelating agent, acetone has a positive effect on the removal percentage.

It can be expected that the residual amount of heavy metals, as can be measured by the sequential chemical extraction procedure, is not available for extraction with water or with a water/acetone mixture to which mall amounts of a chelating agent have been added. We can relate the observed removal percentages in the extraction process to the maximal amount of heavy metals which can be removed by sequential (chemical) extraction. For the metals Pb and Zn these corrected removal percentages are shown in Figure 2. For Pb removal percentages up to almost 80% are found. For Zn these removal percentages are up to 30 %. The figure also confirms the positive influence of a chelating agent and the fact that the influence of acetone on the removal percentage is negligible.

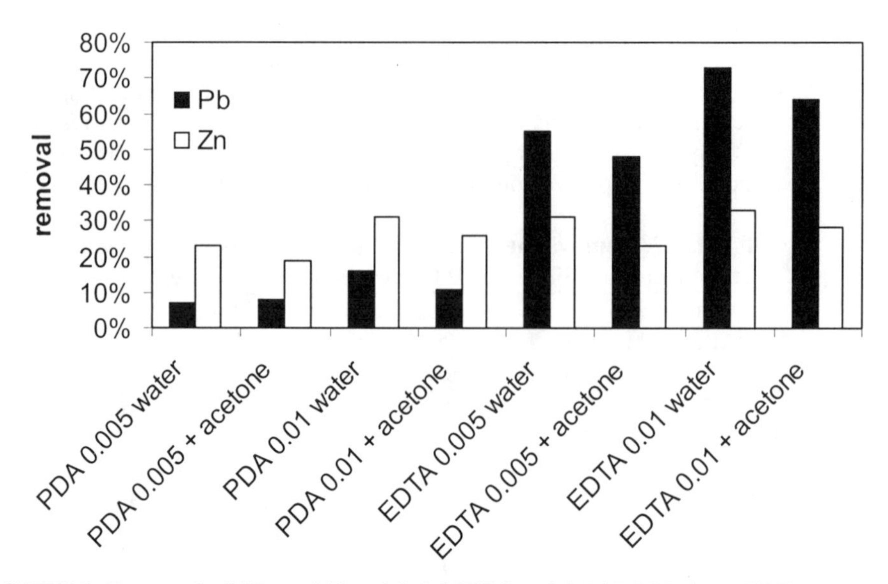

FIGURE 2. Removal of Pb and Zn with 0.005 M and 0.01 M PDA or EDTA in water and acetone/water (40%acetone). Extraction time is 2 hours.

CONCLUSIONS

The exploratory research has shown that removal of heavy metals from sludges and sediments is possible by extraction with a mixture of water and acetone (40% acetone) as solvent to which small amounts of chelating agents such as EDTA have been added. This offers a sound basis for the development of an acetone/water extraction process suitable to treat sediments polluted with both PAH (or other organic pollutants) and heavy metals. In this process three positive effects are combined:

- The positive effect of a mixture of acetone and water as extraction agent on the removal efficiency of PAH (and other organic pollutants) from wet sediments.
- The positive effect of a mixture of acetone and water on the separation of small sediment particles from the extracting agent.
- The positive effect of a chelating agent, added to the acetone/water extractant, on the removal of mildly bound heavy metals.

On basis of this exploratory research the integration of the use of a chelating agent in the solvent extraction process, mentioned in the introduction, is possible. In that case an extra treatment step is necessary in order to remove the heavy metals from the extracting agent and to recover the chelating agent. Assuming the chelating agent has preference for the water phase an ultrafiltration step in combination with a precipitation step using sulphides to remove the heavy metals may be incorporated easily in the process route.

Further research is necessary to optimise the process with respect to treatment time, amount of chelating agent, and the recovery of chelating agent for reuse.

ACKNOWLEDGMENTS

This work is based on the M.Sc. thesis of Léon Voogd, who is kindly acknowledged.

REFERENCES

EPA reports: *EPA540/A5-90/002* 1990, *EPA/540/SR-93/002* 1992[a], *EPA/540/MR-92/079* 1992[b]

Hasselt, H.J. van, A. Costerus, W.H. Rulkens. 1998. *5th International HCH and Pesticides Forum*, Bilbao, Spain 131-139.

Howard, J.L. 1995. "*Sequential extraction analysis of heavy metals using a chelation agent to counteract resorption.*" Department of geology, Wayne State University, Detroit.

Peters, R.W. and K. Young. 1988. "*Evaluation of recent treatment techniques for removal of heavy metals from industrial wastewaters*", Perdue University, AIChE Symposium Series No. 243 Vol. 81

Rulkens, W.H., H. Bruning, H.J. van Hasselt, J. Rienks, H.J. van Veen, J.P.M. Terlingen. 1998. *Wat. Sci. Tech.* 37 411-418.

TECHNOLOGY OF HEAVY METAL ION EXTRACTION AND WATER REUSE

Svetlana Verbych and Mychaylo Bryk (Ecological Research Center at the National University "Kyiv-Mohyla Academy" of Ukraine, Kiev, Ukraine)

ABSTRACT: The comparison of the extraction effectiveness of nickel ions from water solutions by means of the universal ion exchangers and polyampholyte has been investigated. The high value of the ion exchange capacity of the ionate KU-2 comparing with the complex formation amphoteric ion exchange resin ANKB seems to be more attractive and perspective to use for wastewater treatment. The technological circuit and installation for local treatment of rinsing water of nickel plating with subsequent metals utilisation have been developed.

INTRODUCTION

One of the critical pollution problems arising from the electroplating industry is the generation of a large amount of rinsing water for electroplated parts. The used rinsing water consists of many heavy metals and chemicals that can be detrimental to the health of organic bodies when exposed in the environment. Meantime the rinse water has a great possibility for reuse if the valuable heavy metal can be effectively removed and collected.

The traditional method of galvanic wastewater purification, which is based on precipitation of metals, presents certain disadvantages such as: necessity of maintaining specific pH conditions, temperature, and ion concentration; possibility of colloid formation; presence of complexing agents and occurrence of metal in the form of an anion. In addition, the method produces large amount of sludge, several times larger than the initial amount of each metal (Fabiani, 1992). For this reason removing dissolved heavy metals seems to be more attractive by means of ion exchange, phytoextraction, ultrafiltration, reverse osmosis, and electrodialysis (Applegate, 1984; Geselbracht, 1996; Schnoor, 1997; Sengupta and Clifford, 1986). Nevertheless, many of these approaches can be marginally cost-effective or difficult to implement. Therefore the need exist for a treatment strategy that is simple, robust, and that addresses local resources and constraints. In this regard concentration of heavy metals by ion exchange is more available and paying method for rinse water purification (Feksyk and Miller, 1993; Mabel Vaca Mier et al, 2001; Ragosta, 1991). Besides the ion exchange is widely used for water softening and nitrates removing in the potable water processes (Clifford, 1990; Huxster, 1998) as well as in refined technologies. In addition, there is a perspective in use of ion exchange for water systems control of waters with heavy metal pollution (Li Chinhua, 1992). The ion exchange resins are regenerated easily and returned to the cycle, at that from 60 to 100% regeneration of the initial ion exchange capacity can be achieved. But for the regeneration of universal ion exchange resins the consumption of acids and alkalis is three to four times higher than the required stoichiometric quantities. Because of that, application of the complex-forming ion exchangers is considered to be more cost-effective. After saturation of eluate with heavy metals it can be regenerated by electrolysis, which is cost-effective and ecological method for heavy metals removal from solutions (Langfeld, 1992).

New conception of the water treatment management to prevent sediment contamination and to protect environment is declared in the frame of the present work. The proposed approaches to develop of closed loop systems water consumption are based on the suppression of the heavy metals migration into wastewater by ion exchange extraction of the particular heavy metal ions from the rinse water after electroplating process. The purpose of this work was to investigate selectivity of universal cation-exhcange resins and aminocarboxyl ion exchanger for nickel and copper ions extraction, and to determine the influence of diverse factors on the removal and recovery of nickel from simulative rinse water after electroplating.

MATERIALS AND METHODS

The kinetic of nickel ions removed from water solution with the metal concentration of 30 (mg-eq)/l has been studied by means of ion exchangers of different structure, porosity and chemical properties. Solution under investigation was circulated with the velocity 0.65 cm/s through the column containing swollen ion exchange resin under fluidised-bed conditions. For experiment we used strong acidic resins with different binding degree in H^+- form, macroporous resin KU-23 (30/100) containing 30% of binding agent and 100% of pore forming substance, weak acidic ion exchanger KB-4 and polyampholyte ANKB-35, produced in Ukraine, in Na^+-form. Ion exchangers KU-2-4, KU-2-8, KU-2-20 of a gel structure are similar to sulfacation exchanger of Dowex-50 type (USA), and structure and properties of the ampholyte resemble closely to those of Dowex-1-A, Chelex-100 (USA) and ANKB-10, ANKB-50 (Russia) resins. All resins were prepared under method (Polyansky, 1976). The moisture content of the ion exchange resins is diverse in depending on industrial production. Therefore the moisture content and exchange capacity of every used resin were studied (Table 1). To determine these parameters $4 \pm 2 \cdot 10^{-4}$ g of resin was dried up at 90° C till weight became constant and this operation was reproduced at least three times.

The concentration of metal ions was determined with complexometric titration. The degree of adsorption (F) was calculated as the ratio of the equivalent quantity of metal ions adsorbed by the ion exchanger during a fixed period of time, and the equivalent quantity of functional groups:

$$F = \frac{(C_o - C) \cdot V}{E \cdot m},$$

where:

C_o – initial concentration of nickel ion in water solution, [mg-eq/l];

C - current concentration of nickel ions, [mg-eq/l];

V – solution volume, [l];

E – exchange capacity of the resin, [mg-eq/g];

m – mass of the wet resin, [g].

The water sample used in this study was simulated on the basis of rinse water after nickel-plating process at the manufacturing plant.

RESULTS AND DISCUSSION

Industrially produced ion-exchange resins have a gel or macro-pore structure. The functional groups of gel ion-exchangers are distributed over the resin grain. Usually the concentration of the binding agent – divinyl benzene (DVB) - changes from 2 to 20%. The lasting solidity of such grain is caused by high concentration of DVB, and it can change the resin swelling, which is responsible for the ion exchange process. The functional groups of macro-pore resins are accessed more in the exchange process because of large pore size (De Silva, 1995), and on top of that, high concentration of binding agent does not influence kinetic exchange. Unfortunately the capacity of porous resins is lower than capacity of gel resins because of the lesser number of functional groups in the ion exchanger grain. Pores concentration is about 10-30% of polymer frame volume, and it reduces exchange capacity (De Silva, 1995). It was detected that highest extraction rate of nickel is achieved using resin KU-2-8 since its exchange capacity is more than exchange capacity of resin KU-2-4, although the binding degree of this one is two times less (Figure 1). In the meantime, nickel extraction rate is quite high for the resin KU-2-4 from the beginning of process and for next 15 minutes. This could be explained by easy approach of functional sulphur-groups for ion exchanging. It is natural that high degree of swelling as well as a low binding agent concentration led to polymer chains extension that makes the functional groups of resin efficiently involved in exchange process.

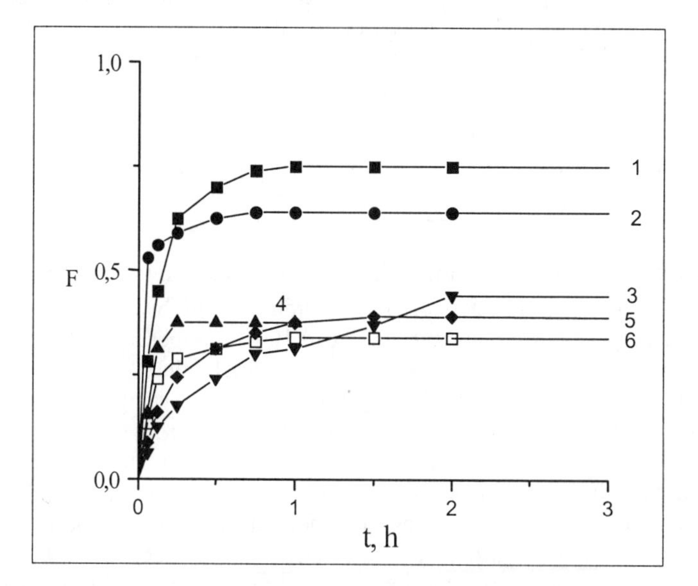

FIGURE 1. Kinetics of adsorption Ni^{2+} ions by ion-exchange resins: 1 – KU-2-8; 2 – KU-2-4; 3 – ANKB-35; 4 – KB-4; 5 – KU-2-20; 6 – KU-23.

The wet KU 2 20 resin have larger static ion exchange capacity than another gel resins, but tight binding (20 % DVB) of this exchanger does not allow to achieve a high rate of nickel sorption, and only 45% of sorption capacity have been used for extraction. With an increase of DVB concentration in the ion exchange resins KU-2-type, moisture content decreased (Table 1), and the steric hindrances in the grain were increased, and that makes the metal ions move toward the functional groups difficult. Therefore the counter-ions mobility decreased with an increase of polymer binding. The ion exchange rate of gel type resins was limited by the diffusion of the metal ions.The highest value of exchange capacity and the moisture content correspond to weak acidic carboxyl-containing cation exchanger KB-4 (Table 1). However the efficiency of nickel extraction by this resin is low. It was assumed that during first 15 minutes, competition of H^+ - ions transport would occur along with the exchange of Ni^{2+} - ions from solution. As a result of this process, hydrolysis of functional groups would occur. This hydrolysis was accompanied by the replacement of Na^+ - ions by H^+ - ions and by the release of OH^- - ions into the solution. The pH of the solution changed from 4.5 to 6.5. The formation of low-dissociated –COOH groups in the ion exchanger led to decrease of extraction rate.

TABLE 1. Physical-chemical characteristics of different types of ion exchange resins in Na^+-form

Type of sorbent	Static exchange capacity of dry exchanger, mg-eq/g	Moisture content, %		Static exchange capacity of wet exchanger, mg-eq/g
		Reference book [*Ion exchange resins*, 1989]	Experimental data	
KU-2-4	5.31	62-70	62	2.04
KU-2-8	5.20	50-60	51	2.57
KU-2-20	4.77	30-40	28	3.44
KU-23	5.10	50-70	64	1.85
KB-4	10.00	55-65	61	3.91
ANKB-35	-	-	54	1.05

It is noteworthy that together with the conversion of the ion exchanger from Na^+ - into Ni^{2+} - form, the formation of hydro-complexes of bivalent cations in the pores of macroporous exchanger KU-23 (30/100) was possible. The kinetic of sorption in this case is accompanied by decrease of pH value.

There are additional co-ordinating binds in the polyampholyte ANKB-35, which led to formation of strong complexes of transition metals. Formation depends also on degree of protonization of the nitrogen in the amino groups. It was known that with increase in acidity of the solution the concentration of complex-forming ionogenic groups decreases and their electrodonor properties and the swelling degree change significantly (Saldadze, 1980). A decrease in the swelling degree and the concentration of complex-forming groups leads to a decrease in stability of ion-exchange complexes. It was found experimentally that Ni^{2+} and Cu^{2+} - ions adsorption decreased with decrease in pH and the extraction degree of Cu^{2+} ions did not exceed 0.1 at pH 1.3 and Ni^{2+} ions sorption did not occur at all from acidic solutions at pH<2 (Figure 2). However, at high pH values complex-formation was limited by the generation of basic salts of the metals and poorly soluble precipitates of hydroxides (Grebenyuk, 1998).

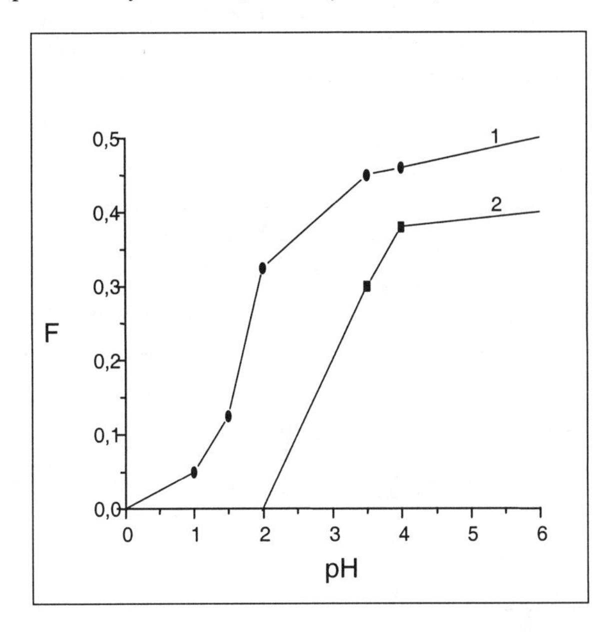

FIGURE 2. Influence of solution pH on adsorption degree (F) of Cu^{2+} (1) and Ni^{2+} (2) ions by exchanger ANKB-35.

But as we can see from Figure 1, the adsorption degree of ampholite ANKB-35 is less than universal resins KU-2-8 and KU-2-4, and the adsorbent filling with Ni^{2+} ions is limited by low exchange capacity of amppholyte ANKB-35 in comparison with resins pointed above (Figure 1). Thus the extraction of heavy metal ions from water solutions can be carried out efficiently, using universal ion exchange resin KU-2-8.

The technological circuit and installation for local treatment of rinsing water of nickel plating and metals utilisation have been developed on the basis of the experiments mentioned above. According to the standard plating technology the parts are moved from the electroplating bath to drag-out (catching) bath and then to the rinsing bath. The volume of drag-out losses and hence the heavy metal concentration in drag-out (catching) and flow type rinsing bathes depend from the composition and concentration of electrolyte in plating bath, geometrical configuration of parts and their conveying speed on the coating line. Heavy metal concentration increases to 1.2-1.7 g/l in the drag-out bath, and exceeds substantially maximum permissible concentration in the rinsing bath during the shift. In this connection from 15 to 30 g of metal are lost per every 1 m^2 of plated parts, and consequently valuable metal is falling into sludge or sewage.

On our technology the ion exchange column is placed directly on the coating line (Figure 3). Solution from a drag-out bath (2) continuously moves by the pump (7) through a column with cation-exhcange resin (4). After absorbent saturation of the first column with metal ions, the flow of solution from a drag-out bath is switched into the second column. The used up column is connected to regeneration path by other pump. The quantity of the ion-exchange resin for the column is selected in such a way that heavy metal ions concentration in the drag-out bath for the period of working shift did

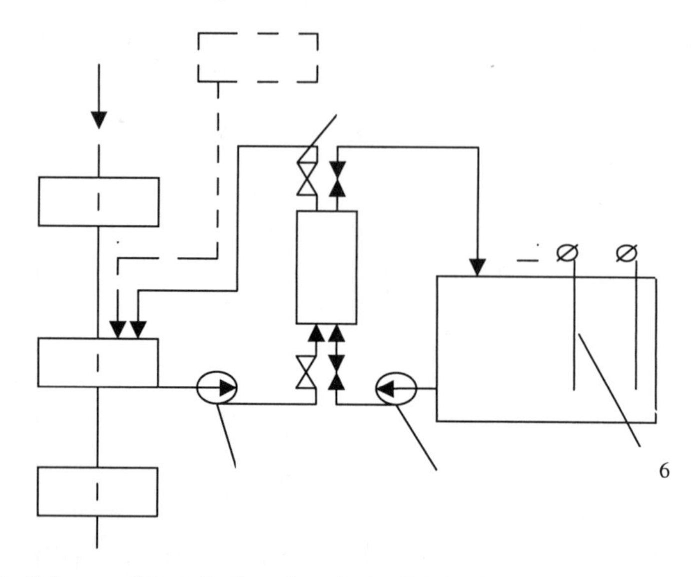

FIGURE 3. Scheme of installation. 1 – plating bath; 2 – drag-out bath; 3 – bath for continuous rinsing; 4 – sorption column; 5 – electrolyser; 6 – cathode; 7 – pump; 8 – valve; 9 – softening bloc; I – solution flow of the drag-out bath; II – recovered solution flow; III – movement of workpieces along the line.

not exceed a certain value. For example, for nickel ions the given concentration is about 70 mg/l under condition of nickel-plating line capacity 11 m^2 /h and drag-out bath volume 1.2 m^3. During multiple solution circulation with low rate the gradual accumulation of nickel ions occurs (Figure 4, curve I). And at flow rate more than 10 v/h, independent of sorption-regeneration cycle, nickel ions concentration in the drag-out bath does not exceed the indicated quantity, while in flow-type rinsing bath will always be lower than the maximum permissible concentration. Regenerated solution, circulating through electrolyzer (5) and the ion exchange column (4), displaces heavy metals ions from it (Figure.3). Under this condition heavy metals ions concentration reaches 3-4 g/l, which allows making deposition of the metal on a cathode of the electrolyzer. The extracted nickel can be used as an anode in the electroplating bath or utilized that depends from quality and thickness of the deposit.

During Ni^{2+} sorption from tap water Ca^{2+} ions are also adsorbed by cation-exchanger and during subsequent regeneration with sodium sulfate, insoluble calcium sulfate are formed blocking functional group of the exchanger. This brings about gradual loss of sorbent's capacity with each sorption-desorption cycle. This can be avoided by using softened water unit, for whose production softening block has been developed. For effective functioning of the unit the block must operate for about 2 days a week. However, the block's cation-exchange capacity gradually falls due to organic compounds and chelates of iron and aluminium hydroxides present in tap water. That is why it is expedient to wash the resin once in 2 month (or after 7 sorption-regeneration cycles) with 3.5 N of sulfuric acid. After that resin should be washed by water and converted in sodium form by salt to reuse in the softened water unit. Thus this adsorption technology allows preventing nickel ions occurrence in the wastewater, reuse treated water into technological process and to obtain pure extracted metal that can be used as an anode in the electroplating process again.

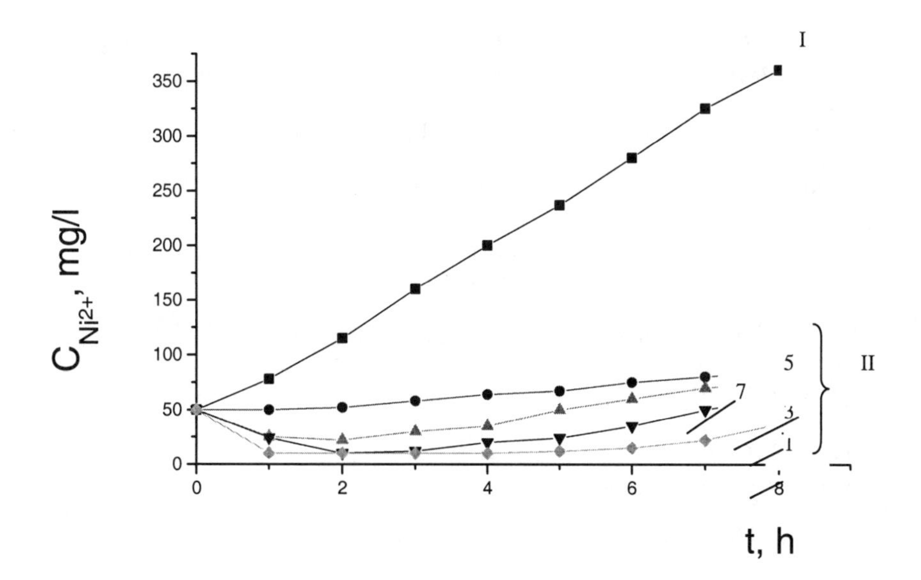

FIGURE 4. Kinetics of change in nickel ions concentration in the drag-out bath on numbers of solution circulation: 2 v/h (I) and 12 v/h (II). Number of curves-number of cycles of sorption-recovery.

REFERENCES

Applegate L. E. 1984. Membrane separation processes. *Chem. Eng*. 91; 64-70.

Cliford, D. 1990. *Ion Exchange and inorganic absorption . Water quality and treatment. A hand-book of community water supplies* 1990 4th addition, Mc Grant Hill, New York.

De Silva, F. 1995. *Essentials of Ion Exchange. Technical Sales Representative, Resin Tech Inc.,* Cherry Hill, NJ.

Fabiani C. 1992. *Recovry of metals ions from waters and sludges*, Report RT/AMB/92/32 ENEA Entre per le Nuove Tecnologie l'Energia e l'Ambiente, Roma.

Feksyk P.U., Miller B.T.1993. *Galvanotechnic and surface treatment*. 2(3); 77-83.

Geselbracht J. 1996. *Microfiltration/reverse osmosis pilot trials for Livermore*, California, Advanced Water Reclamation, 1996 Water Reuse Conference Proceedings, AWWA; 187.

Grebenyuk, V.; Verbich, S.; Linkov, N. & Linkov V. 1998. Adsorption of heavy metal ions by aminocarboxil ion exchanger ANKB-35. *Desalination*. 115; 239-254.

Huxster, J. & Sorg, T. 1998. *Reverse Osmosis. Treatment to Remove Inorganic Contaminants from drinking Water,* EPA/600/52-87/109 Cincinnati, Ohio.

Ion exchange resins. 1989. Reference book. NIITECHIM, Cherkassi.

Langfeld E. 1992. *Galvanotechnik*. 82(11); 3936-3941.

Li Chunhua 1992 Features of ion exchange process for heavy metal pollution control. *Technology of water treatment*. 18(3); 198-201.

Mabel Vaca Mier, Callejas R.L, Gerh R., Cisneros B.E.J. and Alvarez P.J.J. 2001. Heavy metal removal with Mexican clinoptololite: multi-component ionic exchange. *Wat.Res.* 35(3); 373-378.

Polyansky, N.; Gorbunov, G. & Polyanskay, N. 1976. *Investigations methods of ion exchange resins*. Chemistry, Moscow.

Ragosta J.M. 1991. *Metal Finishing*. 89(11); 7-11.

Saldadze, K. & Kopilova-Valova, V. 1980. *Complex - formation ion exchange resins*. Chemistry, Moscow.

Schnoor J. L. 1997. *Phytoremediation*. TE-97-01, Ground-Water Remediation Technologies Analysis Center, Pittsburg, PA.

Sengupta A. K. and Clifford D. 1986. Important process variables in chromate ion exchange. *Environ. Sci. Technol.* 20; 149.

DESIGN OF TREATMENT FOR MERCURY-CONTAMINATED CANAL SEDIMENT USING STABILISATION/SOLIDIFICATION

B.Guha, C.L.Macleod, and C.D. Hills
Centre for Contaminated Land Remediation, The University of Greenwich at Medway, Chatham Maritime, Kent U.K

ABSTRACT: This study was carried out to design an effective treatment method for mercury (Hg)-contaminated canal sediment using stabilisation/ solidification technology. Sediment contaminated with mercury fulminate from the Union Canal in Falkirk, Scotland was mixed with two hydraulic cements, pulverised fuel ash and activated carbon at different dosage rates and water/solids contents. The stabilised samples were cured for 28 days and then leach-tested using the DIN38414-S4 and Toxicity Characteristic Leaching Procedures. The unconfined compressive strength of the stabilised/solidified sediments was also determined. Cold Vapour Atomic Absorption Spectrometry was used to examine the total-Hg concentration in the sediment and the leachates. The results showed that stabilisation/solidification was capable of reducing leachable mercury to within acceptable freshwater-based limits. Strengths of more than 2500 KN/m^2 were observed at 28 days in mixes with and without activated carbon, however, the strengths obtained with this addition were on the whole greater.

INTRODUCTION

The mobility of inorganic Hg in contaminated soils and sediment is very low, but biological and chemical processes can transform the immobile Hg to more toxic and bioavailable methyl-Hg. This form of Hg easily enters the food-chain and biomagnifies itself. Thus, Hg contamination in soils and sediments is a major environmental concern.

Among the present remediation strategies phytoremediation, electrolytic methods, and stabilisation/solidification (s/s) are most popular. Several publications reveal that phytoremediation is the logical approach in terms of treatment cost and size of the contaminated land (Heaton et al., 1998; Pilon-Smits and Pilon, 2000). However, the biggest drawback of this method is that it lengthens the remediation time. The electrolytic method is not feasible where the contaminated area is large and when time and resources are limited. As a result, cement-based stabilisation/solidification may offer a rapidly applied alternative to mercury-contaminated sediments and soils.

Thus, present work investigates the use of s/s in the treatment of Hg contaminated sediment from the Union Canal in Scotland. Because of the gross nature of contamination in the sediment to be studied, organic carbon as a binder additive was also investigated.

Aims/Objectives. The aims of the project were to:

- provide a simple stabilisation/solidification treatment option using hydraulic cements
- yield a strength of more than 1000 KN/m^2 from treated sediments
- reduce leachable Hg to within acceptable limits

Site description. The Union Canal connects the cities of Glasgow and Edinburgh in Scotland. Nine kilometres of this Canal in Falkirk area is heavily contaminated with mercury. The source of this contamination is the former Nobel-ICI factory at Reddingmuirhead. This factory produced over 73 million detonators whose main constituent was mercury fulminate.

MATERIALS AND METHODS

Materials. Rockfast A (RF-A), a proprietary calcium sulphoaluminate cement, pulverised fuel ash (PFA), and ordinary Portland cement (OPC) were used as cementitious binders. All chemicals used were Aristar grade from Fischer Chemicals. The activated carbon (Fisher Chemicals) used in this project was supplied by Aldrich Chemical Company.

Sample storage. The sediment samples were stored in the refrigerator at 4°C after receipt. The DIN leached extracts were preserved in 1% HNO_3. Both the DIN and TCLP leachates were stored in the refrigerator.

Sample pre-treatment. The moisture content of the sediment was determined to be 81% (AWWA). Prior to treatment the moisture content of sediment was reduced by decantation and centrifugation methods where necessary. Ten percent PFA w/w was added to increase the solid content of the sediment.

Binder preparation. The binder was prepared by mixing RF-A, OPC, and PFA as in the ratio 1:1:0.25. The prepared binder was stored in an airtight container.

Digestion procedure for mercury analysis (CVAAS). The sediment was digested (U.S. EPA, Method 7471A) to break down the inorganic and organic form of Hg to elemental Hg prior to determination of total mercury (T-Hg). Triplicate samples of the sediment were weighed (each approximately 0.2 g) and placed in the bottom of BOD bottles. Five ml concentrated H_2SO_4 and 2 ml concentrated HNO_3 were added followed by 5 ml saturated $KMnO_4$ solution. After placing the stopper the BOD bottles were covered with thick aluminium foil to prevent leakage of Hg during digestion. The samples were then heated on a hot plate for 2 hours at 196°C. After heating for 2 hours, samples were cooled at room temperature and the volume made up to 100 ml with deionised distilled water. Then 6 ml of 10% hydroxylamine hydrochloride was added to reduce the excess $KMnO_4$. Digestion of the certified reference materials with this procedure gave more than 70% recovery.

Treatment method. The s/s was carried out at different water/solid contents. Samples were produced at 40%, 50% and 60% solids to assess this important parameter. Mixing was carried out in a 5-l Hobart mixer, and two sets of samples were prepared. One set contained the binder mixture and activated carbon at 1% w/w(solids). Three binder addition rates of 20%, 30% and 40% w/w (total solids) were used. All samples were cured for 28 days. Each treatment was carried out in duplicate to minimise error. Samples were wrapped in cling film to cure at ambient laboratory temperatures.

Leach tests. Treated samples were subjected to DIN38414-S4 (Deutsche Institut for Normung) and TCLP leaching (USEPA, method 1311). The DIN test was carried out at 3, 7, 14 and 28 days after treatment, whereas TCLP leach testing was undertaken at 3 and 28 days only. Duplicate leach testing was carried out on each sample.

Unconfined compressive strength. The compressive strength of triplicate samples was determined at 3, 7, 14 and 2 days. Cylindrical 50-mm x 50-mm samples were tested on an Instron 1195, universal strength-testing machine.

Mercury analysis by CVAAS. The total-Hg concentration of the sediment and the leach extracts were determined using a dedicated mercury analyser (CETAC M6000) at 70°C. The Detection Limit of the Instrument is 0.002 μg/L Hg in solution. The average correlation coefficient of the standard curves obtained was 0.99990.

RESULT AND DISCUSSION

Experimental results revealed that the sediment had an 81% moisture content, based on wet weight. The physical pretreatment methods reduced sediment moisture content to 70%. Further adjustment of solids content by PFA addition at the rate of 10% prior to treatment was successfully employed. Chemical analysis showed that the T-Hg in the sediment was 728 mg/Kg. The T-Hg in the sediment vastly exceeds the upper effects threshold of 560 μg/Kg for fresh water sediment, in the Screening Quick Reference Table developed by the U.S. National Oceanic and Atmospheric Administration (NOAA). The upper effects threshold signifies that Hg concentration equal to or above that value will have toxic effects on the biota.

Analysis of the DIN-leach extracts showed that the 40% solid and 20% binder-only mixtures exceeded the acceptable limit of 1.0 μg Hg/L as defined in Schedule 1 of the U.K. Drinking Water Inspectorates (DWI) Criteria for the Classification of Waters at 28 days (see Table 1).

The TCLP extract from the 50% solid and 30% binder of the non-carbonated treatment also exceeded the regulatory level (the maximum allowable concentration in a TCLP extract) of 0.2 mg/L defined in the TCLP methodology (see Table 1) at days of age. However, a two-sample T-test carried out on data from the 3- and 28-day leaching tests shows that there are no particular differences in the results of the two mix designs described. Thus, it can be concluded that the above values on the 28th day are within their respective criteria.

TABLE 1. Mercury levels in leachates (cement only).

Metal	Solid	Binder	DIN-leach				TCLP-leach	
			3rd day	7th day	14th day	28th day	3rd day	28th day
T-Hg (µg/L)	40%	20%	0.266	0.086	0.514	2.52	N.D*	5.94
		30%	0.064	0.061	0.066	0.069	N.D*	177
		40%	0.049	0.079	0.121	0.011	N.D*	11.44
	50%	20%	0.106	0.296	0.121	0.621	0.014	0.106
		30%	0.076	0.269	0.021	0.086	0.044	340
		40%	0.144	0.424	0.229	0.131	0.001	0.086
	60%	20%	0.051	0.066	0.051	0.086	N.D*	0.089
		30%	0.071	0.066	0.219	0.084	N.D*	0.216
		40%	0.056	0.056	0.056	0.009	N.D*	N.D*

*N.D = Non-detectable and denotes the concentration values <0.001 µg/L.

However, the Hg concentration in the DIN leachates from all mix designs is below the limit set by the UKDWI (Tables 1 and 2). The Hg concentration in the TCLP extracts also meet the limit set for this particular test.

The use of activated carbon appeared to reduce leachable Hg (Table 2, Figures 1 and 2), and an examination of 28-day data, employing the students T-test showed this difference was significant from mixes with and without activated carbon.

TABLE 2. Mercury levels in leachates (cement and activated carbon).

Metal	Solid	Binder	DIN-leach				TCLP-leach	
			3rd day	7th day	4th day	28th day	3rd day	28th day
T-Hg (µg/L)	40%	20%	0.026	0.056	0.046	0.076	N.D*	N.D*
		30%	N.D*	N.D*	0.016	0.016	N.D*	1.71
		40%	0.046	N.D*	0.046	0.166	N.D*	179
	50%	20%	0.016	0.006	0.136	0.096	0.156	0.144
		30%	0.026	0.036	0.016	0.036	0.094	0.166
		40%	0.006	0.086	0.036	0.046	0.086	0.091
	60%	20%	0.056	0.026	0.016	0.126	0.109	0.026
		30%	0.036	0.016	0.116	0.076	N.D*	0.339
		40%	0.036	0.026	0.056	0.046	0.077	1.98

*N.D = Non-Detectable and denotes the concentration values <0.001 µg/L.

A comparison of the DIN leachate values with NOAA ground water criteria (2 µg/L) shows that almost all the leach extracts are within this value. The study also reveals that the DIN leach concentrations are well below the chronic toxic level, of 0.77 µg/L, for fresh surface water criteria set by NOAA.

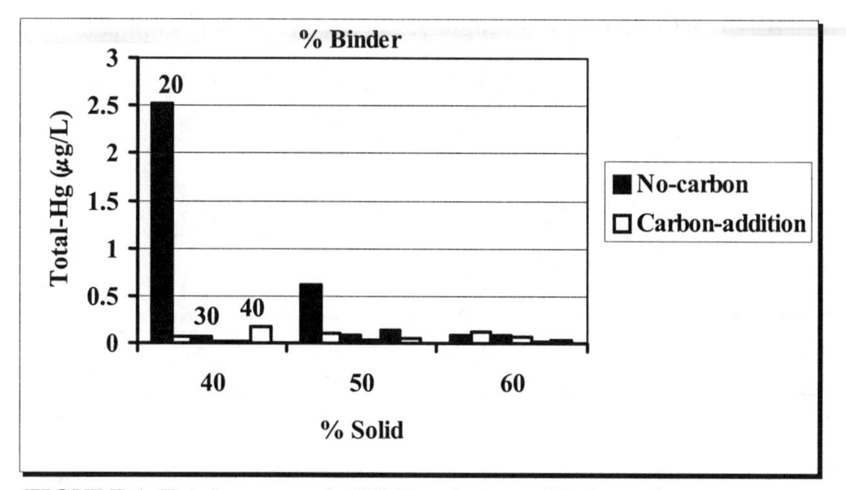

FIGURE 1. Total mercury in DIN leachates at 28 days after treatment.

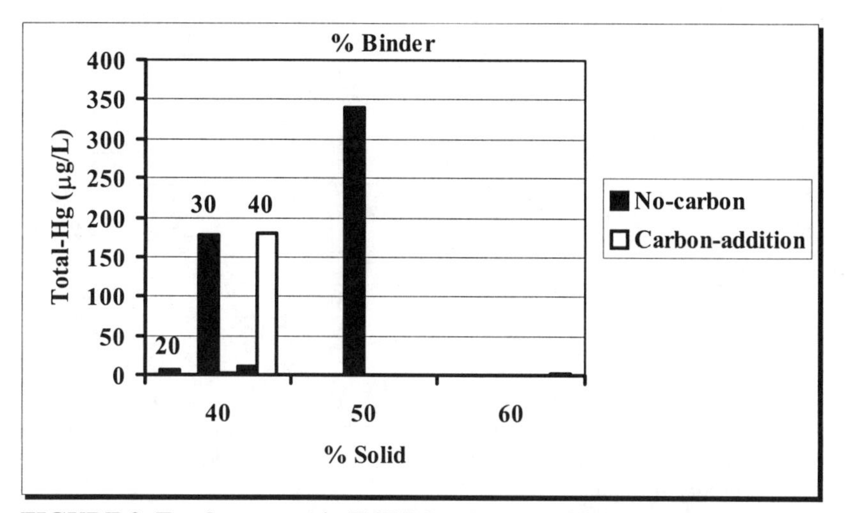

FIGURE 2. Total mercury in TCLP leachates at 28 days after treatment.

The strength of the stabilised/solidified sediment increased with time and with the percentage of binder added to mixes (Table 3 and 4). In mixes without activated carbonation, the 50% w/w solids sediment/30% binder mix met the target strength of 1000 KN/m^2 at 28 days of age. The 40% w/w solids sediment /30% binder mix containing activated carbon also met the target strength at 28 days. The addition of activated carbon was found to enhance strength development at all ages for all mix designs (Table 3 and 4, Figure 3).

TABLE 3. Compressive strength of the cement-treated sediment.

Solid	Binder	Strength (KN/m^2)			
		1st day	7th day	14th day	28th day
40%	20%	95	202	416	452
	30%	159	262	412	540
	40%	210	448	702	908
50%	20%	231	500	682	868
	30%	286	653	825	1281
	40%	494	767	1003	1368
60%	20%	345	657	786	1460
	30%	542	1222	1399	1726
	40%	702	1209	1508	2550

TABLE 4. Compressive strength of the cement-treated sediment with activated carbon.

Solid	Binder	Strength (KN/m^2)			
		1 day	7th day	14th day	28th day
40%	20%	280	423	504	946
	30%	393	677	704	1019
	40%	633	987	1129	1700
50%	20%	321	610	699	997
	30%	697	973	1091	1692
	40%	547	1172	1237	1539
60%	20%	462	718	1144	1436
	30%	747	1117	1400	2363
	40%	1281	1639	2258	2522

CONCLUSION

The present study has shown that:

- Stabilisation/solidification employing hydraulic cement is capable of successfully treating canal sediment grossly contaminated with mercury fulmanate.
- Higher-than-target strengths were realised in the treated sediment at the higher solids contents and higher binder dosage levels examined.
- The addition of activated carbon improved the strength results obtained from samples up to 28 days of age.
- There were significant differences in the levels of mercury obtained from leachates from mixes with and without activated carbon.

ACKNOWLEDGMENTS

The authors gratefully acknowledge support of Land and Water Limited, U.K. We would also like to acknowledge Amanda Beynon for her help and technical support in the laboratory and Dr. Kate Whitehead for technical advice during the programme of work.

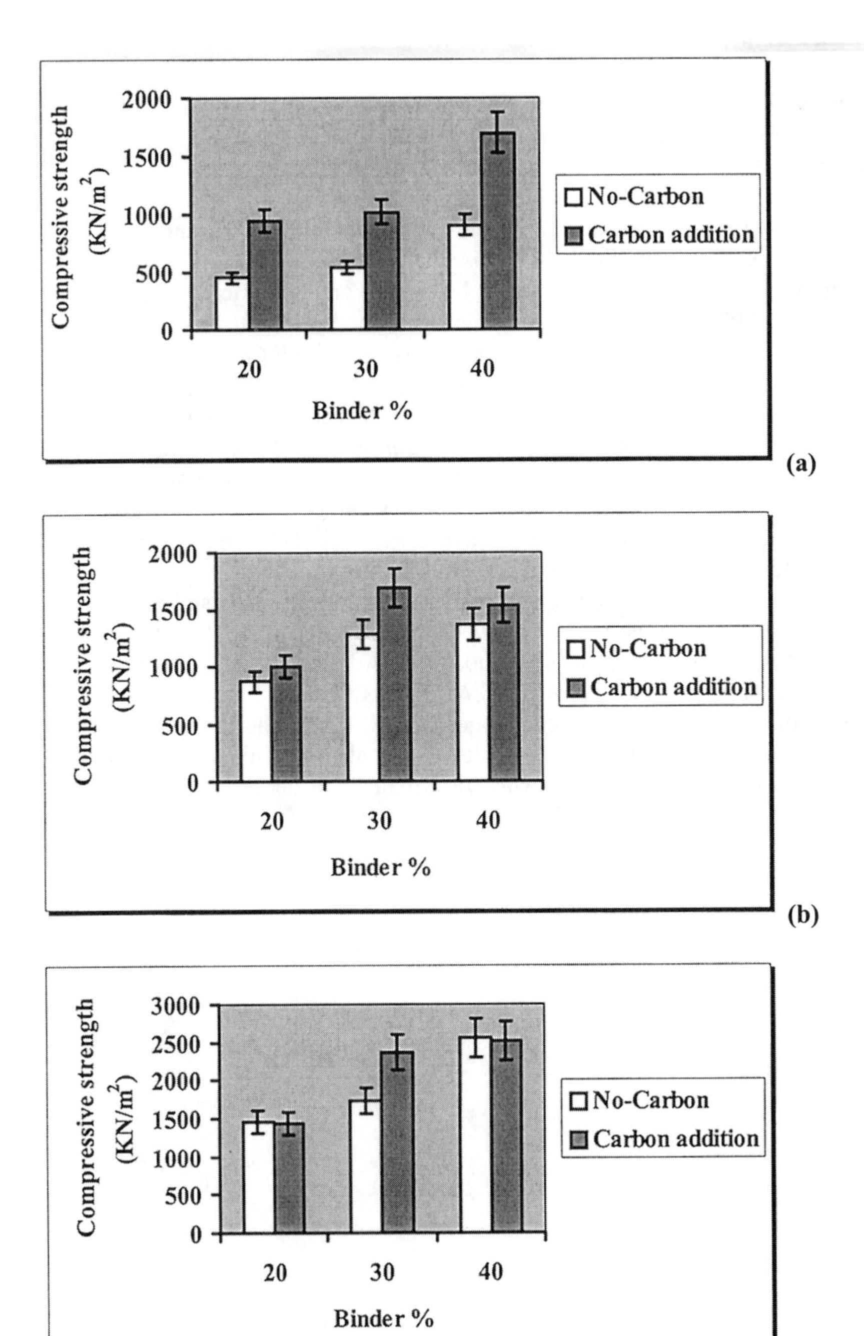

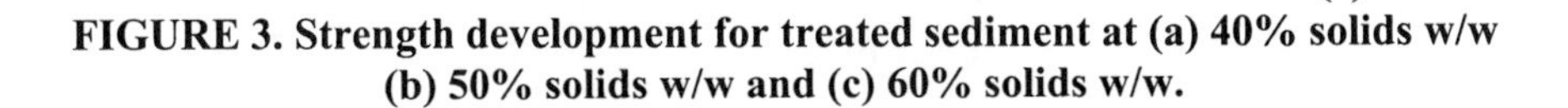

FIGURE 3. Strength development for treated sediment at (a) 40% solids w/w (b) 50% solids w/w and (c) 60% solids w/w.

REFERENCES

Deutsche Institut for Normung. Test Method 38414/S4 German Standard Method for the Examination of Water, Wastewater and Sludge, October 1984.

Drinking Water Inspectorate. Schedule 1 Drinking water standards. www.dwi.gov.uk

Glasser, F. P. 1997. "Fundamental aspects of cement solidification and stabilisation." *Journal of Hazardous Materials.* 52: 151-170.

Heaton, Andrew C. P., Clayton L. Rugh, Nian-jie Wang, and Richard B.Meagher. 1998. "Phytoremediation of mercury- and methyl mercury-polluted soils using genetically engineered plants." *Journal of Soil Contamination.* 7(4):497-509.

Pilon-Smits, Elizabeth, and Marinus Pilon. 2000. "Breeding mercury-breathing plants for environmental cleanup." *Research News.* 5(6): 235-236.

Stein, Eric D., Yoram Cohen, and Arthur M. Winer. 1996. "Environmental distribution and transformation of mercury compounds." *Critical Reviews in Environmental Science and Technology.* 26(1):1-46.

Thoming, Jorg, Bodil K. Kliem, and Lisbeth M. Ottosen. 2000. "Electrochemically enhanced oxidation reactions in sandy soil polluted with mercury." *The Science of the Total Environment.* 261(1-3): 137-147.

United States National Oceanic and Atmospheric Administration (NOAA), Hazmat Report 99-1. *MFB@HAZMAT.NOAA.GOV*

United States Environment Protection Agency (USEPA), Solid Waste 846, Test methods for evaluating solid wastes. Physical and chemical methods. www.epa.gov/epaoswer/hazwaste/test/maina.htm

MOBILITY AND RELATIVE PERMEABILITY EFFECTS ON SURFACTANT-ENHANCED REMEDIATION TREATMENTS

Alessandro Bevilacqua, Matteo Rosti, Giovanni Dotelli and Marinella Levi
(Politecnico di Milano, Milano, Italy)
Steven L. Bryant (University of Texas, Austin, Texas, USA)
Thomas P. Lockhart (Eni Tecnologie S.p.A., San Donato Milanese, Italy)

ABSTRACT: In order to study the influence of mobility and relative permeability effects on surfactant-enhanced remediation processes, carefully-designed porous media experiments have been conducted with trichloroethylene (TCE) as the contaminating phase in sandpacks prepared with Ottawa sand. Injection of an aqueous solution of the anionic surfactant MA-80 into linear sandpacks resulted in the removal of >99% of the residual TCE within 2 pore volumes. Multiple pressure taps along the sandpacks revealed a high pressure transient associated with the movement of the NAPL bank through the column; this is attributed to a relative permeability effect localised at the front of the NAPL bank. Experimental evidence for the formation of viscous NAPL/surfactant phases both within and outside the sandpacks were also obtained. The impact of these mobility and relative permeability effects on SEAR process efficiency was simulated by conducting parallel sandpack floods with one sandpack initially at residual TCE concentration and the other free of TCE. In the higher-dimensional parallel flood, in fact, these factors cause the majority of the surfactant solution injected to enter the clean column.

INTRODUCTION

Contamination of subsurface soil and water by non-aqueous phase liquids (NAPL) is an unfortunate and ubiquitous legacy of man's industrial activity. Current technologies for remediating NAPL- and dense (D)NAPL-contaminated sites encounter severe limitations, particularly where the NAPL has penetrated into an aquifer lying some distance below the surface. The pump-and-treat method is widely employed in such cases and can limit spread of the organic pollutants downstream of the source. However, as it is constrained by the aqueous solubility of the contaminating species, this method constitutes a very long-term approach to the problem. For this reason, considerable interest has been arisen in emerging technologies based on the use of surfactants and other agents that promise to accelerate mass removal of the NAPL source.

In surfactant-enhanced aquifer remediation (SEAR) treatments the surfactant facilitates removal of the NAPL phase through micellar solubilisation and/or mobilisation. Laboratory studies (Pennell et al., 1994) have shown that food-grade surfactants can mobilise chlorinated organics (e.g. tetrachloroethylene, PCE), bringing about the quantitative removal of the DNAPL within a relatively small volume of injected surfactant solution. Pilot studies demonstrate that significant levels of DNAPL removal can also be achieved in the field (Simpkin et al., 1999)

A key issue for the design and economic assessment of a SEAR field process is to accurately estimate the ultimate level of removal and the overall efficiency of the treatment (*i.e.*, how large a volume of surfactant solution must be pumped through the treatment zone to achieve the ultimate level of removal). Among the factors influencing process efficiency, phase mobilities (relative permeability and viscosity) have received relatively little attention. Soil column experiments are essentially one-dimensional and thus do not simulate subsurface situations in which mobility could be important. For example, a heterogeneous distribution of the NAPL phase will tend to impose reduced flow through NAPL-containing matrix compared to flow through the water-saturated matrix. This will occur even in a homogenous medium because the presence of NAPL reduces aqueous phase permeability. In addition, any viscosification accompanying formation of the surfactant/NAPL/brine phase will further reduce flow through the contaminated region. The present work demonstrates the importance of these factors for a model DNAPL system by means of carefully designed porous media experiments.

MATERIALS AND METHODS

Trichloroethylene (TCE) has been employed as the DNAPL phase. The commercial, food-grade surfactant sodium dihexyl sulfosuccinate (MA-80, from Cytec) has been employed as a 4 weight % solution in a 1% NaCl brine.

Sandpacks have been prepared in titanium columns (40 cm length by 2.8 cm diameter) with 120 mesh quartz (Ottawa) sand. Typical permeabilities of about 4 Darcy were measured for the water-saturated columns and the total pore volume (PV) was about 100 mL. Pressure profiles along the column during column flushing were measured by means of four quartz digital transducers placed at equal intervals along the sandpack, in addition to transducers placed just before and after the sandpack. The set-up employed for the dual-column experiment is shown in Figure 1. In all experiments, the sandpacks were kept at constant temperature (22°C) by means of a thermostatted water bath.

Conservative and partitioning tracers (KI for water and phenol or nitrobenzene for TCE) were used to determine the total column PV and residual TCE saturations before and after surfactant flooding. Columns were prepared for surfactant flooding by establishing residual TCE concentrations; these ranged from 11-18% of the total PV.

Mobilisation experiments were conducted by injecting the surfactant at a constant velocity (1 mL/min in the single column, 2 mL/min in the dual column experiment) while measuring pressure continuously. The effluent was collected in 6 mL fractions and the TCE quantified either by weight after physical separation or by gas chromatography (Dwarakanath and Pope, 2000). In the dual column experiment the effluents from the two columns were collected separately at fixed time intervals, allowing flow rates through each column to be accurately reconstructed as functions of time.

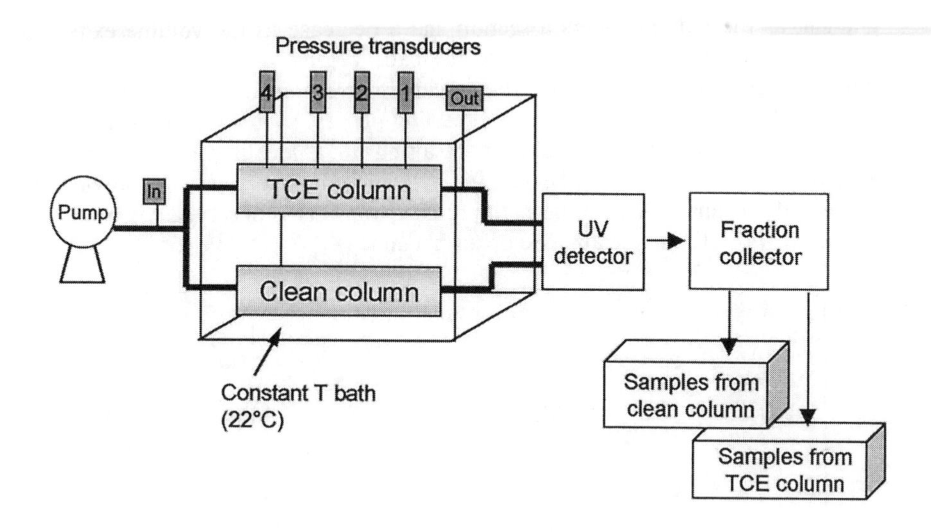

FIGURE 1. Apparatus used for parallel flooding of two columns, showing pressure transducers employed for measuring mobility of phase within TCE-contaminated column.

RESULTS AND DISCUSSION

The surfactant system employed in these experiments has been described in the literature (Pennell et al., 1994). The salinity and surfactant concentrations were chosen so as to obtain a low interfacial tension with TCE and formation of a middle, microemulsion phase (Winsor III) as shown in Figure 2. Within a TCE-contaminated porous medium the surfactant solution employed will promote mobilisation of the residual TCE as the separate phase followed by a TCE/brine/surfactant microemulsion, as illustrated schematically in Figure 3.

A typical effluent profile for a mobilisation experiment (in this case, one PV of surfactant solution was injected, followed by NaCl brine) using this system is shown in Figure 4. Analysis of the effluent showed that about 65% of the TCE present in the column was recovered as the separate phase, while another 35% eluted as the microemulsion. This quantitative recovery of TCE was confirmed with partitioning tracers, which indicated that 0.1% or less of the original TCE (residual to water flooding) remained in the column.

Monitoring pressure across different lengths of the column has revealed, in every experiment, a significant pressure transient localised at the leading front of the TCE bank. Figure 5 illustrates this effect by plotting the pressure drop versus time across each of the successive 8 cm lengths of the sandpack. The pressure increase first appears in the second section of the column after 0.2 PV have been injected. The pressure drop builds up as the surfactant solution progressively invades this section of column. The low mobility volume of fluid continues to propagate through the porous medium, its passage marked by a jump in pressure gradient as the volume enters a section and a decrease as the volume exits that section. This pattern is repeated until the bank begins to exit the column. This timing suggests that the low mobility volume corresponds to a small amount of fluid in which both TCE and brine are flowing phases. In this volume both phases will have low relative permeability, resulting in a localised, higher pressure gradient than elsewhere in the column, where only one phase is flowing. In support of this analysis a similar high pressure transient is observed when water is displaced with TCE in the absence of surfactant.

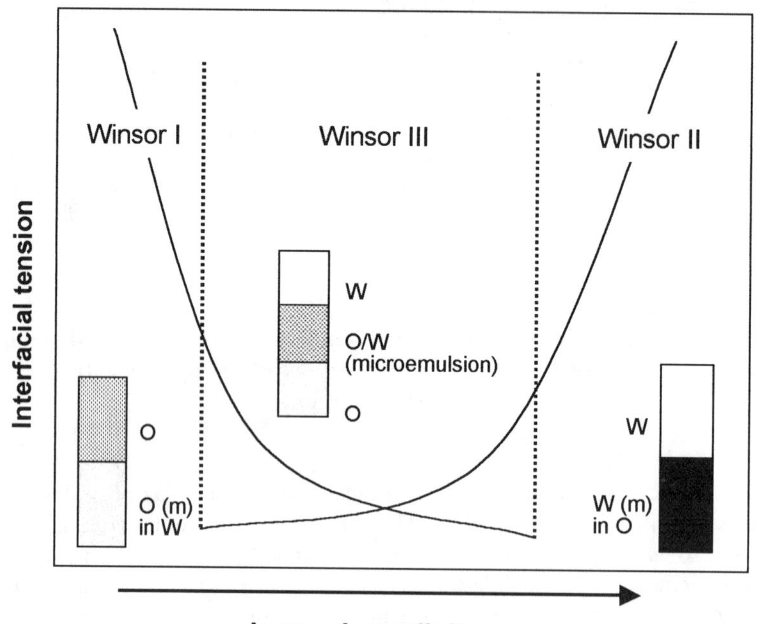

FIGURE 2. Illustration showing surfactant/DNAPL/water phase behaviour and Winsor III microemulsion phase which forms at low interfacial tension (m indicates micellar solution; O, oil and W, water phase).

Further to this point, rheological studies showed that the independently prepared TCE-brine microemulsion phase possesses a viscosity some 2 - 4 times greater than that of the surfactant solution and brine, respectively. These results highlight the importance of taking mobility effects into account in evaluating the efficiency of proposed NAPL field applications. Simple theoretical considerations indicate that these effects will tend to reduce flow through NAPL-contaminated regions once these have been contacted by surfactant, making the process self-limiting and consequently less efficient. A particular concern is that, unless such effects are specifically accounted for, the usual, linear column experiments will tend to over-estimate the efficiency of NAPL removal that will be achieved in the field.

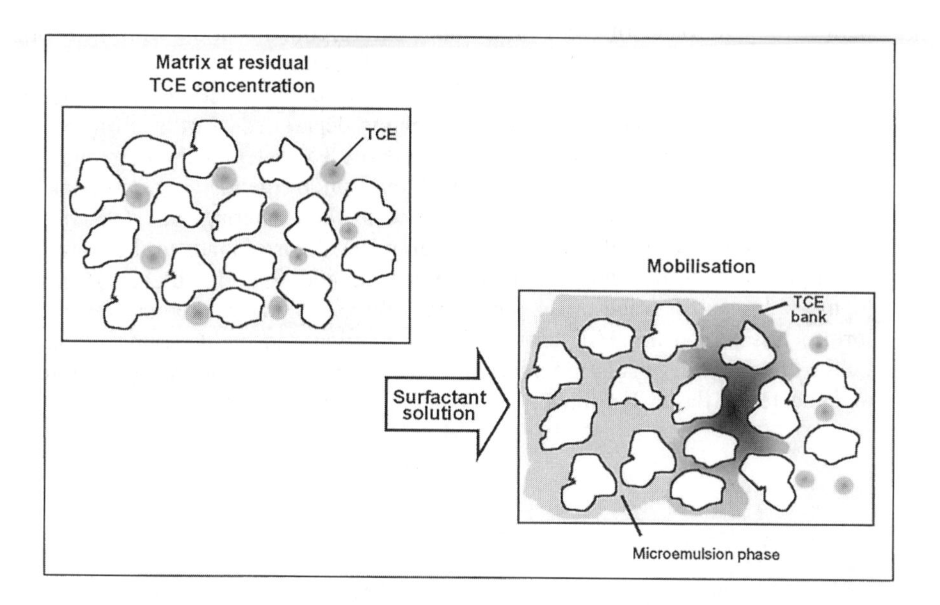

FIGURE 3. Schematic illustration of phase behaviour during surfactant mobilisation of residual TCE trapped by capillary forces in the porous matrix.

In order to demonstrate the importance of these phenomena, flow experiments were conducted with dual columns run in parallel (Figure 1) in which one column contains residual TCE and the other only water. This simple physical model captures an essential element of subsurface treatments: namely, it provides the injected solution with two pathways for flow, one in which NAPL is present, the other free of NAPL. This respects a common situation for polluted groundwater sites, where the NAPL phase will be heterogeneously distributed within the soil matrix. This can be the case following the initial contamination event as well as during intermediate stages of a remediation process.

In the dual column experiment discussed here, multiple pore volumes of surfactant solution were injected in order to simplify interpretation. For the contaminated column, the removal efficiency (cumulative fraction TCE removed per volume of surfactant solution flowed through the column) is essentially identical to that of similar experiments run with a single column. However, inspection of the results reported in Figure 6 tells a different side of the story: getting the necessary volume of surfactant to pass through the contaminated column requires injecting additional surfactant through the rest of the system. Figure 6 plots the instantaneous fraction of surfactant solution entering the two individual columns over the first 250 mL injected. From the outset of the experiment, the TCE-contaminated column receives one-third less surfactant solution than the clean column. This reflects the lower permeability of the TCE column owing to the presence of the residual TCE phase (compare the permeability to water of the TCE column before contamination with TCE, 4.5 darcy, with that at residual TCE saturation, 2.8 darcy).

Secondly, as surfactant injection proceeds, the fraction of surfactant solution entering the TCE-contaminated column departs still further from parity. This may be attributed, at least over the first 0.8 PV entering the contaminated column, to the low mobility associated with the front of the moving TCE bank, as discussed above. In fact, the abrupt increase in fluid entering the TCE column visible in the figure at 230 mL total solution injected corresponds to the arrival of the TCE bank at the end of the column. In addition, the higher viscosity (compared to either brine or the surfactant solution) of the microemulsion phase formed just behind the TCE bank will also contribute to the declining mobility. Of the first 250 mL of solution injected, the ratio of that entering the clean column to that entering the TCE-contaminated column was 2.0.

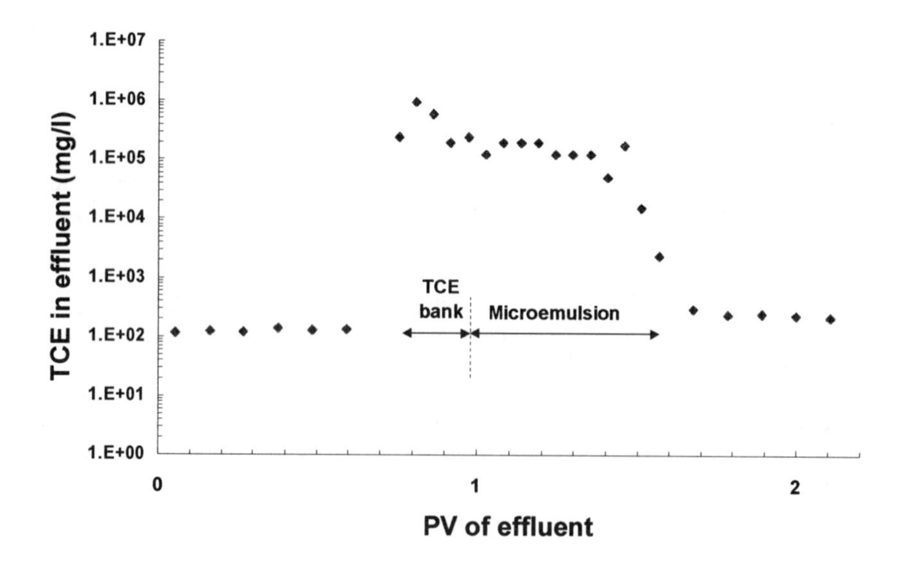

FIGURE 4. TCE in effluent versus total injected volume (1.0 PV surfactant solution injected followed by NaCl brine).

Results are not reported in the figure beyond 250 mL total solution injected because progressive plugging of the flow lines downstream of the TCE-contaminated column occurred after this point. Rheological studies conducted on TCE/brine/surfactant microemulsions revealed a pronounced increase in viscosity at low shear rates such as are calculated for the tubing downstream of the column. Further studies on this shear-dependent behaviour are underway as it could conceivably lead, under some conditions, to trapping of TCE-rich microemulsion in the form of a viscous slug within the subsurface matrix.

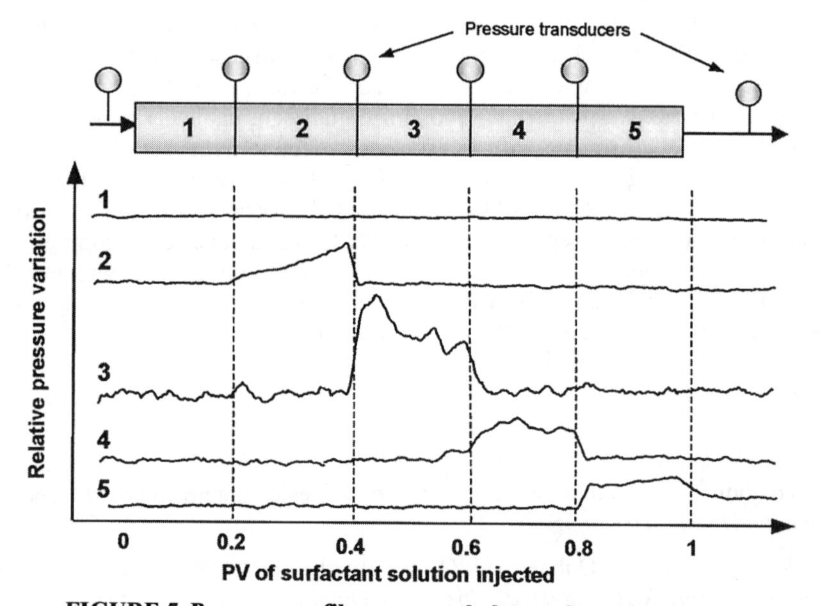

FIGURE 5. Pressure profiles measured along column (tracts 1-5) during surfactant flood show localized low mobility region that moves through column with the front of the TCE bank.

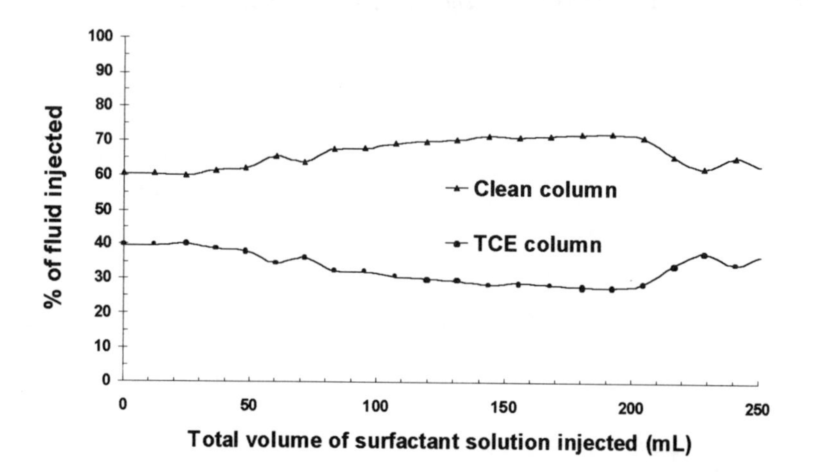

FIGURE 6. Plot showing instanteous splitting of injected solution between clean and TCE-contaminated columns.

CONCLUSIONS

Prior studies of SEAR technology have paid relatively little attention to viscosity and relative permeability effects associated with NAPL mobilisation and/or solubilisation. The present study suggests that such phenomena might strongly influence the sweep, and hence removal efficiency of the process. Unless designed to reveal such effects, laboratory studies using linear sandpacks will provide optimistic estimates of the efficiency of these treatments in the higher-dimensional flow geometry of the subsurface. Finally, while the difficulties of remediating heterogeneous NAPL-contaminated matrices are well known to practitioners, the results reported here show that, even in a perfectly homogeneous matrix, the non-uniform distribution of the organic contaminant will result in uneven flooding of the matrix, with the NAPL-contaminated zone taking less of the treatment solution than would otherwise be expected.

REFERENCES

Pennell, K. D., M. Jin, L. M.Abriola, and G. A.Pope. 1994. "Surfactant enhanced remediation of soil columns contaminated by residual tetrachloroethylene." *J. Contam. Hydrol. 16* (1): 35-53.

Lowe, D. F., C. L. Oubre, and C. H. Ward (Eds.). 1999. *Surfactants and Cosolvents for NAPL Remediation. A technology Practices Manual.* Lewis Publishers, Boca Raton, Florida, USA.

Dwarakanath, V., and G.A. Pope, 2000: "Surfactant Phase Behaviour with Field Degreasing Solvent." *Environ. Sci. Technol. 34* (22): 4842-4848.

REMEDIATION OF PETROLEUM CONTAMINATED LOESS SOIL BY SURFACTANT-ENHANCED FLUSHING ALTERNATIVE

Kun Zhu, Yang Jiantao, Zhang Chunhui, Ma Juan
Department of Environmental Engineering, Lanzhou Railway University,
Lanzhou, Gansu 730070, P. R. China

Abstract: A laboratory study was carried out to evaluate the feasibility of in situ remediation at a diesel contaminated site where an underground storage tank resulted in leakage. The favorable topographic and hydrogeological conditions had surfactant-enhanced flushing alternative available. The average oil residual concentration was 5,000mg/kg at the contaminated site. Through the comparison, aliphatic polyethenoxy ether (AEO_9) and aliphatic polyethenoxy ether sodium sulfate (AES) were optioned for application in the flushing operation. The results proved that AEO_9 was more adapted than AES. Furthermore, the mixing surfactant solution made by 0.8% (v/v) AEO_9 and 0.1% (v/v) AES could significantly enhance the removal rate by 10% as compared with the standard AEO_9 solution. It is estimated that the mixing surfactant solution of 60 pore volumes was able to remove 60% of petroleum remaining in the contaminated unsaturated zone for 9 days.
Key words: Petroleum contamination, remediation, Flushing, Surfactant, Loess

INTRODUCTION

Remediation of petroleum contaminated soils can be accomplished by two types of approaches: excavation or in situ treatment. To date, excavation of contaminated soils is commonly employed since it can be executed quickly with machinery on the site. Excavation, however, can not be desired at the field where a large volume of contaminated soil is to be moved, or contaminant vapors could induce environmental problems if soil was exposed to the surface. In situ remediation offers the advantage of minimum disturbance of a site and infrastructure, as well as minimal exposure hazards. This may lead to lower costs in completing the treatment processes (Davis 1997). The conceptual possibilities of in situ remediation for petroleum contaminated soils include soil venting and vacuum extraction to remove and clean volatile hydrocarbons, bio-remediation to promote breakdown of residual oil by microorganisms, hydraulic methods to collect and remove mobile liquid and dissolved oil, and soil flushing to wash out residual oil from pores of soils. Of all the in-situ remediation technologies, surfactant washing method is a promising candidate for the remediation of a contaminated soil system (Abdul 1992).

In 1996, an environmental observation station detected the occurrence of petroleum contamination within a small section of the Yellow Rive adjacent to Lanzhou City. Then, a comprehensive investigation identified the leakage of an underground storage tank in this area. Originally, a huge underground oil tank with a volume 50m^3 of diesel fuel was installed at the contaminated site by the emergency power workshop attached to a chemical plant. After the petroleum contamination was found and traced to the workshop, the chemical plant was in charge to move the diesel tank away immediately. As a compensating measure, an extraction well was drilled to pump out floating oil that was transported to the wastewater treatment plant. Later, the chemical plant was charged again to clean up the contaminated unsaturated zone that is consist of loess soils 5m thick. We were entrusted to make a possible reclamation draft, thus the flushing – pumping approach was taken as a priority consideration.

The scope of our research involved two aspects: study on the technical aspects of hydrocarbon residual distribution and migration in unsaturated soil; another was the assessment of site-specific characteristics related to the available technical understanding of hydrocarbon cleanup in situ. The objective of our paper is to introduce the selection of the effective surfactants and access their relevant washing competence through the column tests in the laboratory.

HYDROGEOLOGICAL CHARACTERISTICS OF STUDY AREA

The projected remediation field is at the alluvium plan and situated at the suburb of Lanzhou City where geological and hydrological conditions appears likely favorable. The subject site is located in a chemical plant in close proximity to the Yellow River at a distance of approximately 200m. The land surface is covered by loess sediments with the thickness 4.6 - 5.2 m. In the geological structure, bedrock below the remediation site is well-checked Cretaceous sandstone above which Quaternary alluvial sediments mainly consisting of gravel and course sand are overlying. Actually, the stratum of alluvial sediments is vertically sandwiched by the top loess soils and deep bedrock to form a confined aquifer with the average thickness about 6 meters, and its high hydraulic conductivity has a range from 80 to 106m/day. Natural groundwater flow with a porous velocity 1.20m/day crosses the remediation site directly towards the Yellow River. The hydrogeological conditions of the remediation site are illustrated in Figure 1. Therewith, the loess samples were taken from different layers at the remediation field, and were textured. The overall characteristics of loess soils are listed in Table 1. The measurement by using

Darcy's device indicated that loess soils showed comparatively low permeability with a hydraulic conductivity 0.92m/day. Nevertheless, this permeable property was accessible for in situ flushing performance.

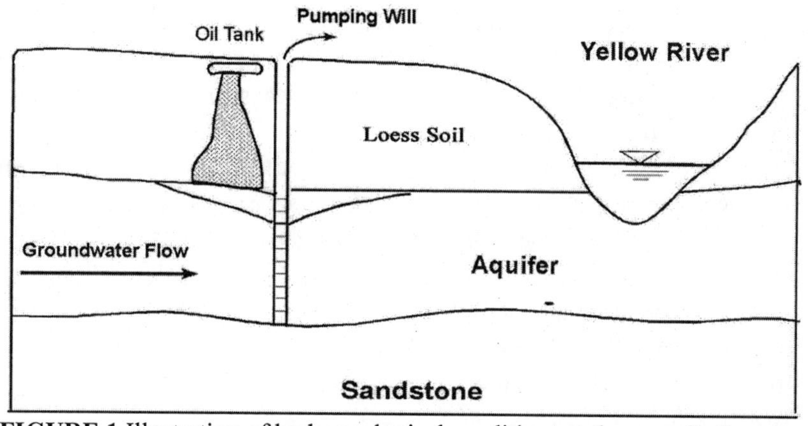

FIGURE 1 Illustration of hydrogeological conditions at the remediation site

TABLE 1 Vertical profile of loess soil at the remediation site

Layer	Sand (%)	Silt (%)	Clay (%)	Bulk density (g/cm^3)	$CaSO_4$ (%)	Organics (mg/kg)	Porosity (%)
Top soil layer (0-2m)	4.6	81.6	13.8	1.10	6.08	1,500	56.1
Middle soil layer (2-3.5m)	6.8	77.2	16.0	1.27	5.73	4,250	50.7
Deep soil layer (3.5-5m)	4.0	73.6	22.4	1.42	6.83	7,500	46.8

Sixteen boreholes crossing overlying loess layer were drilled by hollow stem auger with diameter 1.5 inch along the two vertically crossing lines. Soil samples were taken from different depth in each hole, and the oil content of the soil sample was analyzed individually based on the regular methodology (EPB 1989). The drilling information certified that the hydrocarbons presented in the soil samples below the spot of the former tank were identified to be diesel in character, and the contents varied with depth from 1,500mg/kg in average at the top soil sequentially to 7,800mg/kg at the bottom of vadose zone.

SELECTION OF SURFACTANTS

With respect to hydrophilic properties, surfactants can be divided, in general, into three categories: nonionic, anionic and cationic. The effect of average surfactant hydrophobe length on phase behavior is necessary for surfactant selection for field remediation. A longer surfactant hydrophobe gives higher oil solubility, higher solubilization parameters and lower optimum salinities. Sorption of soil is an important form for the loss of surfactants, which is dependent on the nature of soil surfaces, and pH in soil matrix. Anionic surfactant usually have a low susceptibility for sorption because most soil surfaces are negatively recharged. However, higher adsorption of anionic surfactants to soil usually is observed at lower pH. In contrast, the losses of cationic surfactant often are significant, making their use in remediation practice less favorable compared with anionic and nonionic surfactants (Scamehorn 1982, Harwell 1992, Dwarakanath 2000). In our research project, the essential respect to selection of surfactants was based on the following principles:

- Selection only focused on nonionic and anionic surfactants that must be locally produced and easily be found at market with lower price;
- They must be non-toxic and biodegradable;
- The selected surfactants were of high solubility and strong detergency.

Almost 20 surfactants previously were taken into account among which a sufficient comparison was accessed with respect to the following considerations: critical micelle concentration (CMC), detergency of surfactant for petroleum, potential dispersion of soil colloidal-size particles by the surfactant solution, adsorption by soil surfaces and biodegradability in soil matrix. These concerns are of the critical importance for the selection of surfactants (Abdul 1990). At last, aliphatic polyethenoxy ether (AEO_9) and aliphatic polyethenoxy ether sodium sulfate (AES) were chosen to be the preferable surfactants due to its advantages: strong solubility and detergency (cleaning power), low sorption on soil particle, low dispersion for colloid form structure of soil. The characteristics of AEO_9 and AES are displayed in Table 2 (Lin 1996).

TABLE 2 Characteristics of the selected surfactants in the study

Properties	Surfactants	
	Aliphatic polyethenoxy ether (AEO_9)	Aliphatic polyethenoxy ether sodium sulfate (AES)
Molecular formula	$C_{13}H_{25}O(CH_2CH_2O)_9H$	$C_{12}H_{25}O[CH_2OCH_2]_3OSO_3Na$
Average molecular weight	583	436
CMC (mol/L)	1×10^{-4}	8.7×10^{-3}
HLB	13	11.5
pH	6.5 - 7-5	7 – 8.5
Biodegradable	Yes	Yes

EFFECTS OF SURFACTANTS ON FLUSHING CONTAMINATED SOILS

The optimal concentration of the flushing solution was experimented by the column tests, and the experimental system is shown in Figure 2. The columns were made of thick-walled plexiglass pipes, 8 cm in inside diameter and 30 cm in length. Supported by a rack, a sieve-plate on which a piece of nylon filter screen was placed, was set at the bottom of each column. Before being packed into the column, air-dried loess soil was well mixed with diesel, making the diesel concentration of 5,000mg/kg. The soil was compacted with bulk density about 1.10-1.15g/cm^3 while the porosity was calculated as 0.52. Consequently, the cumulative pore volumes of soil packed in the column were equal to 784.1 cm^3 .

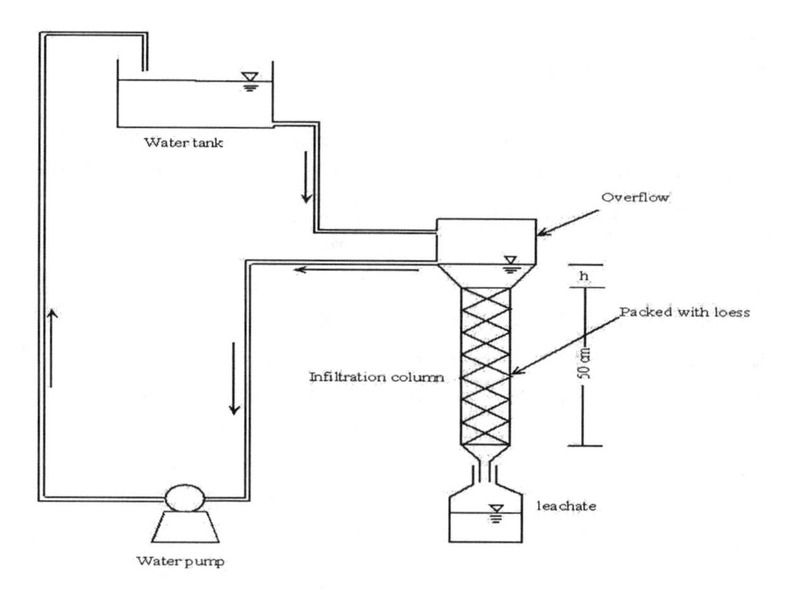

FIGURE 2 The surfactant flushing system for column tests in laboratory

Various surfactant solution consisting of AEO_9 with the concentration (v/v) 0.5%, 1.0%, 1.5%, 2.0%, 2.5%, 3.0%, 3.5% and 4.0% respectively, was put onto the top of the related column. The hydraulic load (h) that was controlled by the overflow outlet, stably maintained 5 cm. After washing with the surfactant solution of 5 pore volumes, the cumulative removal quantity of diesel measured from the effluent of each column is displayed in Figure 3. To be related with the surfactant concentrations distributed into different columns, the alternative tendency of removal quantities appeared to be a parabolic curve on which the point of inflexion approximately matched with the concentration 1.0%. It could be

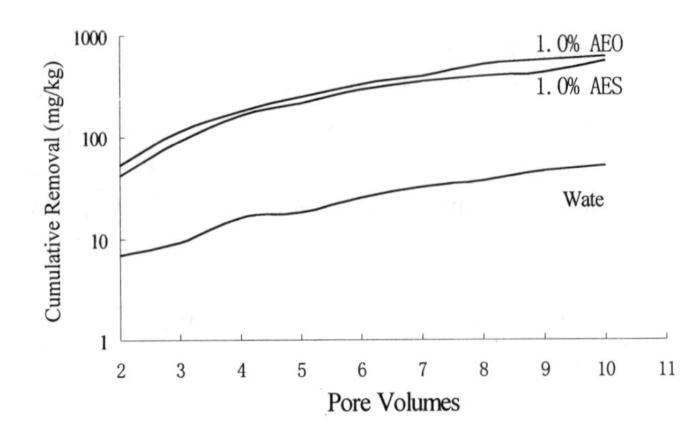

FIGURE 3 Comparison of removal effectiveness between water, AEO_9 and AES solutions in the column tests

understood that flushing effectiveness would not be apparently enhanced if the surfactant concentration continued increasing from the value at point of inflexion. On the economic view, the AEO_9 concentration 1.0% even little less could be considered as the optimal option for the application.

Another comparison was done with three columns those were under the same experimental conditions as above described. The tape water, AEO_9 solution of (v/v)1.0% and AES (v/v) 1.0% were respectively put onto three columns for continuously washing. The increase of the cumulative removal quantity of diesel was proportional with volume of flushing solutions, and the alternative relations are drawn in a semi-logarithmic coordinate as shown in Figure 4. It can be seen that the removal effectiveness of 1.0% AEO_9 solution was higher than AES, additionally at least 10 times higher than fresh water. Surfactants- enhanced soil remediation is one of the high effective technologies for the removal of petroleum residual from soils. Surfactants greatly increase cleanup efficiency by increasing the solubility of petroleum constituents and by lowing the interfacial tension between the hydrocarbons and water (Dwarakanath 2000). Regarding Zhong's suggestion (Zhong 1995) that addition of 0.1% (v/v) AES could significantly enhance flushing effectiveness. A mixing surfactant solution consisting of 0.8% (v/v) AEO_9 and 0.1% (v/v) AES was prepared for the column test, and the washing results are shown in Figure 4. It was verified that the mixing flushing solution was more attainable to clean diesel residual from the soil.

The approximate estimation indicated that the flushing effectiveness by using mixing flushing solution of 0.8% AEO_9 and 0.1% AES would enhance petroleum removal by 10% more than by using the standard 1.0% AEO_9 solution, or by 20% more than using the standard 1.0% AES solution. Based on the above experimental results, the mixing solution consisting 0.8% AEO_9 solution with addition of a cosolution 0.1% (v/v) AES, was suggested to apply for in situ flushing remediation. The first operation period was drafted with a volume of flushing solution equal to 60 pore volumes, which would take 9 days for leaching

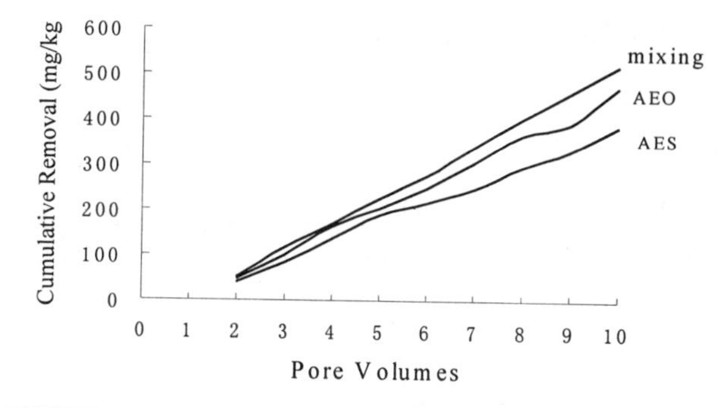

FIGURE 4 Comparison of removal effectiveness by using 1.0% AEO,1.0%AES and the mixing solute

according to a computation by Darcy's law. In this stage, 60% of diesel residual wo uld expectantly be washed out from the vadose zone. The local environmental agency demanded that the flushing operation would not stop unless the oil concentration in the elutriants was less than 5mg/L. However, the surfactant concentration and the proportion of the cosulution should be adjusted in the further flushing process depending on the previous results. As soon as the surfactant flushing processes finished, the fertilizer solution containing N, P, K would be added to the remediated land to promote the biodegradation. A simple calculation by Darcy's law indicated that the flushing leachates would take 5 days to vertically infiltrate the whole loess layer before getting to the groundwater table. After the extraction of groundwater, the drawdown in the well would reach 1.67 m 5 days late meanwhile a stable cone of depression was formed. Thereby, the cone of depression completely controlled leachates and all the mobilized contaminants.

CONCLUSION

We are confident to point out that the above outlined project would present the satisfactory environmental compliance and an appropriate remedial alternative based on the investigation for the soil properties, hydrological and hydrogeological favorable conditions. Both AEO_9 and AES as the selected surfactants were feasible and effective to wash the diesel contaminated loess soil. To be an optimum selection, the mixing flushing solution made by 0.8% (v/v) AEO_9 and 0.1% (v/v) AES could significantly enhance removal rate of oil residual.

It was guaranteed that no detrimental impact would escape from the capture zone created by a pumping well. As such, in this remediation site, no-risk alternative would be the appropriate response.

ACKNOWLEDGMENTS

We sincerely thank the National Natural Scientific Foundation of China for supporting the research (Project No.29977015).

REFERENCES

Abdul, A.D., Gibson,T.L. and Rai, D.N. 1990. "Selection of surfactants for the removal of petroleum products from shallow sandy aquifer." *Ground Water 28*(6): 920-926.

Abdul, A.S.,Gibson,T.L. and Ang, C.C. 1992. "In situ surfactant washing of polychorinated biphenyls and oils from a contaminated site." *Ground Water 30*(2):210-231.

Davis, G. B. 1997. "Site clean-up the pros and cons of disposal and in situ and ex site Remediation." *Land Contamination & Reclamation, Vol. 5* (4): 287-290.

Dwarakanath, Varadarajan and Pope, Gary A. 2000. "Surfactant Phase behavior with field degreasing solvent." *Environ. Sci. Technol. 34*: 4842-4848.

EPB National of China, 1989. Handbook of analysis methodology and observation for water and wastewater analysis (Third Edition). Edit Office of National Environmental Protection Bureau, Beijing, P.R. China.

Harwell, J. H. 1992. "Factors Affecting surfactant performance in groundwater remediation applications, Chap 10, transport and remediation of subsurface contaminants." pp.124-132. ACS Symposium Series 49, American Chemical Society, Washington, DC.

Lin, Qiaoyun and Ge, Hong(Eds.). 1996. *Principle and application of surfactants.* pp.15-28. Petrochemical Industry Publication of China, Beijing, P.R.China.

Scamehorn, J. F., Schecher, R. S. and Wade, W. H. 1982. "Adsorption of surfactants on mineral oxide surfaces from aqueous solutions. I Isopmetrically pure anionic surfactants." *J. Colloid Interface Sci. 85* (2): 463-478.

Zhong Zuoxin. 1995. *Study on petroleum polluted unsaturated zone in a karst region.* Research Report 908. Geological University of China, Beijing, P.R. China.

EXTRACTION OF SMALL PARTICLES FROM PARTICLES BED UNDER BODY OF WATER

Tsutomu Nozaki (Kagishima University, Kagoshima, Japan)
Eiichiro Gombori, Katsuo Kizushi (Kagoshima University, Kagoshima, Japan)

ABSTRACT: As one of the proposals of a classifier, the injection port is mounted at the collecting mouth where the sands (relatively large particles) and the sediments (relatively small particles) mixture, which pile up under a body of water are fluidized before they are removed. The maximum diameter of the sediments to be removed is only determined by the active control of the suction velocity of water at the classifying region of a suction unit. In this device, the horizontal upward velocity is controlled to change the forces acting upon a particle. Whether the particle goes up or down depends upon the balance of forces acting upon the particle. In order to confirm the system, fundamental experiments using flow visualization were carried out. As a result, there exists the combination of the smaller and larger beads, which shows the quickest classification performed for given conditions.

INTRODUCTION

Collecting and removing the sediments under a body of water are very important problems more and more in the future. In the case of removing the sediments especially containing the contaminants from the bottom of water, it is desirable not to diffuse them outside of the cleaning region and is desirable to remove only the contaminants, which are relatively small compared with the sands and stones originally placed there. It is very difficult to only remove the small particles from the mixture by the conventional methods. It was proposed by the authors that by mounting the injection at the collecting mouth collecting and removing of the small particles could be performed more effectively and more smoothly than the other conventional systems.

In the previous papers, the general view of the collector with an injection at the collecting mouth and the relations between the flow ratio of the injection to the suction and the concentration of the beads contained in the suction flow were reported using the spherical glass beads as a collecting material and the water as a carrier (Nozaki et al., 1990). The relation between the bead diameter and the concentration of beads (Nozaki et al. 1991) and the relation between the ratio of inner diameter of injection pipe to that of suction pipe and the concentration of beads were also reported so far (Nozaki et al., 1991).

As one of the applications of such system a classifier removing only small particles from the mixture by the active control of a suction velocity at the entrance region of a suction unit is proposed in this paper. A fundamental experiment of the classifier is carried out using the spherical glass beads as a collecting material again. The suction unit model is fabricated and the flow characteristics are clarified by using the flow visualization technique. From the results of a series of pictures taken by video camera recorder, the effects of the injection on the classification of the beads are clarified quantitatively.

PRINCIPLE OF PROPOSED CLASSIFIER

Principle of a proposed classifier extracting the smaller particles from the various particles mixture under a body of water is illustrated in Figure 1. By mounting the injection unit at the center of the suction unit ②, the particles bed ③ is fluidized by the injection flow before the small particles are extracted. The classifier consists of the injection-suction port, the floating region where the larger particles move from the fluidized region temporally to be able to perform fluidization easier and the classifying region in which the suction velocity is actively controlled at the floating one and only small particles are extracted here. The flow velocity decreases gradually in the floating region, then the particles go up and down according to their sizes. The friction plate ④ is mounted to keep the uniform velocity over the whole entrance section of the classifying region. Then only small particles are extracted in the classifying region where the velocity is controlled actively at the floating velocity of the small particle to be extracted.

The maximum diameter of the deposits to be removed d_{pmax} is determined only by the vertical velocity of water as a carrier at the classifying region of the suction unit. As shown also in Figure 1, the critical velocity V_c is determined by the condition of balance of the buoyancy, the gravity force and the drag force acting upon the particle as follows:

$$\frac{\pi}{6}\rho_p g d_{p\max}^3 = \frac{\pi}{6}\rho_f g d_{p\max}^3 + C_D \frac{\rho_f V_c^2}{2}\frac{\pi}{4} d_{p\max}^2 \quad (1)$$

where C_D is the drag coefficient, pf the density of fluid and pp the density of particle(Bain and Bonnington, 1970). Figure 2 shows the relation between the critical velocity and the maximum diameter of the spherical glass bead to be removed used in this experiment. The larger beads are lifted up together with the smaller beads whereas the suction velocity is kept slightly lower than the critical one. This comes from the fact that as the smaller beads are lifted up in the suction pipe the effective flow area decreases then the suction velocity increases. As the larger beads are lifted vertically upward in the suction pipe, the bead concentration becomes lower, then the larger beads go up and down in the entrance region of the suction pipe during the cleaning. This temporal movement of the larger beads from the cleaning region is so effective as to clean up the particles bed with the injection. When the cleaning finishes, the larger beads which float in the suction pipe go down and pile up in the bottom of water again.

EXERIMENTAL APPARATUS AND METHOD

The experimental apparatus of the classifier-separator system is shown in

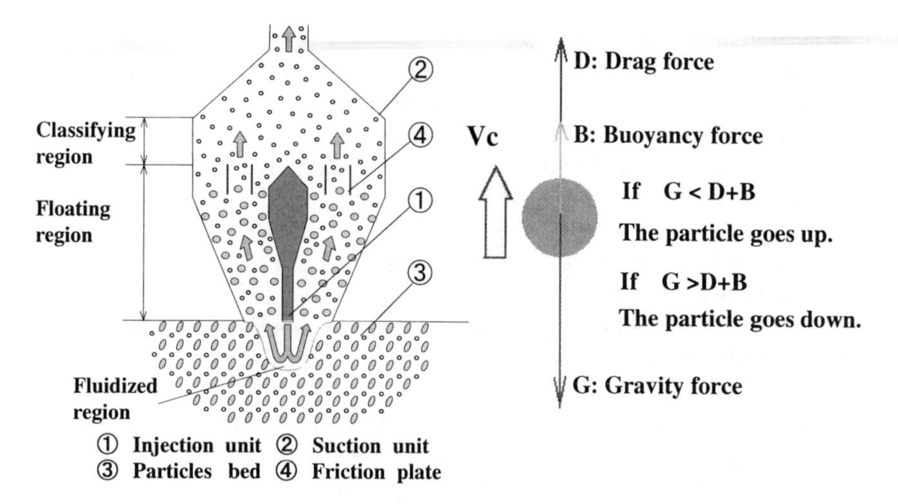

FIGURE 1. Principle of proposed classifier.

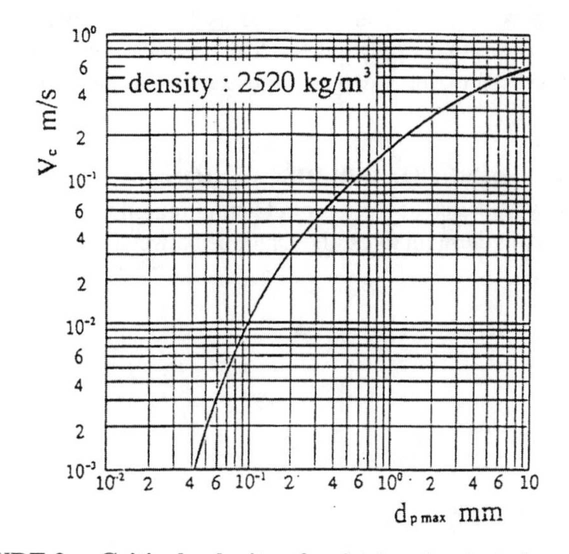

FIGURE 2. Critical velocity of a single spherical glass bead.

Figure 3. The water in the flow tank ① issues vertically in the down direction from the injection pipe ③ to the test vessel ④, where the mixed spherical glass beads ⑤ settle under a body of water. The beads and water two-phase flow entering from the collecting mouth ⑦ flows into the separator ⑧ (Raval Co. 0050LS) through the suction pipe ⑨ and the suction flow control valve ⑩. Then the beads are separated from the water as a carrier and collected in the beads tank ⑪. The cleaned water flows into the water tank ⑫ through the pump for suction (Gear-Ace Industry, SF10) ⑬. In order to visualize the cleaning section, the test section is constructed of a half divided model cut the plane through the axis of suction pipe.

In this experiment, the inner diameter of injection pipe D_i and that of suction D_s are kept constant 10mm and 100mm respectively. The suction velocity v_s is also kept nearly constant at 0.11m/s, which is slightly lower than the critical velocity of larger beads used in this experiment. Two different diameter ranges of the bead are mixed. One is the relatively small bead as a deposit, which is to be removed and the other is the large one as a sand to be left. Five different combinations of the smaller glass beads and larger one are shown in Table 1. The smaller beads are colored in order to recognize the cleaned boundary visually. The volumetric ratio of smaller beads to larger one for each combination is determined by the condition, which gives the most condense particles bed in the case of COM-1.

In order to clarify the effects of injection flow, the experiments were carried out by using the flow visualization technique. A cleaned volume V is obtained by integrating a corrected cleaning boundary $f(r)$ with r as follows:

$$V = 2\pi \int_0^{\infty} f(r) r dr \tag{2}$$

A cleaned up volume V_f is also defined as V finally obtained a picture of cleaned up region is originally taken by the Video camera recorder (EDC-50 Sony Co. Ltd.) and processed by the video copy processor (SCT-CP100, MITSUBISHI ELECTRIC Co. Ltd.).

EXPERIMENTAL RESULTS AND DISCUSSION

Figure 4 shows a series of picture of cleaning region with time in the case of COM-1. By analyzing these pictures, the relations between the cleaned volume

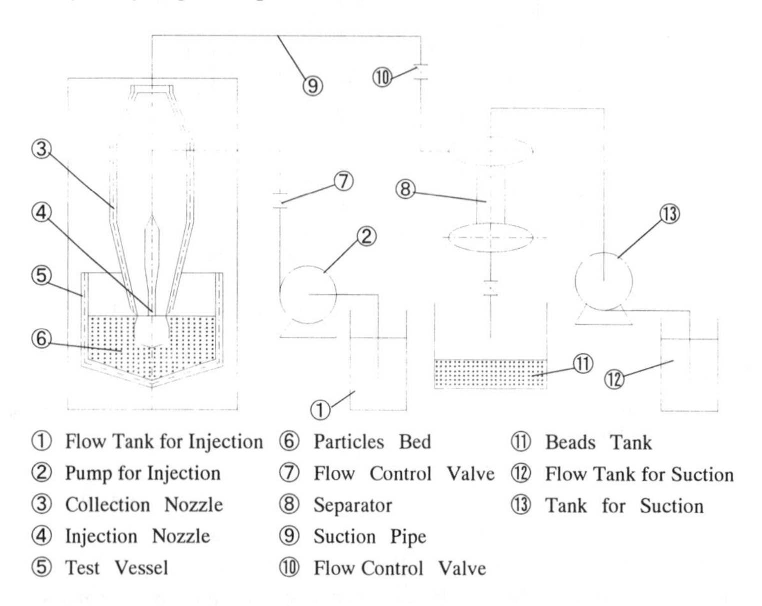

FIGURE 3. Schema of experimental apparatus of classifier.

TABLE 1. Combinations of beads.

	Bead dia. range μ m	Apparent density kg/m³
COM-1	Small beads : d_S = 38 ~ 75 Large beads : d_L = 840 ~1000	1436 1494
COM-2	Small beads : d_S = 75 ~125 Large beads : d_L = 840 ~1000	1517 1494
COM-3	Small beads : d_S = 150 ~ 212 Large beads : d_L = 840 ~1000	1531 1494
COM-4	Small beads : d_S = 212 ~ 300 Large beads : d_L = 840 ~1000	1529 1494
COM-5	Small beads : d_S = 300 ~ 425 Large beads : d_L = 840 ~1000	1533 1494

and the cleaning time are obtained as shown in Figure 5. In the case of the smallest particles of d_p as COM-1, it takes long time to clean the particles bed. As the smaller particle d_p becomes larger, the cleaned volume V increases. In the case of COM-3, the cleaned volume shows the largest value. As the smaller beads diameter is larger than this, the cleaned volume decreases. In the case of the smaller beads diameter being very small, the collisional viscosity is large (Gerdart, 1973 and Saxton et al., 1970). On the other hand, in the case of the smaller particles diameter is larger than COM-3, the gravity force acting the particles increases. Then, the cleaning velocity designated by the gradient of the V-T curve decreases and there exists the particle size which establishes the quickest cleaning as COM-3.

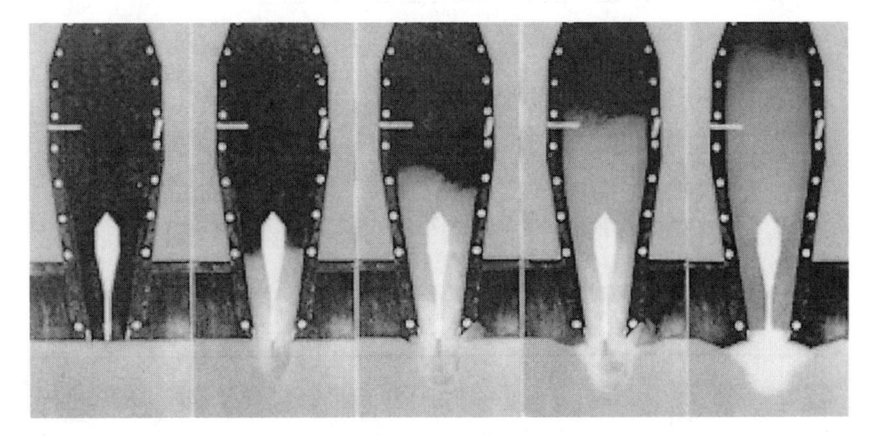

FIGURE 4. Cleaning region(COM-1,Qi/Qs=0.11).

Figure 6 shows the relation between a cleaned volume and a mean bead diameter at 10 sec. As can be seen from this figure, a quick cleaning is established for the bead whose diameter is nearly 150mm. As the mean diameter of a bead becomes smaller than this value, the cleaned volume decreases very rapidly. The actual particles bed consists of numerous sizes of particles. Then it cannot be expected to obtain such clear separation of smaller particles and larger ones as shown in this experiment. However, removement of the sediments by shovel and bucket is not desirable at the point of view of the performance. Furthermore, by using this proposed collector, only cleaning region of the bottom of water especially containing the contaminants is fluidized and these do not diffuse outside. This is one of the most important strong points of this proposed system.

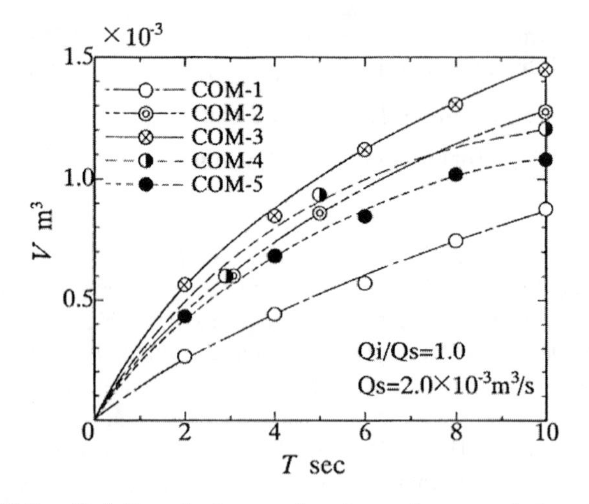

FIGURE 5. Relations between cleaning volume and cleaning time.

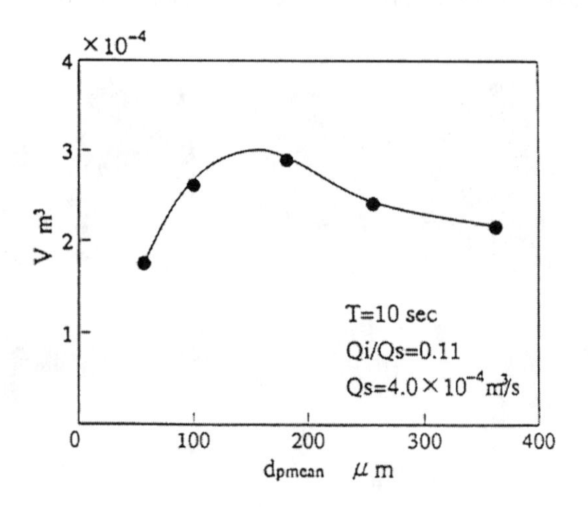

FIGURE 6. Relations between cleaning volume and mean dia. of beads.

CONCLUSIONS

It became clear from the results of flow visualization that in the vicinity of the cleaning region, clear separations are performed for several combinations of the spherical glass beads. By analyzing the pictures of flow visualized section, the effects of the injection on the cleaned up volume and the effects of the collected bead diameter on the cleaning time could be distinguished quantitatively for several combinations of the smaller and larger bead diameters. It was found that there exists the bead diameter, which classifies the particles bed effectively and rapidly for the given condition. Furthermore, by using this proposed classifier, only the classifying region containing the contaminants is fluidized and these do not diffuse outside the classifying region. This is one of the most important and strong feature of this proposed system designed for environmental preservation. As one of the proposals concerning the classifier as a cleaning machine removing the contaminants and many kinds of sediments under a body of water, it was clarified from the fundamental experiments that the classifier is simply performed by the system with an injection at the collecting mouth. From the results of flow visualization of the cleaning region, there exists the smaller particle size, which establishes the quickest cleaning. In the case of smaller and larger particles than this, it takes too much cleaning time as the effect of the collisional viscosity and the gravity force acting upon a particle are predominant, respectively.

This is very useful for the purposes to separate the relatively small particles from the mixture containing the numerous sizes of particles. Furthermore, it can be expected that the proposed system does not diffuse the contaminants outside of the cleaning region. This work was financially supported in part by HARADA Memorial Foundation(2000).

REFERENCES

Nozaki,T., et al., 1990, "Development of the Sand Collector Having an Injection at the Collecting Mouth(Effects of the Flow Rate on the Concentration of the Glass Beads.)." Proc. 3rd Japan-China Joint Con. on Fluid Machinary, 1(1): 339-343.

Nozaki,T., et al., 1991, ASME-FED-Vol.118, Liquid Solid Flows, (1991), 219.

Nozaki,T.,et al., 1991, "Fundamental Experiments of Sand Collector with Injection Port(Variations of Fluidized Volume with Time in the Case of Volcanic Ashes) Proc. 2nd International Conference on Multi-Phase Flow, 4: 9-12.

Bain, A. G. & Bonnington, S. T., The Hydraulic Transport of Solid by Pipeline, Pergamon Press (1970).

Geldart,D., Powder Technology, No.7(1973), 285.

Saxton, J. A. Fitton, J. B. & Vermeulen, T., AIChE Journal, Vol.16, No.1(1970), 120.

REMEDIATION TECHNIQUES APPLIED TO REDUCE PAINT-DERIVED TRIBUTYLTIN (TBT) IN DREDGED MATERIAL

Jacquie Reed (CEFAS, UK), Mike J.Waldock (CEFAS, UK), Bryn Jones (CEFAS, UK), Sylvia Blake (CEFAS, UK), Paul Roberts (CEFAS, UK), Gary Jones (CEFAS, UK), Chris Elverson (University of Nottingham, UK), Steve Hall (University of Nottingham, UK) and ***Lindsay Murray*** (CEFAS, UK).

ABSTRACT: Since the ban on tributyltin (TBT) for small vessels (<25m) in the UK, there has been a general decrease in TBT concentrations at most impacted sites. However, a historic legacy of TBT contamination remains at a number of locations around the UK. Today paint-derived TBT in harbour sediments provides an ongoing source of contamination. DEFRA as the regulatory authority for sea disposal of dredged material in England and Wales has funded CEFAS to investigate the distribution of TBT as sediment-entrained paint particles and initiated research to develop remediation methods to mitigate environmental harm. The aim of this study is to remediate sediments to increase the amount of the sediment that is available for conventional disposal whilst the isolated contaminated material can either be subjected to further remediation (i.e., biological or chemical methods), sent to landfill or placed in a contained site. The study has focused on rapid methods for remediation and a number of different physical and chemical separation techniques have been evaluated. Experiments have been conducted on sediments spiked at two concentrations (1 and 10 mg kg^{-1}) of TBT paint particles and samples taken from contaminated environments (4 mg kg^{-1} TBT). Preliminary results are encouraging and it has proven possible to isolate a heavily contaminated component of the sediment leading to a significant reduction in the volume of grossly contaminated material. Samples were fractionated by size and density using froth flotation and density separation techniques and it was possible to identify paint particles in the large/light samples (250 - 500 μm). The results generally show that the highest concentrations of TBT (i.e., paint particles) are associated with the light fractions. Initial results suggest that it may be possible to enhance isolation of contaminated material by these two techniques. Further method development is ongoing to increase the proportion of TBT-enhanced material removed from the bulk sediment.

INTRODUCTION

Despite the huge amount of research and monitoring that has been carried out during the last 20 years there are still gaps in our knowledge when addressing the issue of the environmental harm of waste paint material entrained in dockyard sediments. TBT occurs in these sediments in three broad categories (i) large paint chippings (ii) small paint particles derived from high pressure hosing of copolymer formulations and (iii) TBT adsorbed to the sediments. Coastal sediments, in particular sediments present in ports and harbours, are contaminated to some degree with TBT from the legacy of use during the last 30 years in vessel antifouling paints. Perhaps the most persistent form of TBT is paint-derived material in dock and harbour sediments.

Currently, there are no proven remediation techniques aimed specifically at TBT contamination. To address this issue, a study was commissioned by DEFRA, the licensing authority, to investigate sediment-TBT contamination in England and Wales, and develop a suitable methodology to remove paint-derived TBT from dockyard material. To reduce the impact of TBT from paint particles when dockyard sediments are disposed offshore, investigations have been carried out on physical and chemical remediation techniques.

Aims. The aims of this study are: (i) to review historic trends in TBT concentrations in dredged material to identify impacted sites around England and Wales and list sites of priority concern; (ii) to determine the behaviour of TBT-contaminated sediment at dredged material disposal sites and (iii) to evaluate whether or not TBT impacted sediments may be improved by remediation techniques before disposal.

HISTORIC LEGACY OF TBT—THE PROBLEM IN ENGLAND AND WALES

Most estuarine sediments in England and Wales are contaminated to some degree compared to offshore sediments. TBT concentrations in dredged material have been measured routinely by CEFAS since 1992 and a total of 1507 locations have been sampled. Sediments at many of these locations (49%) are regarded as suitable for sea disposal. A further 39% of locations have required further investigation to determine the local spatial extent of TBT contamination and additional restrictions have been imposed (e.g. rejection of specific sites within the location) before a licence is approved for sea disposal. Material at the remaining 12% of locations were regarded as unsuitable for sea disposal. Table 1 shows mean annual TBT concentrations (mg/kg) from selected sites in England and Wales.

TABLE 1. A range of TBT concentrations (mg/kg) at sites around England and Wales.

Year	Tyne	Mersey	Falmouth	Thames	Solent
1992	3.02				
1993	0.14 - 1.3		34.88	0.06	0.05 - 0.38
1994	0.13	0.21 - 2.06	15.38		0.13 - 0.48
1995	0.01 - 1.7	0.14 - 8.84		0.15	0.01 - 0.66
1996	0.55 - 8.24	0.15 - 0.44		0.26	0.01 - 0.44
1997	4.47 - 65.25	0.01 - 2.45			0.3 - 0.99
1998	0.02 - 24	0.25		0.17	0.05 -0.24
1999	0.002 - 119	0.01 - 0.22			0.03 - 32.7
2000	0.14 - 8.16	0.01 - 2.7		0.01 - 0.127	0.05 - 1.41

Sediments considered to be grossly contaminated with TBT concentrations (>1 mg/kg) are located close to dockyards at Falmouth, River Mersey, River Tyne, River Orwell, River Humber and Swansea. At several locations sediments are suspected to contain paint chippings due to excessive TBT concentrations (up to 100 mg/kg). Consequently, alternative measures are here sought to assist dock owners and port authorities in managing the paint-derived TBT problem.

TBT AT OFFSHORE DREDGED MATERIAL DISPOSAL GROUNDS

Since gross contamination is patchy when waste paint material is entrained in dockyard sediments, concerns were raised about missing hot-spots when few samples are analysed as part of the risk assessment. There is therefore the potential for redistributing TBT in dredged material at the disposal grounds.

Surveys were conducted to investigate whether TBT concentrations were a problem offshore at selected dredged material disposal grounds. Sediment cores were taken at several sites at each of the disposal grounds (total samples collected *n*=30). As predicted from risk assessments, TBT concentrations were low at all disposal grounds except at one location in the North East of England (Figure 1). Within and outside of this site TBT profiles (TBT:DBT ratios) were determined and even here sites outside the disposal ground showed TBT values (although at relatively low concentrations). A decrease of TBT concentrations with depth has been shown which indicates degradation of TBT. Recent restrictions on disposal at this site will eventually ameliorate the present situation.

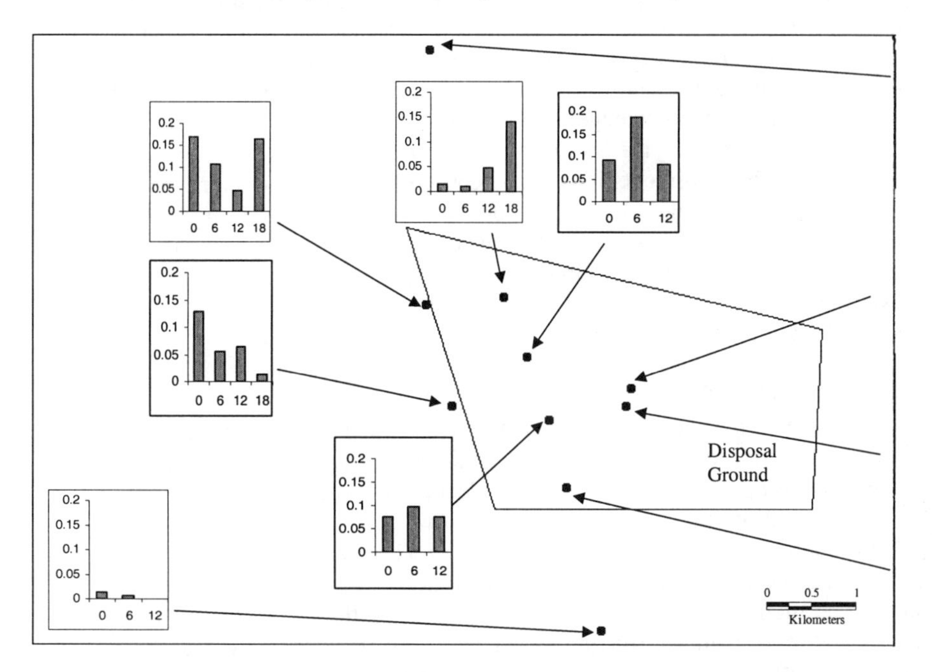

FIGURE 1. TBT concentrations (mg/kg d/w) in sediment cores (cm) at a disposal ground in North East England.

REMEDIATION TECHNIQUES FOR TBT CONTAMINATION

There are several techniques that have been identified for the remediation of TBT from contaminated sediments. These have utilised biodegradation, electrochemistry, physical separation, thermal and chemical extraction processes. Various recoveries have been obtained, but the most successful is up to 80% removal (De Brabandere *et al.*, 1997). Our study concentrates on low-cost rapid solutions based on physico/chemical approaches.

Grossly contaminated dredged material was collected using a Van veen grab and analysed for TBT. Since TBT is highly toxic to aquatic organisms, methods of analysis must be adequate for the determination of trace quantities (nanograms per gram) of analyte. Organotin compounds are therefore extracted from dredged material by sodium hydroxide and methanol, converted to hydrides and partitioned into hexane. The derivatives are then analysed by gas chromatography with flame photometric detection (GC-FPD) (Waldock *et al* 1989). The detection limit for the method is 0.005 μg/g for TBT, 0.010 μg/g for DBT and 0.02 μg/g for MBT.

A subsample of contaminated dock sediment was separated into fractions of different densities and particle sizes using different physical separation techniques to aid in the remediation of contaminated substrates. These included: hydrocycloning, in which a series of hydrocyclones are used to separate increasingly fine fractions of sediment down to ~5 μm (the subsequent fractions are collected and analysed for TBT); and froth flotation, which requires the addition of chemical agents and air to the sediment sample. The choice of chemical depends on the target contaminant. A knowledge of the properties of TBT containing paint flakes was useful in determining the additives.

In addition, sediments have also been prepared for laboratory tests using TBT copolymer paint particles at two concentrations (1.0 and 10 mg kg^{-1}). Paint residues used to spike these sediments were ground to size < 63 μm to mimic slurry from shipyard practices. Similar physical and chemical separation methods above were applied to these spiked sediments to investigate the size fraction where TBT paint became enriched.

RESULTS AND DISCUSSION

TBT enriched samples were identified using analytical methods above and additional isolation of TBT-based paint particles was accomplished using surrogate markers for paint by elemental probe electron microscopy (Figure 2).

Dredged sediments were subjected to two processes to separate the material into fractions by particle size and density. TBT was analysed in different size fractions (Table 2). It was possible to visually identify paint particles in the float sample and the middle fraction (250 - 500 μm) using the froth flotation and density separation techniques, respectively. The results generally show that the highest concentrations of TBT (i.e., paint particles) are associated with the light fractions. This is evident from the field sediment sample. Here, elevated TBT levels within the light fraction were mainly confined to the > 63 μm size fraction. This suggests that Tyne sediments contain larger paint particles than previously suspected. It was possible to identify paint particles in the "large light" fractions

(Figure 3). Results from the spiked sediments suggested that the remediation of the sediments containing paint particles <63 μm shows the reverse contamination pattern (Table 3 and 4). Importantly, paint particles used to spike these sediments were pre-treated to < 63 μm size. This may explain the different contamination patterns shown.

FIGURE 2. Large paint flake identified using electron microscopy.

The most encouraging result from these separation processes is that it has proven possible to isolate a heavily contaminated component of the sediment leading to a marked reduction in the volume of the contaminated sample (60-70%). The aim here is to increase the bulk of the sediment that is then available for conventional disposal or use whilst the isolated contaminated material can possibly be subjected to further remediation (i.e., biological or chemical), sent to landfill or placed in a contained site. Initial results suggest that it may be possible to remove a greater percentage of TBT from dredged material using variants of these techniques.

Further method development is required to increase the proportion of TBT-enhanced material removed from the sediment. A second series of trials is in hand to optimise the fractionation procedure.

CONCLUSIONS

Elevated concentrations of TBT (> 1 mg/kg) are common in UK dock-yards and sediments at many sites contain particulate paint material. TBT at these concentrations are perceived to originate from shipyard practices in the past.

TABLE 2. TBT concentrations in fractions of dredged material from the River Tyne.

density ↓ Size →

	38μm	63μm	125μm	250μm	500μm
light	0.73	0.97	3.25	11.48	8.1
medium	0.44	0.22	2.05	1.85	4.01
heavy	0.21	0.32	0.2	1.84	3.46

The original sample contained a mean concentration of 4 mg/kg (shaded values are higher than those likely to be given approval for disposal to sea). The total mass of sample in the shaded region was above 60% in this trial, but the ability to discriminated heavily and less heavily contaminated material provides a promising way forward for further work.

TABLE 3. TBT concentrations in TBT spiked sediment (<63μm) (1 mg kg^{-1})

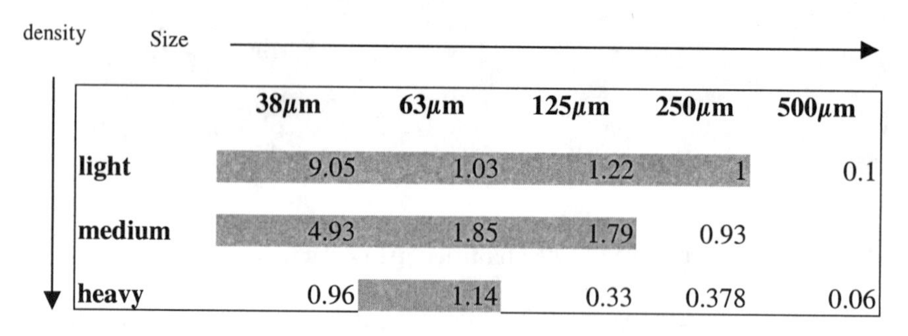

density ↓ Size →

	38μm	63μm	125μm	250μm	500μm
light	9.05	1.03	1.22	1	0.1
medium	4.93	1.85	1.79	0.93	
heavy	0.96	1.14	0.33	0.378	0.06

TABLE 4. TBT concentrations in TBT spiked sediment (<63μm) (10 mg kg^{-1})

density ↓ Size →

	38μm	63μm	125μm	250μm	500μm
light	56.91	24.82	11.45	10.44	1.51
medium	21.92	19.92	1.1	5.71	0.12
heavy	6.09	3.07	0.33	5.9	0.13

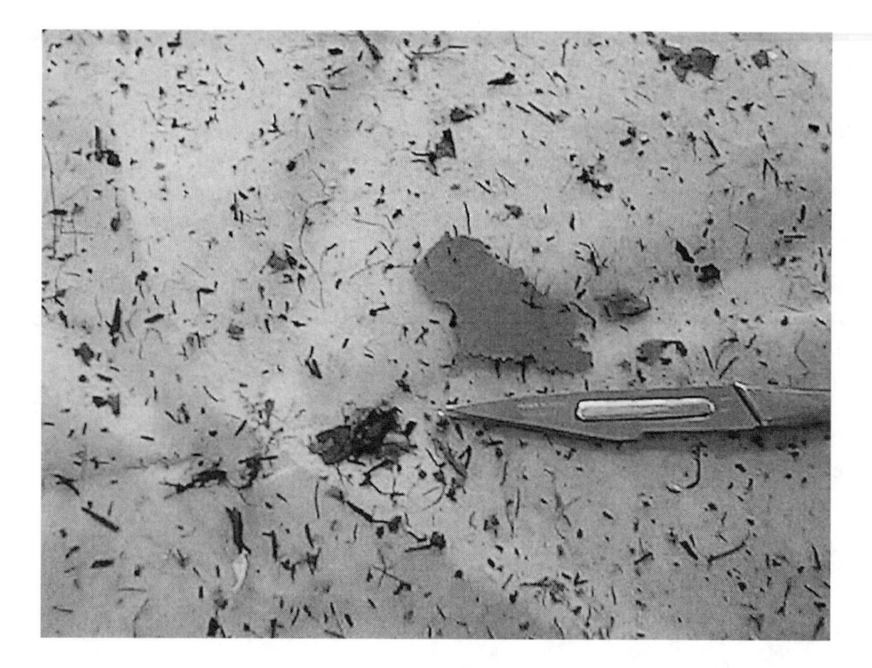

FIGURE 3. A paint flake from the TBT-enhanced sediment fraction using a density separation technique.

Recent surveys of offshore dredge disposal sites show that current restrictions on disposal of such sediments are sufficient to contain any impact within disposal sites. However, it is vital to at least maintain the level of commitment to monitoring that is undertaken at selected disposal grounds to confirm this situation.

The pilot studies utilising TBT-contaminated sediments have identified techniques that provide a promising way forward in removing particulate TBT contamination. A combination of froth flotation and density separation techniques have been effective in reducing >60 percent of paint derived TBT. Further method development is required to increase the proportion of TBT-enhanced material. Accordingly, further samples will need to be analysed so that these results can be validated.

FUTURE WORK

Further optimisation of these methods are required to increase the percentage of TBT from the remediated sediment. Subsequent development requires a large scale experiment to be conducted. In addition, an experimental study should be conducted to investigate the behaviour of paint particles in sediments and studies on the bioavailability of sediment-entrained paint material should also be investigated (Reed *et al.*, 2001).

The authors would like to thank DEFRA, UK for funding this project.

REFERENCES

De Brabandere J., Vanderhaegen B., Dumon G. & Verstraete W. 1997. Bioremediation of Contaminated Harbour Sediment: Development of a Treatment Assessment Strategy. Proc. International Conference on Contaminated Sediments (ICCS), Rotterdam, 1997. Vol 1: 342-345

Reed, J., Thain, J.E., Jones, C., Blake, S.J., James, S., & Waldock, M.J. 2001. Behaviour, fate and bioavailability of paint-derived TBT in sediments. SETAC, Baltimore, November 2001.

Waldock, M.J., Waite, M.E. & Thain, J.E. 1989. The effect of the use of Tributyltin (TBT) antifoulings on aquatic ecosystems in the U.K. Report for Department of Environment.

FIELD SCALE SEDIMENT REMEDIATION USING DRY DREdge™ AND GEOTEXTILE TUBES

Martin L. Schmidt (URS Corporation, Cleveland, Ohio)
Michael L. Duke (WRS Infrastructure & Environment, Inc., Eagleville, Tennessee)
William C. Olasin, (Ashland Inc., Columbus, Ohio)

ABSTRACT: This paper describes implementation of a field scale remediation project that combined innovative dredging technology with geotextiles tubes (geotubes) for dewatering sediments. As part of landfill operations, a concrete basin was constructed to control sedimentation from the landfill. Sediment quantification studies identified approximately 5,500 cubic yards (4205.3 m^3) of sediment. Analytical testing indicated the principal constituents were semi-volatile organic compounds and metals. Geotechnical testing indicated the material was classified as fine-grained plastic clay (CH, CL) soil type. Design of the geotubes was based on site constraints and a sediment bulk density of 1.6 gm/cc. The geotubes were sewn from woven polypropylene fabric and they measured 90 ft (27.432 m circumference), 5 ft (1.524 m height) and 160 ft (48.768 m cubic meters length). A polypropylene liner and fabric were placed beneath the geotubes to collect water passing through the fabric. Approximately, 5,500 cubic yards (4205.3 m^3) of sediment was dredged from the basin and place in geotubes using a patented sealed clamshell dredging technology (Dry DREdge™). Sampling of sediment in the geotubes indicated that after 7 days the dredged material passed the paint filter test. As a result of using the Dry DREdge™, geotube dewatering method and disposal of sediment in the landfill, a project savings of approximately $1 million U.S. dollars was realized.

BACKGROUND INFORMATION

A hazardous waste landfill disposal site was maintained in Catlettsburg, Kentucky since 1976. In September 1998, the Kentucky Division of Waste Management was notified that the 20-acre (80,940 m^2), single cell landfill would no longer be used as a waste disposal site and would be closed.

A wastewater treatment unit had been constructed to control surface water discharges during and fill operations. The wastewater treatment unit was designed to collect and treat surface water runoff and leachate that came from the landfill during operations. The wastewater treatment unit consisted of a concrete sedimentation basin 100 ft (30.48 m) wide x 300 ft (91.44 m) long and water treatment process, which involved chemical precipitation followed by granular activated carbon (GAC) polishing.

As part of the closure process, the Kentucky Department of Environmental Protection (KYDEP) requested that the concrete sedimentation basin be cleared and inspected. It was estimated that approximately 5,500 cubic yards (4205.3 m^3) of sediment was contained in the basin. Since the sediment was from a hazardous waste landfill, the material was considered to be a listed waste. Results from analytical testing indicated the primary chemical constituents were semi-volatile organic compounds (i.e. phenanthrene, chrysene, and naphthalene) and metals (Cr). The KYDEP indicated that it would be feasible to dispose of the sediment from the basin in the on-site landfill during closure, prior to final capping. This procedure provided a cost-effective alternative to off-site disposal. In order for the sediment to be placed in the landfill, the material needed to pass the paint filter test and free liquids would not be allowed.

Since disposal of the sediment from the basin into the on-site landfill was approved, it was necessary to evaluate sediment removal alternatives. Management of surface water runoff from the 40-acre (151,880 m^2) watershed during removal of sediment from the basin was a critical issue. This was important because all surface runoff from the site was required to be treated in the basin until the landfill cap was complete.

As a result of evaluating several techniques, the project team selected Dry DREdge™ technology combined with in-place, geotextile tube dewatering of the wet sediment as the preferred method of sediment dewatering and remediation.

SEDIMENT PROPERTIES

In order to evaluate the use geotubes as a dewatering technique, particle size distribution and hydrometer tests were conducted to characterize the materials in the basin. Three composite samples were obtained from the basin for geotechnical testing. Other geotechnical tests conducted on the samples included Atterberg Limits (liquid limit and plastic limit), natural water content, specific gravity, and geotechnical description. Results of these tests are provided in Table 1.

TABLE 1. Summary of sediment geotechnical characteristics

	Geotechnical Test Parameters							
Sample No.	**Water Content**	**Liquid Limit (LL)**	**Plastic Limit (PL)**	**Plas Ind. (PI)**	**USCS Symb.**	**Sieve Minus ND. 200 (%)**	**Unit Weight**	**Specific Gravity**
1	64.1	45	22	23	CL	89.1	101.9	2.750
2	91.8	57	24	33	CH	99.8	93.5	2.766
3	104.4	60	25	35	CH	99.5	90.8	2.776

The sediment contained in the basin was classified as fine-grained, dark gray, plastic clay (CH to CL) with a trace of sand. Particle size distribution testing indicated that the sediment had more than 90 percent passing a 200 sieve.

Atterberg limit tests indicated the sediment had liquid limits ranging from 45 to 60 and plastic limits ranging from 22 to 25. The plasticity index ranged from 23 to 35. The specific gravity of the sediment varied from 2.75 to 2.78. The natural water content ranged from 64 to 104 percent and void ratio ranged from 1.76 to 2.89. The average saturated wet unit weights for the sediment was 1.41 gr/cc respectively.

In general, the sediment exhibited water content values greater than the liquid limit indicating that the material would act as a fluid mud. The sediment was very soft in consistency and exhibited very low shear strength. The results of this testing indicated the sediment could be handled using the Dry DREdge™ technology.

DESCRIPTION OF DRY DREDGE TECHNOLOGY

Conventional sediment extraction methods, such as, hydraulic dredging and mechanical dredging with clamshells or draglines typically have several serious limitations. The limitations of these techniques include: resuspension of sediments at the point of excavation, inaccurate excavation of "hot spots", and entrainment of freewater within sediments that are excavated.

The United States Army Corps of Engineers (USACE) recognized the many challenges facing conventional dredge technology when dealing with contaminated sediments and established the Construction Productivity Advancement Research (CPAR) Program. Recognizing the potential of the Dry DREdge ™ technology to meet these challenges, the USACE awarded a matching funds contract in 1995. The program consisted of proof testing, prototype design, prototype construction and prototype demonstration. A commercial-scale demonstration of the Dry DREdge ™ was conducted in mid 1996 at the Vicksburg, Mississippi Waterways Experiment Station. Excavating lake sediments produced a discharge stream containing 70% solids by weight, 1.7 specific gravity, and less than 5% water (Parchure, et al. 1997).

The final design of the Dry DREdge ™ (U.S. Patent No. 5,311,682) incorporated a specially designed, sealed clamshell bucket mounted on a rigid, extendible boom. In addition the dredge is intrinsically sound for debris management in that debris can be either removed or shredded.

During operation, the open clamshell is hydraulically pushed into the sediments at low speed, which minimizes sediment disturbance and resuspension. The clamshell is then hydraulically closed and sealed. When the clamshell is removed a plug of sediment remains at its *in-situ* moisture content.

Once brought to surface, the sediment is deposited in the hopper of a positive displacement pump. The hopper can be equipped for debris screening, particle size reduction, vapor emission control, sediment homogenization, and blending of additives to modify flow properties or stabilize contaminants within the sediment. Figure 1 is a photo showing operation of the Dry DREdge ™ in the sedimentation basin. The sediment is pumped in a plastic flow regime through a pipeline to a staging area. Depending on the in-situ moisture content and potential hazards of the sediment, a variety of options for final disposition are available including the following: dewatering, thermal treatment, stabilization, on-site land disposal, and direct vehicle transport.

FIGURE 1. Photo of Dry DREdge ™ operating within sedimentation basin

A key advantage of the Dry DREdge ™ design is the ability to transfer moisture-laden sediments at high solids concentration. High solids content sediment transfer offers major economic advantages because sediment dewatering is reduced or eliminated and the volume of water extracted from the wet sediment is significantly reduced. Since this project involved hazardous waste, the extracted water was required to be treated to NPDES limits. Other advantages include the following:

- Excavation of sediments is accurate and precise. The azimuth, declination, and extension of the clamshell dredge can be monitored continuously by the operator. Therefore, the active excavation area (length, width, and depth) is accurately monitored to minimize over excavation.
- The clamshell-boom design allows the dredge to work around obstructions and in tightly confined areas. This design provides ideal conditions for remediation of "hot spot" or "isolated areas".
- Sediment extraction is achieved with minimal resuspension. Hydraulic dredge cutter heads typically agitate sediment near pump suction. Conventional clamshells free-fall in order to impact the bottom with enough force to penetrate underlying sediment. Draglines are pulled through the sediments and cause significant disturbance. These operations disturb the surrounding sediments, which results in resuspending particles and dispensing contaminants. When dredging is conducted in bodies of flowing water such as estuaries resuspension of sediment is a critical issue to minimize.

GEOTEXTILE TUBE DESIGN

The physical properties of the sediment indicated the material was classified as plastic clay (CH to CL) soil type. Based on the site configuration, the design requirements were for geotubes that had a circumference of 90 ft

(27.432 m), a height of 5 ft (1.524 m) and a length of 160 ft (48.768 m). The final geotube design was determined using a computer program, "Geosynthetics Applications Program (GAP)," (Palmerton,1998). A factor of safety of 5.0 was used to design the tubes, which included factors of safety of 2.0 for seams, 1.5 for creep and 1.5 for biological degradation. When performing this analysis, the program assumes that the geotextile tube is filled with fluid and does not have any shear strength. The strength of the geotextile tube is directly dependent on the available wide tensile strength of the seams. The design analysis indicated that the required seam strength was 259.4 pounds per linear inch (pli). Based on the design analysis, polypropylene geotextile fabric was selected because it had a seam strength of 300 pli.

GEOTEXTILE FABRICATION

In order to dewater the volume of sediment in the basin, three 90-ft (27.432 m) circumference geotubes were manufactured. The geotubes were constructed from 160 ft (48.768 m) long, 15 ft (4.572 m) wide panels of woven polypropylene fabric. The woven geotextile fabric had an ultimate breaking strength in the range of 400 pli. The Area Opening Size (AOS) for the geotube fabric was about 50 (ASTM, 1999).

The geotubes were manufactured in Pendergrass, GA by the TC Mirafi Corporation, and shipped to the project site in a protective covering. Along the top of the geotextile tubes, two rows of inlet ports with 1.5-ft (0.4572 m) diameter, 5-ft (4.572 m) long sleeves were provided every 25-ft (7.62 m). Nylon anchor straps sewn along the perimeter every 10 ft to secure the tubes as they were filled.

The 15-ft (4.572 m) panels of polypropylene were sewn perpendicular to the longitudinal axis of the geotextile tubes. All factory seams were sewn with double stitched butterfly seams and consisted of type 401 double lock stitch that was sewn with a double needle Union Special Model #80200 sewing machine.

Prior to placement of the geotubes on the ground surface, a dewatering area was constructed by grading the ground surface to allow for minimal slope and drainage. A 6-mil polyethylene liner was placed on the ground surface to contain free liquids. A drainage layer consisting of 16 oz. (0.542 kg/m^2) per square yard non-woven polypropylene fabric was placed above the liner to facilitate vertical and lateral drainage of free liquids draining from the geotextile tubes. The water that drained from the geotubes was collected and treated using flocculation and GAC prior to discharge. Figure 2 is a photo of the geotextile tube dewatering area after filling the tubes.

FIGURE 2. Photo of geotubes dewatering on landfill surface.

RESULTS AND DISCUSSION

The field-scale sediment dredging project started in April 1999 and was completed in June 1999. Approximately 5,500 cubic yards (4205.3 m^3) of sediment was dredged from the sediment basin using the Dry DREdge™ technology. The sediment was pumped directly into three geotubes positioned in a dewatering area.

During operation of the dredging process, a very small amount of fine-grained material, less than 5-10 mg/l, was observed initially in the decant water passing through the geotextile tubes. As the tubes filled to the designed height of 5-ft (1.524 m), the decant fluids cleared.

Upon completion of the dredging operation, the geotubes were allowed to dewater without disturbance. Random sampling of the sediment within the geotubes indicated that within 7 days of dewatering, the sediment passed the paint filter test. Measurements taken from the geotubes indicated a free water loss of approximately 20 percent. Observations of the sediment in the geotubes indicated the bulk of this material was interstitial water.

The geotubes were allowed to remain in place for 20 days to allow for additional drainage. At that time the tubes were split open along the top centerline and the dewatered sediments were loaded into trucks and hauled to a disposal area on the landfill where waste was being regraded. The dewatered sediments were spread in lifts less than 0.5 m thick, prior to placement of the final landfill cap.

CONCLUSIONS

As a result of completing this field scale project, approximately 5,500 cubic yards (4205.3 m^3) of sediment were removed and dewatered within 45 days of project initiation. All material was disposed in the on-site landfill, which facilitated transportation and disposal costs. As a result of combining the Dry DREdge™ technology with a geotube dewatering design, a project cost savings of approximately $1.0 million US dollars was achieved.

ACKNOWLEDGEMENTS

Acknowledgements are made to William C. Olasin, Ashland Inc., Roy Ambrose, URS Corporation, Operations personnel of DRE, and Ed Trainer, TC Mirafi for assistance in the successful design and execution of this project.

REFERENCES

Annual book of Standards, Parts 31 and 32, (revised annually). *Textiles: Yarns, Fabrics and General Test Methods,* "Test Method for Determining Apparent Opening Size of a Geotextile,"1987, D 4551 ASTM, 1916 Race Street, Philadelphia, PA 19103.

Duke, M.L., Fowler, J., Schmidt, ML, Askew, AC. "Dredging and Dewatering of Hazardous Impoundment Sediment Using the Dry DREdge™ and Geotubes". Journal of Dredging Engineering. Volume 2, No. 1, March 2000.

Palmerton, J.B., (1998). "Geosynthetics Applications Program (GAP)," Geosynthetics Applications Simulations, Vicksburg, MS.

Parchure, T.M., and Sturdivant, C.N., (1997). "Development of a Portable Innovative Contaminated Sediment Dredge," U. S. Army Corps of Engineers, Waterways Experiment Station, Final Report CPAR-CHL-97-2.

GRAVITY DEWATERING OF DREDGED, CONTAMINATED SEDIMENT TO FACILITATE LANDFILL DISPOSAL

Kendrick Jaglal, Laura Brussel and Stuart Messur
(Blasland, Bouck & Lee, Inc., Syracuse, New York, USA)

ABSTRACT: A bench-scale study was performed to investigate the amount of dewatering and strength gain that would occur in a volume of sediment within 72 hours after being mechanically dredged. Three sediment samples were allowed to sit over a 24-hour period during which supernatant was removed and quantified, following which the sediment from two of the drums (1 and 3) was transferred to two other drums. Drum 2 served as the control in which sediment remained in an undrained condition during the 72 hours after dredging (i.e., only decanting was performed). Drums 1 and 3 both had underdrain systems and the sediment in Drum 3 was mixed twice during the study. At successive intervals during the study, samples were collected for water content, paint-filter, and shear strength testing. Supernatant volumes also were measured at these intervals. For all three scenarios, approximately one-half of the water entrained during dredging was released within the first four hours due to gravity dewatering. The use of an underdrain system, with or without mixing, resulted in the sediment returning close to its average *in situ* shear strength of approximately 46 pounds per square foot (psf) or 2.2 kiloPascals (kPa), and its average solids content of approximately 69 percent, after 72 hours.

INTRODUCTION

Dredging with off-site disposal is one of several remedial options that may be used to address contaminated sediment. Depending upon the material composition and dredging technique employed, the sediment may require dewatering to meet a minimum solids content (e.g. pass the Paint Filter test) prior to disposal in a commercial or project-specific landfill. The required solids content may be met through either dewatering or addition of a bulking agent to absorb the free moisture. Use of the latter technique results in additional cost for the transportation and disposal of the water and associated absorbent bulking agent. Similarly, dewatering by mechanical means may be uneconomical because it is both labor and energy intensive. Consequently, it is beneficial to investigate whether adequate dewatering of particular dredged sediment can be achieved economically through quick, passive techniques, as part of the handling process. Due to space limitations often encountered in many dredging projects and reductions in throughput, extended dewatering times may be both impractical and uneconomical.

Passive dewatering of fine-grained sediment is facilitated by gravity and may result in water rising to the top of the sediment or draining to the bottom. In coarser materials with large pore spaces, water flows through the pore spaces under gravity; towards the bottom of the sediment pile and drainage may be enhanced if underdrainage is provided. In finer or well-graded sediment, the weight of the overlying sediment exerts a pressure that facilitates consolidation of the underlying sediment solids (thereby reducing the permeability at depth) and forces water to the top of the sediment pile, even if underdrainage is provided.

The effectiveness of a sediment dewatering technique may be evaluated based on passing the Paint Filter test, having allowable volume reduction and/or bulking, and increasing shear strength to an acceptable level for equipment mobility and slope stability at the disposal site. Increasing the shear strength of the material makes it easier to landfill in stable geometries using conventional earthmoving equipment and techniques. Materials that cannot be handled or that fail when piled in slopes can create significant cost increases or project delays if additional dewatering or stabilization is required.

A dredging project was proposed in which contaminated river sediment would be placed in scows following dredging, and allowed to sit for 24 hours. After this time period, supernatant overlying the sediment would be removed and the sediment transferred to a dewatering facility to undergo gravity drainage for another 24 to 48 hours (i.e. dewatering over a total of 72 hours after dredging), prior to landfill disposal. The purpose of this study was to investigate dewaterability of this mechanically dredged, contaminated river sediment under passive conditions over the 72-hour period. The dewatering tests were performed both with and without an underdrainage system. Mixing of the sediment was included to evaluate the effect that regular redistribution of water in the sediment may have on dewaterability.

MATERIALS AND METHODS

Field Efforts. Existing geotechnical data for the river sediment targeted for dredging were reviewed to select a location where average, representative conditions may be encountered. Prior to collection of bulk sediment for the tests, a separate "undisturbed" sample was obtained to establish *in situ* conditions. This sample was collected using a 2-inch (5 cm) Lexan® tube that was pushed 3 feet (0.9 m) into the sediment and from which 2.5 feet (0.76 m) of sediment were recovered. The sample was submitted to a geotechnical laboratory for characterization of the *in situ* sediment. Shear strength tests were also performed at various locations and sediment depths (up to 1.0 foot or 0.3 m) in the vicinity of the selected sample location. Next, sediment for use in the dewatering experiment was collected using an Ekman® sampler and transported in 5-gallon (0.02-m^3) pails to the test site. The Ekman® sampler was selected because its operation and function is similar to a mechanical clamshell dredge that is often used in sediment dredging operations. At the test site, the sediment was transferred into 30-gallon (0.11 m^3) drums.

Test Methods. Geotechnical tests that were performed on the sediment included specific gravity, water content, organic content, Atterberg Limits (plastic and liquid), and particle size distribution. These tests were conducted using American Society for Testing and Materials (ASTM) Methods D854, D2216, D2974, D4318 and D422/D1140, respectively. Bulk density was performed using the U.S. Army Corps of Engineers (USACE) EM-1110-2-1906 volumetric method. Vane shear testing to determine shear strength (both *in situ* and in the test drums) was performed using a Geonor® H-60 Inspector Vane Tester, following the manufacturer's instructions. The Paint Filter test was conducted using the U.S. Environmental Protection Agency Method SW846-9095A.

Experimental Procedures. A total of three 30-gallon (0.11-m^3) drums were filled with sediment to a depth of approximately 2 feet (0.6 m) at time zero. The sediment was allowed to stand for 24 hours, which is similar to settling in a scow immediately after dredging. After standing for 24 hours, the sediments from two of the drums were transferred to two additional 30-gallon (0.11-m^3) drums (Drums 1 and 3) containing an underdrain system. This reflects transfer of the sediments from a scow to an onshore dewatering facility. The underdrain system promotes free drainage of water from the sediment. The underdrain system included approximately 8 holes cut in the bottom of each of the two drums, a 3- to 4-inch (7- to 10-cm) thick layer of drainage stone, and a geotextile that was fastened to the drum. An Amoco 4553 nonwoven geotextile was selected to allow for drainage without clogging based on the American Association of State Highway and Transportation Officials (AASHTO) Geotextile Specification for Highway Applications (Designation M 288-96) for filtration.

The sediment in the other 30-gallon (0.11-m^3) drum remained in its original container for the entire 72-hour period of testing and was referred to as Drum 2. This is considered to be similar to just leaving the sediments in a scow for up to 72 hours (3 days).

Bulk sampling of the sediment was performed at time zero and included organic content, grain size analysis, Atterberg Limits and specific gravity. Water content samples (in duplicate) were taken from the top, middle and bottom of each drum at 4, 8, 24, 28, 32, 48 and 72 hours. Supernatant measurements also were made at the same time intervals as the sampling for water content. Paint Filter testing was performed at 0, 24, 48 and 72 hours and vane shear testing was performed at 72 hours.

RESULTS AND DISCUSSION

Sediment Characterization. Laboratory testing indicated that the sediment sampled in the Lexan® tube is silty sand (SM) with trace organics according to the Unified Soil Classification System (USCS) (Holtz, 1981). The wet and dry bulk densities of the sediment were estimated to be 104 and 74 pounds per cubic foot (pcf) or 1666 and 1185 kg/m^3, respectively. Vane shear testing indicated that the *in situ* sediment had very low shear strength; with average undrained shear strength of approximately 46 pounds per square foot (psf) or 2.2 kiloPascals (kPa). The water, solids and organic contents of the *in situ* sediment averaged (for 3 samples) 44, 69 and 5 percent, respectively. The sediments in the drums were classified as organic silts of low plasticity (OL) using the USCS. Based on visual classifications and the laboratory testing, the sediment in each of the three drums appeared similar. The other geotechnical characteristics of the drum sediment are provided in Table 1. The initial round of drum sample testing was used to set the baseline conditions for evaluating changes during the dewatering study.

TABLE 1. Sediment Geotechnical Characterization.

Drum	Plastic Limit	Liquid Limit	Plasticity Index	Specific Gravity	Water Content (%)	Organic Content (%)
1	30	39	9	2.58	105	6
2	26	34	8	2.60	87	6
3	30	39	9	2.60	87	6

Water Content. For each drum, the average water content (Table 2) based on the top, middle, bottom, and associated duplicate samples, decreased with time. During dredging, up to a 100 percent increase in water content due to entrainment into the sample was estimated. Over half of this entrained water was lost during the first 4 hours of settling. This initial water loss is most of the dewatering that occured in the first 24 hours. There was a greater decrease in water content for Drums 1 and 3 (which had underdrain systems) after 48 and 72 hours than for Drum 2.

TABLE 2. Average Sediment Water Content (%) Over Time

Elapsed Time (hrs)	Drum 1	Drum 2	Drum 3
0	105	87	87
4	65	66	74
24	74	70	72
48	61	64	60
72	57	63	55

Solids Content. Using the water content and the specific gravity information obtained from the bulk sampling, the solids content (Table 3) of the sediment was estimated. The data indicated increasing solids content as the experiment progressed. As with the water content, a majority of the change occurred within the first four hours. During the first 24 hours, the average solids content increased from approximately 50 percent to approximately 58 percent. Dewatering during the next 48 hours achieved solids contents of 61 to 65 percent, corresponding to conditions ranging from no underdrain system with supernatant removal to an underdrain system with mixing and supernatant removal. Mixing did not appear to change the dewatering process significantly.

TABLE 3. Average Sediment Solids Content (%) Over Time

Elapsed Time (hrs)	Drum 1	Drum 2	Drum 3
0	49	54	54
4	61	60	58
24	58	59	58
48	62	61	62
72	64	61	65

Extended time to facilitate gravity dewatering is not expected to provide any significant, additional dewatering. Samples taken from the interior of a pile of river sediment that was dredged several months earlier had an average solids content of 65 percent. This suggests that even after several months, the sediment remained at its *in situ* solids content of approximately 65 percent, despite ample time to dewater further.

The resulting volume change from in-situ conditions (assuming no air entrainment) was calculated to range from 10 percent bulking to no change. There were no indications that the potential existed for significant volume reduction during the three days of dewatering.

Shear Strength. The average (based on 6 to 9 samples per drum) shear strength in Drums 1 and 3 after 72 hours was 2.9 kPa (61 psf) and 3.9 kPa (81 psf), respectively, while the average shear strength in Drum 2 (no underdrain) was 0.29 kPa (6 psf). The shear strengths in all three drums tended to increase towards the bottom (Table 4). This may be due to either increasing friction on the rods with depth or additional strength due to small increases in density or confining stress with depth.

TABLE 4. Average Sediment Shear Strength after 72 hours (kPa/psf)

Location	Drum 1	Drum 2	Drum 3
Top	1.5/37	0.13/2	1.7/35
Middle	2.8/58	0.25/5	2.5/52
Bottom	4.5/95	0.5/11	9.3/194

The average *in situ* shear strength of the top foot of sediment was 2.2 kPa (46 psf). The average 72-hour shear strength of the dewatered sediment in all three drums was 3.4 kPa (71 psf). These strengths are similar and indicate that the dewatering process with underdrainage returns the dredged material to approximately its pre-dredging state. Drum 2, which had no underdrainage and therefore had slightly higher water contents and saturation for its full depth, exhibited very low shear strength of less than 0.53 kPa (11 psf). This suggests that settling alone is unable to reestablish even the low *in situ* strength values.

The *in situ* and dewatered sediment shear strengths showed some limited evidence of increasing strength with depth. This apparent strength increase with depth may be due to a reduction in water content due to increased overburden pressure or to a small amount of rod friction on the device for the deeper samples. Due to the potential for some limited rod friction, the measured vane shear strengths should be considered to be upper limits of the potential shear strength that would be available.

Calculations were performed using USACE methods (Willoughby, 1977) of evaluating equipment mobility in confined disposal facilities (CDFs). These calculations indicated that the heaviest equipment that can be used in sustained operations on sediments with strengths similar to those measured in this study would have a ground pressure of 0.5 psi (3.5 kPa or 72 psf). This ground pressure is very low and eliminates the use of all conventional earthmoving equipment which would exert much higher pressures.

Paint Filter Test. The Paint Filter is used to determine the presence of "free liquids" in a representative sample of bulk waste. Since any visible water leaking through the filter causes test failure, even very small moisture losses can be the difference between passing and failing the test. The longer a sample is kept and the more it is handled, the more likely it is that water losses will occur due to evaporation or coating of storage/handling equipment surfaces.

Immediately following dredging, all samples failed the Paint Filter test. After the first 24 hours, only the mixed sediment in Drum 3 passed. After 72 hours, at the end of the study, the uppermost samples in all three drums failed the Paint Filter test while all the samples from the two lower layers passed.

Conclusion. The study suggests that after 3 days of settling and passive dewatering using underdrainage, additional dewatering time is unlikely to achieve a significant improvement without enhanced evaporation/dessication or addition of an absorbent. It appears that the *in situ* sediment conditions represent an approximate endpoint for gravity dewatering. This is because the water that is lost during gravity dewatering is the water that was entrained in the sediment during removal. Therefore, the dredged sediment returns back to approximately its *in situ* water content and shear strength, following gravity dewatering.

REFERENCES

American Society for Testing and Materials (ASTM). 1999. *Annual Book of ASTM Standards - Section 4.0 Construction.* West Conshohocken, PA.

Holtz, Robert D. and W.D. Kovacs. 1981. *An Introduction to Geotechnical Engineering.* Prentice-Hall, Inc., Englewood Cliffs, NJ.

Willoughby, William F. 1977. *Low-Ground-Pressure Construction Equipment for use in Dredged Material Containment Area Operation and Maintenance, Performance Predictions.* Technical Report D-77-7. U.S. Army Engineer Waterways Experiment Station, Vicksburg, MS.

PASSIVE DEWATERING IN A DREDGE MATERIAL REHANDLING FACILITY

Michael D. Dovichi (Earth Science & Technology, Algoma, WI, USA)
Dean Haen (Brown County Port Department, Green Bay, WI, USA)
Donald Miller (Robert E. Lee & Assc. Green Bay, WI, USA)

ABSTRACT: An innovative dredge material rehandling facility (DMRF) has been constructed for the passive dewatering of sediments slightly contaminated with polychlorinated hydrocarbons. The facility utilizes gravity and evapotransporation to dewater dredge material prior to moving the sediments to an onsite storage area. The dewatered sediments, which resemble topsoil, are available for beneficial reuse projects. The DMRF is located near the mouth of the Fox River at Green Bay, Wisconsin, USA. The facility has the potential capacity to store up to 7.4 million cubic yards (5.6 million cubic meters) of sediment in 110 acres (44.5 hectares) at a capital cost of approximately $2,000,000. The annual operating and maintenance (O&M) costs are estimated at $770,000. The combined capital and O&M costs results in a unit cost of less than $6.00/cubic yard ($4.58/cubic meter) of wet sediment. Once dewatered, the soil is available for beneficial reuse projects. The technology incorporated in the Brown County DMRF can be used at any upland confined disposal facility (CDF) and at an in-water CDF if the construction of the CDF allows for suitable subsurface drainage.

INTRODUCTION

The Port of Green Bay, Wisconsin, USA, is managed by the Brown County Harbor Commission and the Brown County Port and Solid Waste Department. In the early 90's, the commission decided that they would begin the process to develop a landfill that would serve the port indefinitely. The US Army Corp of Engineers (USACE) is responsible for dredging the Fox River, and the county is responsible for providing the disposal site. The USACE would construct a conventional upland CDF, but the county wanted a facility that would facilitate beneficial reuse options with an extended site life. The beneficial reuse of all the sediments dredged from the channel was the goal, even though there was no state regulatory process to allow this to happen.

In the early 70's, Brown County provided the USACE with a 400-acre (161.8 hectares) wetland near the mouth of the Fox River to use for the disposal of dredge material. This facility is referred to as Bay Port. A majority of the sediment in the 18-mile (29 km) long shipping channel was dredged hydraulically and pumped to the Bay Port disposal facility. After it was filled to its initial proposed elevation, Brown County received authorization to additionally fill within a 160-acre (64.7 hectares) area of the 400-acre (161.8 hectares) with thc sediments hauled by truck. This additional area was filled by 1998. A nearby in-water CDF was filled to capacity in 1999.

The USACE dredges an average of approximately 150,000 cubic yards (115,000 cubic meters) of sediment annually to maintain the channel at the designed depths of 20-26 feet (6.2-7.9 meters). The federally authorized channel width is not being maintained so there is a current backlog of 3 million cubic yards (2.3 million cubic meters) of sediments. The terminal operators agree that this volume can remain in place without obstructing the normal shipping.

ENGINEERING DESIGN

Because the Harbor Commission wanted to implement beneficial reuse options, they needed to find an economic method to dewater the sediments. The result was a design (Dovichi, and Miller, 1996) that used four cells for dewatering and two for storage or disposal of the dewatered sediments. The facility occupies 110 acres (44.5 hectares) of the original 400-acre (161.8 hectares) CDF. It is designed to accept over 200,000 cubic yards (153,000 cubic meters) of sediment yearly though the normal operation is for the dredging of 150,000 cubic yards (115,000 cubic meters) annually. The dewatered cells range from 7 to 20 acres (2.8 to 8.1 hectares) in size, with perimeter berms as much as 15 feet (4.6 meters) high. The cells were constructed in different sizes, partly because of existing conditions but also to permit the separation of sediment from private dredging projects or “hot spot” dredging.

The design required that the base of the cells, drainage ditches, and ponds had to be excavated in previously deposited sediments within the hydraulically filled CDF. Much of this material was saturated and unstable, but once it was excavated and dried, it was suitable for use in the berms and off-loading ramps.

The perimeter berms were constructed in one-foot thick lifts and compacted to 90% modified Proctor density. The clay for the berms came from new dredgings at private slips and the excavation of the foundation of a new bridge in Green Bay. Previously deposited dredgings that have been partially dewatered were excavated and used to construct the remaining bulk of the berms. The outside slopes of the berms are constructed with a 4 horizontal to 1 vertical slope (4H:1V) and 3 horizontal to 1 vertical (3H:1V) inside slopes. Because the on-site soils consist of dredgings, which is primarily topsoil, it was not necessary to import topsoil for the growth of a vegetative cover. The outside slopes were seeded with a grass/legume mix and the inside slopes became vegetated with the native seeds present in the soil. Mowing controls the growth of weeds and assists in the establishment of a healthy vegetative cover.

There are 1 to 3 off-loading ramps per cell depending upon the size of the cell. The ramps were designed with 10% access slopes. The trucks that deliver the sediment from the off-loading slip are driven up the ramp and backed up to the edge, where the high liquid sediment is dumped into the cell. The off-loading ramps are constructed 3-4 feet (0.9-1.2 meters) higher than the perimeter berms, 10-12 feet (3.0-3.7 meters) above the base of the cell at each location.

The base of each cell has a minimum elevation that is less than 5 feet (1.5 meters) above the average Lake Michigan water level. The cells are generally square or rectangular, with a uniformly sloped base. The base of each cell is constructed at a 0.5% grade to one corner where a gravel dewatering infiltration bed and discharge piping drain off precipitation and carriage water. The off-loading ramps are constructed on the high corners of each cell so that the freshly dumped sediments can flow across the cell. The angle of repose of freshly dredged sediments approaches 0° because of the high water content, but the sediment begins to dewater soon after it is dumped, creating internal friction. This results in a 0.5-1.0% slope on the surface of the sediment that allows precipitation and carriage water to drain to the edges and eventually to the low corner of the cell.

The dredgings typically contain 70%+ moisture (weight moisture/wet weight) when delivered to the facility. Some of this water must be removed from the sediments to enable it to be excavated and transported to the storage cells. A series of perforated pipes have been installed in the base of each cell to assist in dewatering the sediments from below. The 6-inch (15.2 cm) diameter plastic pipes are installed 1-foot (0.3 meter) below the base grades and sloped to the gravel infiltration bed in the low corner of the cell. The infiltration bed is covered with coarse sand to prevent the sediments from draining through the gravel, yet permit the gradual drainage of water. This water leaves the cell through a pipe that drains into ditches that in turn drain to a series of five ponds in series.

The discharge piping in each cell was designed to provide control of the carriage water and precipitation that falls on the cell. It was thought that the carriage water would quickly separate from the sediments and require control so that it could be retained long enough to meet discharge standards. A control structure was designed that could be raised and lowered as the level of the free water in the cell varied.

The ponds are designed with a circuitous water flow path to prevent short-circuiting of the water draining from the cells. The pond system eventually drains to the bay of Green Bay. The state regulatory agency issued a pollution discharge elimination permit which requires monthly monitoring for total suspended solids and phosphorus and quarterly monitoring for PCB's. Any sediment carried in the runoff water would have to migrate through the gravel filter bed, through a minimum of 400-feet (122 meters) of ditches, and through 800 feet (244 meters) of ponds before it would discharged to the bay.

The facility has a design capacity of 2.5 million cubic yards (1.9 million cubic meters) of dewatered and compacted sediments. The final grades are relatively flat because of regulatory concerns with the stability of the dewatered and compacted sediments. The final thickness of the sediments will be approximately 30 feet (9 meters). Theoretically, the facility could hold 10 million cubic yards (7.6 million cubic meters) with 4H:1V slopes and a final thickness of almost 110 feet (33.5 meters). The Brown County Harbor Commission is currently preparing a request to increase the landfill capacity to 7.4 million cubic yards (5.6 million cubic meters).

The annual average of 150,000 cubic yards (115,000 cubic meters) of dredged sediments is measured by the USACE by surveying pre- and post dredging elevations in the shipping channel. The channel has been cut primarily through clay and is relatively stable. The sediment that has to be dredged comes from upriver in suspension, carried along the base of the channel, or washed in from the shallow water adjacent to the channel.

As the dredged sediments undergo dewatering, rehandling, and eventual compaction in the storage cell it consolidate approximately 45%. As a result, the 2.5 million cubic yards (1.9 million cubic meters) capacity will hold 40+ years of dredging if the average of 150,000 cubic yards (115,000 cubic meters) of sediment dredged annually consolidates by 45%. Expanding the facility to hold 7.4 million cubic yards (5.6 million cubic meters) will increase the life of the site to well over 70 years. The facility could last indefinitely if the state allowed the sediment to be used off-site in beneficial reuse projects. Once the concentrations of contaminants are eliminated by upstream remediation projects, the dewatered soil should be available for reuse as topsoil. Green house and field tests have shown that the soil has nutrient value and aids in the growth of plants.

DISPOSAL OPERATIONS

A typical CDF is constructed with berms made out of partially dewatered dredge material to hold the next year's volume of sediment. In order to operate the old CDF at Bay Port, a number of temporary offloading roads were constructed into the fill area from concrete rubble and stone. Besides taking up disposal capacity, these ramps required truckers to back up 50-300 feet (15-91 meters) to get to the end of the ramp to dump. This was a time consuming process, which resulted in delays while trucks waited to access the road.

The DMRF is designed to operate on a three-year cycle, with 150,000 cubic yards (115,000 cubic meters) deposited in the first cell in year one and removed at least three years later before the cell would be used again. The DMRF was first used in 1998 when 250,000 cubic yards (191,000 cubic maters) of sediment was deposited in two dewatering cells (Dovichi, 1999). The sediment was dredged with an "environmental" bucket, barged to an offloading slip, and transferred to 15 and 25 cubic yard (11.5 to 19 cubic meter) dump trucks. The trucks were constructed with a waterproof chute in place of a tailgate. The top of the chute was as high as the sidewalls of the box so sediment would not fall out the back on the road, yet the sediment could flow when the box was raised.

The drivers maintained a 15 minute round trip cycle, from the barge offloading to the DMRF and back, 24 hours a day, six days a week during the dredging operation. The trucks could be driven up the ramp, back up to the edge, dump, and be back down the ramp in less than 60 seconds. Productivity increased significantly from the previous operation, as multiple trucks were now able to use the ramp at a time. An estimated 2000 cubic yards (1529 cubic meters) of sediment were hauled daily with 7-8 trucks.

At the top of the ramp, the drivers stopped the trucks and backed up to a curb at the edge of the ramp. The steel chute on the end of the truck was sufficiently long so that the sediment could drop beyond the curb, almost onto the base of the cell. As filling of the cell progressed, the sediment fell onto previously deposited material and flowed outward. The truckers backed the trucks up to different locations along the edge of the ramp so that the sediment flowed in a relatively uniform manner. The sediment was somewhat cohesive at the advancing front and within a week, desiccation of the surface of the front was noticeable. The sediment advanced as a steamroller, with the surface material progressing over the front onto the base of the cell, to be covered with newer material. There was no evidence of water accumulating at the toe of the sediment and only 2-5 feet (0.6-1.5 meters) of the base of the cell would become saturated from the sediment. The summer of 1998 was extremely dry, with minimal precipitation throughout the dredging period.

The contractor stopped dredging over the Labor Day weekend from Friday September 5 until Tuesday September 9, 1998. By Tuesday morning, the sediment had created a stable front approximately 3 feet (0.9 meters) high along the edge of the sediment. This front remained stationary until September 18, 1998, when fresh sediment broke through the drier edge and started to flow directly downgradient to the lowest part of the cell. The majority of the remaining front showed little movement.

Plastic bottles marked with different colored streamers were tossed from the ramp onto the sediment weekly for four weeks. These bottles were used to get an idea of the relative rate of movement of the sediment. The first bottle thrown on September 14, 1998 was partially buried at the advancing front, a distance of 400 feet (122 meters) within 4 days or approximately 4.5 feet/hour (2 meters/hour).

Two inches of rain fell on September 26, 1998, and the movement of the sediment increased significantly. The bottle thrown on September 28, 1998 traveled approximately 600 feet (183 meters) in 4 days. By October 15, 1998, sediment had migrated to the lowest corner of the cell at the dewatering structure. The sediment rose to the top of the gravel berm constructed in front of the structure by October 21, 1998, after an estimated 130,000 cubic yards (100,000 cubic meters) of sediment was dumped in the cell. Once the sediment covered the gravel-dewatering layer at the dewatering structure, water started to drain from the discharge structure to the perimeter drainage ditch. The surface of the sediment was higher along the axis of the cell than the edges, with the lowest elevation at the farthest end of the cell.

By November 9, 1998, drainage channels started to form on the surface of the sediment. The month of October was dry, so it was apparent that the water was released from the sediment as it consolidated. The drainage channels developed to the low corner as well as to the edges of the cell. The surface water moved to the drainage structure and rapidly drained into the ¾ inch (1.9 centimeter) gravel and out the discharge pipe. The gravel face became blanketed with a gelatinous organic material from the sediment and the blanket had to be removed periodically to maintain flow into the gravel.

There was no visible evidence of solids draining from the discharge structure, in fact minnows were observed swimming into the end of the discharge pipe at the drainage ditch. The sediment dewatered and consolidated sufficiently by November 25, 1998 so that it could be walked on to at least 5 feet (1.5 meters) out from the perimeter berm. Water flowed from the drainage pipe throughout the winter. Open water remained at the end of the pipe even during subzero temperatures. Approximately 10 gallons (37.9 liters) per minute of water drained from the pipe until spring when it slowed to approximately 5 gallons (18.9 liters) per minute. The flow in the winter was only from the subsurface drainage pipes, since the surface of the cell was frozen.

Sediment with higher water content was deposited in an adjacent cell the same year. There was an obvious difference in the operation of the second cell, since the sediment flowed faster and assumed a surface that was essentially flat. Because the surface was flat, drainage channels did not develop as they did on the first cell.

By early spring, vegetation began to sprout on the surface of the sediment in both cells. The sediment was dewatered and developing significant cracks that were 6-8 inches (15-20 centimeters) deep and 1-2 inches (2.5-5 centimeters) wide at the surface. Precipitation drained into the cracks and to the subsurface drainage layer, but excess precipitation would pond in the cracks. The cracks remained open even when filled with water. By June 1999, the cells were completely covered with 6-12 inch (15-30 centimeter) high plants, primarily grasses with some weeds.

DREDGE MATERIAL REHANDLING

The plan was to leave the dredged material in a cell for a minimum of three years and preferably longer if there was enough storage capacity. Because of funding circumstances, Brown County arranged to have the sediment removed during the fall of 2000. A private contractor was hired to move the 130,000 cubic yards (100,000 cubic meters) of sediments deposited in the initial cell in 1998.

The sediment was tested on March 29 and 30, 2000 by drilling 12 hand auger borings to a depth of 8 feet (2.4 meters) (Edlebeck, 2000). Sampling was performed with the use of a 1.25-inch (3.175 centimeter) diameter PVC piston sampler that was pushed into the undisturbed soil within the boreholes. The samples were tested for grain size, Atterberg Limits, and percent moisture in 1-foot (0.3 meter) increments. The moisture content (weight moisture/wet weight) of the samples ranged from a low of 33.9% in sample BP-5 at a 7-8 foot (2-2.4 meter) depth to 63.7% in BP-3 at a 3-4 foot (0.9-1.2 meter) depth. The percentages of moisture of the samples in the 8 borings along with the average percentages by depth are shown on Table 1. The data indicates that generally the sediments are drying from the top as well as from the bottom.

TABLE 1. The percentage of moisture in 2-year-old sediment samples

	BP-1	BP-2	BP-3	BP-4	BP-5	BP-9	BP-10	BP-11	AVE.
0.0-1.0	-	56.9	58.5	58.1	56.1	-	51.5	53.7	55.8
1.0-2.0	53.6	-	57.5	58.0	57.8	-	-	55.1	57.1
2.0-3.0	38.4	62.1	63.9	60.6	60.8	57.9	59.3	-	60.8
3.0-4.0	57.7	62.2	63.7	-	60.7	59.1	61.1	59.9	61.1
4.0-5.0	51.9	60.6	61.9	62.5		57.6	60.9	61.8	60.9
5.0-6.0	54.6	61.8	62.1	62.1	56.9	56.7	61.2	59.4	60.0
6.0-7.0	57.6	61.2	58.6	59.8	56.3	56.1	58.5	48.3	57.0
7.0-8.0	55.1	59.9	63.7	51.4	33.9	-	55.2	55.3	53.2

Access to the cell was made by removing a portion of the interior berm (Haen, 2001). The contractor utilized a tracked backhoe to excavate the soil and off-road dump trucks to transport 1000 feet (304 meters) to the storage cell. The south berm of cell 5 was partially breached to allow haul trucks to exit the cell. Cell 5 was excavated to within .3 foot (10.16 cm) of the original elevation of the base. The excavation was conducted using a backhoe on a floatation pad loading three Volvo off-road articulated haulers. The backhoe was positioned on top of the dredge material and perimeter berms for loading the haul trucks. The trucks traveled through the breached berm or on top of the perimeter berms to the storage cell. Haul trucks dumped the dredge material and a dozer spread the soil to facilitate drainage, prevent ponding, and meet facility design grades.

Survey data indicates that there was a 45% volume reduction from dredged yards to yards compacted in the storage cell. Similar sediments that were handled in the same manner were sampled after they were stockpiled and the moisture content ranged from 57.1 to 70.6% with an average moisture content of 65.7%. The sediment resembled rich topsoil, and was classified as very dark grayish brown elastic silt with sand.

CONCLUSION

The Brown County Port Commission has aggressively pursued unique solutions to dredge material disposal. Their attitude has always been that the sediments are a resource that should be used, not buried. A practical goal of the commission is to reuse every cubic yard of sediment dredged from the river. An ultimate goal would be to eliminate the need to dredge by eliminating upstream soil erosion.

The unique Bay Port DMRF accomplished its purpose by providing the USACE with a long term, relatively inexpensive method of dewatering sediments. The dewatered sediments have physical properties that allow it to be readily transported with conventional trucks for use off-site in beneficial reuse projects. The airspace created by removing the soil allows for the acceptance of additional sediment in the DMRF. Even if the dewatered sediments are never hauled off-site, the community will be left with a property that is available for open space uses, unlike a conventional CDF that requires decades to stabilize.

This type of facility could also be utilized as a first step in the remediation of highly contaminated sediments. Remediation projects do not have to be conducted only during periods of favorable weather or limited in scope because of equipment throughput rates. Sediments can be dewatered passively, stockpiled in a secure facility, and remediated year round in a less costly manner.

REFERENCES

Dovichi, M.D., and Miller, D.L. October 1996. *Brown County Harbor Commission City of Green Bay, Brown County Wisconsin, Bay Port Dredged Material Disposal Facility, Plan of Operation*. Robert E. Lee & Associates, Green Bay, Wisconsin.

Dovichi, M.D., July 1999. *Bay Port Dredge Material Landfill, 1998 Annual Operations Report, July 1999*. Robert E. Lee & Associates, Green Bay, Wisconsin.

Edelbeck, John, 2000. *Report of Subsurface Investigation for Cells 5 and 6 of Bayport Confined Disposal Facility Green Bay, Wisconsin Prepared for: U.S. Army Corps of Engineers-Detroit District CEC Project #G-0066*. Coleman Engineering Co., Iron Mountain, Michigan.

Haen, Dean, R., February 2001. *2000 Annual Report, Bay Port Dredge Material Re-Handling Facility*. Brown County Port and Solid Waste Department, Green Bay, Wisconsin.

INNOVATIVE DEWATERING AND WATER TREATMENT TECHNIQUES FOR HYDRAULICALLY DREDGED SEDIMENT

Martin Wangensteen, PE, PG, Bay West, Inc., St. Paul, MN, USA
Patricia Lafferty, PE, Stone & Webster, Inc., Denver, CO, USA
Jane D. Kobler, U.S. Army Corps of Engineers, Omaha, NE, USA

ABSTRACT: Bay West, Inc., Stone & Webster, Inc., and the U.S. Army Corps of Engineers (USACE) are currently utilizing innovative dredge carriage water dewatering and treatment techniques to remediate more than 66,000 m^3 of mercury-impacted sediment at Grubers Grove Bay (GGB) adjacent to the Badger Army Ammunition Plant (BAAP), Baraboo, Wisconsin, USA. The remedial action consists of hydraulic dredging, treatment, and dewatering of approximately 8.3 m^3/min of sediment-laden dredge carriage water using geotextile tubes, and discharge of filtered water via spray irrigation on agricultural lands. The 12-m circumference, 68-m long geotextile tubes have been particularly effective in dewatering the high water content/low shear strength GGB sediments. A critical element of the sediment dewatering activities was the design and construction of a manifold and pinch-valves system connecting the dredge carriage water pipeline to the geotextile tubes. The manifold and pinch valves provided a means to easily and rapidly redirect flow of the dredge carriage water to the geotextile tubes without impeding the dredging activities. It was also determined that the addition of a cationic flocculating agent was required at concentrations of approximately 60 mg/L, to enhance the geotextile tubes' filtering effect and obtain the required filtrate water quality criteria.

INTRODUCTION

Background Information. Bay West, Inc., Stone & Webster, Inc., and the USACE, herein referred to as the project team, are currently implementing a sediment remedial action (RA) plan at GGB near BAAP, Wisconsin. The RA is unique due to its innovative use of geotextile tubes for sediment dewatering and spray irrigation for final dredge carriage water and interstitial water treatment. This paper will summarize the operational successes and 'lessons-learned' from the sediment removal, geotextile passive dewatering, and spray irrigation water treatment techniques.

BAAP was constructed in 1942 to manufacture small arms and ordnance propellants as part of the United States' military manufacturing effort during the Second World War, Korean Conflict, and Vietnam Era. BAAP discharged treated process and sanitary wastewater to GGB during active manufacturing periods. Process discharges resulted in the addition of wastewater sediments to GGB. The principal wastewater sediments consisted of powdered-activated carbon and aluminum oxide flocculent.

GGB is located on the northwestern shore of Lake Wisconsin, immediately south of BAAP and approximately 1,800 m upstream of the Prairie du Sac Dam.

The southeast end of the bay opens into Lake Wisconsin. GGB was formed following the construction of the Prairie du Sac Dam in 1915 and occupies approximately 101,000 m^2 of water surface area.

Investigation Techniques. To further characterize GGB sediments, the project team completed Phase I and Phase II investigations in 2000. Specific objectives were to:

- Determine volume and variability in the extent of sediments in GGB,
- Determine chemical composition and variability of sediments (for selected chemicals),
- Determine physical characteristics of sediments, and
- Perform a bench-scale treatability study of sediments to evaluate management methods.

The Phase I and Phase II investigations involved the collection of more than 400 sediment thickness measurements and 139 sediment samples for chemical analyses. The time of year the investigations were conducted (January and February during subfreezing temperatures and ice-covered surface water) necessitated the project team's development of an innovative sampling approach to rapidly collect sediment samples while sheltering the sampling crew.

Sediment samples were either collected with an Ekman dredge or a piston-type sediment coring device based on a design developed by the Wisconsin Department of Natural Resources (WDNR) and subsequently modified by Bay West. Surface sediment samples (0 to 0.3 m) were collected with the Ekman dredge while deeper sediments were collected with the piston coring device. Specific analyses included volatile organic compounds (VOCs), semivolatile organic compounds (SVOCs), polychlorinated biphenyls (PCBs), nitrocellulose, nitrogylcerine, ammonia, mercury, methyl mercury, lead, copper, percent solids, particle size distribution, specific gravity, and percent organic material. Sediment thickness was measured using a 0.95-cm-diameter tile probe rod. Measurements were obtained at 428 locates spaced on 15-m intervals.

Investigation Results. The GGB sediments primarily consist of black to brown clayey silts, organic silts, and silty sands. These sediments are thickest on the western end of the bay at depths of up to 4.6 m. The GGB sediment contains concentrations of copper (maximum concentration 277 mg/kg), lead (1,220 mg/kg), mercury (12 mg/kg), methyl mercury (0.149 mg/kg), and ammonia (741 mg/kg) that decrease with depth of sediment and with distance from the BAAP facility's discharge point. The sediments also contained low concentrations of PCBs, trichloroethene, and n-nitrosodiphenylamine.

REMEDY SELECTION PROCESS

Cleanup Goals/Objectives. Data collected during completion of the Phase I and Phase II investigations suggested that the impacted sediments had adversely affected biological communities in GGB. The cleanup criterion for affected sediment used by the WDNR is either the Lowest Biological Effect Level (LEL) for identified constituents or the local background concentration for those contaminants, whichever is greater.

As a result of negotiations with the WDNR, mercury was utilized as the target compound for determining the extent and volume of impacted sediments. The LEL for mercury in sediment is 0.2 mg/kg. The local area mercury background level was determined to be 0.36 mg/kg. In accordance with the WDNR's requirements, the mercury cleanup objective applied to the site was 0.36 mg/kg. Based on the results of the Phase I and Phase II investigations, approximately 66,000 m^3 of sediments were impacted above the 0.36 mg/kg cleanup criteria.

Remedy Screening and Conceptual Design Process. Subaqueous capping and sediment removal were both evaluated for short- and long-term effectiveness in meeting the cleanup objective. Subaqueous capping was not considered a viable option due to questions regarding the permanency of the remedy and geotechnical/physical properties (i.e., high water content and low shear strength) of GGB sediments. As a result of the sediments' physical properties, placement of the capping material would have likely displaced and resuspended the contaminated sediments.

Sediment removal via dredging was determined to be capable of achieving the cleanup objective both in the short-term and long-term. However, it was understood that minimizing the resuspension and redeposition of contaminated sediments would be critical to the success of the project.

It was determined that hydraulic dredging with an experienced dredge team would cost-effectively remove the contaminated sediments to the target depth while minimizing sediment resuspension. The hydraulic dredge selected for the project was the Ellicott International MC-2000, a 0.25-m-diameter hydraulic dredge with a 0.61-m-diameter, 2.6-m-wide auger. Several project-specific modifications to the dredge were completed by Bay West including: 1) adding a shroud to the dredge auger to minimize sediment resuspension; 2) modifying the dredge ladder to allow for a 7.6-m dredge depth while maintaining the use of a hydraulic cylinder for positive crowd; and 3) adding an integrated differential global positioning system (DGPS) to perform positioning and cut monitoring.

During the conceptual design phase, the project team also understood that a significant challenge lay ahead in developing a practical, cost-effective process for managing the sediment dewatering and dredge carriage/interstitial water treatment activities. During the design activities, the following techniques were evaluated for managing the dredge carriage water, sediments, and interstitial water: 1) upland Confined Disposal Facility (CDF) and discharge of treated dredge carriage water and interstitial water back to GGB/Wisconsin River; 2) mechanical dewatering and discharge of treated dredge carriage water and interstitial water back to GGB/Wisconsin River; and 3) passive sediment dewatering utilizing geotextile tubes ('geotubes') and discharge of filtered carriage water and interstitial water overland on adjacent pastureland.

The upland CDF alternative was discarded due to the settling characteristics of the GGB sediments. As a result of the alum and nitrocellulose historically discharged to GGB, the sediments had a tendency to remain in suspension after being disturbed, in particular after being subjected to agitation similar to that anticipated as a result of hydraulic dredging activities. Preliminary design calculations indicated that the land area requirements and corresponding CDF construction, operating, and maintenance costs would result in this option being impractical and cost prohibitive.

Two dewatering technologies were evaluated for both effectiveness and cost. These technologies included a mechanical sediment dewatering system and a passive geotextile dewatering system. A bench-scale test was conducted with the geotextile dewatering system using sediment collected from GGB, both with and without a soluble inorganic sulfide reagent and cationic flocculating agent. During the test, sediments were treated with an inorganic sulfide flocculent and then dewatered. The filtrate was recovered and tested for residual metals.

Both technologies were capable of generating an acceptable filtrate with regard to metals concentration. Total processing cost and reliability favored the passive geotextile filtering technology, which was ultimately selected for this project.

The last challenge associated with the conceptual design was to develop a cost-effective, environmentally sound technique for managing dredge carriage water/interstitial water filtrate generated from the passive geotextile filtering technology. Consideration was initially given to discharging the filtrate back to GGB/ Wisconsin River. However, this alternative was discarded due to the treatment costs required to reduce residual ammonia concentrations to the National Pollutant Discharge Elimination System (NPDES) criteria.

A design was subsequently completed for spray irrigation of the water following the determination that beneficial reuse of the filtrate could be realized by distributing the water over BAAP pastureland. The rate at which the filtered water is discharged to the fields is based on a predefined maximum application rate, water quality testing, evapotranspiration rates, and the amount of naturally-occurring precipitation.

REMEDIATION APPROACH OVERVIEW

The following two paragraphs summarize the remedy selected for the GGB sediments. Subsequent sections of this paper provide a more detailed discussion of the design, construction, and operation of each element of the remediation approach.

A hydraulic dredge is utilized to remove sediment from the bay. The sediment is then pumped via floating and overland pipelines to a dewatering area. A soluble cationic flocculating agent is added to the dredged sediment and carriage water to increase particle size and enhance the filtering effect of the geotextile tubes.

The lined dewatering area consists of a 76-m by 275-m geotextile tube laydown area and a primary catchment basin. Large geotextile filtration tubes are placed in the lined laydown area. Filtered water drains from the geotextile tubes and collects in the south end of the dewatering area in the primary catchment basin. The water is subsequently pumped into a temporary lagoon that will act as the storage reservoir for the spray irrigation system. Irrigation water is applied to selected agricultural fields and forested areas on BAAP. After dewatering, the sediment will be disposed in the dewatering area by constructing a soil cover over the laydown area.

SYSTEM DESIGN, CONSTRUCTION, AND OPERATION

The following sections provide an overview of the major components of the GGB sediment remediation project (Figure 1).

Dredge Operations. Sediment is being removed from GGB using a hydraulic dredge equipped with a horizontal auger. The dredge removes approximately 8.3 m^3/min of sediment and water from the bay. The dredge discharge consists of solids, interstitial water, and carriage water. Interstitial water is water incorporated in the undisturbed sediment layer. Carriage water is additional water removed from the bay in the course of operating the hydraulic dredge. The proportion of carriage water to interstitial water varies with location, time, and depth, based on sediment consistency and dredge operating conditions. Sediment and water are discharged from the dredge through 792 m of floating pipeline that is connected to 823 m of overland pipeline, that transfers the material to the dewatering area. The floating pipeline is constructed in multiple removable sections that vary from 31 m to 124 m long, which permit its length to be adjusted as dredging proceeds. A 275-hp booster pump is located approximately 671 m from the dewatering area in order to maintain the necessary velocity to eliminate sediment settling in the pipeline.

The dredge's cutting head is equipped with a rigid metal shroud to reduce turbidity generated by operation the dredge. The dredge engine is also equipped with a muffler to control noise due to the presence of residential dwellings within GGB.

The dredge pattern was based on a series of lines plotted over the impacted area to attain 100% coverage with overlap. The dredge operator is careful to minimize unnecessary overcutting which would produce an increased volume of carriage water requiring treatment. The dredge utilizes two spud barges equipped with griphoists that are fairleaded to allow one person to accurately index the dredge to the next cut line. In addition, the 0.46-m-diameter spuds are fairleaded to facilitate the use of the dredge tender's winch. To limit the impact to the environment in the event of a hydraulic

FIGURE 1. Aerial Photograph of GGB and Dredge Set-Up.

line break, biodegradable hydraulic oil is utilized in the winch.

A silt curtain was installed across the mouth of GGB. The silt curtain required the closing of GGB to boat traffic during the dredging operation. Warning buoys and signs were placed on the Lake Wisconsin side of the silt curtain to inform boat traffic of the construction activities taking place in the bay.

FIGURE 2. Photograph of Cationic Flocculating Reagent Addition Station.

Metals Removal. Metals are removed from the dredge carriage water and interstitial water during the sediment filtration process utilizing the geotextile tubes. A cationic flocculating agent is added to the dredge carriage water and sediments to enhance the filtering effectiveness of the geotextile tubes. This reagent is introduced in liquid form and helps to coagulate the insoluble metal sulfides it produces with the copper, lead, and mercury ions in the water. The flocculating agent also promotes coagulation of the sediment solids, thus improving dewatering behavior. The reagent is added at a rate of approximately 0.04 m^3/hr to produce a concentration of approximately 60 mg/L in the carriage water (Figure 2).

Dewatering System. The tube laydown area is constructed of an excavated, bermed, and textured 40-mil-lined containment area (Figure 3). The tube laydown area was constructed with a slope of 1 degree transitioning to 2 degrees downward towards the southern end of the laydown area; the centerline of the laydown area slopes 1 degree towards the edges. The southern end of the laydown area is equipped with a transfer and recirculation pump to allow for the filtered water to be transferred to the temporary lagoon for subsequent disposal by spray irrigation on BAAP agricultural fields and forested areas, or recirculation, depending on visual inspection of water quality.

FIGURE 3. Aerial Photograph of Geotextile Tube Laydown Area.

Sediment and water treated with the cationic flocculating agent are pumped into

FIGURE 4. Photograph of Pinch Valve and Tube Manifold System.

the 12-m-circumference, 68-m-long geotextile tubes through a distribution manifold and series of pinch valves (Figure 4) specially fabricated by Bay West for this project. The distribution manifold and pinch valves permit the sediment and water to be delivered to a specific tube or tubes for filtration at controlled rates.

The first time a geotextile tube is filled, it allows some sediment solids to pass through the fabric until a filter cake begins to develop. A primary catchment basin was constructed as part of the dewatering area to trap this initial sediment discharge. All initial, filtered, and rain water follows the constructed gradient to a diversionary wall, built with sandbags covered with polyethylene sheeting, directing flow into the collected solids area of the catchment basin. The transfer side of the catchment basin is isolated from the collected solids area by a gravel-bag weir-wall. Heavy solids can then be recirculated back to the flocculent addition station and geotubes while low turbidity water can continue to pass over the weir and be transferred into the lagoon. All systems run independently, so that dredging, recirculation, and transfer to the lagoon can occur simultaneously.

Sediment and water are pumped into the selected geotextile tube until it reaches a height of approximately 1.8 m, at which time the flow of sediment and water are directed to another tube through the manifold and pinch valve system. Water from the first tube is allowed to filter through the tube fabric into the dewatering area structure.

Geotextile tubes are filled and allowed to drain a number of times. The number of filling and filtration cycles varies from tube to tube with changes in sediment consistency and filtering behavior. Tubes remain in service until they are too full for further use or the accumulated sediment impedes continued use as a filter (Figure 5).

FIGURE 5. Photograph of Filled Geotube.

Dewatered sediment in the filled geotextile tubes will remain in the tube laydown area. When dredging and dewatering activities are complete, the laydown area will be covered with a layer of soil and earth material at least 3 feet thick to make a permanent disposal cell. Material removed during excavation of the laydown area will be used to construct the soil cover upon completion of the project.

The geotextile tubes will continue to drain slowly for an indeterminate period of time, possibly several years. A temporary wetland will be created as part of the disposal cell construction. This wetland will receive any residual seepage from the tubes. When the tubes are finished draining, the temporary wetland will be allowed to dry out and be replaced with upland vegetation by natural colonization and succession.

Temporary Impoundment Lagoon. Filtered water consists of both interstitial water removed by sediment dewatering and carriage water incorporated with the sediment during dredging operations. The proportions of interstitial water and carriage water will vary with the consistency of the sediment and dredge operating conditions. Once water has passed through the geotextile tubes, its identity as interstitial water or carriage water is lost. The term 'filtered water' refers to the product passing through the geotextile tube regardless of its nominal origin as carriage water or interstitial water or their relative proportions.

Filtered water is stored pending disposal in an 11,350 m^3 impoundment lagoon. The lagoon is a 40-mil-lined temporary structure constructed of two temporary berms. The lagoon serves a number of purposes for the project. It provides retention time, allowing any residual suspended solids to settle out of the water prior to disposal. Additionally, the lagoon provides a consistent water supply for operating the spray irrigation system. Spray irrigation rates will be reduced during periods of natural precipitation. The lagoon will provide storage volume, allowing dredging and dewatering to continue during rainy periods that limit disposal of filtered water.

Spray Irrigation. Filtered water from the lagoon is disposed of by spray irrigating agricultural fields and forested areas at BAAP. Three distinct areas for spray irrigation are utilized at BAAP. Water is pumped from the lagoon to spray guns mounted in the irrigation areas (Figure 6). Water flows through a distribution manifold to direct it to a selected irrigation area.

FIGURE 6. Photograph of Spray Irrigation Activities.

The project team has had to carefully balance the dredging and dewatering activities with the spray irrigation activities. Two conditions limit the amount of water that can be discharged onto a given area. There is a regulatory limit of 38 m^3 per 4,047 m^2 per day averaged over a week. In addition to this maximum application rate, the spray irrigation disposal rate must not produce surface runoff from the application areas. During periods of natural rainfall, spray irrigation application rates may be substantially less than the regulatory limit. In order to not limit dredge productivity, additional dewatering at BAAP areas are being evaluated. Operation of the spray irrigation system may be completely precluded in some circumstances. Irrigation activities will be rotated among the designated irrigation areas to allow adequate time for the water to infiltrate the surface soil.

CONCLUSION

The sediment remedial action currently being conducted at BAAP has shown that geotextile tubes can be cost-effectively utilized to dewater and treat carriage water and contaminated sediments generated from hydraulic dredging activities. Prior to implementing this technology, however, bench-scale studies are recommended to determine: 1) the dewatering characteristics of the dredged sediments; and 2) the contaminants levels associated with the filtered dredge carriage water and sediment interstitial water; 3) whether a flocculating reagent should be added to enhance the filtering effect of the geotextile tubes. In addition, it is critical that a manifold and valve system be designed and constructed which can easily and rapidly redirect flow from the dredge carriage water pipeline to the geotextile tubes without impeding dredging activities.

IN SITU CAPPING ON THE PALOS VERDES SHELF, CALIFORNIA

Michael R. Palermo (U.S. Army Corps of Engineers R&D Center, USA)

ABSTRACT: Sediments covering approximately 40 square kilometers of the ocean floor at the Palos Verdes Shelf, near Los Angeles, California, are contaminated with DDT and PCBs. The U.S. Army Corps of Engineers (USACE), working in support of the U.S. Environmental Protection Agency (EPA), has conducted studies investigating the feasibility of in-situ capping all or a portion of the site with a layer of clean sandy dredged material. A study was conducted to determine the feasibility and effectiveness of in-situ capping. This study included the necessary engineering and environmental analyses such as preliminary cap designs, operations plans, and monitoring and management plans for a range of in-situ capping options. The USACE has also recently completed a field pilot study at this site. The pilot study involved placement of approximately 103,000 cubic meters of capping sediments using a split hull hopper dredge. Three 18 hectare capping cells situated at water depths between 40 and 70 meters were capped wholly or in part using both conventional placement methods and special spreading methods. A large scale environmental monitoring effort was conducted before, during and after cap placement using a number of state-of-the-art techniques and specialized equipment. Predictive modeling was also conducted during and following the placements to guide field operations and refine data for design purposes. This paper provides a description of the project setting and conditions and summarizes the results of the feasibility and field pilot studies.

SITE AND PROJECT DESCRIPTION

The Palos Verdes shelf and slope is located off the Palos Verdes peninsula near Los Angeles, California. The shelf is 1 to 6 km wide and very flat with increasing bottom slopes of approximately 1 to 3 degrees out to the shelf break at water depths of approximately 70 to 100 m. Beyond the shelf break, the slope steepens to approximately 18 degrees and extends to water depths of several hundred meters. Three large sewer outfall diffusers discharge onto the shelf approximately 3 km offshore of Whites Point in approximately 60 m of water. Particulate matter discharged through the outfalls has settled out and built up an effluent-affected (EA) sediment deposit on the shelf and slope which contains levels of organic matter and chemical contaminants higher than the native sediments. The contaminants of concern are DDT (and its metabolites such as DDE) and PCBs. The footprint of the EA deposit on the shelf and slope is shown in Figure 1. The highest levels of contamination are on the shelf and extend with the predominant bottom current direction to the northwest from the sewer outfall diffusers.

EVALUATION OF REMEDIAL OPTIONS

The U.S. Army Corps of Engineers (USACE) has played a key role in conducting engineering and environmental studies for the Palos Verdes Shelf in support of the U.S. Environmental Protection Agency (EPA), the U.S. National Oceanic and Atmospheric Administration (NOAA), and the U.S. Department of Justice (DOJ). NOAA acted as the lead agency at this site for a group of Federal and State natural resource trustee agencies to assess damages and injuries to natural resources under the U.S. Comprehensive Environmental Response, Compensation, and Liability Act (CERCLA), also know as the Superfund Act. A feasibility study of sediment remediation options was completed for NOAA and DOJ by the USACE in 1994, which included evaluations of in-situ capping of contaminated sediments and dredging and disposal options to include both in-water capping and diked confined disposal areas. This study determined that in-situ capping was a technically feasible option (Palermo 1994).

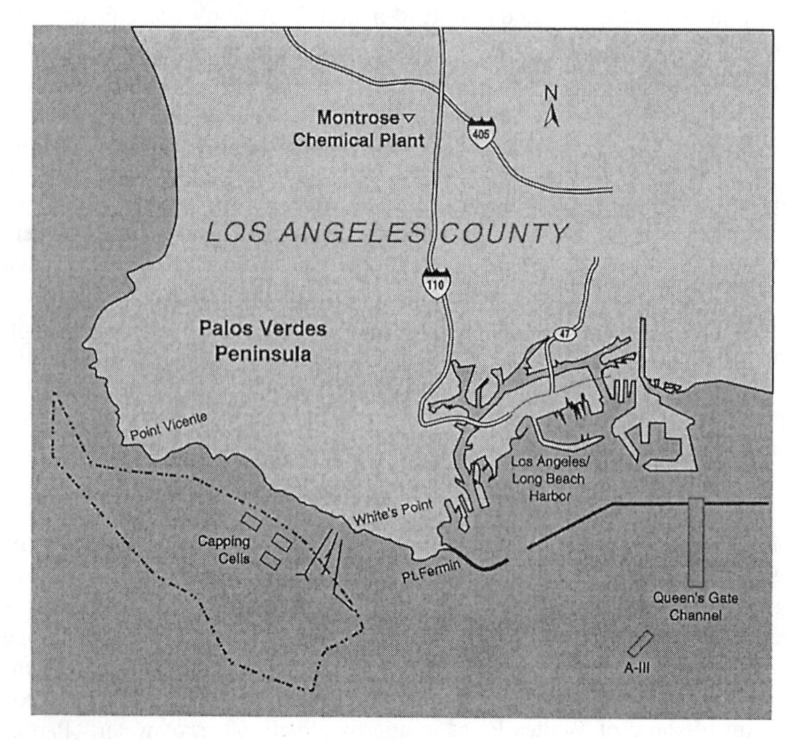

FIGURE 1. Map showing EA sediment footprint on the Palos Verdes shelf and slope, pilot capping cells, Queen's Gate entrance channel and borrow area A-III.

In 1996, EPA began its Superfund investigation of the Palos Verdes Shelf, and requested technical support from the USACE. A detailed evaluation of capping options for the shelf was conducted for EPA by the USACE (Palermo et al 1999). The evaluation included prioritizing areas of the PV shelf to be capped, determining appropriate cap designs, developing an equipment selection and operations plan for placement of the cap, developing a monitoring plan to ensure successful cap placement and long-term cap effectiveness, and developing preliminary cost estimates.

The study identified an area of several square kilometers on the shelf between the 40- and 70-m depth contours as the area which could potentially be capped. If capping is selected as a remedy at the site, the operations would be done in an incremental fashion until the total selected area was capped. Two capping approaches were considered in the USACE report (Palermo et al 1999): 1) placement of a Thin Cap (design thickness of 15 cm) which would isolate the contaminated material from shallow burrowing benthic organisms, providing a reduction in both the surficial sediment concentration and contaminant flux, and 2) placement of an Isolation Cap (design thickness of 45 cm) which would be of sufficient thickness to effectively isolate the majority of benthic organisms from the contaminated sediments, prevent bioaccumulation of contaminants and effectively prevent contaminant flux for the long term.

FIELD PILOT STUDY FOR CAPPING

The USACE has recently conducted a field pilot study of capping on the Shelf for EPA. USACE Los Angeles District planned, designed, and executed the pilot with support from the USACE Research and Development Center, Waterways Experiment Station (ERDC-WES), USACE New England District, and USACE Seattle District. Dredging operations for the pilot were conducted under contract by NATCO Limited Partnership. Field monitoring activities were conducted under contract by Science Applications International Corporation (SAIC).

The overall objective of the field pilot study was to demonstrate that a cap can be constructed on the shelf as intended by the design and to obtain field data on the short-term processes and behavior of the cap as placed. Three 300- by 600-m capping placement cells were selected for the pilot to reflect the range of site conditions including water depth, bottom slope and sediment properties and to allow for different placement methods. The location of the cells within the EA sediment footprint is shown in Figure 1. Pilot cells were located in a comparatively shallow depth with comparatively flat bottom slope (40 to 45 m depth contour with an average slope across the cell of about one degree) and at a comparatively deeper location with steeper bottom slope (60 to 70 m depth contour with average slope across the cell of about three degrees). The cells were separated by a full cell-length in the along-shore direction and by a full cell-width in the perpendicular direction to avoid the potential for interferences during monitoring.

Cap Material Sources. Cap material sources were investigated by the USACE, and dredged sediments from the Port of Long Beach Queen's Gate entrance channel and a sand borrow area designated as A-III were identified as the cap material sources for the pilot. These areas are shown in Figure 1. The Queen's Gate material is a mixture of fine sand and silt, while the A-III material is a coarser sand.

Selection of Placement Equipment. Hopper dredges were earlier identified as a preferable placement equipment type, and use of a hopper dredge was planned for the pilot. A hopper dredge was the equipment of choice because hopper dredges were most readily available equipment for the pilot work, they provide better control of placement in the open ocean environment and allow for more flexibility in placement options to include pumpout capabilities, and they remove material from channels by hydraulic means, resulting in a breakdown of any hardpacked material and addition of water as material is stored in the hopper for transport. Material from hopper dredges is therefore more easily dispersed in the water column, and settles to the seafloor with less energy and less potential for resuspension of the contaminated sediment.

The NATCO hopper dredge *Sugar Island* was used for the pilot placements (see Figure 2). The *Sugar Island* has a split-hull hopper opening mechanism that can be used to control the rate of release. This dredge is also equipped with a hopper pumpout capability over the bow and water jets to aid in pumpout operations. Pumpout can also be accomplished through the lowered dragarm of the hopper dredge.

FIGURE 2. Hopper dredge Sugar Island, and sediment profile photo showing sand cap as placed.

Cap Placement Operations. Dredging and cap placement operations for the pilot project were coordinated with the monitoring efforts. Over 103,000 cubic meters of capping material was placed within three 18 hectare cells on the PV shelf. Single hopper loads were initially placed and monitored. Then, multiple hopper loads were placed to build up the desired cap thickness, and monitoring was conducted at several stages as the cap was constructed. Queen's Gate material was placed using conventional point placements, and A-III sand was placed using spreading methods. In addition, a single load was placed by pumpout through the hopper dragarm.

Monitoring. An extensive field monitoring effort was conducted as a part of the pilot. Major monitoring aspects for the pilot included cap thickness as placed, mixing of cap and contaminated sediments, resuspension of contaminated sediments during cap placement, and short term physical and chemical characteristics of the cap and underlying sediments immediately after capping and following initial sediment consolidation. A number of monitoring technologies were used:

- **sediment profile and plan view images** (photographs) that provide data on cap thickness, disruption of the existing contaminated sediment, and the presence of sediment-dwelling organisms (e.g., worms & other burrowing organisms) before and after capping;
- **sediment cores** that provide data on cap thickness, the degree to which contaminated sediments mixed with clean cap material during construction, and the stability of the cap after construction;
- **water samples and water column measurements** to determine whether contaminated sediment was resuspended during cap construction (and if so, how far and in what direction it traveled before settling to the bottom again) and to assess the magnitude and extent of the turbidity plume created by cap placement;
- **current meters and optical back scatter meters** that helped define both the lateral extent of the surge created when the descending cap material reached the bottom and the variation of current speed and direction at different water depths (which affect the spread of cap material as it fell through the water);
- **side-scan sonar and sub-bottom acoustic profiling** to assess the lateral extent and thickness of the cap;
- **sediment samples from the hopper dredge** used in modeling cap placement and to assess the efficacy of placement methods; and
- **dredge position and load curves** to document where each load of cap material came from and where it was placed.

Modeling. Mathematical models were used to predict the rate of cap material buildup for specific sediment characteristics, various water depths over the shelf and various placement approaches. The models were used to predict cap material dispersion during placement and evaluate the velocities of bottom impact on spreading behavior, respectively. These predictions were initially based on a broad range of assumed properties for the cap material. Once specific cap material sources were selected, refined predictions using the specific site conditions and cap material properties were made. Results of the refined predictions were used to adjust the operational approach and monitoring efforts for the pilot.

Pilot Study Findings. A summary of the pilot study results is presented below:

- Both conventional placement methods and spreading methods proved successful in constructing the pilot cap.

- The actual behavior of the released sediment compared very well to the predictions made by numerical modeling. These predictions included (1) the distribution of cap material on the seafloor, (2) the decrease of surge velocities as a function of distance from the placement location, and (3) dilution of the plume to near background conditions within two to three hours of cap placement.

- Evidence from the sediment profile, coring, and side-scan surveys all support a conclusion that it will be possible to create a cap on the Palos Verdes Shelf that is quite uniform (see photo of in-place sand cap in Figure 2). The caps that were created using both conventional and spreading placement generally vary in thickness by only a few cm across those areas that received what could be considered a full cap application during the pilot.

- Some physical disturbance to the EA sediment was observed, but disturbance was minimized during the pilot study through the management of cap placement points. In addition, it appears that the spreading placement approach has the potential to result in even less disturbance to in-place sediments than conventional placement.

- Water quality measurements determined that suspended solids and contaminant concentrations in the water column showed a rapid return to background levels.

- Contaminant measurements in core samples indicate that a clean cap can be constructed. The process of cap placement resulted in about 3-4 cm of the cap becoming mixed to some degree with the EA sediment. As cap thickness increased beyond this, mixing with the EA sediment became negligible such that the upper portions of the cap had very low levels of contaminants.

- No evidence of cap or EA sediment instability with respect to avalanching or mud flows was observed as a result of operations. Current surge monitoring results indicate that the energy from conventional point cap placement decays with distance and time away from the point of release. Surge energy from spreading placements was much lower than for the conventional placements. No large scale deformations or changes in the seafloor around the cells and in particular down-slope were observed.

- The pilot demonstrated that a cap can be successfully placed on the PV shelf and adequately monitored. The monitoring equipment and techniques proved generally effective in obtaining the desired data, and were generally effective across the range of site conditions encountered in the pilot.

- The pilot study results provided a wealth of data on cap constructability and the effects of site conditions, material type, and placement methods on cap construction. These data will prove useful to decision-makers regarding implementation of any future full scale cap on the PV Shelf.

ADDITIONAL DETAILS AND NEXT STEPS

Additional details of the various modeling and monitoring program components and results are available (Palermo et al 2001; Bratos et al 2001; Pace et al 2001; McDowell et al 2001a and b; Valente et al 2001; Walter and Fredette 2001; and Phillips et al 2001). EPA and USACE are currently reviewing and evaluating the data collected for the pilot project. Complete results for the project and pilot study will be made available in reports prepared for EPA by the USACE and its contractors. Current plans call for the reports and supporting data to be made available both in published form and through a website.

Results of the pilot study will be used by EPA in its decision on any further remedial actions at the site. If additional capping is called for, the results of the pilot will be applied in the design and implementation of any full scale project.

ACKNOWLEDGEMENT

This paper summarizes results of studies conducted by the U.S. Army Corps of Engineers for the U.S. Environmental Protection Agency. Permission to publish this material was granted by the Chief of Engineers.

REFERENCES

Bratos, S., B. Johnson and J. Clausner. 2001. Palos Verdes Shelf Pilot Capping: Dredged Material Fate Modeling for Cap Placement. Proceedings, 21st Annual Meeting of the Western Dredging Association (WEDA XXI) and 33rd Annual Texas A&M Dredging Seminar, Houston, TX.

McDowell, S., E. Tobey, and P. Walter. 2001a. Palos Verdes Shelf Capping: Suspended Sediment Plume Monitoring During Cap Placement. Proceedings, 21st Annual Meeting of the Western Dredging Association (WEDA XXI) and 33rd Annual Texas A&M Dredging Seminar, Houston, TX.

McDowell, S., S. Pace, J. Singer, and P. Hamilton. 2001b. Palos Verdes Shelf Capping: Moored Measurements of Near Bottom Currents and Turbidity During Cap Placement. Proceedings, 21st Annual Meeting of the Western Dredging Association (WEDA XXI) and 33rd Annual Texas A&M Dredging Seminar, Houston, TX.

Pace, S., D. Fischman and S. McDowell. 2001. Palos Verdes Shelf Pilot Capping: Monitoring of Hopper Dredge Operations During Placement of Cap Material. Proceedings, 21st Annual Meeting of the Western Dredging Association (WEDA XXI) and 33rd Annual Texas A&M Dredging Seminar, Houston, TX.

Palermo M.R. 1994. "Feasibility Study of Sediment Remediation Alternatives for the Southern California Natural Resources Damage Assessment," Expert Report, U.S. Army Engineer Waterways Experiment Station, Vicksburg, MS.

Palermo, M., P.R. Schroeder, Y. Rivera, C. Ruiz, D. Clarke, J. Gailani, J. Clausner, M. Hynes, T. Fredette, B. Tardy, L. Peyman-Dove and A. Risko. 1999. Options for *In Situ* Capping of Palos Verdes Shelf Contaminated Sediments. Technical Report EL-99-2, U.S. Army Corps of Engineers, Waterways Experiment Station, Vicksburg, MS. http://www.wes.army.mil/el/elpubs/pdf/trel99-2.pdf

Palermo, M., F. Schauffler, T. Fredette, J. Clausner, S. McDowell, and E. Nevarez. 2001. Palos Verdes Shelf Pilot Capping: Description And Rationale. Proceedings, 21st Annual Meeting of the Western Dredging Association (WEDA XXI) and 33rd Annual Texas A&M Dredging Seminar, Houston, TX.Phillips, C., J. Evans, V. Frank, and P. Walker. 2001. Palos Verdes Shelf Pilot Capping: DDT Concentrations in Baseline and Post-Cap Cores. Proceedings, 21st Annual Meeting of the Western Dredging Association (WEDA XXI) and 33rd Annual Texas A&M Dredging Seminar, Houston, TX.

Valente, R., T. Waddington, J. Infantino and G. Tufts. 2001. Palos Verdes Shelf Pilot Capping: Use of Side Scan Sonar and Sediment Profile Imaging and Plan View Photography for Resolving Cap Material Distribution and Thickness. Proceedings, 21st Annual Meeting of the Western Dredging Association (WEDA XXI) and 33rd Annual Texas A&M Dredging Seminar, Houston, TX.

Walter, P and T. Fredette. 2001. Palos Verdes Shelf Pilot Capping: Geotechnical Results from Sediment Cores. Proceedings, 21st Annual Meeting of the Western Dredging Association (WEDA XXI) and 33rd Annual Texas A&M Dredging Seminar, Houston, TX.

GUIDELINES FOR THE DESIGN, CONSTRUCTION, AND MANAGEMENT OF REVETMENTS AT MARINE WASTE DISPOSAL FACILITIES

Hiroaki Ozasa *(Dr. Eng.)*
Waterfront Vitalization and Environment Research Center
Minato-ku, Tokyo 108-0022, Japan

ABSTRACT: In Japan, incineration residues of solid wastes, residue of industrial wastes, sludge, and other materials are disposed of in both land and marine controlled disposal facilities. More than 80% of the solid waste residue generated in Tokyo Prefecture is disposed of at marine waste disposal facilities. Technical Guideline for Final Disposal Facilities for Solid and Industrial Waste was amended in June 1998. Especially, standards for the structure and maintenance of liners were strengthened and specified clearly. In the light of these circumstances, technical guidelines for the design, construction, and management of revetments at marine waste disposal facilities were formulated.

Role of Marine Waste Disposal Facilities and Appropriate Regulations

In Japan, incineration residues of solid wastes (waste other than industrial waste), as well as industrial wastes are disposed of in both land and marine disposal facilities. Dredged sediment is beneficially utilized as materials for beach nourishment, tidal flat creation, reclamation, and other uses. Some sediment is also disposed of in the ocean. These activities are regulated by Japanese laws based on the London Convention. There are two types of marine reclamation sites; one is stable disposal site where stable materials such as dredged sediments, construction sludge, and glass garbage are filled, and the other is controlled disposal site where incineration residue of municipal solid wastes, residue of industrial wastes subject to intermediate treatment, sludge, and other materials are filled. Figs 1 and 2 show stable disposal site and controlled disposal site respectively. Both types of reclamation sites must be controlled according to government guidelines. Controlled disposal sites are imposed on more strict measures for environment preservation than stable disposal sites. Confined disposal facilities for contaminated sediment do not exist. Contaminated sediment which is defined as 'contaminated' according to Japanese environmental standard is dealt with by procedures for decreasing elution of poisonous agents and is then disposed of at controlled disposal sites.

In large cities such as Tokyo, Osaka, and Nagoya, some controlled final disposal sites are located at marine reclamation areas. In 1997, the final disposal weight of solid wastes in Japan reached 13 million tons. Nineteen percent of this was disposed of at marine waste disposal facilities. Forty-nine percent of the solid waste residue generated in the Tokyo metropolitan area and 84.7% of the solid waste residue generated in Tokyo Prefecture were disposed of at marine waste disposal facilities.

The revetments of marine waste disposal facilities are constructed as social infrastructure subsidized by the government. Every final waste disposal site is regulated by the Waste Disposal and Public Cleansing Law. To obtain permission of the governor of a municipality for the construction of a final waste disposal facility, the facility itself must comply with the Technical Guideline for Final Disposal Facilities for Solid and Industrial Waste. The above Guideline was amended in June 1998. Especially, standards for the structure and maintenance of liners used in final waste disposal sites were strengthened and specified clearly. In the light of these circumstances, the Ports & Harbors Bureau of the then Ministry of Transport (currently the Ministry of Land, Infrastructure, and Transport), which is responsible for revetment construction at marine waste disposal facilities, decided to formulate technical guidelines for the design, construction, and management of revetments at marine waste disposal facilities. They entrusted this task to the Waterfront & Environment Research Center (WAVE).

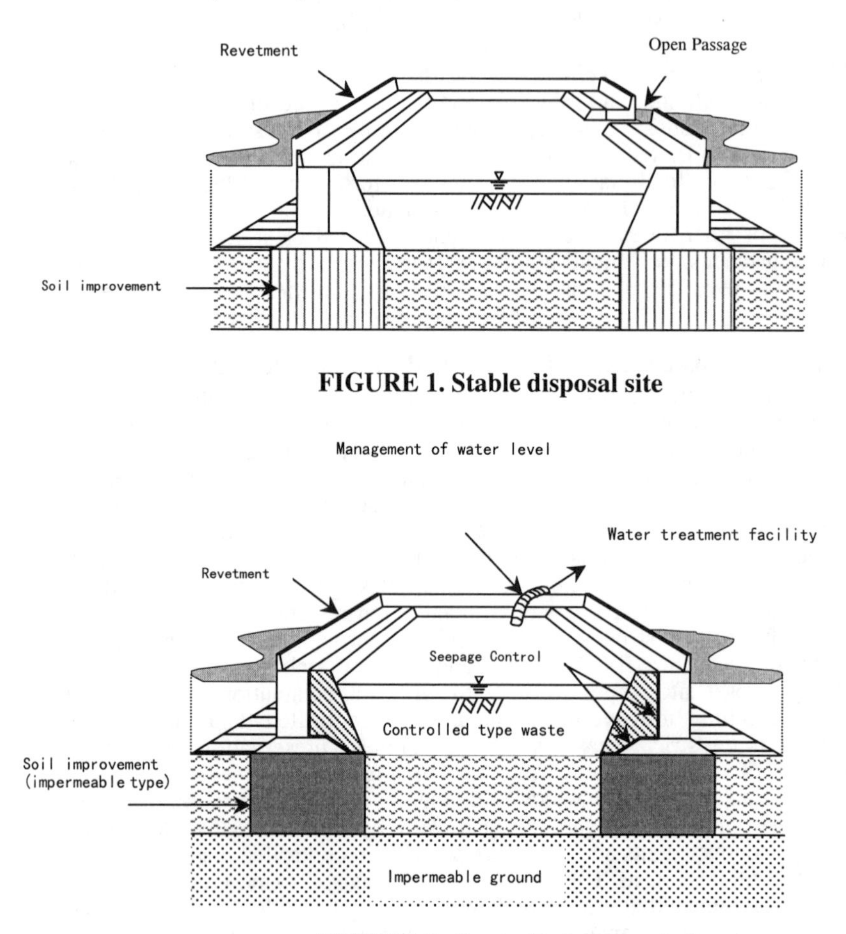

FIGURE 1. Stable disposal site

FIGURE 2. Controlled disposal site

Features of Marine Waste Disposal Facilities

Marine waste disposal facilities (hereafter limited to controlled disposal facilities) possess several features not shared by inland waste disposal facilities. Major differences are as follows:

1. The most important component of the revetment is its liners used to obstruct water movement. Enough consideration must be paid to the factors that affect the liners, such as wave action and the water level difference between the inside and outside of the marine waste disposal facility.

2. Since the revetment of the marine waste disposal facility is usually built on soft ground, measures for reinforcing such soft ground must be implemented in the construction of such a facility while maintaining the seepage control function of the soft clay foundation.
3. Seawater and rain that have accumulated inside the marine waste disposal facility must be processed appropriately and drained out of the facility, once disposal of wastes begins.
4. Dredged sediment and waste soil generated at the construction site are also disposed of at the marine waste disposal facility, in addition to the controlled wastes.

Matters to be Considered in the Planning Stage of Marine Waste Disposal Facilities

The two major points to be borne in planning marine waste disposal facilities are as follows:

1. Natural conditions of a planned site such as ocean waves, storm surges, and tsunami should be taken into consideration.
2. The seepage control function of the ground should be examined when choosing a site for the marine waste disposal facility. Basically, an area having impermeable layers is appropriate for construction of such facility.

Matters to be Considered in Designing Revetments for Controlled Marine Waste Disposal Facilities

The fundamental requirements in designing controlled marine waste disposal facilities are summarized as follows:

1. The revetment should be stable even under the influences of various external forces such as hydraulic pressure, wave force, earth pressure, and seismic force.
2. Even when the marine waste disposal facility is exposed to waves, storm surges, tsunami, and other natural forces, the revetments should function properly to preserve the facility operable.
3. The revetments should exhibit the designed seepage control function to protect public waters around the facility from contamination.

Major points stipulated in the Guidelines with regard to the design of the revetments are as follows:

1. The liners should have a structure that conforms to the intent of the Technical Guideline for Final Disposal Facilities of Solid Waste and Industrial Waste. The liners and the impermeable ground should cooperate with each other to exhibit their expected seepage control functions.
2. The following was specified as a criterion for evaluating impermeability of soft clay grounds: " Layers of equivalent seepage control capacity should have generally the same water penetration time ."
3. The liners that can be used for marine waste disposal facilities include PVC sheet liners , steel sheet-pile liners, joint-sealing liners between caissons, and water-impermeable materials made by mixing soil, cement, water, and other materials. Figs 3 and 4 show examples of seepage control facility at both gravity-type revetment and double steel pipe sheet pile type revetment respectively.

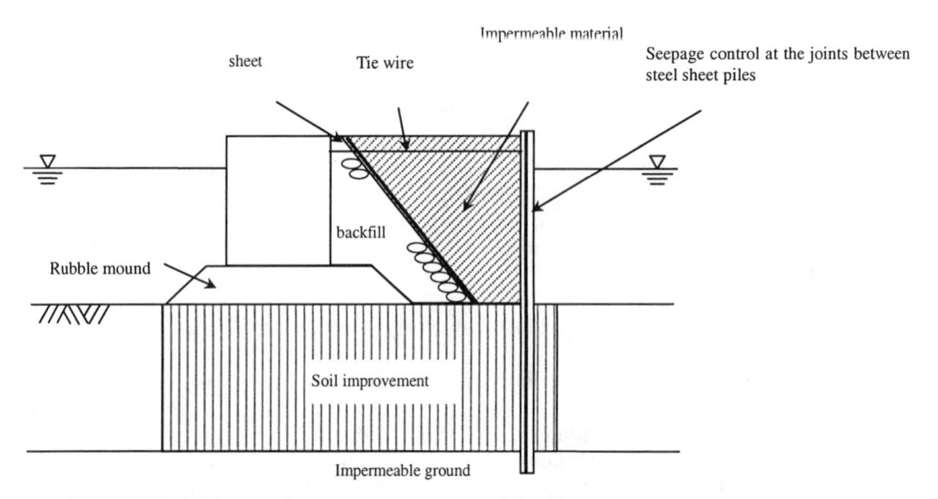

FIGURE 3. Example of seepage control facility at gravity-type revetment

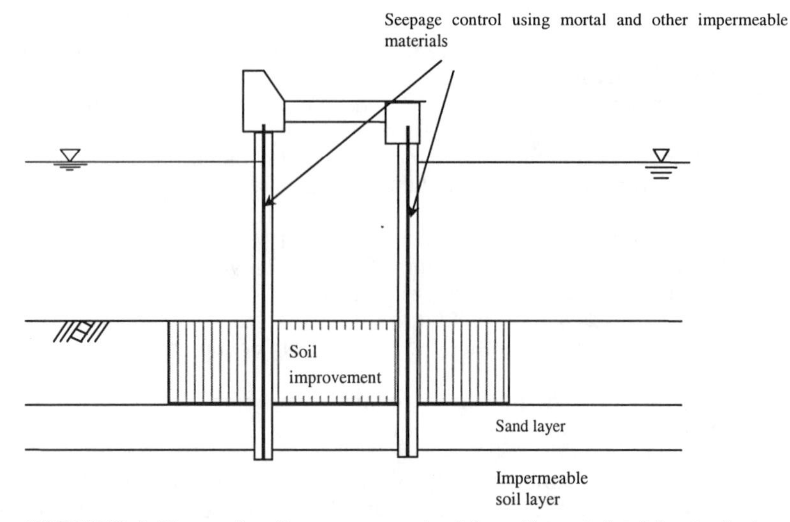

FIGURE 4. Example of seepage control function at double steel pipe sheet pile type revetment

4. To ensure seepage control function by using waterproof sheets only, double layers of them should be employed with non-woven fabric or other suitable material inserted between them. An armor layer should also be installed over them to prevent them from floating and being damaged. Fig 5 shows composition of seepage control facilities on a rubble mound.

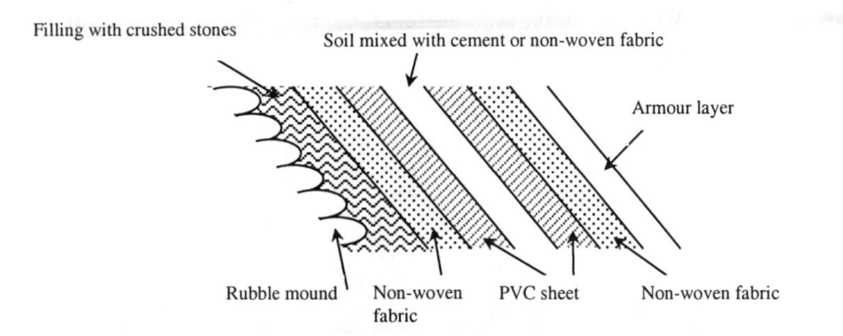

FIGURE 5. Composition of seepage control facilities on a rubble mound

5. The steel pipe sheet piles should be embedded to a proper depth into the impermeable ground so that they exhibit expected liner function. At the joints between steel sheet piles, swelling material liners for permanent work should be used. At the joints between steel pipe sheet piles, on the other hand, mortar and other impermeable materials should be used to maintain sufficient seepage control functions.
6. At the joints between caissons, an appropriate method that can insure sufficient seepage control functions should be employed.
7. The distance between the edge of the waterproof sheets and the foot of the slope of a rubble mound or backfill rubble mound should be at least 5 meters in order to secure a close contact of the sheets with an impermeable foundation. Fig 6 shows this rule.

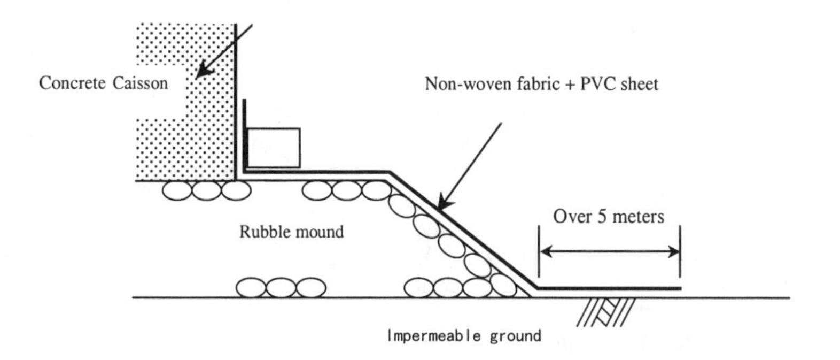

FIGURE 6. Distance between the edge of the waterproof sheets and the foot of the slope of a rubble mound

8. The followings should be paid attention because they are important matters relating stability of the liners : Effects of ocean waves and the water level difference between the inside and outside of the facility should be paid attention; Fig 7 shows pressures by both ocean waves and the water level difference between the inside and outside of the facility. Settlement or deformation of the seabed due to its soft constituents should be coped with; The earthquake resistance of the liners should be insured; Resistance of PVC sheet liners to unevenness in the slope of the

rubble mound as well as their damage-resistance against wastes being disposed should be paid attention.

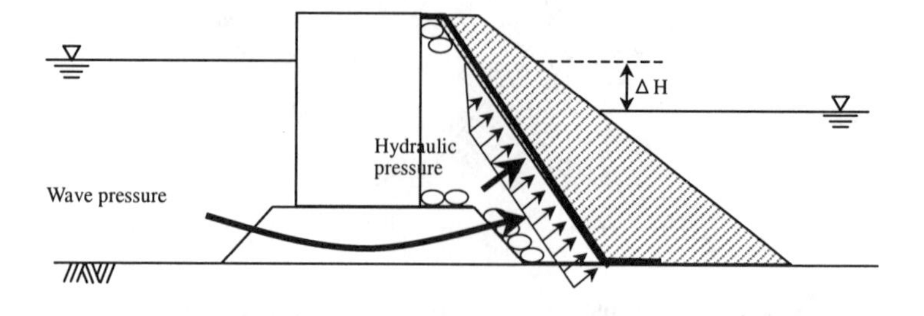

9. As for the structural details of the liners, the Guidelines stipulate that an appropriate construction method can be employed taking into consideration such factors as natural conditions, working conditions, and the characteristics of materials to be used. The thickness of waterproof sheets should be 1.5 mm and over except for asphalt-based ones, and they should be bonded by heat deposition.
10. The coefficient of importance in the seismic design of the revetments at the outer perimeter of the facilities should be 1.2 or over .

Matters to be Considered in Constructing Revetments for Controlled Marine Waste Disposal Facilities

The major points stipulated in the Guidelines concerning the construction of revetments for controlled marine waste disposal facilities are as follows:

1. In the installation of waterproof sheets, the slope of the rubble mound constructed at the required precision should be filled with crushed stones to ensure flatness of the slope (See Fig 5).
2. Laying of waterproof sheets should be carried out under the condition that the water level difference between the inside and outside of the facility is minimized by providing open passages in the revetments.
3. When bonding the waterproof sheets by applying heat deposit, the length of lap should be 10 cm, and bonded length, 3 cm.
4. When installing steel sheet piles together with swelling waterproof materials, appropriate construction method such as coating swelling materials to the joints should be employed. Steel sheet piles should be handled and placed properly.

Matters to be Considered in Managing Revetments at Controlled Marine Waste Disposal Facilities

The Guidelines stipulate the followings:

1. The revetment construction side (port side) should manage the revetments properly. The waste disposal administration side should manage the disposal of wastes at the facility properly. They should consult with each other about methods for final waste disposal.
2. The water level at marine waste disposal facilities during and after disposal of wastes should be set and managed properly taking into consideration natural conditions, structure of the facility, and other

conditions.

3. The revetment construction side should monitor the liners to insure that the seepage control function is working properly. The construction side also should monitor the deformation of the revetment following disposal of wastes. The waste disposal administration side is mandated to monitor the environment in accordance with the amended Technical Guideline for Final Disposal Facilities of Solid Waste and Industrial Waste.
4. The revetment construction side and the waste disposal administration side should implement proper measures when an abnormality is detected during their monitoring activities.

Issues Requiring Further Technological Development

Concerning technologies related to revetments for controlled waste disposal facilities, there are several issues in addition to those mentioned above. They require further technological development. Issues that require further researches have been listed to encourage further technological development.

REMEDIATION OF PCB-CONTAMINATED SEDIMENT IN THE DUWAMISH RIVER, WASHINGTON

Scott J. Mickelson (King County, Seattle, Washington USA)
Diane McElhany (King County, Seattle, Washington USA)

ABSTRACT: Sediment remediation at a King County CSO site in the Duwamish River was the first project undertaken by the Elliott Bay/Duwamish Restoration Program. Remediation of PCB-contaminated sediment was accomplished through mechanical dredging and backfilling with clean sand. Samples collected during dredging showed that PCB contamination was more widespread and severe than originally indicated, resulting in the removal of more sediment than anticipated. Site-wide dredging was completed to a depth of 0.9 meter with two PCB "hot-spot" areas dredged an additional 1.8 meters. Sediment remaining in the hot-spot areas contained PCBs at concentrations exceeding Washington State sediment quality criteria, thus designating the backfill material as a "thick capping layer." Post-remediation monitoring at the site indicates that the backfill material has not been negatively impacted by operation of the CSO. A storm drainage channel that runs through the backfill material, however, has become recontaminated with PCBs. Additional sampling has shown that one potential source of recontamination is sediment located inshore of the remediation area, around the storm drain outfall. King County is attempting to identify a way to eliminate the PCB-contaminated sediment inshore of the site as soon as possible to avoid further recontamination of the backfill.

INTRODUCTION

Combined sewer overflows (CSOs) are discharges of untreated sewage and storm water released directly into surface waters during periods of heavy rainfall. Combined sewers, those that carry sanitary sewage and storm runoff in a single pipe, are found in much of metropolitan Seattle, Washington. All of Seattle's sewers built from 1892 until the early 1940s were combined sewers, following standard engineering practices of the times (King County, 1995). As newer sewers were installed in Seattle, storm water was separated from sanitary sewer systems and additional CSO reduction projects have been undertaken to reduce CSO discharge events and volumes.

CSOs serve as safety valves for the sewage treatment system. In combined systems, the trunk sewers and interceptors have fixed capacities. During periods of heavy rainfall, volumes may exceed the capacity of the sewer pipes to convey the combined sanitary sewage and storm water to treatment plants. To prevent damage to the system and to prevent sewers from backing up into homes and businesses, combined sewers are designed to overflow. Typically, overflows are discharged to rivers and marine waters where the flushing action of tides and currents can ideally disperse pollutants. Several of the CSOs operated by King County discharge into the lower Duwamish River.

The lower Duwamish River is a highly industrialized, salt-wedge estuary influenced both by river flow and tidal effects. The river is considered an estuarine system, exhibiting both marine and fresh water characteristics. During periods of normal river flow, the salt wedge can extend upriver as far as 13 kilometers (km). The lower portion of the Duwamish River, below a navigational turning basin, has been straightened, dredged, and rip-rapped to facilitate navigation and commerce.

A 1991 Consent Decree defined terms of a natural resources damage agreement between King County and federal, state, and tribal natural resources trustees. Representatives of these trustees formed the Elliott Bay/Duwamish Restoration Program (EB/DRP), the goals of which are to remediate contaminated sediments associated with CSOs and storm drains, restore habitat in Elliott Bay and the Duwamish River, and control potential contaminant sources from CSO and storm drain outfalls. King County's Norfolk Street CSO site (Norfolk CSO site) was selected as one of four sites prioritized for sediment remediation under EB/DRP. Discharge volumes at the Norfolk CSO have currently been reduced by 90% to approximately 26,500 cubic meters per year.

SITE CHARACTERIZATION

The Norfolk CSO sediment remediation site is located at river km 10 in the Duwamish River. The site is located upstream of the river reach maintained for commercial navigation and, as a result, retains its natural channel as well as some riparian habitat. The shoreline is characterized by a steeply sloping, erosional bank maintained with large concrete riprap. The bank joins a gently sloping, intertidal mud shoreline that is completely exposed during extreme low tides. Evidence of the salt wedge influence in this portion of the river is indicated by the presence of such marine species as *Mytilis trossulus* (mussel), *Balanus glandula* (barnacle), and *Fucus distichus* (alga).

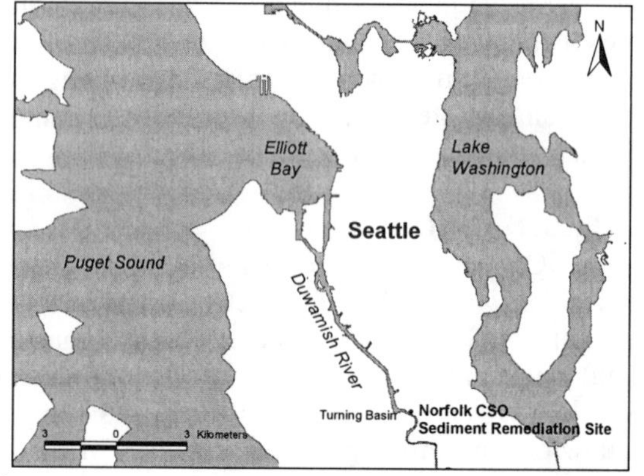

The Norfolk CSO structure has a flap gate over a 2-m discharge pipe and a concrete splash plate that is exposed during normal low tides (EB/DRP, 1996). The remediation site is located adjacent to the CSO outfall structure and is characterized by exposed, intertidal mud habitat as well as subtidal riverbed. The intertidal mudflat had been channelized, both by the discharge of the Norfolk CSO and by a nearby, separate storm drain outfall that drains an industrial complex.

A sediment characterization study was performed in three phases between August 1994 and December 1995 to evaluate the areal extent of contamination at the Norfolk CSO site and determine a preferred remedial alternative.

- Phase 1, conducted in August 1994, included collection of surface sediment grab samples to evaluate the areal extent of surficial sediment contamination around the CSO outfall.
- Phase 2, conducted in August 1995, included collection of additional surface sediment grab samples to refine the boundary of the cleanup area, collection of sediment core samples to evaluate the vertical extent of sediment contamination, and a survey of the river's shoreline to map the topography of the intertidal and upland areas in the vicinity of the CSO outfall.
- Phase 3, conducted in December 1995, was a focused sampling effort to define the boundary of PCB-contaminated sediments in the vicinity of the CSO outfall and several separate storm drain outfalls located downstream of the Norfolk CSO.

The 1991 Consent Decree directed the EB/DRP panel to use Washington State sediment standards to characterize and define the level of contamination including the extent of sediment cleanup required at the site.

Washington is currently the only state that has codified, numeric, chemical criteria for determination of sediment quality. The Washington State Department of Ecology (Ecology) promulgated the Sediment Management Standards (SMS) in 1991 as Chapter 173-204 of the Washington Administrative Code (WAC) (Ecology, 1995). The purpose of the SMS is to reduce and ultimately eliminate adverse effects on biological resources and any significant human health risk from surface sediments in marine, estuarine, and freshwater environments. Currently, numeric chemical criteria have been developed only for marine sediments. Although the Norfolk CSO site is truly an estuarine environment, the marine sediment criteria were applied to assess the levels of contamination and the requisite cleanup.

Three sets of standards were developed under the Washington State SMS: sediment quality standards; sediment cleanup standards; and source control standards. The "no adverse effects level" is the designated Sediment Quality Standard (SQS), the most protective standard for all marine sediments in the state. The SQS corresponds to a sediment quality that will result in no adverse biological effects in the environment (acute or chronic effects to biological resources) and no significant human health risk. To date, no chemical criteria values have been adopted for human health protection, however, 47 chemicals have numeric criteria set to protect against biological harm to the marine environment.

Most of the SQS chemical criteria values are derived from the Apparent Effects Threshold (AET) method (EPA, 1988). This method correlates measured chemistry values with associated biological effects data to arrive at empirically-derived chemical concentrations that predict when adverse biological effects should occur. Chemical concentrations below the SQS values are predicted to have "no adverse effect." The SMS criteria for marine sediment were developed primarily to protect benthic invertebrates with the assumption that such criteria would also protect demersal fish.

The Cleanup Screening Level (CSL) criteria establish minor adverse effects levels, above which station clusters of potential concern are defined as cleanup sites, and also establish minimum cleanup levels (MCULs) to be used in evaluation of cleanup alternatives (Ecology, 1995).

Analytical data from the site characterization study indicated the presence of four chemicals of concern in surficial sediments at multiple stations. Concentrations of mercury at two stations, 1,4-dichlorobenzene at three stations, bis(2-ethylhexyl) phthalate at three stations, and PCBs at four stations all exceeded the CSL. These four chemicals of concern were also detected in multiple core samples at concentrations exceeding the CSL. Coring results indicated sediment contamination to a depth of 60 cm. The highest PCB concentration detected during site characterization was 81.4 milligrams per kilogram dry weight (mg/Kg DW), which exceeds the United States Toxic Substances Control Act maximum level of 50 mg/Kg for hazardous waste disposal.

Results of the site characterization study were combined with evaluations of planned CSO-reduction measures, watershed-wide source control efforts, and recontamination modeling to perform an alternatives evaluation for sediment remediation at the site. Following completion of the alternatives evaluation and interagency discussions with members of the EB/DRP panel, the final footprint of the remediation was set at 3,000 square meters to a depth of 0.6 meter, representing approximately 1,800 cubic meters of sediment. The chosen method of remediation was mechanical dredging and backfilling to grade with clean sand. To obtain complete removal at depth, the cleanup plan called for over-dredging by a depth of 0.3 meter resulting in a total dredge depth of 0.9 meter. Upland disposal was selected but dredged material had to be dewatered before it was transported to the disposal site. It was conservatively estimated that 230 cubic meters of dredged sediment would have to be transported to a hazardous waste landfill for disposal.

SITE REMEDIATION

Remediation of the Norfolk CSO site began in early February 1999 and was completed by late March 1999. Dredging activities employed a mechanical clamshell dredge consisting of a 127-tonne crawler crane equipped with a 3-cubic meter bucket mounted on a 15 by 50-meter barge (EB/DRP, 1999). The dredge started excavation at the location of the highest PCB sediment concentration and expanded out to the dredge boundary at a rate of one barge-load (228 cubic meters) per day. Initial dewatering of the dredged sediment occurred on the barge in six segregated 38-cubic meter compartments. Sediment was then removed to an onshore containment area first arranged to segregate batches about 38 cubic meters in size and later reconfigured to handle larger batches. A total of 590 tonnes of fly ash and saw dust were added to the sediment in the onshore containment areas to absorb excess water, complying with landfill regulations (EB/DRP, 1999).

Composited sediment samples were collected daily from each 38-cubic meter compartment on the barge and analyzed for PCBs within 24 hours to determine the appropriate disposal destination. These samples were extracted according to the United States Environmental Protection Agency (EPA) Method

3550B (ultrasonic extraction). The 24-hour turn around time precluded the use of gel-permeation chromatography as an extract cleanup step, however, preliminary analysis of these sediments indicated that data quality was not adversely affected by this omission. A sulfuric acid cleanup (EPA Method 3665A) was used to eliminate all potential organic interferences and tetrabutylammonium sulfite was used to remove high levels of sulfur from the extracts (EPA Method 3660B).

Initially, 30 grams (g) of sample were extracted and concentrated to a final volume of 10 milliliters (ml). High PCB concentrations in some early samples necessitated a procedural change to an extraction weight of 5 g with the same final volume of 10 ml. Extracts were analyzed on an HP6890 gas chromatograph with dual electron capture detectors (GC/ECD) according to EPA Method 8082. A method blank, spiked blank, and lab duplicate were analyzed with each daily sample batch and a standard reference material for Aroclor® 1254 (National Research Council of Canada #HS-2) was analyzed once every 20 samples.

Sample chromatograms yielded a complex mixture of several weathered Aroclors. The mixture was identified as Aroclors 1016, 1248, 1254, and 1260. Five-point calibration curves were prepared for Aroclors 1016 and 1260 using average response factors. One-point calibrations were used to quantitate Aroclors 1248 and 1254. At least three peaks, unique to each Aroclor, were used to quantitate total PCB concentrations. The largest possible total PCB concentration was reported for each sample as a conservative approach in directing proper disposal of dredged sediment. Sediments with total PCB concentrations greater than 45 mg/Kg DW were transported to a hazardous waste landfill in Arlington, Oregon for disposal. Sediments with total PCB concentrations less than 45 mg/Kg DW were transported to an upland landfill in Bremerton, Washington for disposal. The highest PCB concentration detected during daily testing of dredged sediments was 1,830 mg/Kg DW.

After dredging had removed sediment to a depth of about 0.9 meter over the entire site, composite samples were collected from three locations to verify that chemical concentrations in remaining sediments were below SQS criteria. Analytical results indicated, however, that PCB concentrations were increasing with depth at some of the stations. Two more rounds of dredging were required at two PCB "hot-spot" areas, each round removing an additional 0.9 meter of sediment. A resulting total of approximately 4,000 cubic meters of sediment was ultimately removed during the remedial excavation, of which approximately 1,450 cubic meters were transported to the hazardous waste landfill. Sediment was generally removed to a depth of 0.9 meter over the entire site, however, remediation in the PCB hot-spot areas resulted in removal of sediment up to a depth of 2.7 meters.

The final round of confirmational testing, subsequent to all dredging activities, indicated that, in some of the deepest-dredged areas, sediments were left in place that contained PCB concentrations much greater than the CSL chemical criterion of 1,000 micrograms per kilogram dry weight (μg/Kg DW). Discussions with project oversight personnel from the EB/DRP panel and Ecology determined that, at a depth of 2.7 meters below original grade, these

PCB-contaminated sediments could be left in place and the backfill material would then constitute a "thick capping layer."

Clean backfill material was obtained by the dredging contractor from the Duwamish River navigational turning basin during routine maintenance dredging operations by the United States Army Corps of Engineers. A sample of this material was collected just prior to backfilling activities. Analytical results from this sample confirmed that the material was primarily medium-grained sand and chemically suitable for use as backfill (EB/DRP, 1999). No organic chemicals were detected in the sample and heavy metals were detected at levels indicative of natural, area-wide crustal sediment concentrations (Dexter et al., 1981). Approximately 5,100 cubic meters of turning basin sand were used to backfill the dredged area at the Norfolk CSO sediment remediation site.

POST-REMEDIATION MONITORING

The site hydraulic permit, issued by the Washington State Department of Fish and Wildlife, requires that the Norfolk CSO site be monitored for a period of five years following remediation activities to evaluate possible recontamination of the backfill sediment as a result of continuing CSO or storm water discharges.

Four sampling stations were initially selected to monitor the backfill material for potential recontamination. Two stations were positioned at water depths of 0 and –0.6 meter referenced to mean lower low water (MLLW) along the alignment of the original channel in the mudflat caused by discharges from the Norfolk CSO. A third station was positioned at a water depth of –0.6 meter MLLW along the alignment of the original channel in the mudflat caused by discharges from a nearby storm drain. The fourth station was established as a reference, upstream of the others, at a water depth of –0.6 meter MLLW.

The first monitoring event was performed in April 1999, within a month of remediation activities. This event was used to establish a baseline chemical characterization of the backfill material. Grab samples were collected from the top ten centimeters (cm) of sediment for chemical analysis. As expected, analytical results from the baseline monitoring event were similar to results from analysis of the backfill material prior to placement at the site. The baseline monitoring event verified low concentrations of metals and the presence very few organic compounds.

A topographical survey of the site, conducted during an extreme low tide in June 1999, showed that the channels across the backfill caused by the CSO and storm drain discharges had both changed course compared to conditions prior to remediation. The CSO and storm drain channels had merged and formed a small delta at an approximate depth of –0.9 meter MLLW. Three of the four sampling stations were subsequently relocated to accurately monitor impacts of CSO and storm drain discharges. One sampling station was placed in each of the new channels created by the CSO and storm drain flows across the backfill. The third sampling station was relocated to the delta, "downstream" of the confluence of the CSO and storm drain channels. The upstream reference sampling station remained in the same location.

Subsequent monitoring events occurred in October 1999, April 2000, and April 2001. Each of these monitoring events included the collection of two composite sediment grab samples from each station. During each monitoring event, one sample was collected from the 0 to 2 cm depth stratum to evaluate chemical conditions in recently-deposited sediment and the other sample was collected from the 0 to 10 cm depth stratum to evaluate sediment chemical conditions over the entire biologically-active zone. These samples were routinely analyzed for semivolatile organic compounds, PCBs, metals, grain-size distribution, and total organic carbon.

Based on analytical results over two years, chemical conditions of the backfill material over both the 0 to 2 cm and 0 to 10 cm depth strata can generally be characterized by:

- gradually increasing levels of organic carbon at all stations, with a greater rate of increase evident at the CSO channel and upstream reference stations;
- an associated increase in fine material at the CSO channel and upstream reference stations, with little or no change in grain size distribution at the storm drain channel and combined delta stations;
- little or no change in metal concentrations at all four stations;
- the presence of trace amounts of polynuclear aromatic hydrocarbon (PAH) compounds at all four stations, varying both spatially and temporally, with no indication of a trend toward increasing concentrations;
- the presence of trace amounts of two phthalate compounds (butyl benzyl phthalate and bis(2-ethylhexyl) phthalate) at three of four stations;
- the presence of trace amounts of PCBs at the reference station and the station in the Norfolk CSO channel (see table below); and
- the presence of PCBs at concentrations exceeding the SQS chemical criterion (130 μg/Kg DW) at the two stations downstream of the storm drain outfall, with the highest reported PCB concentrations (exceeding the CSL chemical criterion of 1,000 μg/Kg DW) located at the station closest to the storm drain outfall (see table below).

April 2001 Monitoring Event
Sediment PCB Concentrations (μg/Kg DW)

Sampling Station	0 to 2 cm	0 to 10 cm
Upriver Reference	42.2	30.5
CSO Channel	60.9	35.5
Storm Drain Channel	1,880	1,330
Combined Channel Delta	161	94.1

The concentration gradient indicates that PCBs detected on the backfill material are coming from the direction of the storm drain outfall. A similar pattern of PCB distribution was first observed in samples collected during the October 1999 monitoring event, six months after placement of the backfill material (EB/DRP, 2000). PCB concentrations detected at the station closest to the storm drain outfall during this event, however, were ten times less than in April 2001. This PCB-contaminated sediment, which was unable to be removed during the remediation, could be transported onto the backfill material by discharges from the storm drain. It appears that the source of PCB recontamination on the backfill material is limited to the storm drain channel. King County and the EB/DRP panel are attempting to identify a way to eliminate the contaminated sediment inshore of the Norfolk CSO site as soon as possible to avoid further recontamination of the backfill material.

REFERENCES

Dexter et al., 1981. *Comparison Between Average Crustal Metal Concentrations and Average Metal Concentrations in Puget Sound Regional Soils.* National Oceanic and Atmospheric Administration Technical Memorandum OMPA-13 (Table VIII-1). Seattle, Washington.

EB/DRP, 1996. *Norfolk CSO Sediment Cleanup Study.* Prepared for the Elliott Bay/Duwamish Restoration Program Panel by the King County Water Pollution Control Division. Seattle, Washington.

EB/DRP, 1999. *Norfolk CSO Sediment Remediation Project Closure Report.* Prepared for the Elliott Bay/Duwamish Restoration Program Panel by The EcoChem Team. Seattle, Washington.

EB/DRP, 2000. *Norfolk CSO Sediment Remediation Project, Five-Year Monitoring Program, Six-month Post-construction Monitoring Report, October 1999.* Prepared for the Elliott Bay/Duwamish Restoration Program Panel by The King County Water and Land Resources Division. Seattle, Washington.

Ecology, 1995. *Sediment Management Standards, Chapter 173-204 WAC.* Washington State Department of Ecology. Olympia, Washington.

EPA, 1988. *Sediment Quality Values Refinement:1988 Update and Evaluation of Puget Sound AET.* Prepared for the U.S. Environmental Protection Agency Region 10 by PTI Environmental Services. Seattle, Washington.

King County, 1995. *Combined Sewer Overflow Control Plan 1995 Update.* Prepared for King County Department of Metropolitan Services by Brown and Caldwell, KCM, and Associated Firms. Seattle, Washington.

NATURAL RECOVERY AT A SUBMARINE WOOD WASTE SITE

Joseph D. Germano (Germano & Associates, Bellevue, WA USA)
Lorraine B. Read (EVS Environment Consultants, Seattle, WA USA)

ABSTRACT: Sawmill Cove (Sitka, Alaska) was the receiving point for effluent and storm water discharges from the Alaska Pulp Mill from 1959 to 1993. Operations at the mill resulted in the accumulation of wood solids in some areas up to 20 ft or more in thickness on approximately 100 acres of the seafloor adjacent to the site. The Record of Decision resulting from the RI/FS called for natural recovery with long-term monitoring, with the objective of achieving an equilibrium benthic community by the year 2040. The baseline monitoring survey was carried out in the spring and fall of 2000 using sediment profile imaging (SPI), epifaunal video surveys, benthic community analyses, and sediment chemical analyses. While the AOC is still showing impacts due to organic enrichment (the widespread occurrence of sulfur-reducing bacterial mats, high SOD, high TOC values, presence of methane at depth), there is absolutely no doubt that natural recovery is occurring. Sixteen percent of the AOC has fully recovered and achieved the final management milestone. Based on the positive results from the baseline monitoring, a cost-effective strategy was recommended for future sampling that would save more than 50% of the projected costs of the long-term monitoring program.

INTRODUCTION

For 34 years, Alaska Pulp Corporation operated a pulp mill in Silver Bay outside of Sitka, Alaska. A remedial investigation and ecological risk assessment of the Bay Operable Unit, encompassing Sawmill Cove in Silver Bay, were conducted during 1996 and 1997. The results of these studies were used to delineate an Area of Concern (AOC) of approximately 100 acres in Sawmill Cove. The Remedial Action Objective (RAO) for the AOC in Sawmill Cove, as defined in the Record of Decision (ROD), was to reduce to an acceptable level ecologically significant adverse effects to populations of bottom-dwelling life from hazardous substances, including wood waste degradation chemicals (DEC, 1999). The acceptable level is defined as an observable succession of benthic species that will result in balanced, stable communities as assessed by measures of abundance and diversity at various locations over time. The Alaska Department of Environmental Conservation (DEC) determined that the RAO would best be obtained by natural recovery with long-term monitoring. The ultimate goal is to have 75 percent of the AOC in an equilibrium community by the year 2040. The ecological recovery management milestones are presented in Table 1 (DEC, 1999).

Table 1. AOC recovery milestones

MILESTONE	AREA	TIME (years)	SUCCESSIONAL STATUS
1	>75 % coverage of the AOC	5–10	Decomposers and primary producers
2	>75 % coverage of the AOC	10–20	Primary consumers and detritivores
3	>75 % coverage of the AOC	20–40	Secondary consumers
4	>75 % coverage of the AOC	> 40	Climax (equilibrium) community

Objectives. The Silver Bay Monitoring Program was designed to measure the degree of natural recovery toward the natural resource management milestones outlined in the ROD. The objectives of the monitoring program outlined in the original work plan (Foster Wheeler, 1999) are as follows:

- To compare benthic community abundance and diversity over time to document progress towards natural recovery
- To map natural recovery in the AOC based on changes in benthic community structure over time
- To avoid unnecessary sampling by identifying recovered areas as soon as possible
- To correlate benthic community structure with sediment chemistry parameters to explain spatial or numeric trends in benthic community parameters that deviate from the benthic succession model and the milestones established for recovery

Approach. The intent of the Silver Bay Monitoring Program was to measure progress towards the monitoring objectives using Sediment Profile Imaging (SPI), epifaunal video surveys, benthic community analyses, and sediment chemical analyses in a tiered fashion. The work plan for the monitoring program was prepared by Foster Wheeler (1999) and implemented by EVS Environment Consultants. According to this monitoring plan, the baseline tasks would consist of the following:

- Performing an initial review and assessment of the original remedial investigation (RI) data to determine whether the outlined baseline sampling design should be executed as initially designed
- Conducting Tier 1 sampling using SPI technology and epifaunal video surveys; data would be used to select station locations for Tier 2 sampling
- Conducting Tier 2 sampling for benthic community analysis and associated sediment chemistry
- Evaluating data and characterizing the AOC

The monitoring program recommended that sampling of areas within the AOC be stratified based on types of benthic communities and the anticipated speed of recovery. This approach would allow for the potential reduction of sampling effort in future years by eliminating areas that had been designated as recovered within various strata, thereby allowing efforts to be focused on those remaining areas that are still in the process of recovering.

MATERIALS AND METHODS

The underwater video and sediment profile imaging surveys were conducted between April 18-23, 2000. Underwater video images were obtained using an underwater color camera (SeaCam 2000, DeepSea Power and Light, San Diego, CA) mounted in a "down-looking" orientation on a heavy towfish. Two lasers (DeepSea Power and Light, San Diego, CA) mounted 10 cm apart were attached to the camera (to provide a reference scale for sizing objects), and a 250 watt underwater light provided illumination. The towfish was deployed directly off the stern of the vessel using the cargo boom and boom winch. The heavy weight of the towfish kept the camera positioned beneath the DGPS antenna, thus ensuring that the position data accurately reflected the geographic location of the camera.

A laptop computer equipped with a video overlay controller and data logger software (Discovery Bay Software, Port Townsend, WA) integrated DGPS data (date, time, latitude, longitude) and the video signal. Date, time and position data were stored on a floppy disk at one second intervals. DGPS data (updated every second) and transect identification numbers were stored directly onto the videotape using a four head video cassette recorder (VCR). The video signal also was fed into a digital camcorder (Sony TRS 310) and recorded on Sony Digital-8 tape. A television/VCR (two head) combination unit located in the pilothouse assisted the helmsman control the speed of the towfish over the seabed and was used to record a backup VHS tape. A second television monitor on the work deck assisted the winch operator control the vertical position of the towfish.

Eleven line transects spaced 200 ft apart through the AOC were defined for the video survey. Wind and current conditions coupled with the area's complex topography made it difficult to follow the defined straight-line transects, and most of the tracks meandered back and forth over the intended straight survey lines.

The SPI survey was conducted with a Benthos (North Falmouth, MA) Model 3731 sediment profile camera at a total of 97 stations throughout the AOC located on an orthogonal grid with 200 ft. spacing. SPI was developed almost two decades ago as a rapid reconnaissance tool for characterizing physical, chemical, and biological seafloor processes and has been used in numerous seafloor surveys throughout the United States, Pacific Rim, and Europe (Rhoads and Germano, 1982, 1986). The sediment profile camera works like an inverted periscope (Figure 1). A deep-sea 35mm camera mechanism is mounted horizontally inside a water-tight housing on top of a wedge-shaped prism. The prism has a Plexiglas® faceplate at the front with a mirror placed at a 45° angle at the back. The camera lens looks down at the mirror, which is reflecting the image from the faceplate. The prism has an internal strobe mounted inside at the back of the wedge to provide illumination for the image; this chamber is filled with distilled water, so

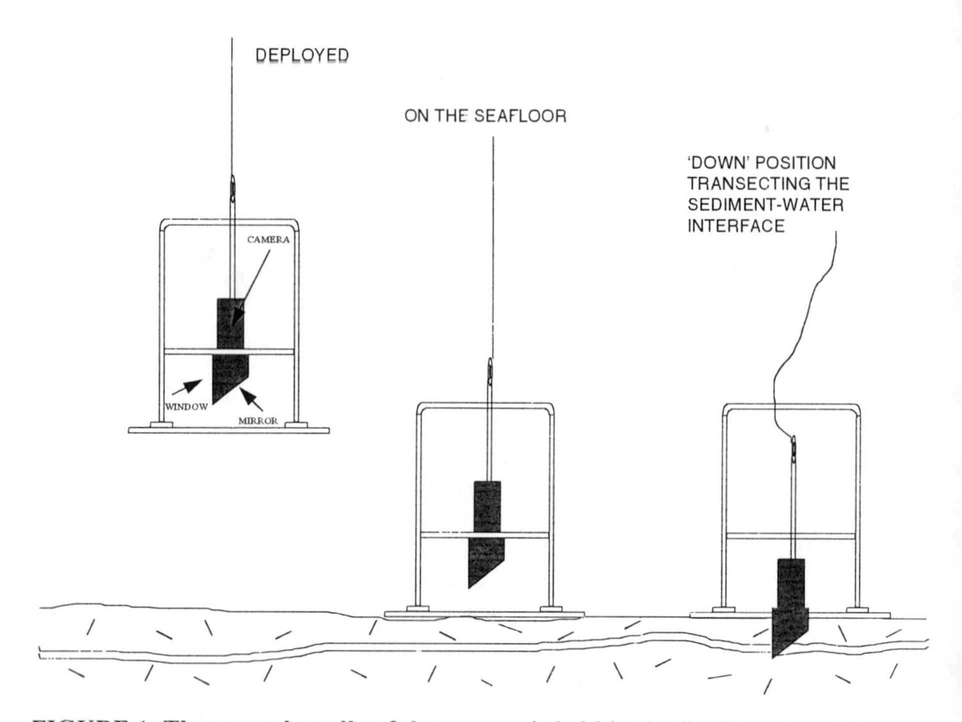

FIGURE 1. The central cradle of the camera is held in the "up" position by tension on the winch wire as it is being lowered to the seafloor (left); once the frame base hits the bottom (center), the prism is then free to penetrate the bottom (right) and take the photograph

the camera always has an optically clear path to shoot through. This wedge assembly is mounted on a moveable carriage within a stainless steel frame. The frame is lowered to the seafloor on a winch wire, and the tension on the wire keeps the prism in its "up" position. When the frame comes to rest on the seafloor, the winch wire goes slack (see illustration) and the camera prism descends into the sediment at a slow, controlled rate by the dampening action of a hydraulic piston so as not to disturb the sediment-water interface. On the way down, it trips a trigger that activates a time-delay circuit to allow the camera to penetrate the seafloor before any image is taken. The knife-sharp edge of the prism transects the sediment, and the prism penetrates the bottom. The strobe is discharged twice with each lowering to obtain two cross-sectional images of the upper 20 cm of the sediment column. After the two replicate images are obtained at the first location, the camera is then raised up about 2 to 3 meters off the bottom to allow the strobe to recharge. The strobe recharges within 5 seconds, and the camera is ready to be lowered again for another two images. Surveys can be accomplished rapidly by "pogo-sticking" the camera across an area of seafloor while recording positional fixes on the surface vessel. The resulting images give the viewer the same perspective as looking through the side of an aquarium half-filled with sediment.

Sediment samples were collected with a van Veen grab (0.1 m^2), sieved to 1mm, and all taxa identified to the family level; wet-weight biomass was recorded for each major taxonomic group from each station.

RESULTS AND DISCUSSION

Video Analysis. A total of 20,616 time/position records were examined during the video analysis, of which 15,933 (77 percent) were deemed to have acceptable images. The unacceptable records did not yield useful information due to excessive turbidity, debris covering the lens, the camera not yet reaching the bottom on initial deployment, the camera off-bottom during a transect, or the camera dragging in the sediment. The visibility on the tapes varied considerably from clear views to extremely turbid footage. This survey took place soon after the spring plankton bloom in Silver Bay, and the dead phytoplankton cells had settled to the bottom, producing a floccular detrital layer over a large portion of the site. This organic-rich layer of decomposing plant material was visible in the video survey either as a near-bottom nepheloid layer that obscured details on the sediment surface or as marine “snow” in the water column that obstructed the camera’s view. The acceptable records represent a total area of 5,338 m^2 surveyed. Approximately 24% of the bottom surveyed was covered by sulfur-reducing bacterial mats (*Beggiatoa* spp.), and 23% of the frames showed evidence of the site being utilized as a resource for large macrofauna (fish, shrimp, anemones, and crabs). Many of the occurrences of macrofauna were coincident with the presence of the bacterial mats, indicating that dissolved oxygen was not a limiting factor.

SPI Analysis. The sediment grain-size major mode was fairly uniform throughout the entire site. With the exception of four stations that had large rocks and cobble on the sediment surface preventing prism penetration, the remaining stations within the AOC were all primarily fine-grained sediments ($\geq 4\ \Phi$) consisting mainly of decomposing wood particles. Wood fibers and fragments of various sizes were evident in the cross-sectional profile at most of the stations. The wood fragments ranged in size from distinct chips/chunks 2–4 cm in diameter to fine sand- or silt-sized particles. Evidence of wood fibers/waste was present in all but five of the stations throughout the site. Not unexpectedly, sediment shear strength was relatively low in most areas, with the camera prism over-penetrating the bottom at about 20 percent of the stations sampled. With this large volume of both labile and refractory organic matter at the sediment surface, the SOD is relatively high compared with other nearshore embayments.

Evidence of organic loading contributing to the high SOD is prevalent throughout the site and documented in the profile images in a number of different parameters: the absence of an apparent Redox Potential Discontinuity (RPD) boundary, low reflectance values of dark sediments, the presence of methane gas bubbles, and a seasonal input pulse of labile organic matter from dead phytoplankton (detectable even at relatively thin layers in the sediment profile images). Well-developed redox boundaries were encountered in the eastern and southern areas of the site.

Not surprisingly, the distribution of infaunal successional assemblages followed a pattern similar to that described for the mean apparent RPD. Recolonization and recovery has definitely started within the site at more than half the stations sampled. As one moves to the south and east within the AOC, Stage I (*sensu* Rhoads and Germano, 1982) polychaete assemblages appear at 41 percent of the stations. Seven percent of the stations exhibited evidence of the start of a transitional (Stage II) deposit-feeding community developing, and 9 percent primarily at the southern end of the AOC had well-developed Stage III assemblages present. In addition to visual evidence of macrofaunal recolonization, the sediment profile images also revealed evidence of microfaunal colonization in the form of anoxic, sulfur-reducing bacteria (*Beggiatoa* or other similar species). The diagnostic white, string-like colonies of anoxic bacteria were readily visible in SPI photographs, even in low densities. Sulfur-reducing bacterial colonies were found at almost half (42 out of 91) of the locations sampled.

The overall pattern of habitat "stress" due primarily to excess organic loading from the profile images showed those stations closest to the coastline experiencing the most severe stress; as with RPD values, conditions improve as one moves to the southern and eastern edge of the AOC, away from the heaviest accumulation of wood waste.

Strata for benthic grab sampling were determined by conducting statistical cluster analysis with the Tier 1 SPI data. Because the Tier 1 data were a combination of continuous data variables and ordinal or categorical variables, a dissimilarity metric that allows for different types of variables was applied. For this method, the dissimilarity value differs depending on the type of variable, but all variables are scaled to the same interval and hence have equal weight. The multidimensional distance between two points is the sum of the dissimilarity for each variable, scaled to the interval [0,1]. The clustering algorithm known as *partitioning around medoids* (Kaufman and Rousseeuw, 1990) was used for the statistical analysis. Partitioning cluster methods such as this one divide the stations into a pre-specified number of groups based on minimizing the dissimilarities within each group. The cluster analysis identified three clusters with spatially contiguous stations.

The three identified strata had clear separation in terms of depth and successional stage, with the exception of one area in the southwest boundary area of the AOC. This area was on a very steep incline and showed elements of two strata without fitting neatly into either one. The stations in this area were used to define a fourth stratum, so that conditions in this area could be documented separately.

Benthic Community Analyses. At the 37 stations, a total of 3,543 individuals were found, belonging to 62 families in 3 phyla. Of the three phyla, annelids were the most abundant group present, representing 75 percent of the total abundance and 89 percent of the total biomass. Molluscs represented 23 percent of the total abundance and only 10 percent of the total biomass. Arthropods were the smallest group, representing 2 percent of the total abundance and less than 1 percent of the total biomass. A more thorough analysis of community composition among stations was done using cluster analysis (Bray-Curtis metric of family abundance). A total of six sets of stations were identified from the cluster analysis at a distance of 0.67. A trophic functional analysis of the cluster groups integrated with the SPI results showed that the geographic regions identified as "stressed" and "recovered" coincided with one another.

CONCLUSIONS

The results from the video survey, profile images, and grab samples allowed the AOC to be divided into three final strata coinciding with the progress toward benthic recovery. While the AOC is still showing impacts due to organic enrichment (the widespread occurrence of sulfur-reducing bacterial mats, high SOD, high TOC values, presence of methane at depth), there is absolutely no doubt that natural recovery is occurring. The first two RAO recovery milestones (Table 1) have already been achieved: 81 percent of the site is covered with decomposers (sulfur-reducing bacterial colonies, most likely *Beggiatoa* spp.) (**Milestone 1**), and primary consumers (Stage I polychaetes) were present at all but 2 grab stations and in sufficient densities at 33 (89 percent) of the grab stations sampled (**Milestone 2**). Approximately 16 percent of the AOC (Stratum 3) has fully recovered and achieved the final management milestone, and 22 percent of the AOC (Stratum 2) is in transition to the final recovery stage with notable abundances of deposit-feeding taxa. Sixty-two percent of the AOC (Stratum 1) is still considered seriously impaired in regard to benthic community status. The excellent agreement between the SPI and grab samples will allow future monitoring to rely primarily on the more cost-effective SPI technology, eliminating the need for the more labor-intensive grab sampling and faunal identification.

The important conclusion for the state regulatory agencies in charge of oversight of this area was to recognize that the successional recovery of a soft-bottom infaunal community is not a linear progression that cannot be reversed. Once any benthic community gets established in an area after a disturbance, there are deviations from the idealized representation of recovery, and the never-ending ecosystem "dance" of disturbance and recovery is a part of the natural landscape, due to the frequency of natural disturbances on a wide variety of spatial and temporal scales (Sousa, 1984; Hall, 1994). If one monitors any area of seafloor, the results will show a patchwork-quilt pattern of bottom in different stages of successional recovery (Rhoads and Germano, 1982). In fact, it is quite possible that there will never be a point when monitoring coincides with the AOC having more than 75~percent of its area ***simultaneously*** covered by a Stage III infaunal assemblage (the final recovery milestone) because of localized or small scale disturbances (physical or biological) that will be constantly occurring. The key realization for the regulatory community from these results was a re-definition of what "full recovery" means for this site. Any evaluation of future progress of the

AOC toward achieving the RAO management milestones needs to consider that "full recovery" in an area is really the **capacity** of a mapped region to support a Stage III assemblage; one should not assume that once a Stage III community is established, it is there for eternity. For example, it is quite likely that if the stations in Stratum 3 (the fully recovered area) were sampled in the following year, some of them might show a Stage I or Stage II assemblage. This does not mean that they can no longer be considered "recovered" because just as reaching the endpoint of the continuum is not an immovable constant, so a temporary retrograde does not represent a permanent condition. Once the SOD is not at a level where only hypoxic-tolerant species can survive (*sensu* Diaz and Rosenberg, 1995) and the area has been documented to have had the sediment reworked to a sufficient depth by Stage III taxa at one point in time, then that area will be designated as recovered, having achieved the final RAO management milestone.

REFERENCES

DEC. 1999. *Record of Decision, Mill and Bay Operable Units, Alaska Pulp Corporation Mill Site, Sitka, Alaska*. Alaska Department of Environmental Conservation, Juneau, AK.

Diaz, R. J. and R. R. Rosenberg. 1995. "Marine Benthic Hypoxia: A Review of Its Ecological Effects and the Behavioural Responses of Benthic Macrofauna." *Oceanography and Marine Biology: an Annual Review* 33: 245-303.

Foster Wheeler. 1999. *Monitoring Program, Long-term Benthic Monitoring Program and Bioaccumulation Survey, Sawmill Cove*. Alaska Department of Environmental Conservation and the City and Borough of Sitka. Foster Wheeler Environmental Corporation, Bellevue, WA.

Hall, S. J. 1994. "Physical Disturbance and Marine Benthic Communities: Life in Unconsolidated Sediments." *Oceanography and Marine Biology: an Annual Review* 32: 179-239.

Kaufman, L., and P.J. Rousseeuw. 1990. *Finding Groups in Data: An Introduction to Cluster Analysis*. Wiley, New York, NY.

Rhoads, D.C., and J.D. Germano. 1982. "Characterization of Benthic Processes Using Sediment Profile Imaging: An Efficient Method of Remote Ecological Monitoring of the Seafloor (REMOTS™ System)." *Marine Ecology Progress Series* 8:115-128.

Rhoads, D.C., and J.D. Germano. 1986. "Interpreting Long-term Changes in Benthic Community Structure: A New Protocol." *Hydrobiologia* 142:291-308.

Sousa, W.P. 1984. "The Role of Disturbance in Natural Communities." *Annual Review of Ecology and Systematics* 15: 353-391.

FIELD MEASUREMENTS OF PCB VOLATILIZATION FROM CONTAMINATED LAKE SEDIMENTS AND SURFACE WATERS

Eric Foote, Victor Magar, James Abbott, Carol Peven-McCarthy
(Battelle Memorial Institute)
Makram Suidan, Shuang Qi (University of Cincinnati)
Paul de Percin, Craig Zeller (U.S. Environmental Protection Agency)

Between 1955 and 1978, capacitors were manufactured at the Sangamo-Weston plant, South Carolina. PCB releases into the environment resulted in sediment contamination at the Sangamo-Weston/Twelvemile Creek/Lake Hartwell Superfund Site. Sangamo-Weston reported using Aroclors 1016, 1242, and 1254.

This study consisted of a field investigation that was conducted at Lake Hartwell to determine measurable levels of volatilized PCBs occurring from PCB-contaminated shoreline sediments and from the lake-water surface. Flux chambers were designed, constructed, and deployed at locations corresponding with the United States Army Corps of Engineers (USACE) transect locations T-12, T-L, and T-O to trap volatilized PCBs over sediments and the lake surface. Sediment flux sampling was conducted for periods of 24- and 72 hours over surface- and deeper more contaminated sediments. Lake surface flux sampling was conducted over a 72-hour period. Volatilized PCBs were trapped onto pre-cleaned, certified, polyurethane foam cartridges (PUFs). Sediment and water samples were also collected at the sediment flux sampling locations. Other operational measurements included airflow rate, ambient and internal flux sampler temperatures, and barometric pressure. The PUF, water, and sediment samples were analyzed by GC/MS using EPA SW-846 8270M, from which 51 PCBs congeners were identified and resolved.

Total PCB data was used to calculate net fluxes for the various sampling locations. PCB flux rates for surface sediments were 0.6 to 1 $\mu g/m^2$-day; deeper sediments (after removing the upper 15 cm of sediment) were 6 to 9 $\mu g/m^2$-day; and surface waters were 55 to 140 ng/m^2-day. Congener specific distributions for sediment, water, and air samples were analyzed and showed that PCB volatilization favored lower-chlorinated, and hence lower molecular weight, PCBs.

The work presented herein summarizes this research and does not represent policy of the United States Environmental Protection Agency (U.S. EPA).

INTRODUCTION

The site chosen for this investigation was the Sangamo-Weston/Twelvemile Creek/Lake Hartwell Superfund Site (hereafter referred to as Lake Harwell). Lake Hartwell is an artificial lake located in northwest South Carolina. It was created between 1955 and 1963 when the USACE constructed

Hartwell Dam on the Upper Savannah River, 7 miles from its confluence with the Seneca and Tugaloo Rivers. The Sangamo-Weston plant was used for capacitor manufacturing from 1955 to 1978 (U.S. EPA, 1994). The plant used a variety of dielectric fluids in its manufacturing processes, including fluids containing PCBs. Waste disposal practices included land-burial of off-specification capacitors and wastewater treatment sludge on the plant site and at six satellite disposal areas. PCBs were discharged with effluent directly into Town Creek, a tributary of Twelvemile Creek, which is in turn a major tributary of Lake Hartwell. Between 1955 and 1977, the average quantity of PCBs used by Sangamo-Weston ranged from 700,000 to 2,000,000 lb/year. An estimated 3% of the quantities received and used by the plant were discharged into Town Creek, resulting in an estimated cumulative discharge of 400,000 lb of PCBs. Sangamo-Weston reported using Aroclors 1016, 1242, and 1254.

In 1994 the U.S. EPA selected natural recovery for restoring the sediments at Lake Hartwell. The decision to employ natural recovery rather than more intrusive environmental remediation efforts (such as dredging) have ensured limited sediment disturbance. As a result there exists a relatively homogenous distribution of PCBs in the surface sediments, overall providing a unique opportunity in which to study PCB volatilization.

MATERIALS AND METHODS

Two flux chamber designs were developed for measuring PCB volatilization; one was developed for trapping volatilized PCBs over contaminated sediments, and the other for trapping volatilized PCBs on top the lake surface. The flux chambers were deployed at transect locations previously established by the USACE and U.S. EPA Region 4 as shown in Figure 1.

Sediment Flux Chambers incorporated an inverted 208 L drum and battery-operated diaphragm pump for continuous air sampling. The lid of each drum was fastened and the bottom portion of the drum was cut off to create a sharpened edge that was embedded into the sediment, creating a confined air space above the sampling area. The flux chambers were operated to pull entrapped headspace air through two PUF cartridges placed in series. After the air was pulled through the cartridges, it passed through a moisture trap to minimize the accumulation of water in the pump. A flowmeter was located on the backside of the pump to monitor the flowrate. Airflow was returned inside the drum to maintain a closed and circulated system.

PUF cartridges were factory-cleaned and pre-encapsulated in glass tubes (SKC, Inc.). Each PUF cartridge consisted of two PUF sections that "sandwiched" a band of Tenax™ granules. Two PUF cartridges were connected in series to ensure full capture of all PCBs, should breakthrough beyond the first cartridge occur. All ancillary tubing consisted of 0.5 cm inner diameter (I.D.) Teflon™ tubing. Each sampler was equipped with a "K"-type thermocouple that was connected to a Fluke™ meter for monitoring internal temperatures during sampling. Ambient temperatures were measured in parallel using a separate thermocouple.

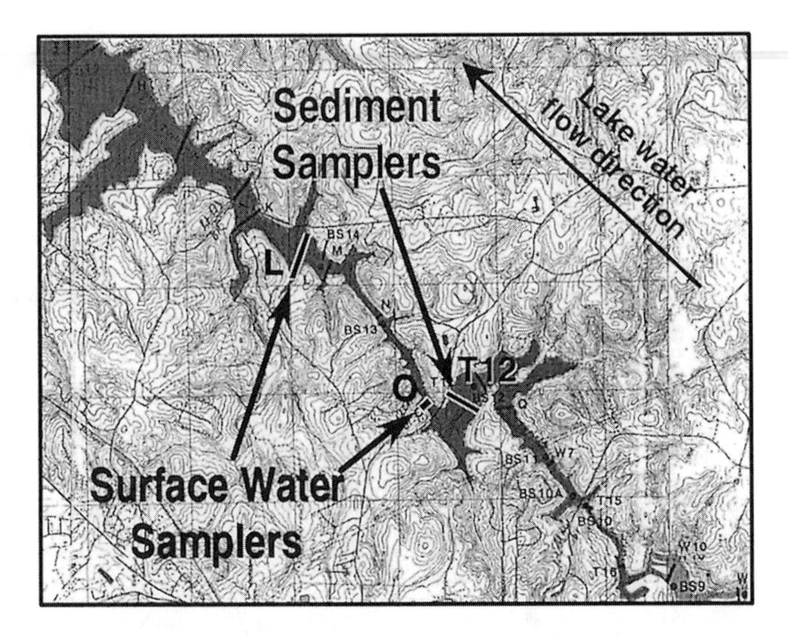

FIGURE 1. Location of Flux Chamber at Sangamo-Weston/Twelvemile Creek/Lake Hartwell Superfund Site

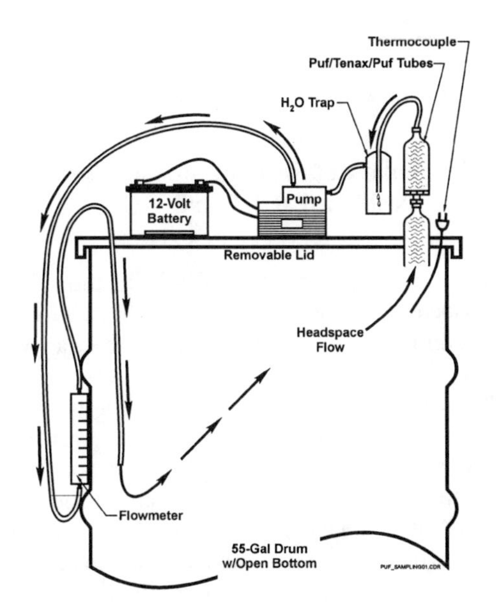

FIGURE 2. Schematic of Sediment Flux Chambers Positioned at T-12

Lake Surface Flux Chambers consisted of an inverted galvanized steel chamber that was equipped with a similar pump and ancillary equipment, as described above. The steel chamber was kept afloat with the use an inflated innertube. The innertube was inflated to maintain an inner diameter of approximately 60 cm, and the steel chamber was fixed inside the inner circle of the innertube in an inverted fashion. The chamber was anchored into position and weighted so that the outer edging of the chamber remained approximately 30 cm below the water surface, to prevent the wave action on the lake from breaking the seal between the entrapped air and the atmospheric air. Figure 3 shows a schematic of the lake surface flux chamber deployed at T-L and T-O.

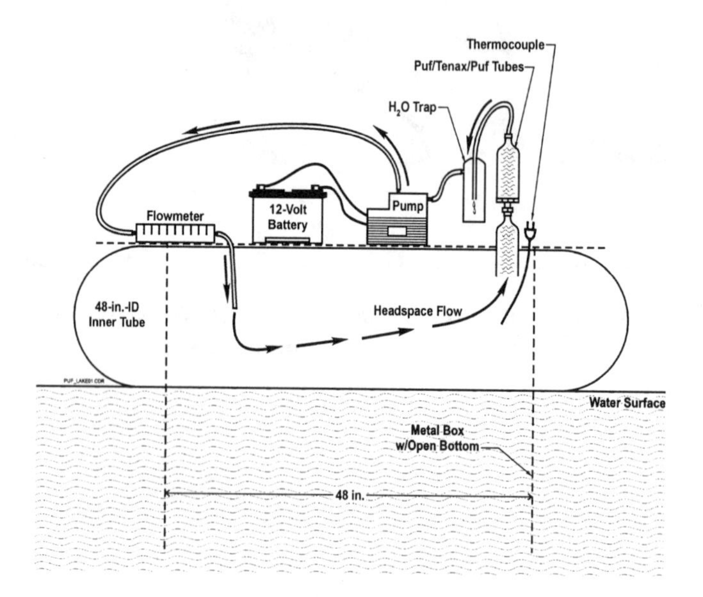

FIGURE 3. Schematic of the Lake Surface Flux Chambers Positioned at T-L and T-O

Table 1 outlines the number and type of volatilization flux tests conducted at Lake Hartwell. To determine the volatility of PCBs above surface shoreline sediments, duplicate sediment flux chambers were placed approximately 60 cm apart, in approximately 15-20 cm of water overlying undisturbed sediments. Two sets of samplers in duplicate trapped volatilized PCBs for 24 and 72-hour periods, respectively. A second sediment volatilization flux test was conducted at T-12, using the same volatilization flux chambers. However, for this second test, the samplers were repositioned after approximately 15 cm of the overlying sediment was removed. Overlying sediment was removed to freshly expose buried contaminated sediments where, based on results of previous investigations, buried sediments are known to contain higher levels of PCBs than the surface sediments (Battelle, 2001). Additional ambient air samples were collected during each of the 24-hour flux tests conducted over contaminated sediments to determine the concentration of PCBs in ambient air. To determine the volatility of PCBs from the lake surface, the lake surface flux chamber was deployed at transects T-O and T-L. Each of the surface water flux chambers operated for 72 hours.

All air-sampling activities used modified-Method TO-10A for acquiring samples on PUF cartridges at a flowrate of approximately 9 to 10 liters per minute (LPM). Water and sediment samples were collected after the completion of each sediment flux sampling. The PUF, water, and sediment samples were analyzed by gas chromatograph/mass spectrophotometer (GC/MS) analysis using EPA SW-846 8270M, from which 51 PCBs congeners were identified and resolved.

TABLE 1. Number and Type of Volatilization Flux Tests Conducted at Lake Hartwell

Condition	Flux Measurements	Test Duration	
1	Surface Sediment (~1 mg/kg)	24-Hour	72-Hour
2	Deeper Sediment (~6 mg/kg)	24-Hour	72-Hour
3	Lake Surface	-------	72-Hour

Operational parameters were monitored approximately 2-3 times per day to ensure that the flux chambers functioned properly throughout the testing period. Flow measurements were made and recorded to ensure the pumps maintained a flowrate of approximately 9 to10 LPM during continuous operation. The energy indicator on each rechargeable battery was monitored. Batteries that indicated a marginal battery life were replaced with a fully charged battery. The internal and external temperature of each chamber was measured and recorded to the determine temperature variance during continued operation. The time and barometric pressure also were recorded.

EXPERIMENTAL RESULTS

Tables 2 and 3 show the results for the surface sediment and deeper sediment flux tests, respectively. The tables show the test durations (hrs), measured sediment and water total PCB (t-PCB) concentrations associated with each flux measurement, PUF PCB measurements for the two PUFs in series and their total PCB mass accumulated (sum of two PUFs in series), and the calculated flux for each test. Flux was calculated as the total measured PCB mass (sum of the two PUFs in series) divided by the flux chamber surface area and the test duration. The data generated from these tests indicates that PCBs volatilized from the shoreline sediments and surface water on the lake.

Two anomalies existed in the data and are difficult to explain. The first anomaly was that t-PCB concentrations in water were much higher than expected, based on previous investigations (Battelle, 2001) where the measured lake water t-PCB concentration was 3.52 ng/L in the vicinity of Core L; t-PCB concentrations in water shown in Tables 2 and 3 were approximately 1,000 times greater than previously measured. The higher aqueous concentrations reported in these tests are attributed to the introduction of sediment fines in the water samples during sampling. PCBs preferentially adsorb to fine particles, so the presence of fines could have increased the apparent t-PCB concentrations in water. Fines likely were introduced from sediment disturbances during setup.

TABLE 2. Results from Surface Sediment Flux Chambers Positioned at T-12

Flux Chamber	Test Duration (hrs)	Total PCBs			Flux ($ng/m^2/day$)
		Sediment (ng/g)	Water (μg/L)	PUF (ng)	
1	24	**761 ± 188**	**0.043 ± 0.018**	280	**1047**
				0.81	
Total				**281**	
2	24	**1114 ± 64**	**1.72***	287	**1078**
				2.38	
Total				**289**	
3	72	**1046 ± 79**	**4.97***	524	**652**
				0.39	
Total				**524**	
4	72	**1098 ± 15**	**7.05***	474	**590**
				0.39	
Total				**474**	

* No duplicate sample
Flux chamber surface area = 0.27 m^2

TABLE 3. Results from Deeper Sediment Flux Chambers Positioned at T-12

Flux Chamber	Test Duration (hrs)	Total PCBs			Flux ($ng/m^2/day$)
		Sediment (ng/g)	Water (μg/L)	PUF (ng)	
1	24	**6105 ± 2060**	**1.05 ± 0.07**	1666	**6490**
				73.5	
Total				**1740**	
2	24	**7888 ± 417**	**1.05 ± 0.10**	5.69	**150**
				34.4	
Total				**40**	
3	72	**4142 ± 304**	**0.52 ± 0.003**	7033	**8763**
				13.5	
Total				**7047**	
4	72	**6208 ± 746**	**1.02 ± 0.09**	152	**190**
				NA	
Total				**152**	

NA = Not analyzed, sample broken in transit
Flux chamber surface area = 0.27 m^2

The second anomaly was that flux rates from Chambers 2 and 4, over deeper sediments, were much lower than expected and did not follow the expected pattern of higher t-PCB mass measured from sediments of much higher t-PCB concentrations. For comparison, the flux rates reported for Chambers 1 and 3 were much higher above deeper sediments, as shown in Table 3.

Duplicate PUF data from above undisturbed sediments showed that there was minimal breakthrough (<1% based on Table 2 and <4.2% based on Table 3) into the second PUF cartridge, indicating that the first PUF cartridge trapped greater than 95% of the PCBs that passed through the system. Data among experimental duplicates for both sediments and PUF samples were similar, resulting in low standard deviations. Table 2 suggests that more volatilization occurred in the first 24 hours of testing as the flux rates above surface sediments were approximately 40% greater than for the 72-hour tests. However, comparison of Chambers #1 and #3 in Table 3 suggests that the fluxes from the 24-hour and 72-hour tests from the deeper sediments were similar. The freshly exposed contaminated sediments resulted in elevated flux levels for chambers #1 and #3, as expected.

The results from the lake surface flux chamber tests are shown in Table 4. Table 4 shows the individual and total PCB mass captured on the two PUF cartridges in series. Neither water nor sediment samples were collected from the locations where the surface water flux chambers were placed, but previous studies have shown the lake water PCB concentrations to be in the low ng/L range. Total flux per location was calculated as described previously. PCB volatilization was observed on the lake surface at both transect locations (T-O and T-L). Flux rates at T-O and T-L were 140 ng/m^2-day and 55 ng/m^2-day, respectively.

TABLE 4. Results from Water Surface Flux Chambers

Flux Chamber #	PUF t-PCBs (ng)	Flux (ng/m^2-day)
1	494	
	N/A	
Total	**494**	140
2	131	
	63.3	
Total	**194**	55

N/A = not available. Sample extract broke during transit to analysis location.
(a) Flux chamber surface area = 1.2 m^2.

For comparison, ambient air samples collected during 24-hour operation during surface- and deeper sediment testing resulted in 11.7 ng and 2.15 ng of t-PCBs, respectively.

PCB Composition by Congener Distribution Figure 4 shows an example of the congener distribution relationship between sediment, water, and air samples collected from flux chamber 1 during the 24-hour surface sediment PCB volatilization test and is representative of the typical distribution observed for each test conducted. PCB congeners were normalized to the t-PCB concentration in each sample, respectively, and are plotted as percent of t-PCBs.

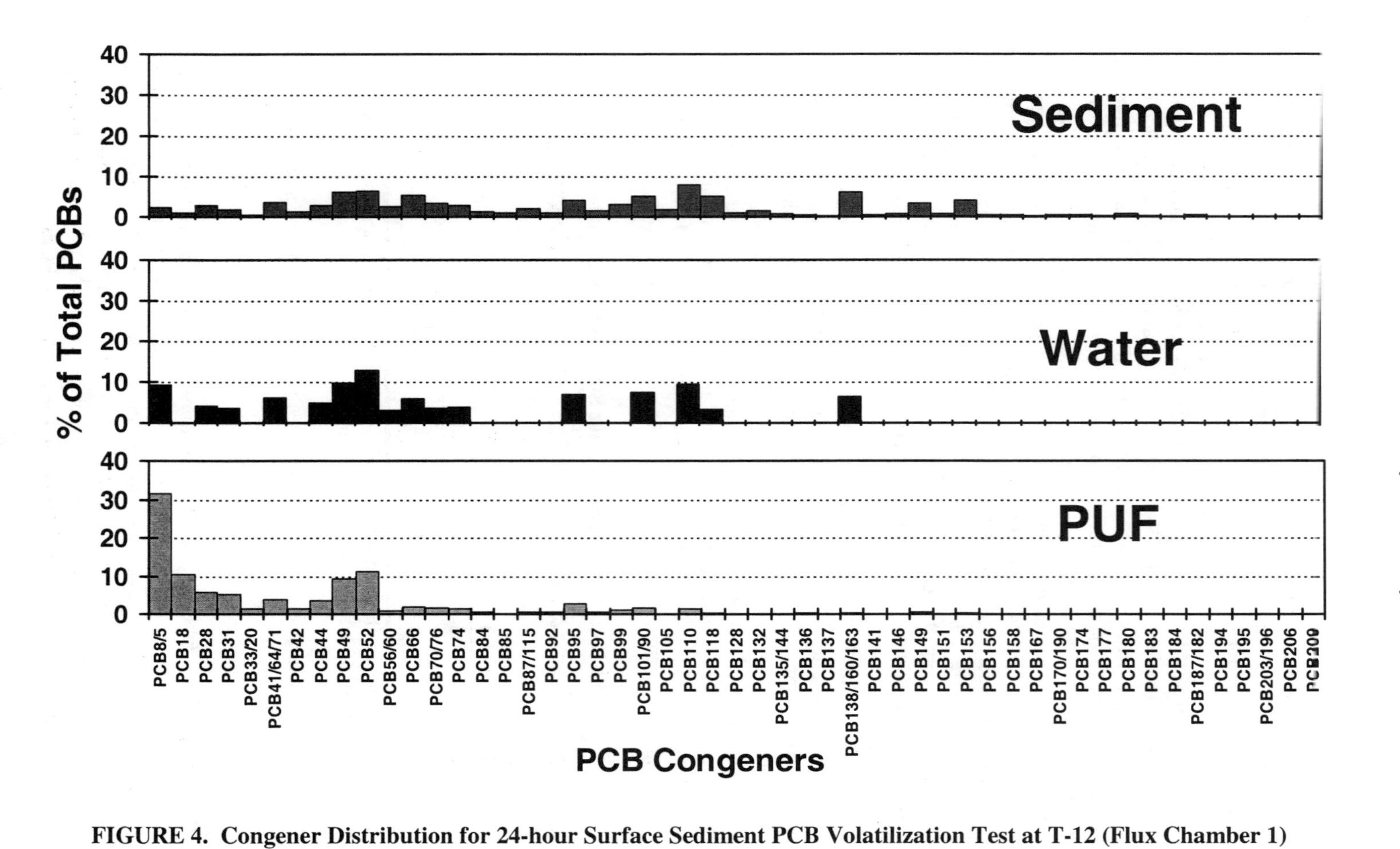

FIGURE 4. Congener Distribution for 24-hour Surface Sediment PCB Volatilization Test at T-12 (Flux Chamber 1)

The figure illustrates that the lower-molecular-weight, less chlorinated PCBs make up the largest fraction of PCBs that volatilized over the test periods. A similar trend was seen in 72-hour tests that were performed over surface sediments, and in the 24- and 72-hour tests conducted over deeper sediments.

DISCUSSION

The results produced from this study indicate that PCB volatilization can be measured using controlled volatilization flux chambers, over short time periods of 24 to 72 hours. There is a paucity of literature information on the volatile loss of PCBs from contaminated sediments and surface waters. Others have investigated the volatility of PCBs in the environment, and conflicting information on the rate of volatility from solid matrices exist. Variability in experimental data can be a result of the natural complexities of each unique site. These complexities are governed by the types of PCBs released into the environment, the capacity of sediments to adsorb PCBs, and chemical properties of individual PCB congeners (including diffusivity, solubility, partitioning coefficients, and most importantly their volatility, expressed as Henry's law coefficients)..

This investigation utilized equipment that was designed and constructed to aggressively, sample PCB volatilization, because the volatility of PCBs was unknown and it was anticipated that volatilized PCBs would be difficult to measure. Each sampler was constructed with a closed system that recirculated the sampled air through dual PUF traps in series and back into the chamber headspace, above the contaminated sediments or surface water.

In spite of this perceived success, future work is recommended to improve the flux chamber method and to further quantify the volatilization of PCBs at Lake Hartwell. Issues regarding internal temperature control, varied flow rates, and flow designs are anticipated for future investigations.

REFERENCES

Battelle. 2001. *Natural Recovery of Persistent Organics in Contaminated Sediments at the Sangamo-Weston/Twelvemile Creek/Lake Hartwell Superfund Site*. Prepared for the U.S. EPA, Contract No. 68-C5-0075, Work Assignment 30. September 29.

United States Environmental Protection Agency. 1994. *Superfund Record of Decision: Sangamo-Weston/Twelvemile Creek/Lake Hartwell Site, Pickens, GA: Operable Unit 2.* EPA/ROD/R04-94/178.

NATURAL RECOVERY OF PCB-CONTAMINATED SEDIMENTS AT THE SANGAMO-WESTON/LAKE HARTWELL SUPERFUND SITE

Victor S. Magar[(a)], Jennifer A. Ickes, James E. Abbott (Battelle, Columbus, Ohio)
Richard C. Brenner (U.S. EPA, Cincinnati, Ohio)
Greg S. Durell and Carole Peven-McCarthy (Battelle, Duxbury, Massachusetts)
Glenn W. Johnson (University of Utah, Salt Lake City, Utah)
Eric A. Crecelius and Linda S. Bingler (Battelle, Sequim, Washington)

Abstract: The goal of this study was to investigate and identify natural recovery processes responsible for the recovery of polychlorinated biphenyl (PCB)-contaminated sediments at Lake Hartwell, SC. Ten sediment cores were collected from predetermined transect locations. Cores were analyzed by depth for ^{210}Pb, ^{137}Cs, and PCBs. ^{210}Pb and ^{137}Cs were used to date sediments in order to determine sediment accumulation rates and sedimentation rates. Upgradient transect cores were impacted significantly by sediment built up in the headwaters of Lake Hartwell, due to historical sediment releases from upstream impoundments. The highest PCB concentrations were associated with buried silt/clay layers. Downgradient transect cores consisted primarily of silt and were not noticeably impacted by the impoundments. Surface sediments showed decreasing t-PCB concentrations, which approached the target t-PCB cleanup goal of 1.0 mg/kg. The PCB congener compositions became increasingly dominated by lower-chlorinated congeners with sediment depth and corresponding age, resulting in a relative accumulation of *ortho* chlorines and loss of *meta* and *para* chlorines. End member (EM) patterns had characteristics of Aroclors 1248 and 1254, Aroclor 1248 dechlorination via *Process C* (Bedard and Quensen, 1995), and *Process H'* dechlorination (Bedard and Quensen, 1995).

INTRODUCTION

The U.S. EPA's NRMRL is interested in developing effective, inexpensive remediation technologies for contaminated sediments. This study focuses on the recovery of sediments contaminated with PCB compounds. It has been determined that anaerobic and aerobic biotransformation/biodegradation of PCBs reduces the overall contaminant mass (Quensen and Tiedje, 1997), sorption onto organic particles reduces bioavailability (McGroddy et al., 1996), and sediment containment through natural capping can act as a natural barrier to protect the aquatic environment from contaminated sediments (Cardenas and Lick, 1996). The goal of this study is to achieve a better understanding of the natural mechanisms that contribute to the recovery of PCB-contaminated sediments. The study focuses on the natural recovery of PCB-contaminated sediments at the Sangamo-

---

[(a)] This proceedings paper was previously published in Leeson, A., Foote, E.A., Banks, K., and Magar, V.S., Phytoremediation, Wetlands, and Sediments, The Sixth International In Situ and On-Site Bioremediation Symposium, San Diego, California, June 4-7, 2001 (pp. 231-236).

Weston/Twelve-mile Creek/Lake Hartwell Superfund site in South Carolina (i.e., the Lake Hartwell site).

In general, natural attenuation of contaminated sediments relies on two primary mechanisms: (1) burial of contaminated sediments with clean sediments, and (2) contaminant weathering. Sediment burial (capping) acts both to protect the water column from the vertical diffusion and advection of contaminants from near surface sediments, and to reduce contaminant transport into the food chain that can occur through bioturbation and bioaccumulation in surface or near-surface sediments. PCB weathering, which includes such mechanisms as dilution, volatilization, biodegradation, and sequestration, can provide a permanent reduction in levels of PCB contamination and of PCB chlorination. Biological reductive dechlorination under anaerobic conditions preferentially dechlorinates higher-chlorinated PCB congeners, transforming them to lower-chlorinated congeners. Other weathering processes such as dilution and volatilization preferentially remove lower-chlorinated PCBs that are more mobile.

This study focused on evaluating natural sediment capping and contaminant weathering at the Lake Hartwell site. The contributions of the various potential weathering mechanisms were not examined separately, except by careful analysis of PCB congener profiles in aged sediments.

Site Description. The 730-acre Sangamo Weston/Twelve-Mile Creek/Lake Hartwell site is located in Pickens County, SC (U.S. EPA, 1994). The Sangamo-Weston plant was used for capacitor manufacturing approximately from 1955 to 1978. The plant used a variety of dielectric fluids in its manufacturing processes, including ones containing PCBs. Waste disposal practices included land burial of off-specification capacitors and wastewater treatment sludge on the plant site and at six satellite disposal areas. PCBs also were discharged with effluent directly into Town Creek, a tributary of Twelve-Mile Creek, which is a major tributary of Lake Hartwell. The site was divided into two operable units (OUs). OU 1 addressed the land-based areas, which included the Sangamo-Weston land, and six satellite disposal areas. The source of PCB contamination has been cleaned as part of the remediation of OU 1. OU 2 consists of sediments contaminated with PCBs that lie at the bottom of Lake Hartwell (Figure 1). This study pertains to the recovery of sediments in OU 2.

MATERIALS AND METHODS

The field studies included the collection of sediment cores and subsequent chemical analyses at 10 sample locations at the Lake Hartwell site. Cores were collected at predetermined river transects already established by the U.S. EPA Region 4 and United States Army Corps of Engineers (USACE) under the site's ongoing annual monitoring program (Figure 1). Sediment cores were collected using either 5-cm by 183-cm Lexan™ tubes or 5-cm by 244-cm transparent polyvinyl chloride (PVC) pipe. Triplicate sediment cores were taken at each of the sample locations. Each core was segmented vertically into 5-cm samples and sealed in glass containers. The three cores were used as follows: one core was used for contaminant analyses (Battelle Ocean Sciences Laboratory, Duxbury,

Massachusetts), which included PCB concentrations with congener distributions for the Lake Hartwell samples; one core was used for lead-210 (^{210}Pb) and cesium-137 (^{137}Cs) isotope concentrations (Battelle Marine Sciences Laboratory, Sequim, Washington) for sediment age-dating analyses; and one core was used for PCB partitioning experiments (Battelle Columbus, Ohio). Additional analyses from each sample location included total organic carbon (TOC), grain-size/particle-size distribution, and moisture content (Soil Technology Inc., Bainbridge, Washington).

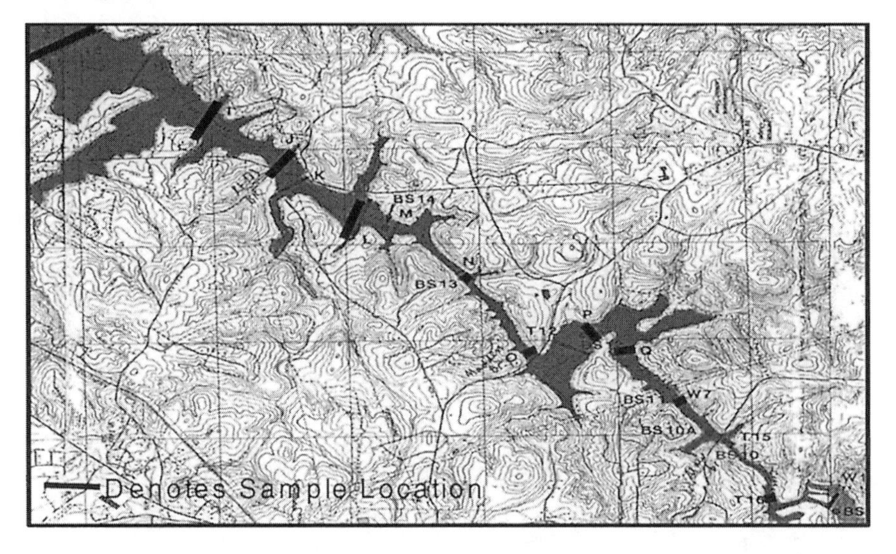

PCB analyses were conducted using high-resolution gas chromatography with mass spectrometry (GC/MS). The PCB method included the separation and analysis of 107 PCB congeners. ^{210}Pb was analyzed using the alpha counting technique described by Koide et al. (1973); ^{137}Cs concentrations were determined by gamma counting a known amount of sediment (~100 g dry) on a Ge-diode detector.

RESULTS AND DISCUSSION

Sediment cores were obtained from 10 transect locations in Lake Hartwell. As much as possible, sample cores were taken from the deepest portion of the river. PCB congener analyses were conducted; total PCB concentrations were determined by the sum of individual congeners. Vertical PCB profiles were established at each transect location. Typical PCB profiles are shown for Transect Q (Figure 3) and T6 (Figure 4), from upgradient and downgradient locations in Lake Hartwell.

The most upgradient transects (T16, W7, Q, and P) were impacted significantly by historical releases of sediment built up in the headwaters of a dam located upgradient of the lake. As expected, the highest PCB concentrations were associated with silt layers while sand layers contain very low PCB concentrations. In Transect Q (Figure 2), the top 25 cm of sediment contains less than 93 μg/kg total PCBs, well below the cleanup goal of 1 mg/kg for surface sediments (U.S.

EPA, 1994). Transect T6 (Figure 3), consisting primarily of silt, was not noticeably impacted by the release of silt and sand from the dam, and has a more typical PCB profile, where the highest PCB concentrations in the middle of the core are associated with the period of maximum PCB release into the lake. The top 5, 10, and 15 cm of the core contained 0.93, 2.6, and 8.9 mg/kg. While surface sediment concentrations appeared to be decreasing, at the time of sampling they continued to exceed the 1 mg/kg PCB goal.

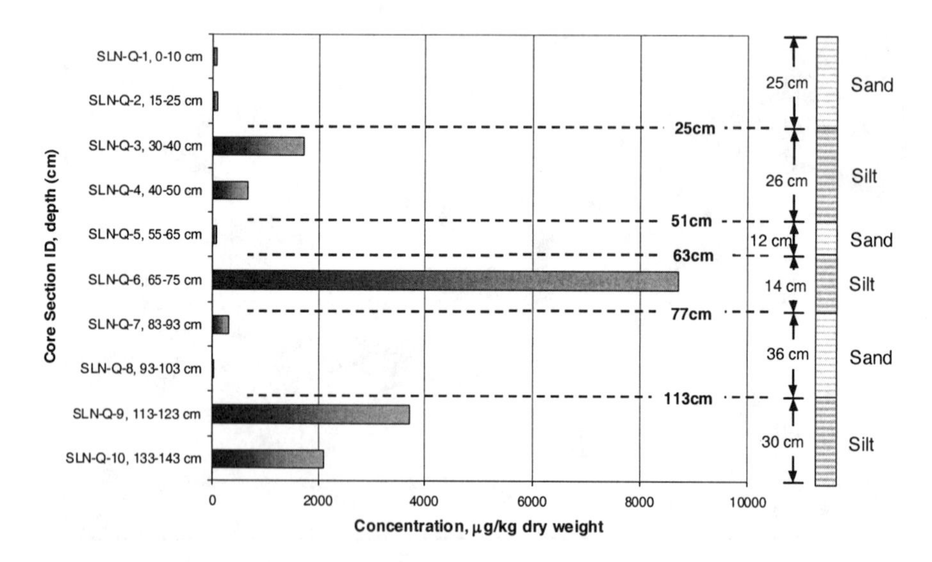

FIGURE 2. Vertical PCB profile from upgradient transect Q. Silt/sand layers are formed in the upgradient portion of the lake due to planned releases of sediment built up behind a dam located upstream of the lake.

PCB Composition Data. PCB composition (i.e., the relative concentrations of PCB congeners) was examined based on PCB homologues (i.e., level of chlorination) and congener data. The PCB homologues and congener compositions became increasingly dominated by lower-chlorinated congeners with sediment depth and corresponding age of the deposited sediments, which is consistent with changes observed in the homologue composition. A significant loss (–45%) of tetra-, penta-, and hexachlorobiphenyl congeners and accumulation of mono-, di-, and trichlorobiphenyls dominated at greater depth. In surface sediment samples, tetrachlorobiphenyls were most abundant, followed by the tri- and pentachlorobiphenyls, then by di- and hexachlorobiphenyls, and mono- and heptachlorobiphenyls.

Polytopic Vector Analysis. The PCB data generated for this study were modeled using the multivariate statistical method known as Polytopic Vector Analysis (PVA) to identify fingerprint (also known as end-member) compositions from the data generated for Lake Hartwell. End-member (EM) patterns then were compared to source patterns reported in the literature (e.g., Aroclor compositions and known PCB dechlorination or weathering patterns).

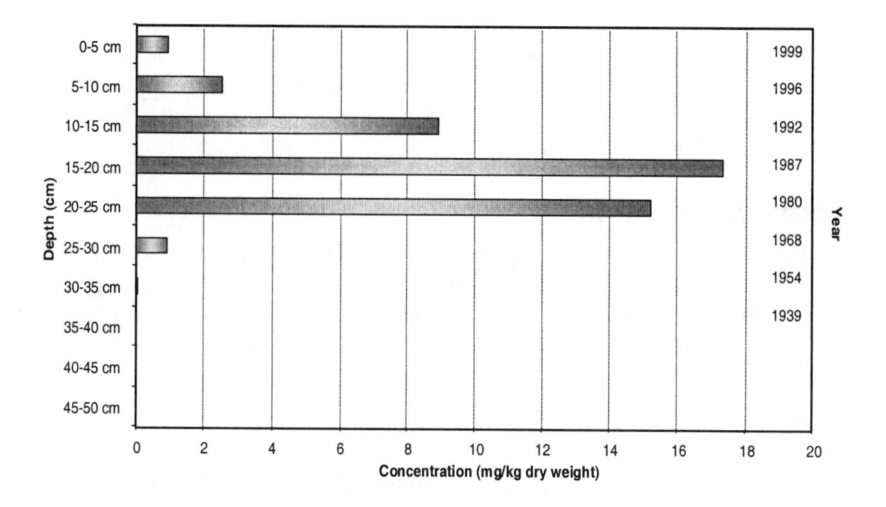

FIGURE 3. t-PCB vertical concentration profile in downgradient Core T6

Initial results suggest that PVA appeared to resolve three EM patterns. The composition and distribution of EM-1 had characteristics of a mixture of Aroclors 1248 and 1254.

EM-2 was dominated by low-chlorinated congeners, including mono-, di-, and trichlorobiphenyls. The congeners that made up EM-2 preferentially exhibited chlorines in the 2 (*ortho*), 4 (*para*) and 6 (*ortho*) positions. The dominance of *ortho*-chlorines suggested that EM-2 was a result of a microbial dechlorination process.

EM-3 (End-Member 3) was characterized by di-, tri-, and tetrachlorobiphenyl congeners, but also showed an increase in *ortho*-chlorinated congeners relative to EM-1. EM-3 appears to be an intermediate between EM-1 and EM-2. EM-1 was most abundant in surface sediments, and EM-2 and EM-3 dominated deeper sediment segments.

CONCLUSIONS

Vertical PCB profiles indicate that the sediments are recovering by the slow deposition of increasingly less-contaminated sediment, but that the downgradient reaches of the lake have not yet recovered to the target concentration of 1 mg/kg. The presence of *ortho*-halogenated compounds found in high concentrations provides indications of reductive dechlorination of higher-chlorinated PCBs in deeper sediments. PVA provided an excellent tool to evaluate end-member PCB patterns.

The coring approach demonstrated in this research provides a uniquely effective approach to understanding the history of contaminated sediment deposition, weathering processes, and contaminant sources in the interest of studying natural recovery of PCB-contaminated sediments. This research summarizes just one method of studying recovery of contaminated sediments at Lake Hartwell. In order to fully study PCB-contaminated sediments it is necessary to study other mechanisms that may play a role in sediment recovery and the fate and transport of sediment contaminants, such as bioaccumulation, bioturbation, and volatilization to name a few. In addition, a better understanding is needed of how major events such as 100- or 500-year storms, sea level rise, and human activity impact sediment erosion and deposition.

ACKNOWLEDGEMENTS

Funding for this study was provided by the U.S. EPA National Risk Management Research Laboratory under Contract No. 68-C5-0075, Work Assignment 4-30. This extended abstract summarizes recent research and does not represent U.S. EPA policy.

REFERENCES

Bedard, D.L., and J.F. Quensen. 1995. "Microbial Reductive Dechlorination of Polychlorinated Biphenyls." In: L.Y. Young and C.E. Cerniglia (Eds.), *Microbial Transformation and Degradation of Toxic Organic Chemicals*, pp. 127-216. Wiley-Liss, New York, NY.

Cardenas, M., and W. Lick. 1996. "Modeling the Transport of Sediments and Hydrophobic Contaminants in the Lower Saginaw River." *Great Lakes Res.* *22*(3): 669-682.

Koide, M., K.W. Bruland, and E.D. Goldberg. 1973. "Th-228/Th232 and ^{210}Pb Geochronologies in Marine and Lake Sediments." *Geochim. Et cosmochim. Acta, 37*: 1171-1187.

McGroddy, S.E., J.W. Farrington, and P.M Gschwend. 1996. "Comparison of the In Situ and Desorption Sediment-Water Partitioning of Polycyclic Aromatic Hydrocarbons and Polychlorinated Biphenyls." *Environmental Science and Technology 30*(1): 172-177.

Quensen, J.F. III, and J.M. Tiedje. 1997. "Evaluation of PCB Dechlorination in Sediments." In D. Sheehan (Ed.). *Methods in Biotechnology; Bioremediation Protocols*, pp. 257-273.

United States Environmental Protection Agency. 1994. *Superfund Record of Decision: Sangamo-Weston/Twelvemile Creek/Lake Hartwell Site, Pickens, GA: Operable Unit 2*. EPA/ROD/R04-94/178.

NATURAL RECOVERY OF PAH-CONTAMINATED SEDIMENTS AT THE WYCKOFF/EAGLE HARBOR SUPERFUND SITE

Victor S. Magar[a], Jennifer A. Ickes, James E. Abbott (Battelle, Columbus, Ohio)
Richard C. Brenner (U.S. EPA, Cincinnati, Ohio)
Scott A. Stout and Richard M. Uhler (Battelle, Duxbury, Massachusetts)
Eric A. Crecelius and Linda S. Bingler (Battelle, Sequim, Washington)

Abstract: Eagle Harbor is a shallow marine embayment of Bainbridge Island, WA (~10 miles west of Seattle, WA) and was formerly the site of the Wyckoff wood-treatment facility. The facility became operational in the early 1900s and contributed large quantities of creosote to the polycyclic aromatic hydrocarbon (PAH) contamination of harbor sediments. The site was partially capped to control PAH migration into the water column and surrounding sediments. However, residual surface sediment contamination remains. This investigation focused on the natural recovery of the PAH-contaminated sediments.

TPH fingerprinting provided information on the nature and source of hydrocarbon contamination in Eagle Harbor. The distribution of the 50 PAH analytes provided insight into the sources and degree of weathering in the sediments. With decreasing sediment depth, the creosote signature showed increasing weathering and an increasing influence of an urban runoff signature. These results provided information on the ability for Eagle Harbor sediments to recover under natural conditions, to distinguish between PAH sources (i.e., between creosote contamination and urban runoff), and to identify the occurrence of creosote-derived PAH weathering in surface sediments. More extensive discussions and more conclusive evaluations will be made in an upcoming full-paper publication.

INTRODUCTION

The U.S. EPA's National Risk Management Research Laboratory (NRMRL) is interested in developing effective, inexpensive remediation technologies for contaminated sediments. This study focuses on the recovery of sediments contaminated with PAH compounds. In general, natural attenuation of contaminated sediments relies on two primary mechanisms: (1) burial of contaminated sediments with clean sediments, and (2) contaminant weathering. Sediment burial (capping) acts both to protect the water column from the vertical diffusion and advection of contaminants from near surface sediments, and to reduce contaminant transport into the food chain that can occur through

---

(a) This proceedings paper was previously published in Leeson, A., Foote, E.A., Banks, K., and Magar, V.S., Phytoremediation, Wetlands, and Sediments, The Sixth International In Situ and On-Site Bioremediation Symposium, San Diego, California, June 4-7, 2001 (pp. 237-246).

bioturbation and bioaccumulation in surface or near-surface sediments. PAH weathering, which includes such mechanisms as dilution, volatilization, biodegradation, and sequestration, can provide a permanent reduction in PAH contamination, but tends to preferentially remove the lower-molecular-weight PAH compounds. This study focused on evaluating natural sediment capping and contaminant weathering at the Eagle Harbor site.

Site Description. Eagle Harbor is a shallow marine embayment of Bainbridge Island, WA. The island is located approximately 10 miles due west of Seattle, WA, and was formerly the site of the Wyckoff wood-treatment facility. The facility, which became operational in the early 1900s, used large quantities of creosote, contributing to the contamination of Eagle Harbor sediments with PAHs. In 1987, the Wyckoff/Eagle Harbor site was placed on the National Priorities List as a Superfund site. The site has been partially capped to control PAH and mercury migration into the water column and surrounding sediments. However, residual surface sediment PAH contamination was reported to exist, generally to the west and south of the existing cap. During this investigation, which focused on the natural recovery of PAH-contaminated sediments, mercury was not measured.

FIELD SAMPLING AND ANALYTICAL METHODS

Sediment cores were collected from 10 locations in Eagle Harbor. Coring locations are shown in Figure 1. A vessel from Battelle's Marine Research Laboratories in Sequim, WA (Battelle-Sequim) was used as a platform for sediment coring. Sediment cores were collected by manually driving the cores into the sediment using a conventional sand driver equipped with a vibrator. Sediment cores (28 to 78 cm long) were collected in 5-cm-diameter Lexan™ tubes. Four sediment cores were collected west of the cap, labeled E01, E02, E03, and E04; three cores were collected southwest of the cap, labeled X01, X02, and X03; and three cores were collected south of the cap, adjacent to the former Wyckoff facility, labeled B01, B02, and B03.

Three parallel cores were drawn from each location for further analysis; one core was dedicated to each of three Battelle laboratories. Suitable cores were extruded on shore and extruded samples were sent to their respective laboratories; unsuitable cores, such as those with poor sediment recoveries, were discarded. PAH and TPH analyses were conducted at Battelle's Ocean Sciences Laboratories in Duxbury, MA (Battelle-Duxbury); ^{210}Pb and ^{137}Cs analyses were conducted at Battelle Marine Sciences Laboratory, in Sequim, WA (Battelle-Sequim); and partitioning experiments were conducted at Battelle's Columbus Operations (Battelle-Columbus).

Sediment samples were extruded vertically upward through the cores using a 4.5-cm-diameter rubber gasket connected to a metal dowel, which was used to push the sediments from the bottom through the top of each coring tube. The water column was pushed over the top of the core tube and discarded until the sediment-water interface reached the top of the tube. Then the sediment core was pushed through in 5-cm segments, which were collected in 8-oz glass sample jars.

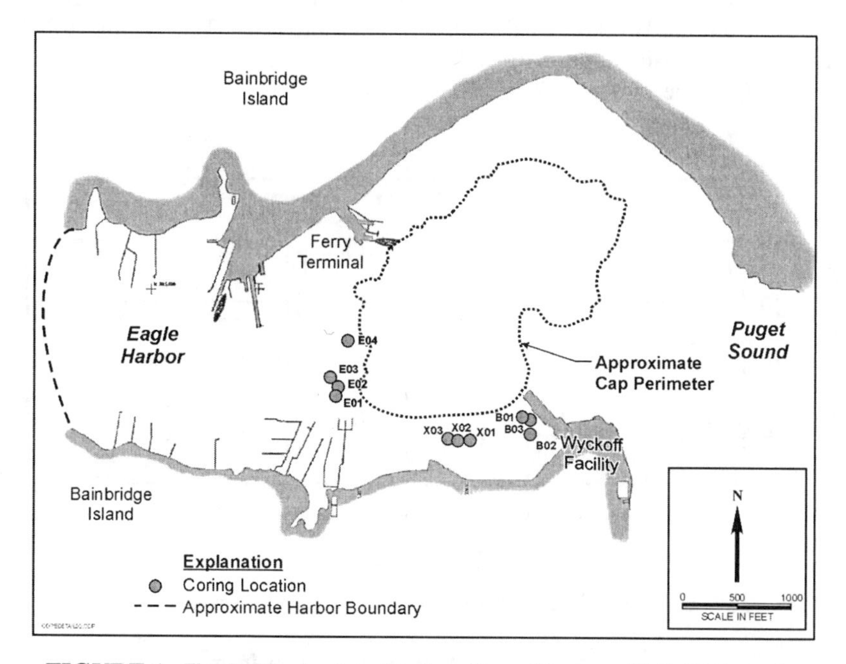

FIGURE 1. Eagle Harbor coring locations (Source: U.S. EPA and USACE, 2000).

RESULTS AND DISCUSSION

TPH Analysis: TPH concentrations were determined for the sediment core segments. The distribution of TPH concentrations among the 97 sediments analyzed is depicted in Figure 2. TPH values ranged from <10 mg/kg to >142,000 mg/kg. Most samples contained between 10 and 1,000 mg/kg TPH. The highest concentrations occurred near the bottom of the B02 core (B02-7 and –8) where more than 13-14 weight percent (dry) of the sediment was comprised of extractable hydrocarbons. Concentrations of this magnitude indicated the presence of free-phase hydrocarbons in the pore space of these sediments, which was consistent with the physical appearance of these sediments and the close proximity of the B02 sample location to the former Wyckoff facility. In fact, all 26 of the sediments containing more than 1,000 mg/kg TPH were collected from the bottom portions of the B0 and X0 cores; these cores were located closest to the former Wyckoff creosote manufacturing facility. It is equally notable that, with one exception (X02-9), all of the sediments containing less than 100 mg/kg

TPH were collected from the bottom portions of the E0 cores; these cores were located furthest from the former Wyckoff facility.

Sediment Age-Dating. Battelle-Sequim conducted sediment dating on 10 cores collected from the Wyckoff/Eagle Harbor Superfund Site. This sediment dating was conducted using lead and/or cesium isotope analyses (^{210}Pb or ^{137}Cs, respectively) as described by Koide, et al. (1973).

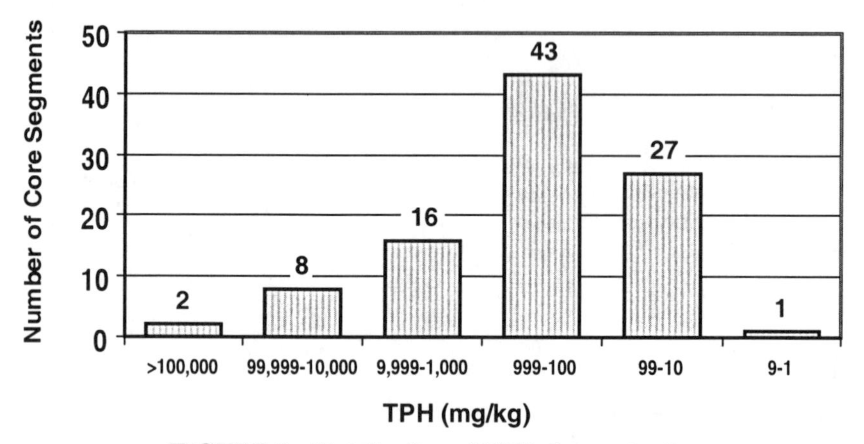

FIGURE 2. Distribution of TPH Concentrations

^{210}Pb and ^{137}Cs isotopes are relatively common in marine sediments and typically are used to determine the age of the sediments over years, decades, or centuries. This type of dating relies on two simplifying assumptions: (a) that the sediments to be aged have a relatively uniform grain distribution with depth, and (b) that the sediments have a relatively constant annual deposition rate. As much as possible, these conditions were confirmed visually during the initial inspection of the sediment cores and with total organic carbon and grain size distribution analyses. Excess sediment core material not used in the actual age-determination process was analyzed for PSD to determine the percentage of clay, silt, sand, and gravel in the sediments; PSD was used to assess the uniformity of the sediment samples.

^{210}Pb and ^{137}Cs analyses were conducted in accordance with the Quality Assurance Project Plan (Battelle, 2000). One-hundred-gram aliquots of wet sediment were used for the ^{137}Cs sediment age-dating analysis; approximately 3.0 g of sediment were used for the ^{210}Pb analysis. Approximately 3.0 g of dry sediment underwent acid digestion followed by plating onto silver disks for ^{210}Pb analysis using the alpha-counting technique described by Koide et al. (1973). The remaining sediment was used for ^{137}Cs analysis using gamma counting on a Ge-diode detector. These two measurements were used to age-date sediment samples.

Although all of the sediment core segments were analyzed for ^{210}Pb and ^{137}Cs, not all core segments could be age-dated. Cores whose PSD analyses showed nonuniform historical sediment deposition could not be dated. Based on the observed sedimentation patterns, cores B01, E01, E02, E03, and E04 could be dated and cores B02, B03, X01, X02, and X03 could not. It is likely that sediment deposition patterns in the B and X cores were more significantly impacted by off-shore runoff due to their close proximity to the shore, resulting in erratic sediment deposition patterns and confounding the age-dating analysis for these cores.

The year 1900 fell in the depth segment of 50-55 cm in core E0-1 and at a depth of about 30-40 cm in cores E0-2, E0-3, and E0-4. This was the approximate depth at which the hydrocarbon concentrations in the E0 cores began to increase dramatically. The history of hydrocarbon contamination in the greater Puget Sound sediments is very similar to that in the E0 cores from Eagle Harbor, where hydrocarbon concentrations increase significantly at the turn of the 20th century, presumably due to regional urbanization and industrialization. Not surprisingly, due to its proximity to the shore, the sedimentation rate for core B0-1 was significantly greater than for the E0 cores. The bottom section of core B0-1 (41-46 cm) has an estimated date of 1943. For comparison, the 45-cm interval in the E0 cores would have been dated between 1848 and 1915.

PAH and TPH results were used to characterize contaminated sources. Three characteristic TPH and PAH profiles (i.e., end-member patterns) were identified and compared with patterns taken from the literature. These patterns were used to identify three primary PAH sources and mixtures of sources, including a creosote source (from the Wyckoff facility), an urban runoff source, and a natural background source. The urban runoff and natural background end-member patterns dominated the EO cores, while creosote and urban runoff end-member patterns dominated the BO and XO cores.

Creosote-contaminated sediments were studied to evaluate contaminant weathering. Creosote weathering showed a marked decrease in lower-molecular-weight PAHs, particularly 2-, 3, and 4-ring PAHs.

CONCLUSIONS

The results of this study provided information on the ability of Eagle Harbor sediments to recover under natural conditions, to distinguish between PAH sources (i.e., between creosote contamination and urban runoff), and to identify the occurrence of creosote-derived PAH weathering in surface sediments. The coring approach demonstrated in this research provides a uniquely effective approach to understanding the history of contaminated sediment deposition, weathering processes, and contaminant sources in the interest of studying natural recovery of PAH-contaminated sediments.

This research examined just one method of evaluating recovery of contaminated sediments at Eagle Harbor. In order to fully study PAH-contaminated sediments, it is also necessary to study other mechanisms that may play a role in sediment recovery and the fate and transport of sediment contaminants, such as bioaccumulation, bioturbation, and volatilization, to name a few. In addition, a better understanding is needed of how major events such as 100- or 500-year storms, sea level rise, and human activity impact sediment erosion and deposition more extensive evaluation of the results of this work will be made in an upcoming full-paper publication.

ACKNOWLEDGEMENT

Funding for this study was provided by U.S. EPA's National Risk Management Research Laboratory under Contract No. 68-C5-0075, Work Assignment 4-30. This technical note summarizes recent research and does not represent U.S. EPA policy.

REFERENCES

Battelle. 2000. *Level III Quality Assurance Project Plan: Natural Attenuation of Persistent Organics in Contaminated Sediments at the Wyckoff/Eagle Harbor Superfund Site*. Prepared for U.S. EPA, NRMRL, Cincinnati, OH, under Contract No. 68-C5-0075.

Koide, M., K.W. Bruland, and E.D. Goldberg. 1973. "Th-228/Th232 and ^{210}Pb Geochronologies in Marine and Lake Sediments." *Geochim. Et cosmochim. Acta, 37*: 1171-1187.

U.S. Environmental Protection Agency and USACE. 2000. Final *1999 Environmental Monitoring Report: Wyckoff/Eagle Harbor Superfund Site, East Harbor Operable Unit, Bainbridge Island, Washington*. Prepared by Striplin Environmental Associates, Inc., for U.S. EPA Region 10, Seattle, WA. May 15.

U.S. EPA, see .U.S. Environmental Protection Agency.

SILT REMOVAL FROM A RESERVOIR IN HEEL, THE NETHERLANDS

Ir. P.A. Appel (MTI Holland B.V., Kinderdijk, the Netherlands)
Mr. M.J. Jeroense (Van den Biggelaar/Conetra, Kerkdriel, the Netherlands)

ABSTRACT: A former gravel pit is used as a reservoir for a nearby water treatment plant. A layer of fine silt has been deposited over the gravel that remained in the pit and this significantly reduces the permeability of the gravel. For properly operating the plant, this silt had to be removed with as less impact on the environment as possible. To investigate the impact of dredging this material, a trial project was executed first. A wheel dredger with a specially designed hood around the wheel was put to use. The hood functioned extremely well; no turbidity was observed outside the hood. The silt needed to be dumped back in the reservoir at 30 m depth, and in order to avoid too much contamination of the reservoir and spreading during dumping, a silt screen was utilized. Especially the use of the screen was innovative since no experience with such a function and length existed. The silt screen caused initially some operational problems, but by making adaptations to the screen it functioned very satisfactory. After successfully completing the trial test, the permeability of the gravel layer was found to be increased by 40%.

INTRODUCTION

At the village of Heel, in the southern part of the Netherlands, a former gravel pit known as "De Lange Vlieter" is used as a reservoir for a nearby water treatment plant. Through the gravel layers the water is pumped to the plant. However, a layer of fine silt has been deposited over the gravel that remained in the pit. This silt is problematic since it reduces the permeability of the gravel and by doing so the flow to the plant's waterpumps. To optimize the efficiency and capacity of the pumps the silt needs to be removed by dredging.

Since the water in the basin will be used as future drinking water and therefore should be as fresh and clean as possible, the basin is closed to the public and any recreation or other polluting activity is prohibited. For the same reason, the "Waterleidingmaatschappij Limburg" (waterworks of the province Limburg), who gave the order for removing the silt, set very strict operational requirements that needed to be met while dredging. As a consequence, some special adaptations had to be made to the dredger in order to diminish the impact on the environment.

In order to investigate the effects of dredging the silt on the pollution and condition of the reservoir, even with the application of the suction hood and silt screen, a trial project had to be executed first. Having started on the 15th of March 2001, this trial project lasted till half June.

In this paper, the development and execution of this project is described, while special attention is being paid to the concept of the silt screen and the suction hood.

Operational requirements. The first requirement was that, by absence of other suitable dumping areas, the dredged material should be dumped back in the deepest part of the reservoir. This deepest part is located in the center of the reservoir at a depth of 28 to 30 m. Outside this dumping area no increase in solid matter, measured in mg/l, was allowed to take place. In order to achieve this, the dredged material was to be dumped back in the reservoir via a silt screen.

Another important requirement was that the increase in solid matter at a distance of 20 m from the dredger was not allowed to exceed 20 mg/l. Covering the dredge wheel with a suction hood was therefore necessary for reducing the turbidity caused by dredging.

Apart from the silt-screen and the suction hood, several other regulations were in order. For example, the dredge, as well as the support vessels, had to be fuelled and lubricated by environmental friendly petrol and oils.

Site Description. The reservoir "De Lange Vlieter" contains approximately 150.000 m³ of silt that needs to be removed, distributed over 4 locations. Most of this material originated from the dredging activities in the former gravel deposit (i.e. it was in-situ material), although also the decay of organic material, rain and natural flows of water contribute to the accumulation of silt.

The material that is to be dredged is found between 4 to 28 m water depth and has a thickness varying between 0.5 to 5 m. Since this material is found not only on top, but also in the upper gravel layer itself the first 5 cm of the gravel should be removed as well. The allowable horizontal and vertical accuracy of the dredge wheel was set to only 10 cm, because excavating more gravel would endanger the stability of the pit.

During the trial project 27.000 m³ has been removed from one of the four locations. Only those sediment layers with a thickness of less than 1.5 m were removed, since dredging thicker layers would at this point be too costly. A schematic overview of the operation is shown in figure 1.

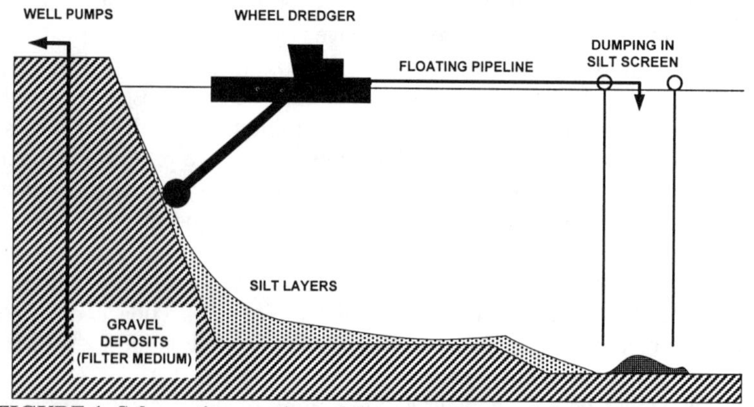

FIGURE 1. Schematic overview of the excavation and dumping process.

DESIGNING THE PROJECT

The contractor "Van den Biggelaar/Conetra", specialized in dredging and civil engineering tasks, was called in to remove the silt layer. "MTI Holland BV", the research and development department of the dredge building company "IHC Holland NV", gave advice and comments with respect to the operational possibilities within the given constraints. "MTI" paid special attention to the design of the silt-screen and the development of the suction hood covering the dredging wheel.

Dredging wheel. Since as much turbidity had to be prevented during this project, the choice which type of dredging unit had to be used was of prime importance. The most suitable unit was considered to be a dredging wheel (Figure 2), because such units have a high clean up efficiency (less spoil than e.g. a cutter suction dredger), can dredge more or less in situ material and can achieve the required high level of precision dredging.

FIGURE 2. An IHC dredging wheel.

Design of the Hood. Since the waterworks required that the increase in solid matter at a distance of 20 m from the dredger may not exceed 20 mg/l, a suction hood covering the dredging wheel was designed to diminish thc turbidity around the wheel. The basis for the design of the hood was a symmetrical, completely covering hood, as shown in figure 3. This hood covers most of the dredge wheel but has an open area in the front en sides to allow for easy excavation of the deposit.

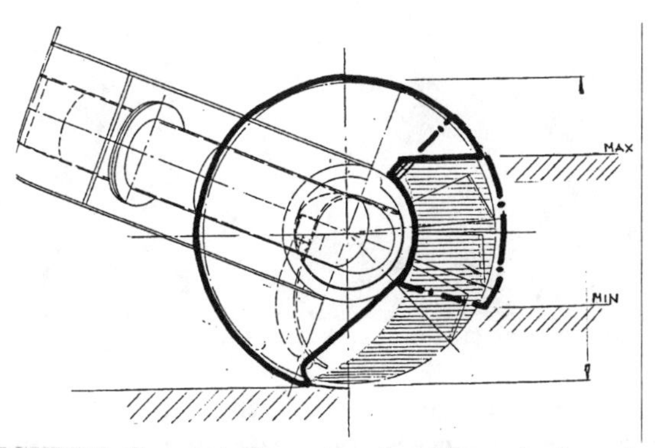

FIGURE 3. Concept of a suction hood for a dredge wheel.

Three different layouts of the hood have been considered. Two big hoods that cover both the wheel itself and the drive unit at the side, and a smaller one that only covers the wheel. The big hood can have a rectangular or trapezium shape. A trapezium shaped hood has the advantage that it's construction is stronger and it has a relatively smaller open area in the front. A small hood, only covering the wheel has the least open area. The dimensions of the big hoods were taken so that they fit around the wheel easily (approximately 2 m wide, 2 m long and 1 m high for the big hood). Because the wheel had to work at different depths and thus at different angles, the hood had enough open area to allow the wheel to excavate the silt at any given angle.

The open area has been determined for each of the wheels. By combining this open area with the given production capacity of the dredgepump (maximum of 1500 m^3/hr), and a certain concentration of the pumped mixture (ranging from 5% to 20%) the velocity of the material flowing into the hood could be calculated. These velocities ranged from 0.1 m/s for the big rectangular hood, to 0.4 m/s for the small hood. At a velocity of 0.1 m/s particles smaller than 1 mm that become suspended by the cutting action of the wheel are dragged into the hood. Larger particles may remain out of the hood. However, such particles are likely to settle in a few seconds and will not cause any problems with regard to the turbidity restriction that was set. At higher velocities, even the larger particles are sucked into the hood. By decreasing the open area of the hood, the velocity of the water flowing into the hood is increasing, causing almost all particles to be sucked up in the hood. This reduction of open area can be achieved by using a visor on the hood, which can be lowered via a cylinder or manually.

The concepts of the suction hoods, as proposed by "MTI" have been technically worked out by "Van den Biggelaar/Conetra". Out of constructional-technical considerations, the actual hood was based on the big, rectangular hood design. This hood (Figure 4) was equipped with a visor that could be adjusted by hand when necessary.

FIGURE 4. The actually used suction hood, with the visor adjustment mechanism.

To this hood, a connection for a possible pipeline extension has been added. This was thought to be necessary in case the dredgepump itself was not sufficient to completely remove the mixture within the hood. Using this connection has not been necessary.

Design of the Silt Screen. The waterworks required the use of a silt-screen since the dredged-up silt has to be dumped back in the deposit. In this way, it was thought that the fine material would not pollute the whole reservoir. Usually, silt screens are used only to confine a contaminated material that lies under water. This allows for using e.g. a grab dredger to excavate that material without contaminating the surrounding environment. Such screens are often used in harbors and are limited to only a few meters length. So far, they have not been used to confine material that had to be deposited back into an area and they were certainly not as long as the finally installed screen, which was about 30 m long.

The first step in designing the screen was determining several properties of the material, like the grainsize distribution, the density and the settling velocities. Hydrometer analysis indicated that most of the material consisted for more than 70% out of clayey and silty material. Density measurements of the material showed that its (wet) density was about 1500 to 1600 kg/m³. The settling properties have been determined with the aid of measuring cylinders and various amounts of sample. It turned out that the bulk of the material settled in about 30 minutes and that even the very fine fraction was settled after 90 minutes. Taking the height of the measuring cylinder into account, the settling velocity of the fine material was 0.2 m/h. The area where the dredged material was allowed to be dumped was located between 28 and 30 m depth. With the given settling velocity a particle would have needed 150 hours to reach this depth.

At "MTI", four different types of screen layout have been looked at (Figure 5). Since not even the fine material was allowed to be suspended outside the dump area, this velocity formed the most important design criterion for the silt screen. To keep the dimensions within feasible range, the calculations made have been based on a square screen with sides of 12 m (open area around the discharge pontoon 144 m²), and a dredged capacity of maximum 1500 m³/h.

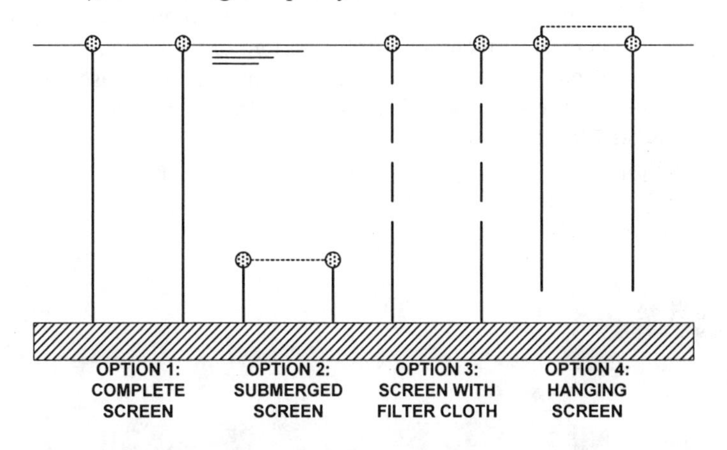

FIGURE 5. Four different types of screen layout.

Option 1: complete screen. This screen reaches completely from the surface to the bottom. The only way water can escape this screen is at the surface. This means that with the given screen dimensions and dredged capacity the upward velocity in the screen would be 10 m/h. Particles smaller than 100 μm cannot settle under these circumstances and will be discharged over the upper part of the screen, causing contamination of the surface water. Since it takes about 150 hours for the fine fraction to reach the bottom, and wave action and currents at the bottom of the reservoir might transport these particles outside the designated dumping area, this option is not regarded to be feasible.

Option 2: submerged screen. This is a screen that rests at the bottom, but it is only a few meters high. The discharge area within the screen is fed by a long diffusor. Since a continuous stream of material is fed to discharge area, some of the mixture will be pushed over the top of the screen with, again, a velocity of 10 m/s. Assuming a screen of 5 m height, the fine fraction will still need 25 hours to reach the bottom.

Option 3: screen with filter cloth. This is a submerged screen, but atop of the actual screen that rests on the bottom filter cloth is attached that reaches to the surface. Through this cloth the water has the chance to exit the discharge area while the fine particles remain behind. However, this material may clog the cloth, rendering it useless. In the worst case the whole cloth is clogged, resulting in a screen like the one described under option 1.

Option 4: hanging screen. This screen hangs above the water surface so that no water can flow over the top and ends a few meters above the bottom of the reservoir. The dredged mixture is discharged at the top of the screen, resulting in a settling column of about 30 m length. Since the water in the column can only escape at the bottom, all material, even the fine fraction, is released at a minimum height above the reservoir bottom. Some fine material may escape with this water out of the screen, but outside the screen the velocity reduces quickly to zero, resulting in settlement of all material near the bottom of the screen.

The screen that was actually used during the trial was based on option 4 and is shown in figure 6. It was hanged on a frame that was supported by pontoons. The dredged mixture was discharged within the screen via several openings in a pipe end. By shifting the pontoons by its anchors, the screen spread out the settling material evenly in the discharge area.

FIGURE 6. The discharge pontoon, surrounded by the silt screen.

EXECUTION OF THE TRIAL PROJECT

The trial project started on the 15th of March, 2001 and lasted till half June. During this period no significant downtime was recorded. Throughout the project an overall efficiency of 76% was achieved. The remaining time was primarily necessary to remove debris like steel cables from the wheel, and to reposition the anchors of the dredger.

Observations. During the project, an underwater video filmed the hood in operation. It turned out that it worked exceptionally well in keeping any loosened material from escaping to the surroundings. Although some clouds of suspended material escaped when it moved along the deposit, these were immediately sucked back into the hood.

Such satisfying results were not experienced with the silt screen. Because of it length of almost 30 m, it was very difficult to handle. During some rough weather, the screen ripped and broke in two. It took a week to recover the remnants of the screen from the bottom of the reservoir. Because such problems were not wanted to be experienced again, a new but shorter screen (10 m length, reaching from the surface) was installed. Although this screen did not guide the dredged mixture directly towards the dump area, it was observed that the material flowed almost as one column to the bottom. Only because of light current action could some fine material suspend and cause turbidity.

Screen squeezing. A problem that was encountered while operating both the long and the short screen was that the screens were squeezing together. This was undesirable because it resulted in the cloth coming to the surface, where the impact of the discharging mixture could have caused damage to the screen itself.

This unforeseen effect only took place when dredged material was being dumped into the screen. When pumping was ceased the screen returned slowly but surely to its original position. Therefore, the most likely cause for this is that the velocity of the settling mixture created an underpressure within the screen. As a result the screen squeezed together, which in turn caused the velocity of the settling mixture to increase, thereby intensifying the effect.

Several measures have been undertaken to cope with this problem. Large pieces of conveyor belts were hung on the screen in order to protect it from being damaged by the discharging mixture. On the short screen, a weight of in total 1900 kg was attached but even this could not prevent the occurrence of squeezing.

Turbidity measurements. Throughout the project, the turbidity was measured by taking samples around the dredger and the silt screen, as well as by visual inspection with a camera. As stated before, the suction hood worked very satisfactory in reducing turbidity. Also the turbidity near the silt screen was negligible. Even with the shorter screen, no increase in solid matter content outside the dumping arca was measured. This proves how effective the taken measurements have been in keeping the turbidity in the whole reservoir below acceptable values.

EVALUATION OF THE TRIAL PROJECT

The trial project as described above is successful in two ways. First of all, because of the direct improvement in permeability of the reservoir and, second, because of the usability of the suction hood and silt screen in other dredging projects where any impact on the environment should be avoided as much as possible.

Improvement of Permeability. Prior to the trial project the permeability and waterflow from the reservoir through the gravel towards the waterpumps was measured. This test was also executed after completion of the project. It turned out that after dredging an increase in permeability of 40% was achieved. Apart from that, the requirements that were set with regard to the allowable turbidity formed no real obstacles during the execution of the project.

These two factors proved that the trial project was such a success that the contractor is allowed to dredge the remaining locations in the reservoir as well. At the moment that this paper was written, the outcomes of the trial project were further evaluated. In November 2001 the waterworks will designate exactly which future areas are to be dredged in the same way as the trial project.

However, due to decomposition of (organic) material and influx of fine material via rain and small streams, fresh silt is continuously brought into the reservoir anew. At this moment it is not sure at which rate the quantity of silt increases but it is likely that in the future freshly deposited material has to be removed from the areas that are now being cleaned. This cannot be achieved by standard dredging techniques since these involve the removal of some of the remaining gravel. As a consequence the slopes of the reservoir may become unstable. Cleaning and/or removal projects in the future should therefore preferably be directed towards "non-destructive" dredging. Studies are now underway to identify suitable techniques.

Applicability of Dredging Method. There are various locations in the Netherlands where fresh water is extracted from surface reservoirs. All such reservoirs have, to a certain degree, problems with silting up of the gravel layers through which they extract their water. Although some drastic measurements can be undertaken, like temporary reclamation of the reservoir to remove this material, these are not very time and cost effective. However, the current project has proven that by using a silt screen to dump back the dredged spoil and a protective hood around the excavation unit, it becomes possible to remove silt and other fine material from areas that are very sensitive to turbidity and contamination of the surrounding water without harming the environment.

As a matter of fact, the concept of a silt screen to contain dredged material within a limited area is now used at a project in the "Loosdrechtse Plassen", a lake district in the middle of the Netherlands. So far, the results are very promising as well.

MASS BALANCE MODELS OF SEDIMENT REMEDIATION

Christopher S. Warren and Donald Mackay (Canadian Environmental Modelling Centre, Trent University, Peterborough, Ontario, Canada)

ABSTRACT: A common problem is the contamination of sediments in lakes and river systems as a result of past discharges of hydrophobic organic substances such as polychlorinated biphenyls (PCBs), hexachlorobenzene (HCB) and metals such as mercury. Remediation programs usually involve reduction or elimination of discharges or interventions such as dredging or capping. Allowing decontamination by natural attenuation is also a feasible option. We suggest that a valuable strategy when addressing these options is to develop a dynamic mass balance model of the system, which describes the disposition of the chemicals over a period of time, during which actual, or proposed remediation actions take place, as well as natural remediation or attenuation. A periodic (e.g. annual) mass balance description of the entire system can assist in establishing the effectiveness of past efforts and proposed future efforts. In addition, the inclusion of a bioaccumulation model can provide further information on how the remediation actions may affect contaminant levels in aquatic biota. This approach is illustrated by applying a multi-segment QWASI (Quantitative Water Air Sediment Interaction) fugacity model to the fate of HCB in the sediments and biota of the St. Clair River, which connects Lake Huron to Lake St. Clair and Lake Erie in the Laurentian Great Lakes.

INTRODUCTION

The St. Clair River, which borders Canada on the east and the United States on the west, flows from Lake Huron, past the city of Sarnia and a number of chemical and petroleum industries on the Canadian side, into Lake St. Clair. Figure 1 highlights the area near Sarnia where the chemical industries are located. As a result of an assessment of impairments to usages of the river, including wildlife habitat reduction, spills to the area and fishing restrictions, the river has been designated by the International Joint Commission as an Area of Concern and has been subject to a Remedial Action Plan (IJC, 1988).

Among the contaminants of concern in the St. Clair River are mercury (Hg), hexachlorobenzene (HCB), hexachlorobutadiene (HCBD), octachlorostyrene (OCS) and polychlorinated biphenyls (total PCBs). They show an affinity for solid particles when present in water and once associated with sediment may contaminate aquatic and benthic organisms (IJC, 1988). There have been reductions and ultimately elimination of the loadings of these contaminants, however, as occurs frequently in similar situations, sediment and biota concentrations have been slower to respond.

FIGURE 1. Sarnia Industrial Area, St. Clair River.

Several remediation actions have been taken or may apply in the future. First is the elimination of mass flux of contaminants from industrial outflows, which previously entered the St. Clair River from point source discharges. Second, a source of highly contaminated sediment near a former industrial outfall was removed by dredging in 1996. Third is natural remediation as chemical is slowly removed from the sediment by degrading reactions and through transfer to the flowing water. This process requires no human intervention. Fourth is a possibility of future capping or bulkheading in which the sediment and associated chemical is essentially buried to an inaccessible depth. This has not been attempted to date.

The object of this project is to document quantitatively the changes in contaminant levels in sediments, water and biota since 1995, including an evaluation of the effects of past remediation actions. This is done using a mass balance model. The model can then be used in a predictive mode to determine likely changes in contamination levels as a result of proposed remedial actions, including natural attenuation or remediation. By quantifying past behavior and possible future behavior it can assist decision making.

MODEL STRUCTURE

The model treats the river as a number of segments of defined location and area, connected by flows of water and suspended sediment and exposed to the atmosphere. Each segment consists of a defined volume of water underlain by a volume of sediment. To simplify the equations, the water and sediment compartments are treated as being well mixed, thus a single concentration prevails in each. The QWASI unit in Figure 2, which shows the fate and transport processes treated, illustrates a representative segment. All significant processes are quantified, as follows:

- Emission into the water columns from shore sources (dependent on segment)
- Inflow into water from tributaries (dependent on segment)
- Inflow from upstream and adjacent segments of both water and suspended sediments
- Input from the atmosphere
- Volatilization to the atmosphere
- Degradation in the water column and sediment

- Water to sediment transport by deposition and diffusion
- Sediment to water transport by resuspension and diffusion
- Sediment burial
- Outflow in water and suspended sediments to downstream and adjacent segments

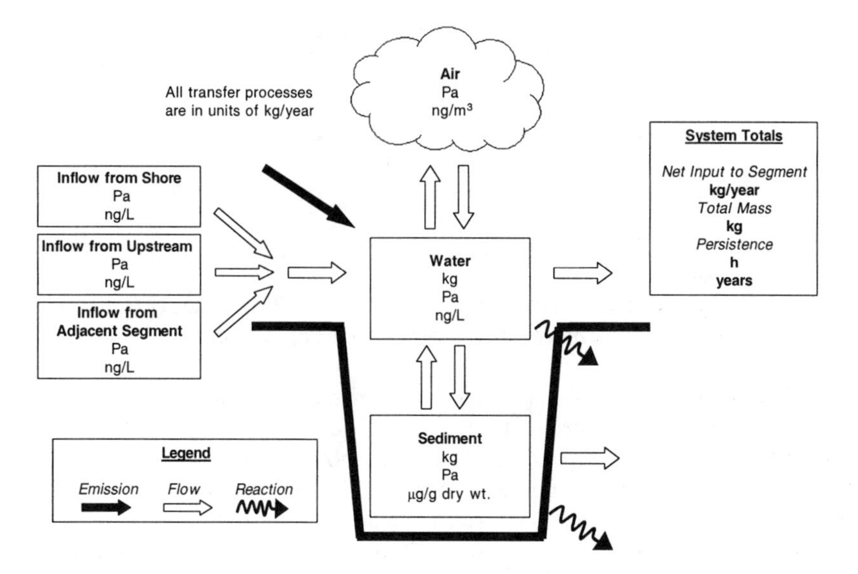

FIGURE 2. Segment mass balance diagram for the St. Clair River Multi-segment QWASI model.

The model is designed to calculate, for the subject chemical of known input rate, the masses and concentrations in the water column and sediment, and the above prevailing fluxes (e.g. kg/year). These data are reported at a specified time and totals are calculated for the entire river. A time history of the quantities and locations of contaminants is thus deduced.

The effects of remediation actions are included by interrupting the calculations at appropriate times to simulate the following actions:

- Changing or eliminating emission rates from the shore.
- Simulating dredging by removing a mass of sediment and associated chemical from the model. This results in a step-change in chemical mass, it being assumed that the sediment returns to its original condition.
- Simulating bulkheading or permanent capping by causing a mass of chemical to be buried permanently.

Natural remediation is also quantified as occurring continuously as a result of loss of chemical by reaction and by transport from the sediment to the water and then to the outflow water or the atmosphere.

The river is treated, for the present purposes as two parallel sets of connected compartments as shown in Figure 3.

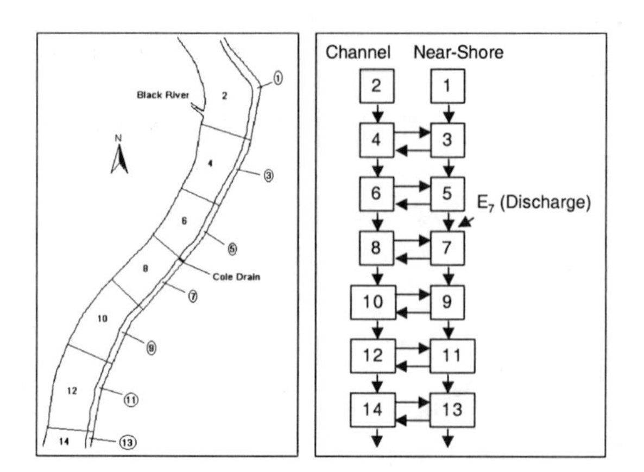

FIGURE 3. Connected river segments of Multi-QWASI model (a) Actual location in the river. (b) Schematic representation.

Figure 3(a) shows the actual location of the segments and 3(b) displays their connections schematically. It was decided necessary to treat near-shore and main channel segments separately. The near-shore segments extend from the shoreline a typical distance into the river of 70 m, have a mean depth of 5 m and contain most of the contaminants. The remainder of the river, consisting of the main navigation channel with a mean depth of 10 m and the United States near-shore region, is lumped together. The main channel is not as contaminated and dredging is not feasible. In any event, the bottom is heavily scoured and is subject to episodic disturbance by vessel traffic. The total number of segments is 14, i.e. 7 each of near-shore and river channel, but this can be modified.

The average annual water flow is approximately 5000 m^3/s of which 98% is assumed to flow in the main channel and 2% in the near shore segment (Aleshin and DePinto, 1997). Inter-segment exchange is included i.e. near-shore to main channel and vice versa. Average annual conditions are used and greatly simplify the calculations. Seasonal or episodic changes in flows, temperature or sediment mobilization are not treated. The model thus describes year to year changes, not day to day or month to month changes. This is regarded as satisfactory because long term conditions are of primary interest.

In addition to the fate and transport model a separate bioaccumulation model is provided. This model accepts as input the concentration of the chemical in the sediment and water column at any location and calculates the concentrations in selected benthic and pelagic organisms comprising a typical food chain or web. This has the benefit that biomonitoring data can be used to confirm the validity of the model calculations. Future concentrations in biota can be estimated and compared with levels of concern from the viewpoints of human or wildlife consumption or the risk of toxic effects. It is noteworthy that the primary reason for remediation is the restoration of the biota in the river to a satisfactory state at which adverse effects are, or will be, virtually eliminated.

The food web model is based on work by Haffner and Morrison and consists of a one compartment model with a series of equations that determine the rate of uptake and accumulation of contaminants in an aquatic organism, the freshwater clam (Haffner and Morrison, 1997). In their work, the concentrations of a variety of organic contaminants, i.e. PCBs, PAHs, OCs and chlorinated pesticides, were measured in freshwater clams. From this the rate of uptake was determined, and along with the rate of contaminant elimination from the mussel and the contaminant water concentrations, the rate of accumulation was calculated.

MODEL FORMULATION AND EQUATIONS

The model is written in fugacity format as described fully by Mackay (1991). It is emphasized that an analogous conventional concentration-based model would give identical results. Fugacity is used primarily because it simplifies the coding and it is particularly convenient for linking the fate and transport model to the bioaccumulation model.

For a single segment as depicted in Figure 2, the differential mass balance equations are as follows:

Water Column

$$V_{Wi} Z_{Wi} \frac{df_{Wi}}{dt} = E_W + f_S (D_R + D_T) + f_A (D_M + D_C + D_Q + D_V) + \Sigma f_{WB} (D_i + D_X) - f_{Wi} (D_T + D_D + D_W + D_V + D_J + D_Y) \quad (1)$$

Sediment

$$V_{Si} Z_{Si} \frac{df_{Si}}{dt} = f_{Wi} (D_T + D_D) - f_{Si} (D_B + D_S + D_R + D_T) \quad (2)$$

where the parameters are listed and defined in Table 1.

TABLE 1. Parameter definitions for the mass balance equations.

Term	Definition
Subscripts *A, W & S*	Subscripts for *V*, *Z* and *f*, referring to Air, Water and Sediment
V	Volume of each phase
Z_{Wi} & Z_{Si}	Bulk phase Z values
f	Fugacity
f_{WB}	Water fugacity from an adjacent segment to segment *i*
E_W	Chemical discharge or input into the water column (mol/h)
$\Sigma f_{Wi}(D_I + D_X)$	Sum of advective inflows from adjacent water compartments

Two solutions are derived and used. First is a steady state solution, which is used to confirm that the model equations satisfy a mass balance constraint. It is also used during the parameterization process to select reasonable parameter values and test the sensitivity of the results to changes in parameter values.

Second is the dynamic or unsteady state solution, which gives a simulation of the actual time course of remediation. It is of most relevance in this context.

Included within the sediment was a small volume of non-aqueous phase liquids or tar, which contained a substantial concentration of contaminant.

Parameterization. As with any model which treats a complex environmental situation, a large number of parameters must be defined. These include hydrological data, especially rates of deposition and resuspension, chemical data including partition coefficients and degradation rate constants and data defining chemical inputs from discharge, Lake Huron (upstream of the site) and the atmosphere.

RESULTS

Steady State Conditions. The model was first run in steady state form for two conditions. The first was typical of the years up to 1995 in which there was a constant discharge of HCB approximately 2 g/h (Kauss, 1999). This rate resulted in concentrations in the river and sediments, which were typical of those existing about 1995. This simulation represents conditions, which would apply after prolonged (decades long) discharge at the above rate. This is viewed as a reasonable initial condition from which to deduce future conditions. A feature of this period was the presence in the sediment of oily material, which was later dredged. Introducing an in-place non-aqueous phase contaminant source to segment seven of the model included this oily material. This has the effect of increasing the mass of contaminant in the individual segment.

Secondly, the model was run for a steady state condition with zero discharges and the only inputs were from the atmosphere and Lake Huron. This represents the likely condition some decades in the future to which the river will approach. It is noteworthy that the concentrations are not zero because of the "background" inputs from the atmosphere and Lake Huron. Table 2 is a summary of the steady state results.

TABLE 2. Summary of Steady State concentrations for the Near-Shore Segments, 5 (background segment) and 7.

Steady State Results				
Emission Conditions	**2 g/h (1996)**		**0 g/h (2000)**	
	Segment 5	Segment 7	Segment 5	Segment 7
Sediment Conc. (ug/g)	0.0040	0.124	0.0040	0.0040
Benthic Conc. (ug/g)	0.0213	0.659	0.0213	0.0212
Pelagic Conc. (ug/g)	0.0049	0.151	0.0049	0.0049
Measured Environmental Parameters (a-1995, b-1994)				
	Segment 5		Segment 7	
Sediment Conc. (ug/g)[a]	0.004		0.124	
Benthic Conc. (ug/g)[a]	0.0013 – 0.0036		0.1067-0.4323	
Pelagic Conc. (ug/g)[b]	0.0012 – 0.0056		0.037 – 3.6	

Table 2 depicts results in only two near-shore segments, No. 5 (representing the upstream background, with inputs only from the atmosphere and Lake Huron) and No. 7 (which is the most contaminated segment, having received the direct discharges) for each of the conditions described above.

Dynamic Conditions. The dynamic results are of most interest and relevance. A dynamic model yields a large quantity of data including fugacities, masses, concentrations and all fluxes listed earlier for all 14 compartments at regular intervals each year for some decades. The quantity of data generated by the computer vastly exceeds the capacity of any user to assimilate it. As a result a strategy was adopted of printing out a complete statement of the prevailing conditions at any desired time. This is essentially a "report card" of the conditions in the river. By comparing "report cards" at various times the progress of remediation can be assessed. The model can also be used to test various remediation strategies and compare their effectiveness.

An initial condition was selected of conditions typical of 1995. At this time there was a discharge into Segment 7 of 2 g/h, and there was a relatively high concentration of HCB in the vicinity of the discharge. This condition was allowed to run until 1996. There was then a simulation of dredging, which took place in 1996, removing a total of 200 kg of HCB from zone 1. There was simultaneous removal of oily material. The model was then run with decreasing discharges in 1998 to 2 x 10^{-5} g/m^3. In 2000 the discharge was set at zero, its current (2001) rate. The simulation was then continued until 2002, representing current conditions and those, which may apply prior to any dredging in 2002. If dredging were to occur in 2002 to zone 1 sediments, 100 kg of HCB would be removed. For interest, the model was then run until 2029.

Illustrative results of the dynamic program for segment 7 are represented in Figure 4.

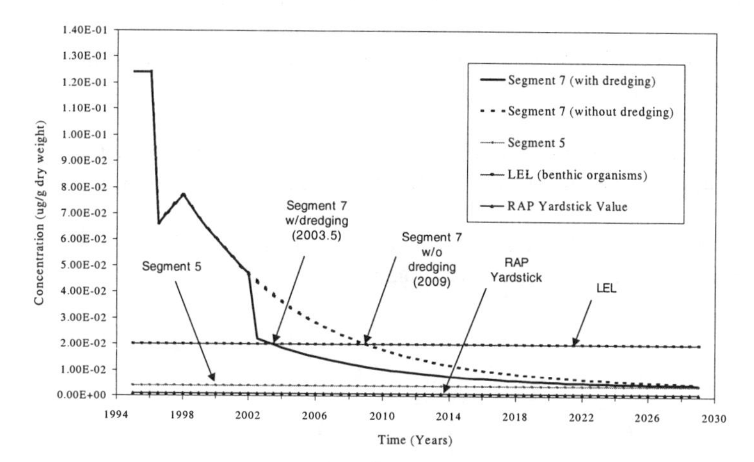

FIGURE 4. Summary of sediment dredging effects on segment 7 in the St. Clair River Multi-QWASI model.

CONCLUSIONS

Figure 4 suggests that remediation of sediments in 2002 could affect the river system in such a way that in the next few years the sediment concentration in segment 7 should be at the Lowest Effect Level (LEL) for possible deleterious effects on benthic organisms. Without human intervention in 2002, the LEL should not be reached in segment 7 until approximately 2009. This particular remediation scenario thus advances restoration by about 5-6 years.

This model provides predictions about the effects of remediation actions on contaminated systems, which can then be used to help assess the most appropriate approach to be taken. Be it capping, dredging or simply natural attenuation, the most cost effective ecologically beneficial method can be suggested. As a result this model would be very effective in remediation strategies and could be applied in many other of these possible situations. It has the potential to quantify the conditions in the past present and future, including the benefits of various remediation actions. It can estimate the time required to reach target concentrations and determine if indeed these concentrations are feasible given background inputs. In total, by quantifying in all the remediation processes it can, we believe, contribute to more rational decision making about the restoration of such systems.

REFERENCES

Aleshin, K. and DePinto, J. 1997. *Mass balance model of sediment and HCB transport in the St. Clair River.* Draft report prepared for St. Clair River Remediation Action Plan.

Haffner, G.D. and Morrison, H. 1997. *Environmental assessment in the St. Clair River biomonitoring and foodweb models 1995.* Great Lakes Institute for Environmental Research. University of Windsor, Ontario.

International Joint Commission. *Upper Great Lakes Connecting Channels Study.* Final Report, VII, 1988.

Kauss, P.B. 1999. *Cole Drain (Sarnia) contaminant concentrations and loadings 1995.* A report prepared for St. Clair River Remediation Action Plan.

Mackay, D. 1991. *Multimedia Environmental Models: The Fugacity Approach.* Lewis Publishers, Inc., Chelsea, Michigan.

SURVEY OF CONTAMINATED SEDIMENT RESUSPENSION DURING CAPPING

Victor S. Magar, Jennifer Ickes, Lydia Cumming (Battelle, Columbus, Ohio)
Wayne Trulli, Carl Albro (Battelle, Duxbury, Massachusetts)
Terrence Lyons (U.S. EPA, Cincinnati, Ohio)

ABSTRACT: Capping is a common remediation technology for the containment/stabilization of contaminated sediments. During capping activities, capping material is released from a barge at the water surface and falls through the water column to the sediment, thus providing a clean sediment layer over the contamination. Little information exists on the potential release of contaminated sediments during and after the capping operation. This paper focuses on the measured release of contaminants during capping events at Eagle Harbor, Washington (creosote-contaminated from a wood treating facility) and Boston Harbor, Massachusetts (confined aquatic disposal [CAD] cells for contaminated sediment). The U.S. EPA and Battelle sampled the water column during these two capping events to evaluate whether cap placement resulted in the release of TPH and PAH contaminated sediments at Eagle Harbor, or TPH, PAH, PCB, and RCRA metal-contaminated sediments at Boston Harbor. Results at both sites showed relatively minor contaminant resuspension. The most significant releases occurred when previously uncapped sediments were capped; the magnitude of contaminant resuspension decreased with successive capping layers.

INTRODUCTION

U.S. EPA's NRMRL is interested in whether the placement of a marine cap results in the release of contaminants into both the surrounding water column. The goal of these studies was to achieve a better understanding of the amount of contaminants released into the surrounding water column before, during, and after capping. For this study, the Battelle Ocean Sampling System (BOSS) was used to acquire high-resolution field data in order to determine the extent of the release of any contaminated sediment into the water column at two sites: Wyckoff/Eagle Harbor Superfund Site, Bainbridge Island, WA. and the Mystic River/Boston Harbor, Boston, MA

Eagle Harbor is a shallow marine embayment of Bainbridge Island, Washington, located approximately 10 miles due west of Seattle, WA. The former Wyckoff wood-treatment facility, located on Bainbridge Island, became operational in the early 1900s. During its operation, large quantities of creosote were used as a wood preservative, resulting in PAH contamination of Eagle Harbor sediments. The Wyckoff/Eagle Harbor site was placed on the National Priorities List (NPL) in 1987 as a Superfund Site. The site was partially capped to control PAH migration into the water column and surrounding sediments; however, residual surface sediment contamination remained, generally to the south and west of the existing cap (Figure 1). Capping of the remaining sediments continued October 2000 to February 2001 by the U.S. Army Corps of Engineers (USACE), transporting sand to the site in a flat-topped barge and using a high-pressure hose to wash the sand overboard to gradually produce a uniform cap. This method was used to minimize potential sediment resuspension during capping; a portion of the capping operation (Figure 1) was monitored. Final capping is scheduled between November 2001 and February 2002.

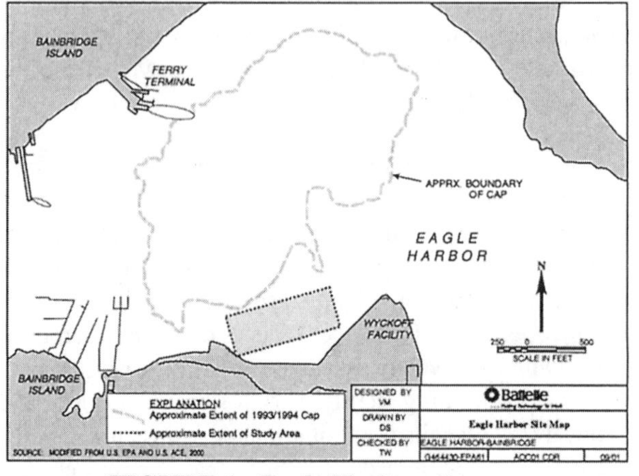

FIGURE 1. Eagle Harbor site map

CAD cells in ***Boston Harbor*** were used as disposal sites for silty, fine-grained sediments, that were determined to be unsuitable for ocean placement, from the Boston Harbor Navigation Improvement Project (BHNIP). Two of the CAD cells (Cells M8 and M19), located in the Mystic River portion of Boston Harbor (Figure 2), were filled with dredged material in January 2000. Sand dredged from Cape Cod Canal was used to cap CAD cells M8 and M19 (Fredette et al., 2000). The sand cap was "sprinkled" on the cells using a partially opened hopper dredge, which was maneuvered using a tug rather than utilizing the dredge's engines. This method of capping was expected to minimize disturbance of the silt material in the CAD cells.

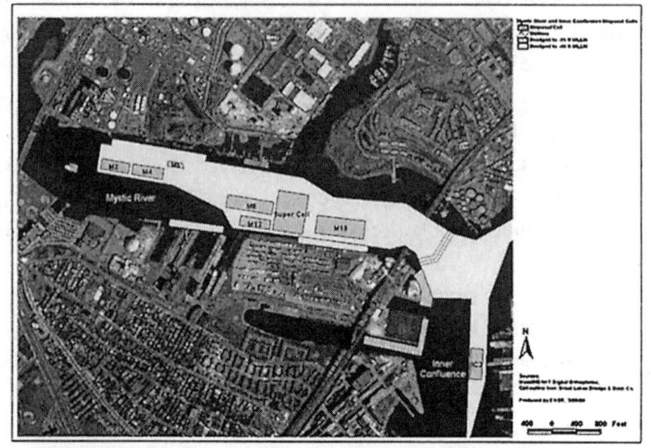

FIGURE 2. Boston Harbor/Mystic River Site Map

FIELD SAMPLE COLLECTION AND ANALYSIS

Sampling and Navigation. Monitoring of the capping events at Boston Harbor and Eagle Harbor was conducted using the BOSS deployed from survey vessels that towed the BOSS underwater during survey operations. A photo of the BOSS prior to deployment is shown in Figure 3. The BOSS in situ sensor package included: conductivity, temperature, depth, and turbidity sensors, and a Teflon™/titanium pumping system for sample collection. Additionally, an Acoustic Doppler Current Profiler (ADCP) was deployed to obtain vertical profiles of horizontal currents at Boston Harbor. Vessel positioning during sampling operations was accomplished with the BOSS navigation system, which consists of a Northstar Differential Global Positioning System (DGPS) interfaced to the BOSS computer.

FIGURE 3. Battelle's Ocean Sampling System

***Eagle Harbor*:** The first sampling event at Eagle Harbor was completed on October 13, 2000 prior to the initiation of the USACE capping operations. Marine water samples, manually collected within the uncapped area of Eagle Harbor and analyzed for TSS, TPH and PAH as part of a separate site investigation, were used as baseline reference information for this study. The BOSS was not used for these baseline samples; instead, underwater divers carried the appropriate sample bottles to the bottom of the harbor and collected the samples approximately one meter from the bottom at each of three locations.

During the capping and post-capping surveys at Eagle Harbor, real-time BOSS sensor data was collected continuously. The survey vessel towed the BOSS at a distance of 1 meter from the bottom. The capping survey was conducted from November 1 through 3, 2000. On each survey day, samples were collected at three targeted locations within the study area and at background locations before and after daily capping operations. During the capping operations, samples were collected at multiple locations within the study area at varying distances and directions from the capping barge. A total of 90 water samples were collected for TPH, PAH, and TSS analyses during the three-day survey. On June 13, 2001, approximately 4 months after the USACE capping operations ceased, water samples were collected at fourteen locations within the target study area and at background locations. Only the PAH and TSS data for the samples will be discussed in this paper. TPH concentrations were very low in all the water samples and did not change significantly during capping operations.

***Boston Harbor*:** The field survey operations at Boston Harbor consisted of 10 separate sampling events conducted over a 22-day period (approximately two events per day plus background samples before and after capping). All 10 events were conducted using the BOSS (Figure 3). During the first event, sensor data and water samples were collected prior to initiation of capping activities at CAD cells M8 and M19 for baseline/reference data. During the next eight events, sensor data and water samples were collected while capping activities were conducted at either CAD cell M8 or M19. During the last (10th) event, sensor data and water samples were collected four days after completion of all capping activities at both cells. All water samples collected at Boston Harbor were analyzed for PCB, PAH, TPH, TSS, and RCRA 8 metals. Only the PCB, PAH, and TSS data for samples collected from CAD cell M19 are discussed in this paper. Samples for CAD cell M8 showed similar patterns to those from CAD cell M19.

During these monitoring activities, real-time BOSS sensor data was collected continuously along several transects within the designated CAD cells. The BOSS was towed at a distance of 2 m above the river bottom for all transects (unless obstructed) in order to optimize the potential to detect resuspended sediments.

EAGLE HARBOR RESULTS

Polycyclic Aromatic Hydrocarbons. During capping operations, t-PAH concentrations ranged from 20 to 3,872 ng/L and TSS concentrations ranged from 1 to 191 mg/L. Figure 4 shows the average t-PAH concentrations measured for each sampling event (run) conducted during the three-day capping survey. Run 1, Run 2, and Run 3 occurred successively during daily capping operations; each run included six to nine samples. "Pre-run" and "post-run" include three samples each that were collected before and after daily capping operations, respectively. Averaged baseline and post-capping survey results are included in Figure 4 for comparison. Elevated contaminant concentrations were observed during capping operations; they appeared to decrease with each successive capping day and dissipated after capping was completed.

TSS vs. Organics. To analyze the relationship between t-PAH and turbidity, the t-PAH was plotted against TSS concentrations to create x-y (scatter) graphs for samples collected during capping operations (Runs 1-3 for every capping survey day). The correlation coefficients (r^2 value) for the best-fit linear regression lines were calculated. During the first survey day, r^2 values ranged from 0.71 through 0.95, showing a relatively strong correlation between turbidity and PAH. However, r^2 values decreased significantly during the capping survey, and by the third day r^2 values were less than 0.3. Resuspension may have become less significant as the capping progressed, covering the surface sediments with a clean sediment layer. Consequently, with each successive capping run, turbidity from suspended capping material increasingly overshadowed resuspended contaminated sediments. Figure 5 presents the results of the linear regression analysis for the first survey day.

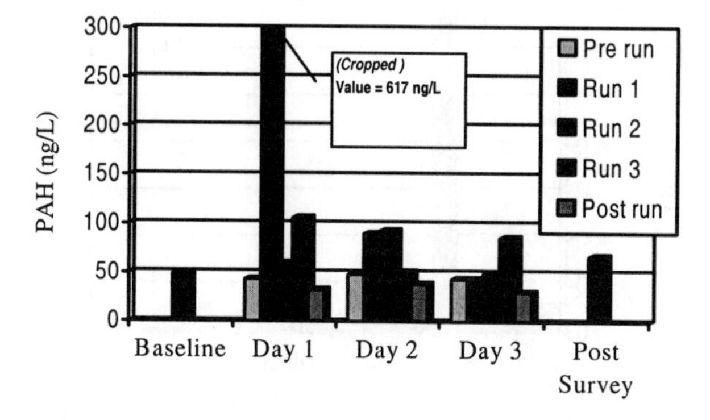

FIGURE 4. Average PAH Concentrations – Eagle Harbor

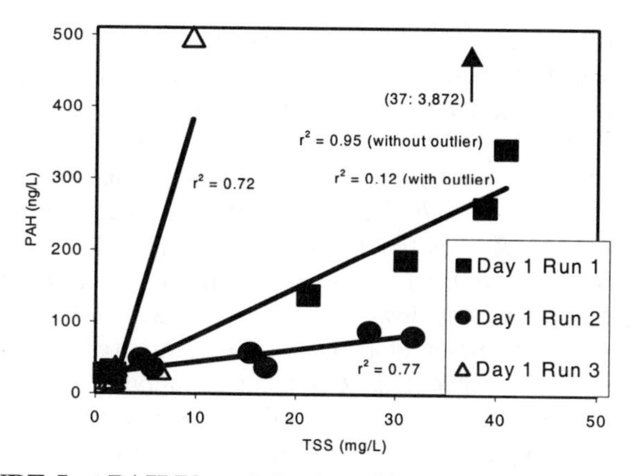

FIGURE 5. t-PAH Plotted Against TSS - Day 1 Eagle Harbor

BOSTON HARBOR RESULTS

Polycyclic Aromatic Hydrocarbons. PAH concentrations in samples taken during Run 1 averaged 1,372 ng/L ± 1,569 ng/L, with a maximum of 5,242.30 ng/L (Figure 6). The total PAH concentrations collected during capping for Run 1 were much greater than background, Pre Run 1, and Post Run 1 sample concentrations. Post Run 1 samples collected one hour after capping stopped for Run 1 remained elevated above the Pre Run 1 samples, but decreased significantly from samples collected during Run 1.

Pre Run 2 samples had higher PAH concentrations compared to the samples collected during Post Run 1 and Run 2. These samples also had elevated TSS, TPH, and PCB concentrations. Further resuspension after Run 2 was not observed at cell M19. PAH concentrations during the capping runs decreased over the remaining capping events at cell M19. By the end of sampling at cell M19, concentrations had returned to background levels measured during the Pre Survey before capping began.

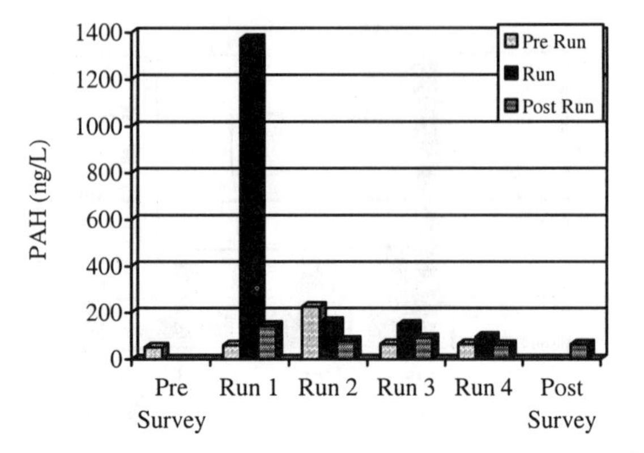

FIGURE 6. Average PAH Concentrations – Boston Harbor Cell M19

Polychlorinated Biphenyls. PCB concentrations in background samples and Pre Run 1 samples were all BDL for cell M19. The first PCB detection occurred during Run 1,where concentrations average 19.7 ng/L ± 25.6 ng/L, with a maximum value of 84.24 ng/L. The 84.24 ng/L sample collected during Run 1 was the highest PCB detection for all sampling events, and was the only sample that exceeded the water quality criteria for PCBs of 30 ng/L (USACE, 1999). Samples taken 1 hour after Run 1 dropped back to <4.7 ng/L. The first Pre Run 2 sample had a PCB concentration of 18.02 ng/L, but the other two Pre Run 2 samples had concentrations <5 ng/L.

PCB concentrations were lower during Run 2 than Run 1, and averaged 1.8 ng/L ± 1.6 ng/L with a maximum of 4.25 ng/L (Figure 7). Average concentrations for Run 3 and Run 4 were higher than Run 2, but much lower than Run 1, and Pre Run 4 samples returned to BDL. The average PCB concentrations for Runs 3 and 4 were 3.3 ng/L and 4.6 ng/L, respectively; variation in the Run 4 samples was high with one elevated detection at 23.79 ng/L, and three Run 4 samples measured BDL.

TSS vs. Organics. In general, high TSS concentrations did not correlate to high organic concentrations at cell M19 (Figure 8). At cell M19, four water samples taken during Run 1 had both high TSS concentrations and relatively high organic concentrations; however several water samples also had high TPH levels and very low TSS concentrations. Cell M19 also had one sample in Run 4 with a high PCB concentration (23.79 ng/L), but low TSS (7.54 mg/L) (Figure 9).

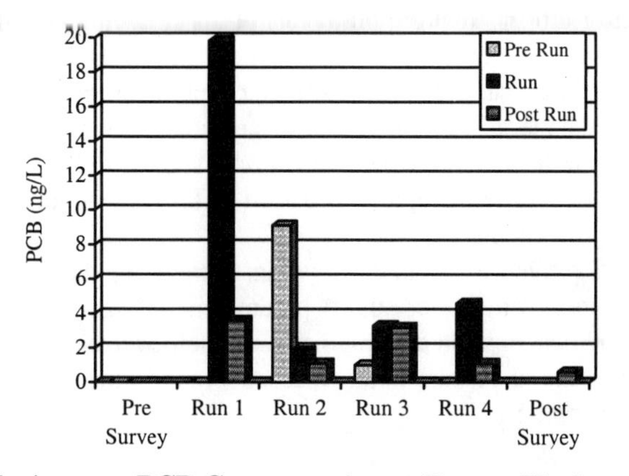

FIGURE 7. Average PCB Concentrations – Boston Harbor Cell M19

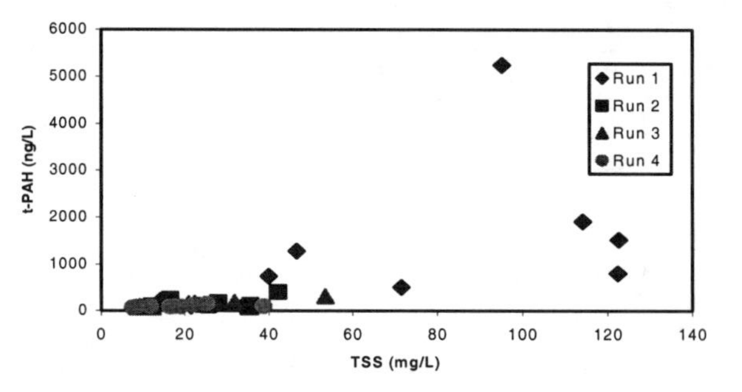

FIGURE 8. PAH Plotted Against TSS for Runs 1 Through 4 – Boston Harbor

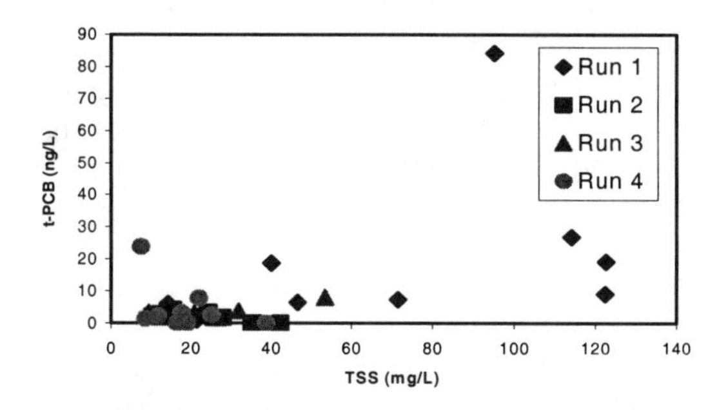

FIGURE 9. PCB Plotted Against TSS for Runs 1 Through 4 – Boston Harbor

CONCLUSIONS

The capping material at Eagle Harbor was washed off of a barge with a high-pressure hose, while a "sprinkling" method used at Boston Harbor. These sites used different methods for capping contaminated settlements, but both sites showed similar results from the monitoring during the capping events.

Table 1 compares sampling results at Boston Harbor to results from the parallel study at Eagle Harbor. Background levels of the contaminants of concern were below detection limits or at very low levels, at both sites. The highest resuspension of contaminant material was seen during the first capping event at each CAD cell and at Eagle Harbor. In general, contaminant resuspension was relatively low for all capping events during both surveys, where contaminant concentrations were in the low ng/L range for most samples. Rapid dissipation of TSS and contaminants was seen following all capping events.

The BOSS proved to be an effective instrument for the collection of water samples and for monitoring water quality parameters before, during, and after capping. Samples and data collected using the BOSS helped to achieve a better understanding of the amount of contaminants released into the surrounding water column as a result of the capping events.

TABLE 1. Comparison of Results from Boston Harbor and Eagle Harbor.

Monitoring Event	Boston Harbor		Eagle Harbor	Observations
	PCB (ng/L)	PAH (ng/L)	PAH (ng/L)	
Before Capping	Non detect	46 - 59	46 - 73	Non-detect to low level background at both sites
During Capping	Max. = 84	Max. = 5,242	Max. = 3,872	Elevated suspended contaminants during capping; contaminant resuspension decreased with successive capping layers
After Capping	0.4 – 1.5	41 – 83	38 – 159	Rapid settling of suspended contaminants at both sites

ACKNOWLEDGEMENT

Special thanks to Ms Olga Quirin (U.S. EPA, Region 1) and Dr. Thomas Fredette and Mr. Michael Keegan (USACE) for their assistance and access to the Boston Harbor/Mystic River Site; and to Mr. Ken Marcy (U.S. EPA, Region 10), Ms. Brenda Bachman (USACE), and Mr. Travis Shaw (USACE) for their continued assistance and access to the Wyckoff/Eagle Harbor Superfund Site. Funding for this study was provided by the U.S. EPA National Risk Management Research Laboratory.

REFERENCES

Fredette, T.J., P.E. Jackson, C.J. Demos, D.A. Hadden, S.H. Wolf, T.A. Nowak Jr., and E. DeAngelo. 2000. The Boston Harbor Navigation Improvement Project CAD Cells: Recommendations for Future Projects Based on Field Experience and Monitoring. Western Dredging Association Conference June 2000, Warwick, RI.

United States Army Corps of Engineers (USACE). 1999. Water Quality Certification, Boston Harbor Navigation and Berth Dredging Project, Letter Report. December 28.

KEYWORD INDEX

This index contains keyword terms assigned to the articles in the three books published in connection with the First International Conference on Remediation of Contaminated Sediments (Venice, 10–12 October 2001). Ordering information is provided on the back cover of this book.

In assigning the terms that appear in this index, no attempt was made to reference all subjects addressed. Instead, terms were assigned to each article to reflect the primary topics covered by that article. Authors' suggestions were taken into consideration and expanded or revised as necessary to produce a cohesive topic listing. The citations reference the three books as follows:

1(1): Pellei, M., A. Porta, and R.E. Hinchee (Eds.), *Characterization of Contaminated Sediments*. Battelle Press, Columbus, OH, 2002. 357 pp.
1(2): Porta, A., R.E. Hinchee, and M. Pellei (Eds.), *Management of Contaminated Sediments*. Battelle Press, Columbus, OH, 2002. 309 pp.
1(3): Hinchee, R.E., A. Porta, and M. Pellei (Eds.), *Remediation and Beneficial Reuse of Contaminated Sediments*. Battelle Press, Columbus, OH, 2002. 463 pp.

D

E

F

O

P

Q

R

S

T

AUTHOR INDEX

This index contains names, affiliations, and book/page citations for all authors who contributed to the three books published in connection with the First International Conference on Remediation of Contaminated Sediments (Venice, 10–12 October 2001). Ordering information is provided on the back cover of this book.

The citations reference the three books as follows:

1(1): Pellei, M., A. Porta, and R.E. Hinchee (Eds.), *Characterization of Contaminated Sediments.* Battelle Press, Columbus, OH, 2002. 357 pp.
1(2): Porta, A., R.E. Hinchee, and M. Pellei (Eds.), *Management of Contaminated Sediments.* Battelle Press, Columbus, OH, 2002. 309 pp.
1(3): Hinchee, R.E., A. Porta, and M. Pellei (Eds.), *Remediation and Beneficial Reuse of Contaminated Sediments.* Battelle Press, Columbus, OH, 2002. 463 pp.